Waste Not! How Nanotechnologies Can Increase Efficiencies Throughout Society

Waste Not!
How Nanotechnologies Can Increase Efficiencies Throughout Society

By

Michael Berger
Nanowerk LLC, USA
Email: michael@nanowerk.com

ROYAL SOCIETY OF **CHEMISTRY**

Hardback ISBN: 978-1-83767-487-9
EPUB ISBN: 978-1-83767-730-6
PDF ISBN: 978-1-83767-729-0

A catalogue record for this book is available from the British Library

The Royal Society of Chemistry is a charity, registered in England and Wales, Number 207890, and a company incorporated in England by Royal Charter (Registered No. RC000524), registered office: Burlington House, Piccadilly, London W1J 0BA, UK, Telephone: +44 (0)20 7437 8656.

For further information see our website at www.rsc.org

For general enquiries, please contact books@rsc.org

For EU product safety enquiries, please email books@rsc.org or contact Royal Society of Chemistry Worldwide (Germany) GmbH, Römischer Hof, Unter den Linden 10, 10117 Berlin.

Printed in the United Kingdom by CPI Group (UK) Ltd, Croydon, CR0 4YY, UK.

Preface

When you hear the word *nanotechnology*, you probably picture something like tiny robots, futuristic medical procedures, or miniature electronics embedded in high-tech devices. But what if you were told that nanotechnology shows promise in transforming something much more familiar—your food?

Food production has already been revolutionized by various technologies, from genetic modification to industrial processing, and nanotechnology represents the next frontier—one that could be particularly transformative in addressing one of the industry's biggest challenges: food waste. Globally, nearly 40% of all food produced never makes it to our plates. This waste amounts to an estimated 2.5 billion tons of food each year. The problem is immense and touches every stage of the supply chain—from farms where crops are lost due to pests and disease, to transportation and storage where poor conditions lead to spoilage, and finally in stores, food services, and homes where food expires unused. This inefficiency doesn't just waste food—it's responsible for nearly 10% of global greenhouse gas emissions, straining our environment and resources.

> *"OK, that is a massive problem! But what does that have to do with nanotechnology?"*

Well, a lot actually.

In the food industry, nanotechnology research and development is advancing in three critical domains: safety, packaging, and processing (see Figure 1). From detecting pathogens and allergens to improving packaging with bio-based materials and enhancing food texture and nutritional value, nanotechnology offers innovative solutions at every step of the supply chain. These developments reduce waste while enhancing food quality and safety, helping us tackle one of the world's most pressing challenges—food waste.

Waste Not! How Nanotechnologies Can Increase Efficiencies Throughout Society
By Michael Berger

Published by the Royal Society of Chemistry, www.rsc.org

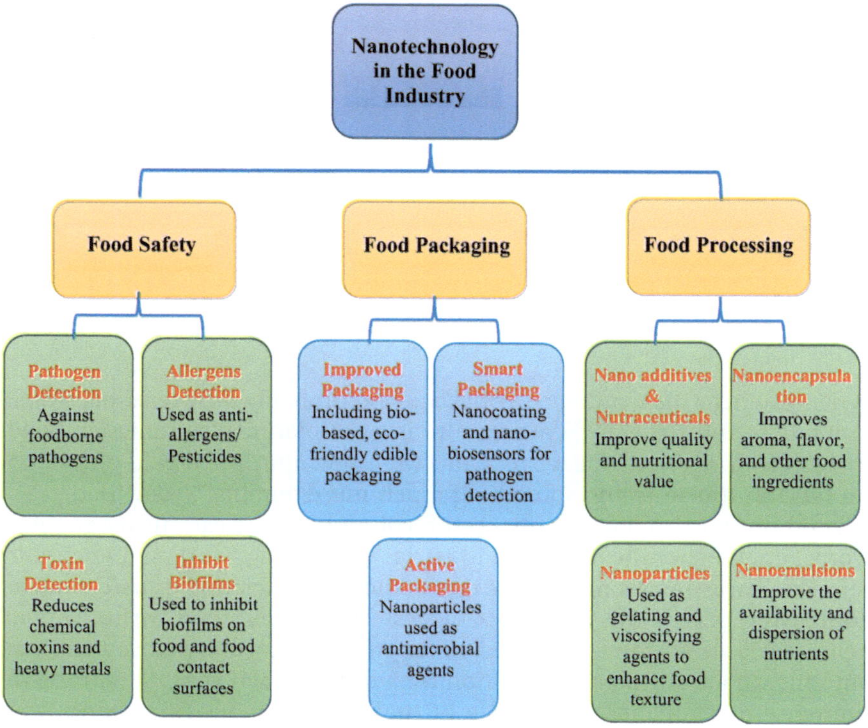

Figure 1 The application of nanotechnology in food safety, food packaging and food processing. Reproduced from ref. 1, https://doi.org/10.1007/s44187-022-00013-9, under the terms of the CC BY 4.0 license, https://creativecommons.org/licenses/by/4.0/.

For example, nanotechnology-enhanced precision agriculture is showing promise in allowing farmers to reduce waste at the very start of the process. Laboratory studies of nanosensors have demonstrated that their microscopic size and enhanced sensitivity could potentially enable more precise monitoring of soil and crop conditions. While several promising prototypes have been developed, these technologies are still primarily in the research and development phase, with ongoing work needed to address challenges in durability, cost-effectiveness, and regulatory approval before widespread agricultural implementation. Their small size allows them to be distributed throughout fields without disrupting crop growth, while their heightened sensitivity enables earlier detection of nutrient deficiencies or the first signs of plant disease. This more precise and distributed monitoring helps prevent the unnecessary overuse of water, pesticides, or fertilizers. This technology reduces environmental harm while ensuring healthier crops with fewer losses.

As food moves from the farm to the market, nanotechnology continues to play a vital role. Research on nanoparticles in smart packaging suggests the potential for improved monitoring of food freshness and signalling when products are nearing spoilage. Unlike traditional freshness indicators that only show temperature exposure, nanoscale sensors can detect specific molecular markers of spoilage, such as gases released by bacterial growth or changes in pH levels, providing more accurate and earlier warnings of food degradation. By using these packaging systems, supermarkets can better manage stock, reducing the volume of expired food that ends up discarded.

Research on nanoscale antimicrobials in packaging has demonstrated potential for superior protection against food contamination. Their extremely small size creates a vastly larger surface area for interaction with microorganisms compared to conventional antimicrobials. This increased contact area, combined with their ability to be uniformly distributed throughout packaging materials, results in more effective and longer-lasting antimicrobial protection while using fewer active ingredients.

Studies of nanomaterials like nanoemulsions suggest they could extend the shelf life of perishable items through superior preservation. Their extremely small droplet size allows them to spread more uniformly throughout food products and penetrate deeper into food structures than conventional preservatives. This results in more effective preservation while using less additives and maintaining the food's natural taste and quality. These innovations could not only help fight spoilage but also combat foodborne illnesses, protecting consumers and reducing waste from spoiled, unsafe food.

The promising research results of nanotechnology in addressing food waste illustrate a broader pattern: when we can manipulate and engineer materials at the nanoscale, we gain powerful new tools for solving complex systems-level problems. Just as nanoscale innovations help us monitor food freshness or deliver nutrients more effectively, similar principles are transforming other crucial sectors. Consider how nanoscale engineering approaches that preserve food freshness share common scientific foundations with those improving energy storage efficiency, enabling targeted drug delivery in medicine, or creating more sustainable manufacturing processes.

In each case, precise control at the smallest scales leads to significant real-world impacts. However, as illustrated by our discussion of food applications, nanotechnology's journey from basic research to market varies significantly across industries and applications. Some nanotechnologies, like those used in chemical catalysis or solar cells, are already commercially implemented and creating value. Others are in late-stage development or pilot testing, showing promising results as they move toward commercialization. Many applications remain primarily in the research phase, where scientists are still working to overcome technical, economic, or regulatory challenges. And some represent theoretical possibilities whose potential is yet to be fully explored. This diversity in development stages is typical of platform technologies like nanotechnology, where different applications mature at different rates depending on their complexity, market readiness, and regulatory requirements.

This book explores how nanotechnologies provide solutions to drive sustainable solutions across these diverse sectors. Through real-world examples and cutting-edge research, you'll discover the common threads linking these applications—from the nanoscale structuring of materials to the smart monitoring systems that reduce waste and increase efficiency. Whether it's preventing food spoilage, improving solar cell performance, or developing cleaner manufacturing processes, nanotechnology is quietly but powerfully transforming industries in ways that align technological innovation with environmental stewardship.

From nanoengineering to global sustainability solutions, this book reveals how innovations at the nanoscale have the potential to reshape our approach to some of humanity's most pressing problems. These innovations are not confined to one domain; they represent a myriad of small but impactful innovations across various sectors, each contributing to larger systemic improvements.

Nanotechnology presents a diverse array of opportunities to drive sustainability and innovation across multiple sectors. Unlike many technological fields that rely on a few broad approaches, nanotechnology's impact is characterized by its granularity—the vast array of small-scale applications and incremental improvements that together create significant progress.

When we refer to "nanotechnology", it is important to recognize that this term encompasses a wide range of techniques and innovations, making it a platform technology. As a platform technology, nanotechnology serves as a foundational tool that can be applied across various industries and scientific fields, enabling a multitude of other technologies and solutions. It not only facilitates the creation of new products and processes but also enhances existing ones, contributing to a more sustainable and efficient future. This versatility allows nanotechnology to address challenges in fields as diverse as medicine, energy, environmental science, and manufacturing, among others.

Nanotechnology can involve the development of novel nanomaterials like graphene, carbon nanotubes, and quantum dots, which possess unique electrical, thermal, and mechanical properties. It also includes the nanostructuring of surfaces to create coatings with specific functionalities, such as anti-corrosion, self-cleaning, or antimicrobial properties. Additionally, nanotechnology encompasses the manipulation of nanoscale objects, such as nanoparticles for targeted drug delivery in healthcare, nanosensors for detecting pollutants in environmental monitoring, and nanocomposites for creating stronger and lighter materials in aerospace and automotive industries.

Nanotechnology can be applied in the development of nanoelectronics, which are essential for the miniaturization and enhancement of electronic devices, as well as in nanophotonics, which deals with the interaction of light with nanostructures to develop advanced optical devices and sensors. The field also extends to nanomedicine, where nanomaterials are engineered to interact with biological systems at the molecular level to deliver drugs more effectively or detect diseases earlier.

This broad applicability and adaptability make nanotechnology uniquely positioned to provide highly targeted solutions to specific challenges across multiple domains. Each application—whether enhancing the efficiency of energy storage with nano-engineered materials, improving the recyclability of plastics through nanoscale additives, or developing smart packaging that uses nanomaterials to reduce food spoilage—represents a precise, often minuscule improvement. Yet, when these innovations are viewed collectively, their cumulative effect becomes profound, driving significant advancements in sustainability and resource efficiency.

Throughout this book, we will explore numerous opportunities where nanotechnology can contribute to sustainability, demonstrating that these myriad small-scale applications and innovations collectively offer a powerful toolset for advancing circular economy principles and addressing global environmental challenges.

This book is written for a diverse audience of professionals seeking to understand and harness nanotechnology's potential for sustainable solutions. Whether you're a researcher exploring new applications, an industry professional looking to innovate processes, a policymaker developing science-informed regulations, an educator teaching about emerging technologies, or an environmentalist interested in technological solutions to ecological challenges, you'll find valuable insights into how nanoscale innovations are reshaping our approach to global sustainability. While the concepts discussed are sophisticated, they are presented in an accessible manner that allows readers from various backgrounds to grasp both the theoretical foundations and practical applications of nanotechnology across different sectors.

The journey of nanotechnology from basic research to commercial solutions varies across applications, but its potential to address critical sustainability challenges is clear. As climate change, resource depletion, and environmental degradation become increasingly urgent, supporting the responsible development and implementation of nanotechnology solutions across sectors is not just an opportunity—it's an imperative. Through understanding and advancing these technologies, we can accelerate their contribution to building a more sustainable and resilient future.

Michael Berger

Reference

1. Z. H. Mohammad, F. Ahmad, S. A. Ibrahim and S. Zaidi, Application of nanotechnology in different aspects of the food industry, *Discover Food*, 2022, **2**, 12.

Contents

Waste Not! How Nanotechnologies Can Increase Efficiencies Throughout Society
By Michael Berger

Published by the Royal Society of Chemistry, www.rsc.org

Section II: An Essential Overview of Nanotechnology

CHAPTER 1

Introduction

Waste generation touches every aspect of modern society, from industrial production, food supply chains, consumer products, and public infrastructure to widespread inefficiencies in energy use, building design, and resource consumption. In the context of this book, 'waste' refers to a broad spectrum of unwanted or unused materials, byproducts, and inefficiencies resulting from human activities. This encompasses various forms, including solid waste (such as discarded consumer goods and packaging), liquid waste (like industrial effluents), gaseous waste (such as emissions from factories and vehicles), agricultural waste from farming and food production, and importantly, energy waste (see Figure 1.1). Energy waste includes inefficiencies in production processes, transmission losses, and unnecessary consumption in various sectors. Additionally, we consider other inefficiencies in resource use as forms of waste.

Industrial processes significantly contribute to waste and inefficiency, from raw material extraction to the disposal of used products. Food production generates waste throughout its cycle, from fields to processing plants to supermarket shelves and household waste. Packaging materials and end-of-life electronics are major sources of solid waste, while 'vampire energy' consumption (the energy consumed by electronic devices when they are plugged in but not actively in use, such as when in standby mode or turned off) exemplifies energy waste in consumer products. Energy-inefficient buildings, along with public infrastructure such as transportation systems, street lighting, and the heating and cooling of public spaces, pose challenges for resource management. Improving fuel efficiency and reducing wear and tear in transportation are critical steps toward minimizing waste.

In the energy sector alone, nearly 70% of the energy produced globally is lost as waste heat, with industrial processes wasting 55% to 68% of their input energy. In transportation, only 12% to 41% of fuel is used efficiently,

Waste Not! How Nanotechnologies Can Increase Efficiencies Throughout Society
By Michael Berger

Published by the Royal Society of Chemistry, www.rsc.org

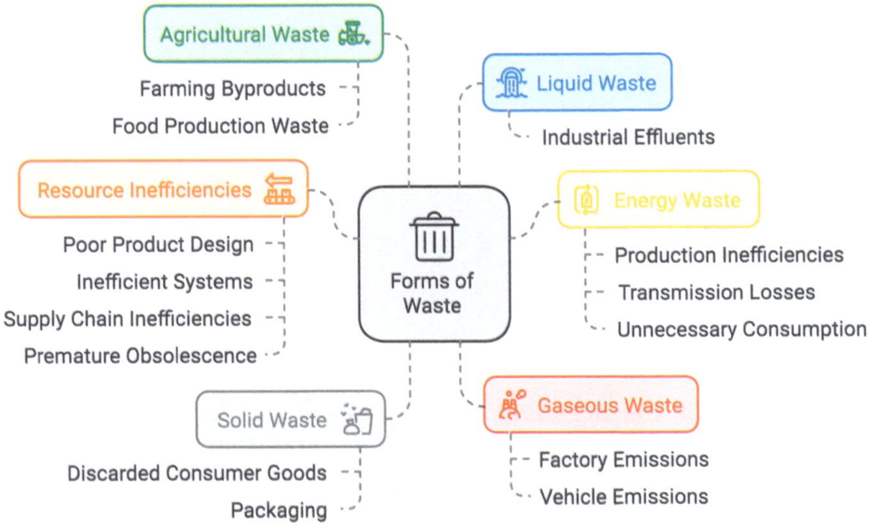

Figure 1.1 Major categories of waste in modern society.

and power plants convert just 38% of primary energy into electricity.[1] These inefficiencies contribute significantly to global warming, underscoring the urgent need for sustainable energy solutions.

Waste has always been an unavoidable byproduct of human consumption. From early civilizations to modern society, humans have generated various forms of waste. As populations grew and urbanized, waste management became a pressing issue. Ancient civilizations disposed of waste by dumping it into rivers or landfills, which eventually became unsustainable, leading to environmental degradation.

Around the 5th century B.C.,[2] Athens likely introduced the first known regulation on waste disposal, prohibiting the dumping of waste into streets or water supplies. This law required citizens to dispose of waste in designated public dumps outside the city walls and keep their properties clean. Over time, societies developed more advanced waste management techniques, leading to the concept of municipal solid waste management in the 19th century, which spurred cities to organize waste collection and disposal systems.

Today, waste management has turned into a critical global issue. The world grapples with unprecedented levels of waste generation, including solid, liquid, gaseous, radioactive, and agricultural waste. Landfills, incineration facilities, and other waste collection systems are increasingly strained, and the environmental impact of waste disposal continues to grow. Modern waste management strategies emphasize waste reduction at the source, the promotion of recycling and composting, and the adoption of new technologies such as waste-to-energy systems and zero-waste initiatives.

A related concept is "the end of abundance", which highlights resource scarcity and depletion as global populations grow and consume more

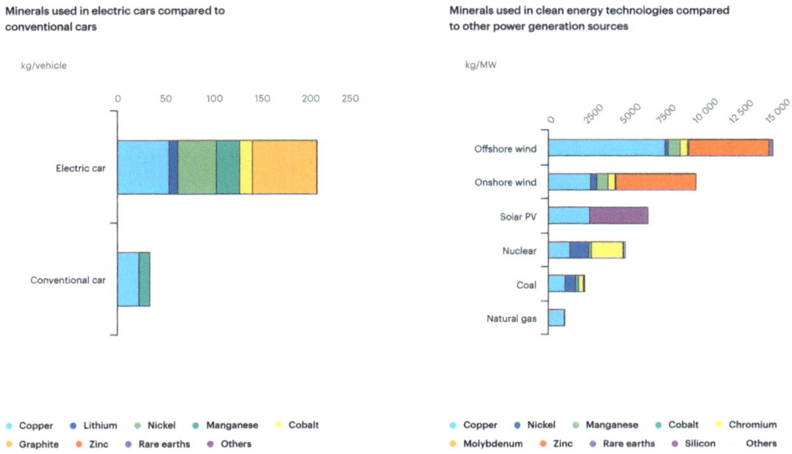

Figure 1.2 The Role of Critical Minerals in Clean Energy Transitions, IEA, Paris, 2021. Reproduced from ref. 3, https://www.iea.org/reports/the-role-of-critical-minerals-in-clean-energy-transitions, under the terms of the CC BY 4.0 license, https://creativecommons.org/licenses/by/4.0/deed.en.

resources. Minimizing waste and maximizing resource efficiency have become more critical than ever, as the previous perception of abundant resources is no longer sustainable.

However, efforts to address one problem often create new challenges. Consider the transition to clean energy, which is driving increased demand for minerals such as lithium, manganese, and rare earth elements. While solar and wind energy systems are environmentally friendly, they require significantly more minerals than fossil fuel-based systems (see Figure 1.2). An average electric vehicle requires six times more mineral inputs than a conventional car, and an onshore wind plant requires nine times more mineral resources than a gas-fired plant.

Current estimates suggest that if the transition to green energy systems continues as forecasted, supply from existing mines and projects under construction will meet only half of the projected lithium and cobalt requirements and 80% of copper needs by 2030.[3]

Or take biofuels: the production of biofuels, derived from renewable organic materials like corn-based ethanol, was initially seen as a more sustainable alternative to fossil fuels. While biofuels offer potential benefits such as reduced greenhouse gas emissions, decreased reliance on exhaustible resources, and increased farm income, they also present challenges. These include potential land use changes (such as large-scale deforestation), pressure on water resources, pollution, and increased food costs due to displacement of food crops. Some biofuels may even emit more greenhouse gases than fossil fuels, depending on the feedstock and production process. Moreover, biofuels often require subsidies and interventions to compete in the marketplace, leading to economic inefficiencies.[4,5]

In yet another example, desalination, the process of removing salt and impurities from seawater to produce fresh drinking water, has been heralded as a solution to global water scarcity. However, desalination plants require substantial amounts of energy, often from fossil fuel sources, and produce large volumes of concentrated brine as waste. This waste can have negative impacts on marine ecosystems when discharged back into the ocean, as it increases salinity and may contain residual chemicals from the desalination process.

According to the International Desalination Association (IDA), there were 15 906 operational desalination plants in 2019, producing more than 95 million cubic meters of water per day. These desalination plants also produce about 142 million cubic meters of brine daily.[6]

Or take the ban on single-use plastics: In an effort to reduce plastic pollution, some governments have implemented bans on single-use plastic items such as bags, straws, and cutlery. While this policy helps decrease plastic waste, it has led to an increased demand for alternative materials like paper and biodegradable plastics. The production of these materials also has environmental impacts, such as deforestation for paper production or the use of land and resources for growing bio-based plastics feedstocks. Additionally, the disposal of biodegradable plastics can be problematic, as they often require specific industrial composting conditions to break down effectively.

The increasing scarcity of raw materials will likely raise the cost of producing goods and providing services, affecting industries and consumers alike. Consequently, governments, businesses, non-governmental organizations, and society as a whole are shifting their attention and prioritization to sustainable practices and circular economy models. These models aim to minimize waste at every stage of the product lifecycle and maximize resource use through recycling, reusing, repurposing materials, as well as better product design and innovative solutions to avoid waste in the first place (see Figure 1.3).

The shift in focus to a circular economy is motivated by a combination of environmental, economic, and social factors. These include the need to preserve finite resources, mitigate climate change, reduce pollution, and create new economic opportunities through innovation and job creation in emerging industries.

Governments around the world are implementing policies and regulations to promote sustainable practices and circular economy models, driven by the need to protect the environment, mitigate climate change, and ensure long-term economic growth. Regulations that promote recycling, waste reduction, and resource efficiency, as well as incentives for businesses that adopt circular practices, are examples of government-driven initiatives.

Companies are increasingly adopting circular economy models to reduce their environmental footprint, save on material and energy costs, and

Figure 1.3 The circular economy is a sustainable economic model that emphasizes reducing waste and resource consumption by designing products for reuse, repair, and recycling. It involves a closed-loop system where raw materials are efficiently utilized through sustainable design, production, distribution, and consumption, followed by waste collection, management, and transformation into new resources, minimizing residual waste.

respond to consumer preferences for sustainable products. Businesses may also see opportunities for innovation, creating new products and services that align with the principles of the circular economy.

Non-governmental organizations (NGOs), environmental organizations, and community groups play a crucial role in raising awareness about the importance of the circular economy and sustainable practices. They also work with governments and businesses to develop and promote initiatives that support the transition to a circular economy.

Finally, consumers are increasingly seeking out sustainable products and services as awareness grows about the environmental impacts of resource depletion and waste generation. This demand creates a market incentive for businesses to adopt circular economy principles.

The "3Rs"—reducing, reusing, and recycling—remain essential strategies for managing the diverse types of waste generated by society. However, a truly circular economy encompasses a wider range of approaches, including resource efficiency, recovery, and eco-design. These strategies aim not only to minimize waste production but also to reduce the environmental impact of disposal and maximize resource conservation across the product lifecycle. Current targets are primarily concentrated in areas like recycling and recovery, but the transition towards a circular economy necessitates embracing additional solutions, such as reusing, refurbishing, and enhancing resource efficiency (see Figure 1.4).

The focus of this book—nanotechnology—has the potential to transform not only the 3Rs but also the broader strategies of a circular economy, such as resource efficiency, recovery, and eco-design. By making these processes more efficient, cost-effective, and sustainable, nanotechnology can play a crucial role in reimagining waste management and resource conservation.

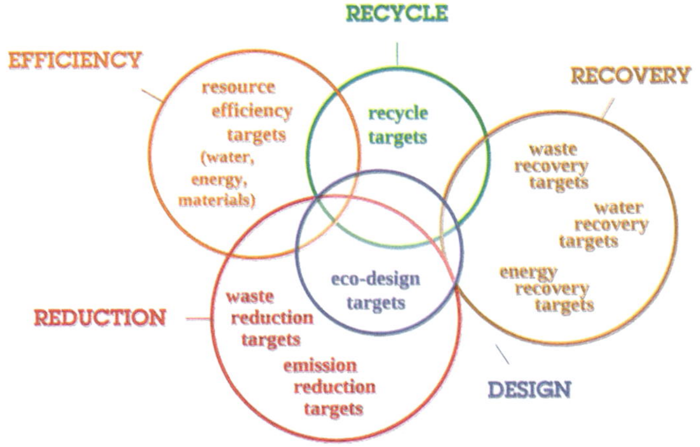

Figure 1.4 Main existing circular economy targets by areas of application. Reproduced from ref. 7, https://doi.org/10.1016/j.resconrec.2019.104553, under the terms of the CC BY 4.0 license, https://creativecommons.org/licenses/by/4.0/.

However, major technological shifts don't happen overnight, and they are rarely the result of a single breakthrough. This is especially true for the wide-ranging capabilities encompassed by nanotechnology.

Nanotechnologies manipulate matter at the molecular and atomic levels to create materials and devices with unique properties, often significantly different from their bulk counterparts. They offer great potential for addressing waste management challenges and revolutionizing how we approach waste reduction, reuse, and recycling.

It is important to understand that nanotechnology is not an industry; nor is it a single technology or a single field of research. What we call nanotechnology consists of sets of enabling technologies applicable to many traditional industries, therefore it is more appropriate to speak of nanotechnologies in the plural. These technologies serve as a versatile foundation, offering a broad range of tools and applications that can be adapted and integrated into numerous sectors. This versatility is what makes nanotechnologies a quintessential example of a platform technology.

Unlike specialized technologies that are confined to specific uses or industries, nanotechnologies provide a base upon which diverse and multifaceted applications can be built, ranging from healthcare and electronics to environmental science and beyond. This platform nature allows for unprecedented cross-industry collaborations and innovations, fostering a multidisciplinary approach to problem-solving. As industries continue to adopt and adapt various aspects of nanotechnologies, we are likely to witness a sweeping transformation in how materials and processes are developed and utilized, underscoring the pivotal role of nanotechnologies as foundational tools for modern industrial and scientific endeavors.

The potential applications of nanotechnology in sustainability efforts can be best understood by examining how they align with and enhance established waste reduction strategies. Let's explore how nanotechnology can contribute to each component of the waste hierarchy, starting with the most important:

1.1 Reducing

This solution focuses on reducing the amount of waste generated—ideally by preventing it altogether. It involves using fewer materials, energy, and resources to produce goods and services. Businesses can reduce waste by minimizing packaging, using energy-efficient appliances, and employing renewable energy sources. Nanotechnologies can contribute to this effort by producing new materials that are more durable, lightweight, and resistant to degradation, such as nanocomposites made of nanoparticles and biodegradable polymers for packaging materials. Additionally, nanotechnologies can be used to develop more efficient and durable batteries and energy storage devices, reducing electronic waste.

1.2 Reusing

This solution involves finding new uses for products and materials that would otherwise be discarded. Repairing and refurbishing products, donating them to charity, or repurposing them for other uses are some ways to implement reusing. Nanotechnologies can help by creating new materials and products with enhanced properties that can be reused multiple times, such as self-healing materials that repair themselves when damaged, extending their lifespan and reducing the need for replacement. Nanotechnologies can also be used to develop smart sensors and devices that monitor the condition of products and materials, identifying opportunities for reuse.

1.3 Recycling

Recycling involves collecting and processing waste materials to create new products. Nanotechnologies can improve the efficiency and effectiveness of recycling by creating new materials and processes that separate, purify, and transform waste materials into new products. For example, nanotechnologies can be used to develop new methods for separating and purifying metals from electronic waste, which can then be reused in the production of new electronic devices. Nanoparticles can also be used to improve the properties of recycled materials, making them stronger, lighter, and more durable.

Throughout this book, we explore how various aspects of nanotechnology can be applied to different types of waste and inefficiencies, with an initial focus on strategies to reduce waste generation. For unavoidable waste, we envision a future where waste is no longer merely a byproduct of human activity but a valuable resource to be recovered and reused. This fundamental shift in perspective and practice represents a crucial step toward a more sustainable and waste-resilient society.

In addition to exploring the technology aspects, it is vital to consider the global context in which these technologies will be implemented. International cooperation and policies are key to standardizing, ensuring safety, and promoting the ethical deployment of these innovations. Global standards and regulations are crucial for the safe production, handling, and disposal of nanomaterials, given their environmental and health risks. Collaborative international research can establish shared safety protocols and best practices.

Cross-border transfer can accelerate the adoption of effective technology solutions. International partnerships can bridge gaps between developed and developing countries by facilitating knowledge exchange and advancing technological progress.

Finally, global policies can drive investments and funding for nanotechnology research. International initiatives ensure these technologies are accessible and beneficial worldwide.

While nanotechnology is not a panacea for the waste problem, it should be seen as a powerful set of tools and approaches that can contribute to a broader suite of solutions, working in tandem with policy interventions, behavioral change, and infrastructure development. The examples and case studies presented throughout this book demonstrate how nanotechnology can significantly contribute to reducing waste, increasing resource efficiency, and supporting the transition towards a circular economy.

Realizing this potential requires more than isolated technological advances. The insights and possibilities explored in this book point toward the need for a comprehensive strategic framework. This book introduces NanoLEAP—a proposed vision for orchestrating the next decade of nanotechnology development in waste management and sustainability. This envisioned 10-year strategic initiative would combine targeted funding, collaborative efforts, and cross-industry partnerships to leverage nanotechnology's full potential in critical areas such as carbon capture, energy efficiency, and circular economy practices.

The NanoLEAP framework charts a path toward a sustainable, resource-efficient future, where the technologies developed today will play a pivotal role in addressing tomorrow's most pressing environmental challenges.

References

1. Q. Bian, Waste heat: the dominating root cause of current global warming, *Environ. Syst. Res.*, 2020, **9**, 8.
2. F. A. UriarteJr, Solid waste management: principles and practices: an introduction to the basic functional elements of solid waste management, with special emphasis on the needs of developing countries, 2008, [cited 2024 Sep 16]. Available from: https://www.ukdr.uplb.edu.ph/books/28.

3. The Role of Critical Minerals in Clean Energy Transitions, IEA, [cited 2024 Sep 16]. Available from: https://www.iea.org/reports/the-role-of-critical-minerals-in-clean-energy-transitions.
4. United States Environmental Protection Agency (EPA), Economics of Biofuels, [cited 2024 Sep 15]. Available from: https://www.epa.gov/environmental-economics/economics-biofuels.
5. T. J. Lark, J. Meghan Salmon and H. K. Gibbs, Cropland expansion outpaces agricultural and biofuel policies in the United States, *Environ. Res. Lett.*, 2015, **10**(4), 044003.
6. E. Jones, M. Qadir, M. T. H. van Vliet, V. Smakhtin and S.-M. Kang, The state of desalination and brine production: A global outlook, *Sci. Total Environ.*, 2019, **657**, 1343–1356.
7. P. Morseletto, Targets for a circular economy, *Resour. Conserv. Recycl.*, 2020, **153**, 104553.

Section I: Understanding Waste and Its Impact on Society

CHAPTER 2

Where Is All the Waste Coming from?

Waste, in its many forms, is an inevitable byproduct of human activity, permeating every aspect of our lives. From industrial manufacturing to household consumption, humans generate waste in solid, liquid, gaseous, and energetic forms. This waste creation occurs across all sectors of society – industrial, agricultural, commercial, medical, and residential.

The sources and types of waste are diverse. Manufacturing processes generate solid waste, hazardous materials, and emissions. Energy production creates radioactive waste, ash, and significant heat losses. Healthcare facilities produce biohazardous materials and pharmaceutical waste. Agricultural activities result in chemical runoff, animal waste, and food loss. Even our daily activities contribute to multiple waste streams through energy inefficiency, water waste, and discarded products.

Poor design and inefficient practices play a significant role in exacerbating these problems across industries. This includes not just product design, but also process design, system design, and infrastructure design. For instance, inefficient building design leads to energy waste, poor manufacturing processes result in material and water waste, and inadequate infrastructure design contributes to transmission losses and resource inefficiency.

Understanding the origins, categories, and volumes of waste across these different domains is crucial for identifying systemic issues and potential areas for intervention. This understanding highlights inefficiencies in our current systems and serves as a call to action for developing sustainable practices and policies across all sectors and waste types.

While waste takes many forms—from gaseous emissions to liquid effluents to energy losses—solid waste remains one of the most visible and challenging aspects of our waste crisis. Figure 2.1 illustrates the composition of municipal

Waste Not! How Nanotechnologies Can Increase Efficiencies Throughout Society
By Michael Berger

Published by the Royal Society of Chemistry, www.rsc.org

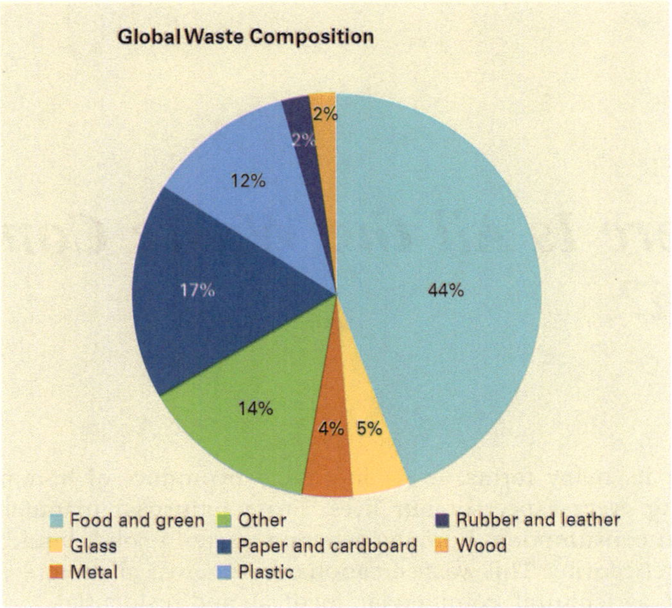

Figure 2.1 Categorization of types of materials in municipal solid waste in percent. Reproduced from ref. 44, https://hdl.handle.net/10986/30317, under the terms of the CC BY 3.0 license, https://creativecommons.org/licenses/by/3.0/igo/.

solid waste, showing how food and green waste makes up the largest portion. This breakdown helps us understand one important aspect of our waste challenge, though it represents just one piece of the broader waste puzzle discussed throughout this chapter.

Recognizing the staggering amounts and varieties of waste generated emphasizes the environmental, economic, and public health challenges we face. This realization should drive us towards implementing more sustainable solutions and fostering responsible practices in all aspects of human activity.

The comprehensive overview presented in this chapter underscores how waste generation and improper management permeate every sector of modern society, resulting in detrimental impacts on the environment and public health. Key highlights include:

- Billions of tons of waste are generated globally each year, with significant contributions from sectors like manufacturing, agribusiness, retail, healthcare, and households. This puts immense strains on ecosystems, infrastructure, and resources required to handle accumulating waste.
- A significant portion of the waste produced is hazardous or challenging to manage, containing heavy metals, plastics, chemicals, and other

toxic substances. Improper handling of these waste streams causes pollution, biodiversity loss, and increased risks to human health.
- A throwaway culture and products designed for obsolescence are major factors exacerbating waste, especially plastic, e-waste and textile waste. This promotes disposability over sustainability.
- Beyond solid waste, significant challenges include emissions and air pollution from industry and transport, water waste and contamination from agriculture and manufacturing, energy losses across production and transmission systems, and wastewater from various sectors. These less visible forms of waste often have far-reaching environmental and health impacts.
- Innovative technologies and circular economy approaches can help mitigate various waste issues through recycling, recovery, remanufacturing, and waste-to-energy processes. However, infrastructure and implementation remain insufficient.
- Sustainable product and process design, such as packaging-free goods, reusable alternatives, energy-efficient systems, and toxic-free manufacturing, can reduce waste at the source while benefiting business through cost and resource savings.

2.1 Industrial Manufacturing

The industrial manufacturing sector is a significant contributor to global waste production, resource depletion, and environmental pollution. Manufacturing processes, driven by resource-intensive practices and often outdated technologies, lead to the generation of solid waste, hazardous materials, and emissions, alongside inefficient energy use. For example, the cement industry accounts for around 7% of global CO_2 emissions. Traditional cement production methods release approximately 0.8–1.0 tons of CO_2 per ton of cement, while newer, more efficient technologies can reduce CO_2 emissions by up to 30%.[1]

Manufacturing processes involve complex supply chains and the use of raw materials, chemicals, and energy, all of which lead to substantial waste production. The primary forms of waste in industrial manufacturing include solid waste, hazardous waste, and emissions (both air and water).

2.1.1 Solid Waste Generation

The global manufacturing sector produces substantial amounts of solid waste.[2] The industrial sector generates approximately 11 billion metric tons of solid waste annually (that is in addition to the 2 billion tons of municipal waste[3]), with construction, metal fabrication, and automotive production being key contributors.[4] Waste from manufacturing processes often includes metals, plastics, and chemicals. Over 20% of this industrial waste is discarded without proper recycling, leading to increased pressure on landfills and environmental degradation.[3]

2.1.2 Hazardous Waste

Hazardous waste is another major byproduct of industrial manufacturing. This type of waste poses substantial or potential threats to public health and the environment due to its characteristics. Hazardous waste can be toxic, reactive, ignitable, or corrosive. It includes materials such as chemical solvents, heavy metals, and certain types of sludge from industrial processes. Over 400 million tons of hazardous waste is generated annually worldwide, much of it originating from manufacturing sectors such as electronics, chemicals, and pharmaceuticals.[5] Improper handling of this waste can result in contamination of soil and water systems, with dangerous substances like lead, cadmium, and mercury being released into ecosystems.

2.1.3 Water Usage and Wastewater

Manufacturing is also a major consumer of freshwater, accounting for roughly 19% of global freshwater withdrawals.[6] This water is often used for cooling, cleaning, and material processing. However, inefficient water use and improper wastewater management lead to pollution and resource depletion. The textile industry, for example, consumes around 93 billion cubic meters of water annually (including cotton farming), contributing to water pollution through the discharge of untreated wastewater, which contains dyes, chemicals, and microplastics.[7]

2.1.4 Energy Inefficiency

The energy-intensive nature of industrial processes exacerbates waste generation. The International Renewable Energy Agency (IRENA) estimates that the industry sector accounted for 36% of the global final energy consumption in 2020.[8] Moreover, production processes are carbon intensive, making industry responsible for one-fourth of the global energy-related CO_2 emissions (the second-largest emitter after the power sector). A significant portion of this energy is wasted due to inefficiencies in production methods, outdated technologies, and reliance on fossil fuels.

2.2 Agribusiness

Agribusiness, encompassing large-scale agricultural production, faces significant environmental and public health challenges due to waste generation from the excessive use of pesticides, synthetic fertilizer runoff, and animal manure. This waste not only contaminates water sources but also contributes to environmental degradation. Precision agriculture offers a sustainable solution by employing high-tech equipment and innovative practices to optimize farm inputs, minimize waste, and reduce greenhouse gas emissions. Consequently, these practices enhance farm productivity and economics, benefiting both the environment and the agricultural sector.

2.2.1 Excessive Pesticide Use

Agribusinesses commonly use large quantities of pesticides to protect crops from pests and diseases. However, excessive pesticide use can lead to waste and environmental harm. According to the US Environmental Protection Agency (EPA)[9], about 40–60% of applied pesticides can end up in non-target areas due to factors such as drift or runoff, posing threats to water sources, ecosystems, and wildlife. For instance, the widespread use of neonicotinoid pesticides has been linked to the decline of pollinators, such as bees and butterflies, which are essential for maintaining biodiversity and crop production.

2.2.2 Synthetic Fertilizer Runoff

Agribusinesses often rely on synthetic fertilizers to boost crop yields. However, improper application and poor management practices can result in nutrient runoff into water sources. One study[10] found that up to 50% of nitrogen and 40% of phosphorus from fertilizers are not absorbed by crops, instead entering water sources and contributing to harmful algal blooms, eutrophication, and dead zones in bodies of water. For example, the Gulf of Mexico's "Dead Zone," an area with low oxygen levels that cannot support marine life, has been linked to excessive nutrient runoff from the Mississippi River Basin, where large-scale agriculture is prevalent.

By addressing these waste-generating practices in agribusiness, such as adopting integrated pest management strategies and optimizing fertilizer application, the sector can mitigate its environmental and public health impacts.

Consider the concept of precision agriculture. This is a farming management concept that uses technology to optimize farm inputs, such as water, fertilizer, and pesticides, to maximize crop yields while minimizing waste and environmental impact. Precision agriculture practices[11] employ high-tech equipment to minimize agricultural inputs and target them to specific needs of fields, reducing greenhouse gas emissions. By utilizing technologies like GPS-guided machinery, remote sensing, and variable-rate application equipment, farmers can apply the precise amount of inputs needed for specific areas of their fields, reducing waste and preventing pollution.

These practices also positively impact farm productivity and economics by maintaining or increasing yields while lowering production costs. This work presents precision agriculture technologies that can mitigate greenhouse gas emissions, offering brief descriptions and their reported impacts. Technologies include variable rate sowing/planting, fertilizing, spraying, weeding, and irrigation.

2.3 Consumer Products

Key waste issues in the consumer industry include excessive packaging, contributing to landfill buildup and pollution; single-use plastics, exacerbating the plastic pollution crisis; and fast fashion waste, promoting a throwaway culture and increased textile waste. By adopting sustainable materials and recycling innovations, companies can help reduce waste and its environmental impacts.

2.3.1 Excessive and Non-recyclable Packaging Materials

Packaging waste significantly contributes to environmental pollution. In 2021, the countries in the European Union generated an estimated 84.3 million metric tons of packaging waste.[12] This equates to about 188.7 kg per inhabitant in the EU, with numbers varying from 73.8 kg per capita in Croatia to 246.1 kg per capita in Ireland. The total consisted of 40% paper and cardboard, 19% each for plastic and glass, 17% wood, and 5% metal.

One example of excessive packaging is the use of multiple layers of plastic and non-recyclable materials for single-use products. For instance, individually wrapped snack items may be further packed in a larger plastic bag and then placed in a cardboard box for shipping. Similarly, toothpaste typically comes in a plastic tube, which is often placed inside an additional cardboard box. By opting for sustainable, reusable, and recyclable packaging materials, companies can help reduce packaging waste and its associated environmental impacts.

2.3.2 Fast Fashion

The fast fashion industry is characterized by the rapid production and turnover of inexpensive clothing items, contributing to a throwaway culture and increased textile waste. Consumers inevitably participate in a fashion market that produces garments at ever-lower costs (such as $4.99 supermarket t-shirts), encouraging impulse buying or shopping as entertainment.

Now let this figure sink in for a moment: 100 billion garments are produced globally each year, with 33% ending up in landfills within the first year of purchase.[13]

Fast fashion items are often made from low-quality materials that do not last long, further exacerbating the waste problem. Technological solutions, such as recycling innovations and the development of sustainable materials, can help mitigate textile waste. Companies like Evrnu and Worn Again Technologies have developed processes to recycle discarded textiles into new, high-quality fibers that can be used to create new clothing items, reducing waste and promoting circularity in the fashion industry.

2.3.3 Single-use Disposable Items

Single-use disposable items, like plastic cutlery, straws, and food containers, generate considerable waste and environmental pollution. Single-use plastics account for 40% of the total plastic produced globally, and less than 10%

of all plastic waste is recycled.[14] Technological innovations can help address this issue by offering alternatives to single-use items.

Companies like Notpla and Stora Enso have developed biodegradable and compostable materials made from seaweed and plant fibers, respectively, which can be used to create sustainable packaging and disposable products. By replacing single-use plastics with eco-friendly alternatives, we can reduce waste generation and the environmental impacts associated with plastic pollution.

2.4 Food Waste

Strictly speaking, food waste would largely be part of agribusiness waste on the production end and consumer waste on the consumption end, but this category is worth mentioning on its own because it is so significant.

One of the main issues with food waste is the inefficient use of resources, such as water, land, and energy, which contributes to environmental degradation. Food waste exacerbates food insecurity, as food that could have been used to feed the hungry is discarded.

Food waste is an increasingly pressing global issue, with serious environmental, economic and social implications. The 2024 UNEP Food Waste Index Report[15] estimates that in 2022, around 1.05 billion tons of food waste was generated by households (60%—which amounts to 132 kg per capita wasted per year globally), food services (28%) and retail (12%). Overall, this food waste is responsible for 8–10% of global greenhouse gas emissions. This is in addition to the estimated 1.2 billion tonnes of food waste generated at the production stage on farms and fisheries. Add to that losses occurring between farm and retail—transport, storage, manufacturing and processing—of about 436 million tonnes, the total amount of food waste generated globally is estimated to be well over 2.5 billion tonnes annually.[16]

This suggests that as much as 40% of all food produced is never eaten when both farming and post-farming losses are considered.

Many valuable resources go down the drain as well—literally—including water used during production/harvesting stages plus energy used throughout transport & storage phases prior to eventual disposal by landfill sites.

In developed countries, food waste often occurs due to overproduction combined with a culture that values presentation over taste or nutrition—leading consumers to discard large quantities of perfectly edible produce because it doesn't look 'perfect' enough for sale or consumption at home. Other factors contributing towards household wastage are confusion about best before dates on packaging labels; poor portion control; inadequate storage practices; and a lack of meal planning skills.

Food waste ends up in various destinations, including landfills, composting facilities, or anaerobic digestion plants. Landfills are the most common destination, where food waste decomposes and generates methane, a potent greenhouse gas.[17]

Composting is a more sustainable alternative, as it converts food waste into a valuable nutrient-rich material that can be used to enrich soil and support plant growth. However, not all food waste is suitable for composting, and the process requires appropriate infrastructure and management.

Anaerobic digestion is another option for managing food waste, wherein microorganisms break down organic materials in an oxygen-free environment, producing biogas (a mixture of methane and carbon dioxide) that can be used for energy production. This method not only helps reduce methane emissions from decomposing food waste but also provides a renewable source of energy.

2.5 Automotive Industry

The automotive sector generates waste throughout the manufacturing process, vehicle lifespan, and maintenance operations. These wastes range from packaging materials, metal shavings, and scrap materials to solvents, wastewater, and chemicals. Additionally, the energy consumed during the production and assembly of vehicles contributes to greenhouse gas emissions and climate change. Manufacturing an average passenger vehicle produces between 5.6 (gasoline) and 8.8 (electric) metric tons of CO_2 emissions.[18]

During the manufacturing process, waste materials may include metal shavings, scrap materials, and packaging materials, which result from cutting, shaping, and assembling vehicle components. Additionally, the use of solvents, chemicals, and wastewater is common in automotive production as they aid in processes such as cleaning, painting, and surface treatment.

Throughout a vehicle's lifespan, waste continues to be generated. Replacement parts such as batteries, tires, filters, and various fluids contribute to this ongoing waste stream. Vehicle maintenance often involves the use of corrosive materials like phosphoric, hydrochloric, and hydrofluoric acids for cleaning and degreasing, which can pose environmental risks if not properly managed. Daily vehicle operations also contribute to air pollution, with emissions influenced by factors such as fuel efficiency and maintenance practices.

End-of-life vehicles (ELVs) represent a significant waste management challenge. In 2020, the EU scrapped 5.4 million vehicles, generating 6.3 million tons of waste,[19] while the U.S. scrapped over 15 million vehicles in the same year.[20] The rise of electric vehicles (EVs) presents new waste management challenges, particularly regarding battery disposal and recycling. By 2030, an estimated 145 million EVs will be on the road globally, necessitating specialized recycling processes for their batteries, which contain hazardous materials.

The rapid growth of electric vehicles has brought significant environmental challenges associated with batteries to the forefront. As EVs gain popularity, the International Energy Agency estimates[21] that by 2030, there would be around 145 million electric vehicles on the road, across all transport modes. This increase will have considerable implications for battery waste and recycling. Battery recycling, especially for lithium-ion batteries used in many EVs, is essential for reducing waste and managing the demand for finite resources used in battery production.

To address these challenges, the automotive industry is adopting various waste reduction strategies. The use of lightweight materials such as carbon fiber composites, advanced high-strength steel, and aluminum alloys can improve fuel efficiency and reduce emissions. Improved recycling processes for ELVs and their components, especially batteries, are being developed by companies like Redwood Materials and Li-Cycle.

Manufacturers are also focusing on designing for sustainability, creating vehicles with longer lifespans, easier maintenance, and more recyclable parts. Energy efficiency is another key area of focus, with investments in more fuel-efficient designs and alternative power sources like electric and hydrogen fuel cells. The industry is also exploring the use of eco-friendly materials, including bio-based materials and nanomaterials, to reduce its environmental impact.

By addressing these waste challenges through innovative technologies and sustainable practices, the automotive industry can significantly reduce its environmental footprint and contribute to a more sustainable future. This holistic approach to waste reduction aligns with broader efforts across various sectors to minimize environmental impact and promote circular economy principles.

2.6 Retail Sector

The retail sector is a significant contributor to waste generation, primarily through excessive packaging, single-use products, and inefficient inventory management. These issues not only lead to environmental problems but also result in economic losses for businesses.

Excessive packaging is one of the most visible forms of waste in retail. Around 40% of plastic packaging waste ends up in landfills, while only 14% is recycled.[22] Common examples include plastic clamshells for fruits and vegetables, individually wrapped items, and layers of packaging for shipping. This overuse of packaging materials, especially plastics, contributes to pollution and resource depletion.

To address this issue, retailers are exploring eco-friendly packaging alternatives. Biodegradable materials like polylactic acid (PLA) and bio-based polyethylene are becoming more common. Some innovative companies are pushing the boundaries even further. For instance, Indonesian startup Evoware has developed seaweed-based packaging for food items, which is not only biodegradable but also nutritious.

Single-use products represent another significant source of waste in the retail sector. Disposable cutlery, cups, and bags are ubiquitous in many retail environments, contributing to the growing problem of plastic pollution. Retailers are increasingly adopting sustainable alternatives made from materials like bagasse (a byproduct of sugarcane processing) or bamboo.

Moreover, there's a growing trend of encouraging customers to use reusable items such as cloth bags and metal or glass containers.

Inefficient inventory management is a less visible but equally important contributor to retail waste. When retailers overstock or fail to sell products before they expire or go out of fashion, perfectly good items often end up in landfills. This not only wastes resources but also represents a significant economic loss for businesses.

To combat these issues, retailers are leveraging technology solutions. Advanced inventory management systems using artificial intelligence and machine learning can help predict demand more accurately, reducing overstocking. Some retailers are also adopting "just-in-time" inventory practices to minimize waste.

An increasing number of retailers are embracing the circular economy concept. This involves designing products for longer life, easier repair, and eventual recycling or upcycling. Some companies are even experimenting with rental or resale models to extend product lifecycles.

By addressing these waste issues through technological innovation, sustainable materials, and improved management practices, the retail sector can significantly reduce its environmental footprint. These efforts not only contribute to sustainability goals but can also lead to cost savings and improved customer loyalty, as consumers increasingly prioritize eco-friendly practices.

As we move forward, the challenge for the retail sector will be to balance consumer convenience and product protection with sustainability. This will require ongoing innovation in materials science, logistics, and business models. The retail sector's efforts to reduce waste are crucial in the broader context of global waste management and environmental protection.

2.7 E-waste

Electronic waste, or e-waste, has emerged as a significant global environmental challenge in our increasingly digital world. This category of waste encompasses discarded electronic devices such as smartphones, computers, and televisions, which often contain hazardous materials that can harm both the environment and human health.

The scale of the e-waste problem is staggering. According to the Global E-waste Monitor 2020,[23] the world generated 53.6 million metric tons of e-waste in 2019. Alarmingly, only 17.4% of this waste was officially documented as properly collected and recycled, leaving a vast amount of potentially harmful materials unaccounted for.

Several factors contribute to the growing e-waste crisis. Chief among them is the rapid advancement of technology, coupled with the continuous release of new devices. The average lifecycle of a smartphone, for instance, is approximately 2.5 years and is expected to decrease further.[24] This short lifespan results in millions of discarded devices annually. Similarly, the constant improvements in computer performance and capabilities lead consumers to frequently upgrade their hardware, further exacerbating the e-waste problem.

The environmental implications of improper e-waste disposal are severe. When electronic devices end up in landfills, toxic substances such as lead, mercury, and cadmium can leach into the soil and groundwater. This contamination poses significant risks to local ecosystems and, by extension, human health.

The e-waste challenge also presents opportunities for innovative solutions. Technological advancements are paving the way for more efficient and environmentally friendly recycling methods. For example, researchers are exploring hydrometallurgical processes, which use aqueous chemistry for the recovery of metals from e-waste.[25,26] These methods can be more efficient and less polluting than traditional pyrometallurgical processes.

Another promising avenue is the use of biological methods for metal recovery. This approach, known as bioleaching, employs bacteria, fungi, or other microorganisms to extract valuable metals from the dust created during the mechanical recycling of electronic waste. These biological processes can be more energy-efficient and environmentally friendly than conventional methods.[27]

The concept of the circular economy is also gaining traction in the electronics industry. This approach emphasizes designing products for longevity, repairability, and eventual recycling. Some manufacturers are exploring modular designs that allow for easy upgrades and repairs, potentially extending the lifespan of electronic devices and reducing e-waste generation.

Policy measures also play a crucial role in addressing the e-waste challenge. Many countries are implementing extended producer responsibility (EPR) programs, which make manufacturers responsible for the entire lifecycle of their products, including disposal and recycling. These policies incentivize companies to design more sustainable products and establish effective recycling systems.

As we move forward, addressing the e-waste problem will require a multifaceted approach. This includes advancing recycling technologies, promoting sustainable design practices, implementing effective policies, and raising consumer awareness about responsible electronics disposal. By tackling this issue comprehensively, we can mitigate the environmental impacts of e-waste and work towards a more sustainable electronics industry.

2.8 Fashion Industry

Fast fashion is a business model that emphasizes rapid production cycles, aiming to capitalize on the latest trends by delivering high volumes of affordable clothing to consumers. This strategy, while catering to the ever-changing demands of customers, has led to a surge in textile waste, as garments are quickly deemed unfashionable and discarded. The short life cycle of these garments, combined with the utilization of low-quality materials, exacerbates the waste generation problem faced by the fashion industry.

The issue of textile waste has emerged as a significant contributor to global waste and environmental pollution. For instance, the Environmental Protection Agency (EPA) reported[28] that in 2018, the United States generated approximately 17 million tons of textile waste, with only 15% being recycled, 19% combusted for energy recovery, and a staggering 67% ending up in landfills. This increase in waste not only strains landfills but also contributes to greenhouse gas emissions, as decomposing textiles release methane and other harmful gases.

Synthetic fibers, such as polyester, play a critical role in this problem. As non-biodegradable materials, they persist in the environment and release microplastics when washed, which eventually enter aquatic ecosystems and accumulate in the food chain. In fact, a study by researchers at the University of California, Santa Barbara, found that a single polyester garment could release around 1900 microplastic fibers per wash.[29]

Additionally, the textile dyeing process is a significant source of water pollution and the fashion industry is considered the second-largest consumer of water globally. Harmful chemicals from dyes and finishing agents are often released into water bodies, leading to adverse ecological and health consequences.

In contrast, technology-driven solutions have emerged to help mitigate the environmental impact of the fashion industry. For instance, the use of closed-loop water recycling systems in textile dyeing can significantly reduce water consumption and pollution. A notable example is the ColorZen technology, which has demonstrated the ability to decrease water use by up to 90% and energy consumption by 50% during the dyeing process.[30]

Another promising development is the incorporation of biodegradable materials, such as Tencel lyocell fibers in garment production. Derived from sustainably sourced wood pulp, Tencel fibers are extracted from sustainably grown wood using a unique closed loop system which recovers and reuses the solvents used, minimizing the environmental impact of production.[31]

The textile industry is notorious for combining outdated technologies and inefficient resource use, generating significant waste, including offcuts, trimmings, and dye byproducts. The dyeing sector of the fashion industry is particularly harmful, depleting essential resources. Lifecycle assessments show that it takes 3781 liters of water to make one pair of jeans, considering cotton production, manufacture, transport, and washing. The overall environmental cost of the fashion industry is staggering:[32–34]

- The fashion industry uses 93 billion cubic meters of water annually—enough to meet the consumption needs of five million people.
- Dyeing 1 ton of fabric requires 200 tons of water. Furthermore, 1 ton of fabric barely represents 0.5% of the estimated 80 billion garments produced annually by the fashion industry.

- Around 20% of wastewater worldwide comes from fabric dyeing and treatment.
- Of the total fiber input used for clothing, 87% is incinerated or disposed of in a landfill. Every second, one truckload of clothing is landfilled or burned.
- Every year a half a million tons of plastic microfibers are dumped into the ocean, equivalent to 50 billion plastic bottles.

2.9 Healthcare and Medical Facilities

A study[35] analyzing the carbon footprint of products used in the five most common surgical operations in the NHS in England revealed that 68% of carbon contributions come from single-use items, such as gowns and drapes. The research identifies significant carbon contributors as the production and waste disposal of single-use items, as well as decontamination processes for reusable products.

The study found that relatively few products (23%) were responsible for over 80% of the product carbon footprint. Applying the carbon dioxide equivalent (CO_2-eq), the standard unit for measuring carbon footprints, to each of the operations, the researchers showed that the operation with the highest product carbon footprint was knee replacement (85.5 kg CO_2-eq). This was followed by gall bladder removal (20.3 kg CO_2-eq), carpal tunnel decompression surgery (12.0 kg CO_2-eq), hernia repair (11.7 kg CO_2-eq) and tonsillectomy (7.5 kg CO_2-eq).

Across the five operations, the researchers found that relatively few products (23%) were responsible for more than 80% of the product carbon footprint. The researchers suggest focusing on strategies to reduce emissions from high-emission products, typically large single-use items, to mitigate the carbon footprint of surgical operations.

Proposed strategies include eliminating or finding low-carbon alternatives for high-emission products, avoiding non-sterile gloves when handwashing is sufficient, not opening gauze swab packs unless required, and asking suppliers to remove rarely used items from single-use pre-prepared packs. The researchers note that reusable surgical textiles can significantly reduce carbon footprints without being clinically inferior. By eliminating single-use items, switching to reusable alternatives, optimizing decontamination processes, and improving waste segregation and recycling, the product carbon footprint could be reduced by one third.

Outside the operating room, healthcare facilities generate a variety of waste, including biohazardous waste, sharps waste, pharmaceutical waste, and general waste. The improper disposal of this waste can pose risks to public health and the environment. The improper management of medical waste can result in the spread of infections, water contamination, and air pollution. Proper handling, segregation, and disposal of medical waste are essential to minimize the environmental and public health impacts.

Biohazardous waste, which includes blood, body fluids, and tissues, may contain pathogens that can cause diseases. For instance, the improper disposal of biohazardous waste, such as contaminated needles or syringes, can lead to the transmission of bloodborne pathogens, like hepatitis B, hepatitis C, and HIV. In addition to direct transmission, these pathogens can also contaminate water sources and soil, leading to indirect exposure to infections.

Sharps waste, such as needles, syringes, and lancets, can pose a risk of injury and infection if not handled and disposed of properly. Pharmaceutical waste, which includes expired, unused, and contaminated medications, can have detrimental effects on the environment and public health if not managed appropriately. The presence of pharmaceuticals in water systems has been linked to the development of antibiotic resistance in bacteria and negative impacts on aquatic life[36] and that exposure to pharmaceuticals in waterways led to changes in fish behavior and physiology.[37]

General waste, such as paper, plastic, and food waste, generated in healthcare facilities can also contribute to environmental pollution if not properly managed. The improper disposal of general waste can lead to air pollution from open burning or methane emissions from landfills, contributing to climate change and public health issues, such as respiratory problems and cardiovascular diseases.

Several technology options exist to prevent some of this waste from accruing. For instance, healthcare facilities can adopt green chemistry principles to reduce waste generation at the source by opting for pharmaceuticals with fewer hazardous components or utilize alternative, environmentally friendly chemicals for cleaning and sterilization processes.

Laboratories can implement strategies to minimize waste such as microscale techniques, which involve using smaller amounts of chemicals for experiments and substituting hazardous chemicals with less toxic alternatives. Moreover, the sharing of equipment and reagents between labs can help minimize the generation of waste.

2.10 Construction and Demolition

The construction industry is a significant contributor to waste generation, particularly in the form of construction and demolition waste. Construction and demolition (C&D) activities are a major contributor to solid waste generation worldwide. In fact, the construction industry is responsible for generating more than a third of all waste in the European Union.[38]

C&D waste encompasses a wide range of materials discarded during the construction, renovation, and demolition of buildings, roads, and bridges. This category of waste includes but is not limited to concrete, bricks, wood, glass, metals, plastics, asphalt, and excavated soil.

C&D waste is a critical environmental concern for several reasons:

- **Volume:** It represents a significant portion of the total waste generated, with varying figures globally but notably high in developed regions like the EU.

- **Resource consumption:** The generation of C&D waste signifies a substantial consumption of natural resources, given the materials involved.
- **Landfill space:** A significant portion of C&D waste ends up in landfills, occupying valuable land space and contributing to pollution.
- **Carbon footprint:** Many construction materials, especially concrete, have high carbon footprints due to energy-intensive production processes.

These materials, if not properly managed, can contribute to landfill overflow and environmental pollution. Additionally, improper waste management can lead to issues such as soil contamination, water pollution, and greenhouse gas emissions.

Some other waste materials generated during construction include concrete, bricks, wood, glass, metals and plastic. For instance, the disposal of gypsum-based materials, such as drywall, can release harmful hydrogen sulfide gas if not managed correctly.[39]

Sustainable construction practices, such as using recycled materials, implementing waste reduction strategies, and employing green building techniques, can help minimize waste generated during construction and demolition activities. These practices not only reduce the environmental impact of construction projects but also promote resource efficiency and cost savings.

2.11 Energy Production and Distribution

Waste generation in the energy production and distribution sector arises from several sources. The most common types of waste generated in the energy sector include:[40]

- **Fly ash and bottom ash**: These residues, generated from the combustion of coal in power plants, contain heavy metals and other pollutants. In the United States alone, coal-fired power plants generate approximately 110 million tons of coal ash annually, with about 40% being recycled and the rest disposed of in landfills or impoundments.
- **Flue gas desulfurization sludge**: Produced during the removal of sulfur dioxide from flue gases, this waste can contain hazardous substances. A typical 500-megawatt coal-fired power plant can produce around 125 000 tons of gypsum from flue gas desulfurization annually.
- **Waste energy**: Generated in the form of heat loss during energy conversion processes and losses during transmission and distribution due to resistance in power lines and transformers. It is estimated that up to 60% of the energy generated in traditional power plants is lost as waste heat during conversion processes.
- **Radioactive waste**: Generated by the nuclear energy industry, such as spent nuclear fuel, which poses environmental and public health risks. The nuclear industry worldwide produces over 2000 metric tons of spent nuclear fuel annually, which requires safe long-term storage

solutions. The International Atomic Energy Agency (IAEA) reported that the current total global inventory of solid radioactive waste is approximately 38 million m^3, of which 30.5 million m^3 (81% of the total) has been disposed of permanently and a further 7.2 million m^3 (19%) is in storage awaiting final disposal.[41]

- **Decommissioning waste:** Arising from the decommissioning of energy facilities like power plants and oil rigs, including hazardous substances like asbestos and lead.
- **Renewable energy waste:** While renewable energy sources are far less waste-intensive than traditional energy production, they still have waste considerations. For example:
  - Solar panel waste: End-of-life solar panels, while recyclable, can contain some hazardous components. Currently, it is estimated that over 90% of discarded solar panels end up in landfills, though this is starting to change with increased recycling efforts. The International Renewable Energy Agency (IRENA) estimates that by 2050, there could be around 78 million tonnes of cumulative PV panel waste globally.[42]
  - Wind turbine waste: Older wind turbine blades can be challenging to recycle. In some locations, decommissioned wind turbine blades are being disposed of in landfills due to the lack of viable recycling options. According to the Global Wind Energy Council (GWEC), the accumulation of wind turbine blade waste is expected to reach approximately 43 million tonnes worldwide by the year 2050.
  - Battery waste: Large-scale energy storage systems rely on batteries, which present end-of-life disposal concerns. While lithium-ion batteries have valuable components, the recycling process is complex, and infrastructure is currently insufficient to handle the expected volume of waste.

These waste types highlight the environmental challenges associated with energy production and distribution, emphasizing the need for sustainable waste management practices in the sector.

Technical solutions to reduce waste and waste energy in this sector include transitioning to renewable energy sources like solar, wind, and hydroelectric power, implementing carbon capture and storage technology to capture CO_2 emissions from power plants and industrial facilities, and recycling and reusing waste materials like fly ash in construction materials. To address energy loss during transmission and distribution, upgrading power grids with advanced technologies, such as high-voltage direct current (HVDC) systems, smart grid solutions, and superconducting materials, can improve energy efficiency and reduce waste energy.

2.12 Information and Communication Technology

Waste generation issues in the Information and Communication Technology (ICT) sector are multifaceted. The rapid growth of ICT and the short lifespan of electronic devices contribute to a significant increase in electronic waste, or e-waste. This includes discarded computers, smartphones, and other electronic equipment.

Improper disposal of e-waste can lead to environmental pollution and health risks due to hazardous substances like lead, mercury, and cadmium (see Section 2.7).

Additionally, data centers and other ICT facilities consume large amounts of energy for operations and cooling systems, leading to waste energy in the form of heat loss and contributing to greenhouse gas emissions. Furthermore, the production of electronic devices requires the extraction of scarce and non-renewable resources, such as rare earth metals and minerals, which contributes to resource depletion and environmental degradation.

Technical solutions to address waste generation in the ICT sector include implementing extended producer responsibility programs, which require manufacturers to take responsibility for the collection, recycling, and disposal of their electronic products at the end of their lifecycle.

To tackle energy consumption and waste energy, it is essential to incorporate energy efficiency in the design and production of ICT equipment and devices. Manufacturers can prioritize energy efficiency during the design phase by using components that consume less power, incorporating power management features, and optimizing device performance to minimize energy usage.

Additionally, promoting a circular economy approach in the ICT sector, which emphasizes the recycling, refurbishment, and reuse of electronic devices and components, can help conserve resources and reduce waste generation.

2.13 Energy Waste

Energy waste refers to the inefficient use of energy resources, leading to unnecessary energy consumption and environmental impacts. This issue cuts across all sectors of society.

2.13.1 Residential Sector

The residential sector accounts for a considerable share of energy waste in modern society. Various factors contribute to this waste, including inefficient heating, ventilation, and air conditioning systems (HVAC). In many homes, outdated and poorly maintained HVAC systems result in significant energy waste. These systems often consume more energy than necessary to maintain comfortable indoor temperatures and air quality, leading to higher energy bills and increased greenhouse gas emissions.

Electronic devices, such as televisions, computers, and chargers, often continue to consume energy even when they are not in use, contributing to what is known as "vampire power" or "standby power" waste. This can account for a substantial portion of a household's overall energy consumption.

Poor insulation and inefficient building design can result in significant energy waste, as they necessitate increased heating and cooling efforts to maintain comfortable indoor temperatures. Older buildings, in particular, are often characterized by air leaks and inadequate insulation, leading to higher energy consumption.

2.13.2 Industrial Sector

Many industries continue to rely on outdated and inefficient machinery and equipment, resulting in excessive energy consumption. This not only increases operational costs but also contributes to higher greenhouse gas emissions.

Some production processes are inherently energy-intensive and wasteful. By adopting more energy-efficient processes and technologies, industries can significantly reduce energy waste, lower costs, and decrease their environmental footprint.

Energy losses occur during the transmission and distribution of electricity from power plants to end-users. These losses can be attributed to factors such as resistance in transmission lines, transformer inefficiencies, and theft. Reducing these losses can help improve overall energy efficiency in the industrial sector.

2.13.3 Transportation Sector

Many vehicles, especially those with internal combustion engines, are highly inefficient, converting only a small fraction of the energy in fuel into useful work. This inefficiency leads to significant energy waste and higher greenhouse gas emissions.

Traffic congestion and excessive idling waste large amounts of fuel and contribute to air pollution. Encouraging carpooling, improving traffic flow, and implementing congestion pricing can help reduce energy waste in this area.

In many cities, public transportation systems are underdeveloped, inefficient, or insufficiently accessible, leading to increased reliance on private vehicles and higher energy consumption. By expanding and improving public transportation networks, cities can reduce energy waste and promote more sustainable transportation practices.

Energy waste contributes to several pressing issues, including climate change, resource depletion, and air pollution. Energy waste can be attributed to a significant portion of global CO_2 emissions. Furthermore, wasted energy represents a significant economic burden, as resources are used inefficiently, and money is spent on unnecessary energy consumption.

Energy efficiency is called the "first fuel" in clean energy transitions, as it provides some of the quickest and most cost-effective CO_2 mitigation options while lowering energy bills and strengthening energy security. According to the IEA,[43] energy efficiency is the single largest measure to avoid energy demand in the Net Zero Emissions by 2050 Scenario.

References

1. A. Morgado, P. Hugues and T. Vass, Cement, IEA, [cited 2024 Aug 16]. Available from: https://www.iea.org/reports/cement.
2. S. Nath, K. K. Singh, S. Tangjang and S. Das, Industrial solid wastes and environment: An overview on global generation, implications, and available management options, in *Springer Water*, Springer Nature Switzerland, Cham, 2024, pp. 221–246.
3. Trends in solid waste management, Worldbank.org. [cited 2024 Aug 16]. Available from: https://datatopics.worldbank.org/what-a-waste/trends_in_solid_waste_management.html.
4. Fortune Business Insights, [cited 2024 Aug 16]. Available from: https://www.fortunebusinessinsights.com/industry-reports/industrial-waste-management-market-100116.
5. The world counts, [cited 2024 Aug 16]. Available from: https://www.theworldcounts.com/challenges/planetearth/waste/hazardous-waste-statistics.
6. AQUASTAT, *FAO's Global Information System on Water and Agriculture*, [cited 2024 Aug 16]. Available from: https://www.fao.org/aquastat/en/overview/methodology/water-use.
7. Ellen MacArthur Foundation, A new textiles economy: Redesigning fashion's future, 2017.
8. World Energy Transitions Outlook 2023:1.5°C Pathway, International Renewable Energy Agency (IRENA), Abu Dhabi.
9. US Environmental Protection Agency (EPA), (n.d.), Pesticides' Impact on the Environment. Retrieved from https://www.epa.gov/pesticide-science-and-assessing-pesticide-risks/pesticide-exposure-human-and-environmental-health.
10. L. Lassaletta, *et al.*, *Environ. Res. Lett.*, 2014, **9**, 105011.
11. A. Balafoutis, B. Beck, S. Fountas, J. Vangeyte, T. Wal and I. Soto, *et al.*, Precision agriculture technologies positively contributing to GHG emissions mitigation, farm productivity and economics, *Sustainability*, 2017, **9**(8), 1339.
12. Packaging waste statistics, Europa.eu, [cited 2024 Aug 17]. Available from: https://ec.europa.eu/eurostat/statistics-explained/index.php?title=Packaging_waste_statistics.
13. J. Boulton, A. McCallion and J. Curtis, *Transitioning to a circular textile economy in Australia – Summary of Findings*, Monash University, 2020.
14. World Wildlife Fund (WWF), *Solving Plastic Pollution Through Accountability*, 2019.
15. UNEP Food Waste Index Report 2024, UNEP – UN Environment Programme, 2024 [cited 2024 Oct 3]. Available from: https://wedocs.unep.org/handle/20.500.11822/45230.
16. Worldwildlife.org. Driven to waste, The Global Impact of Food Loss and Waste on Farms. WWF-UK 2021 [cited 2024 Oct 15]. Available from: https://www.worldwildlife.org/publications/driven-to-waste-the-global-impact-of-food-loss-and-waste-on-farms.

17. US EPA, Inventory of U.S. Greenhouse Gas Emissions and Sinks: 1990–2019, 2021 [cited 2024 Aug 17]. Available from: https://www.epa.gov/ghgemissions/inventory-us-greenhouse-gas-emissions-and-sinks-1990-2019.
18. LowCVP, Lifecycle emissions from cars, [cited 2024 Aug 17]. Available from: https://www.zemo.org.uk/assets/workingdocuments/MC-P-11-15a%20Lifecycle%20emissions%20report.pdf.
19. Eurostat, End-of-life vehicle statistics (2023), [cited 2024 Aug 11]. Available from: https://ec.europa.eu/eurostat/statistics-explained/index.php?title=End-of-life_vehicle_statistics.
20. Reuters. Average age of U.S. vehicles hit record 12.1 years in 2020-IHS Markit. [cited 2024 Aug 11]. Available from: https://www.reuters.com/article/us-autos-research-age-idTRNIKCN2DQ05V
21. IEA, *Global EV Outlook 2021*, IEA, Paris, 2021.
22. Ellen MacArthur Foundation, *The New Plastics Economy: Rethinking the future of plastics*, 2016.
23. V. Forti, C. P. Baldé, R. Kuehr and G. Bel, *The Global E-waste Monitor 2020: Quantities, flows and the circular economy potential*, United Nations University (UNU)/United Nations Institute for Training and Research (UNITAR) – co-hosted SCYCLE Programme, International Telecommunication Union (ITU) & International Solid Waste Association (ISWA), Bonn/Geneva/Rotterdam, 2020.
24. Statista, Replacement cycle length of smartphones in the United States 2013–2027, in years, [cited 2024 Aug 11]. Available from: https://www.statista.com/statistics/619788/average-smartphone-life/.
25. Ž. Kamberović, M. Ranitović, M. Korać, Z. Andjić, N. Gajić and J. Djokić, *et al.*, Hydrometallurgical process for selective metals recovery from waste-printed circuit boards, *Metals*, 2018, **8**(6), 441.
26. C. Bishnoi, S. V. Daware and B. Rai, Recent and novel leaching processes for recovery of metals from spent lithium-ion batteries: A review, *Trans. Indian Inst. Met.*, 2024, 77, 3139–3152.
27. A. I. Adetunji, P. J. Oberholster and M. Erasmus, Bioleaching of metals from E-waste using microorganisms: A review, *Minerals*, 2023, **13**(6), 828.
28. EPA, Textiles: Material-Specific Data, [cited 2024 Aug 11]. Available from: https://www.epa.gov/facts-and-figures-about-materials-waste-and-recycling/textiles-material-specific-data.
29. M. A. Browne, P. Crump, S. J. Niven, E. Teuten, A. Tonkin and T. Galloway, *et al.*, Accumulation of microplastic on shorelines worldwide: sources and sinks, *Environ Sci Technol*, 2011, **45**(21), 9175–9179.
30. ColorZen® – environmentally friendly solution to cotton dyeing. ColorZen®, [cited 2024 Aug 11]. Available from: https://www.colorzen.com/.
31. Tencel.com, [cited 2024 Aug 11]. Available from: https://www.tencel.com/b2b/product/tencel-lyocell.
32. UNEP. Cleaning up couture: what's in your jeans? 2018, [cited 2024 Aug 11]. Available from: https://www.unep.org/news-and-stories/story/cleaning-couture-whats-your-jeans.

33. Ellen MacArthur Foundation, Fashion and the circular economy, [cited 2024 Aug 11]. Available from: https://ellenmacarthurfoundation.org/fashion-and-the-circular-economy-deep-dive.
34. The World Bank, Accelerating Change To Solve The Water Crisis, [cited 2024 Aug 11]. Available from: https://www.worldbank.org/en/news/immersive-story/2023/03/22/accelerating-change-to-solve-the-water-crisis.
35. C. Rizan, R. Lillywhite, M. Reed and M. F. Bhutta, The carbon footprint of products used in five common surgical operations: Identifying contributing products and Processes, *J. R. Soc. Med.*, 2023, 014107682311661.
36. K. Kümmerer, Antibiotics in the aquatic environment – a review – part I, *Chemosphere*, 2009, **75**(4), 417–434.
37. T. Brodin, *et al.*, Dilute concentrations of a psychiatric drug alter behavior of fish from natural populations, *Science*, 2013, **339**(6121), 814–815.
38. Construction and demolition waste, Environment, [cited 2024 Aug 12]. Available from: https://environment.ec.europa.eu/topics/waste-and-recycling/construction-and-demolition-waste_en.
39. Calrecycle.ca.gov, [cited 2024 Aug 12]. Available from: https://calrecycle.*ca*.gov/condemo/wallboard/.
40. A. Karaeva and E. Magaril, Considering waste generation in the energy sector during the transition to a circular economy, *Recycling*, 2023, **8**(2), 42.
41. IAEA Nuclear Energy Series No. NW-T-1.14 (Rev. 1), Status and Trends in Spent Fuel and Radioactive Waste Management, 2022.
42. IRENA and IEA-PVPS, *End-of-Life Management: Solar Photovoltaic Panels*, International Renewable Energy Agency and International Energy Agency Photovoltaic Power Systems Programme, 2016, ISBN: 978-92-95111-99-8 June 2016.
43. IEA, Energy Efficiency, IEA, Paris, 2022, [cited 2024 Aug 12]. Available from: https://www.iea.org/reports/energy-efficiency.
44. S. Kaza, L. C. Yao, P. Bhada-Tata and F. Van Woerden, *What a waste 2.0: A global snapshot of solid waste management to 2050*, World Bank, Washington, DC, 2018.

CHAPTER 3

Where Is All the Waste Going?

While waste piles up in landfills, contributing to unseen health crises and environmental disasters, many regions struggle with the toxic fallout of our consumer-driven lifestyles. As global consumption rises, so too does the waste it leaves behind. Economic growth has spurred improvements in living standards for poorer societies and driven increased luxury consumption in wealthier ones. Global GDP surged from $18.9 trillion in 1970 to $92.8 trillion in 2023 (in constant 2015 US dollars).[1] But this growth has come at a cost: more people with greater purchasing power means more goods, more packaging, and ultimately, more waste.

This trend has significant implications for waste generation and resource consumption. The relationship between economic growth and waste generation is influenced by multiple factors, such as population growth, urbanization, technological advancements, and changes in consumption patterns.[2] Nonetheless, it is clear that the increasing global economic output has generally been accompanied by an increase in waste generation over the past five decades.

According to the United Nations, the world population is projected to reach 9.7 billion by 2050,[3] up from just over 8 billion in 2023. This population growth, combined with rapid urbanization and the expansion of the global middle class, has led to an increase in the overall demand for goods and services. As more people gain access to higher incomes and improved living standards, their consumption patterns change, often resulting in increased resource use and waste generation.

The United Nations Environment Programme (UNEP) projects that global solid waste generation will increase from 2.1 billion to 3.8 billion metric tons by 2050. While waste generation is strongly correlated with factors like the Human Development Index, urban population share, gross national income, and adult literacy rates, UNEP analysis indicates that the most accurate

Waste Not! How Nanotechnologies Can Increase Efficiencies Throughout Society
By Michael Berger

Published by the Royal Society of Chemistry, www.rsc.org

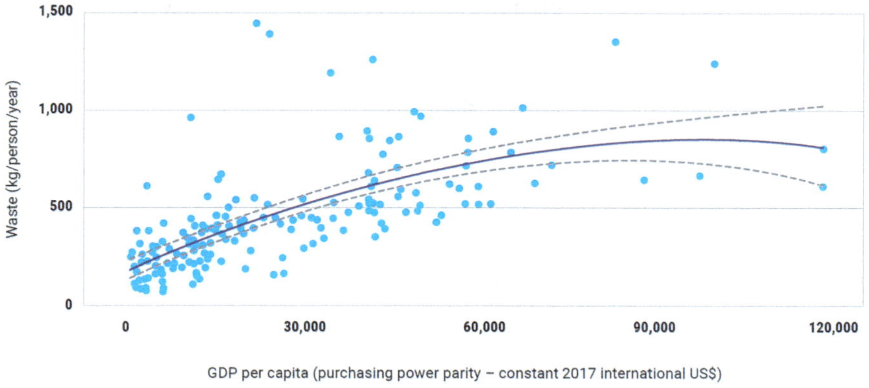

Figure 3.1 Relationship between gross domestic product (GDP) and waste generation in the most recent year available between 2010 and 2020. Note: Each dot represents a country, with GDP data for the corresponding year converted to international dollars using purchasing power parity rates. Data is in constant 2017 international US dollars, corresponding to the World Bank International Comparison Program 2023. The line of best fit is shown in purple. Reproduced from ref. 4 with permission from United Nations Environment Programme, Copyright 2024.

model for predicting waste trends is a linear regression based solely on GDP per capita.[4] As GDP rises, so does consumerism—a cultural and economic phenomenon where the consumption of goods and services becomes a priority. Economic development and globalization have accelerated consumerism, leading to increased demand for products with shorter lifespans, such as electronics, clothing, and single-use items like food and drink containers. This surge in consumption, closely tied to rising GDP, further drives waste generation (see Figure 3.1).

3.1 Landfills

Landfills are the antithesis of sustainability, and they are among the most toxic places on the planet. They generate greenhouse gases, primarily methane (CH_4) and carbon dioxide (CO_2), as a result of anaerobic decomposition of organic waste materials.[5] Methane is a particularly potent greenhouse gas, with a global warming potential 28–36 times greater than carbon dioxide over a 100-year period. In terms of hazardous materials, landfills can contain various toxic substances, including heavy metals (*e.g.*, lead, mercury, and cadmium), volatile organic compounds (VOCs), and other pollutants. These contaminants can leach into the surrounding soil and groundwater, posing risks to human health and the environment.

Incinerators, on the other hand, burn waste to reduce its volume and mass. However, this process emits harmful substances into the atmosphere, such as dioxins and furans, which are toxic and persistent organic pollutants. Dioxins and furans can cause a range of health problems, including cancer, reproductive and developmental issues, and damage to the immune system. Incineration can also release heavy metals such as lead, mercury, and cadmium, which can cause neurological and developmental disorders, as well as respiratory, cardiovascular, and renal problems.[6]

The World Bank provides comprehensive data on global waste generation and management, broken down by geography, including the quantities of waste that end up in landfills. According to a World Bank report,[7] approximately 55% of municipal solid waste globally was landfilled. The latest data for the EU shows 32% of solid municipal waste was landfilled.[8]

3.2 Waste Treatment and Recycling

Solid waste that is not landfilled is typically managed through various other methods, which may include recycling, composting, incineration, and waste-to-energy conversion. The choice of waste management method depends on the type of waste, local infrastructure, and environmental regulations. Here is an overview of the main alternatives to landfilling:

Recycling involves collecting, sorting, processing, and converting waste materials into new products. Commonly recycled materials include paper, cardboard, glass, metals, and plastics. Recycling helps conserve resources, reduce the need for raw materials, and decrease energy consumption and greenhouse gas emissions.

The most straightforward approach involves recycling products for identical uses by different consumers, such as reusing items through second-hand markets. Alternatively, products may find new life in different applications. Frequently, products reaching their end-of-life undergo reprocessing to extract essential raw materials, like recycled paper or plastics. When products contain valuable substances, targeted recycling efforts focus on extracting these components (see Figure 3.2). Advanced recycling technologies often leverage synergies among multiple industrial processes, utilizing the waste from one process as raw material for another.

The recycling rate varies significantly by country and region, with higher-income countries generally having higher recycling rates. Data for the EU shows a recycling rate of 39% for 2020.[8]

Composting is the process of breaking down organic waste, such as food scraps and yard waste, into a nutrient-rich soil amendment. Composting can occur at an individual or community level and helps divert organic waste from landfills, reducing methane emissions and providing a valuable resource for agriculture and gardening.

Composting rates also vary by country and region, with higher rates typically found in countries with strong policies and infrastructure for organic waste management.

Incineration is the process of burning waste at high temperatures to reduce its volume and weight. The combustion process generates heat, which can be used to produce steam and generate electricity. Incineration can reduce the volume of waste by up to 90% and is often employed for waste

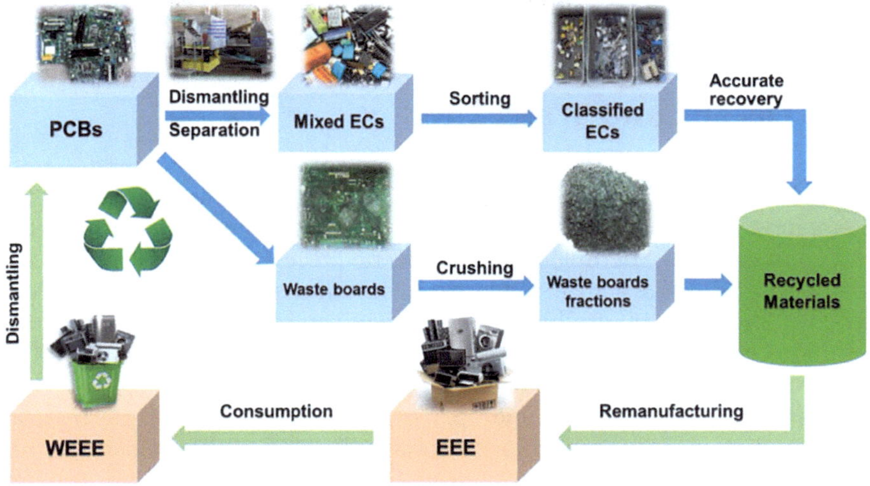

Figure 3.2 Recycling flow chart of a printed circuit board (PCB) and its electronic components (ECs) in waste electrical and electronic equipment (WEEE). Reproduced from ref. 16 with permission from the Royal Society of Chemistry.

that cannot be recycled or composted. However, as previously mentioned, incineration can release harmful substances into the atmosphere if not properly managed.

The prevalence of incineration and waste-to-energy facilities varies significantly by country, with some European countries, such as Denmark and Sweden, relying heavily on these methods for waste management.

Waste-to-energy (WTE) facilities use incineration systems that convert the energy stored in waste into electricity, heat, or fuels through combustion, gasification, or anaerobic digestion. These facilities help reduce the volume of waste, lower greenhouse gas emissions, and provide an alternative source of energy.

Anaerobic digestion is a process that breaks down organic waste in the absence of oxygen, producing biogas (a mixture of methane and carbon dioxide) and nutrient-rich digestate. The biogas can be used for heating, electricity generation, or transportation fuel, while the digestate can be used as a fertilizer.

Mechanical biological treatment (MBT) facilities process mixed waste streams through a combination of mechanical and biological processes to separate recyclable materials, stabilize organic waste, and reduce the volume of residual waste. MBT can be used as a pre-treatment for waste destined for landfilling or incineration to reduce the environmental impact of these disposal methods.

The choice of waste management method depends on various factors, including the type of waste, available infrastructure, economic considerations, and local regulations. Ideally, a combination of these methods should be employed in an integrated waste management system to minimize the environmental impact of waste disposal.

3.3 Oceans and Waterways

In 2019, 6.1 million tons of plastic waste are estimated to have ended up in aquatic environments, of which 1.7 million tons flowed into the ocean.[9] This figure most likely has increased since then due to growing plastic production and consumption. Microplastics, tiny plastic pieces less than 5 mm in size, represent a secondary form of plastic pollution that stems from the breakdown of larger plastic products and waste in the marine environment. It is estimated that 82–358 trillion plastic particles weighing 1.1–4.9 million tons are suspended in the world's oceans.[10]

Like larger plastic debris, microplastics accumulate up the marine food chain and negatively impact organisms through physical blockages and leaching toxic chemicals. The tiny size of nanoplastics (smaller than 1 micron) also allows them to cross cell walls and bioaccumulate to higher concentrations, posing greater risks of toxicity as they reach human seafood diets.

The ubiquity of nano- and microplastics symbolizes the need for intervention not just at the waste disposal stage but also upstream policies to curb excessive plastic usage. Reducing single-use plastics could significantly mitigate this pernicious byproduct that evades quantification and detection.

This plastic waste originates from various sources, including land-based activities and maritime industries. Around 80% of marine plastic pollution comes from land-based sources, such as littering, improper waste management, and stormwater runoff. The remaining 20% is attributed to maritime activities, including fishing gear, shipping, and offshore oil and gas platforms.

The waste entering the oceans is, to a large extent, a subset of the waste sources discussed earlier. For example, single-use plastics and packaging from the food and beverage industry, plastic components from electronics, and synthetic fibers from the fashion industry can all end up as marine debris. Mismanaged waste and lack of proper waste disposal infrastructure in certain regions contribute to the problem. Additionally, waste generated by maritime activities, such as abandoned fishing nets and gear, forms a distinct category of ocean waste.

Marine plastic pollution poses severe threats to marine ecosystems, wildlife, and human health. Ingestion of microplastics by marine organisms can lead to physical harm, reduced reproductive success, and even death. As microplastics move up the food chain, they accumulate in larger predators, including fish and shellfish consumed by humans, potentially transferring harmful chemicals and posing risks to human health.

Efforts to address marine plastic pollution require a multi-pronged approach, including improved waste management practices, increased recycling rates, reduction of single-use plastics, and innovations in product design to minimize plastic waste.

While plastic pollution represents the most significant and persistent threat to marine environments, other pollutants also impact oceans and waterways. These include industrial effluents containing heavy metals, agricultural runoff with fertilizers and pesticides, oil spills, untreated sewage discharge, and thermal pollution from power plants. Mining waste, including acid mine drainage and tailings, can severely impact local water systems. Additionally, excess nutrients from agricultural and urban runoff can cause algal blooms and oxygen-depleted "dead zones". However, unlike many of these pollutants which may dissipate or break down over time, plastic pollution's persistence and accumulation in marine food chains makes it particularly concerning for long-term ecosystem health.

3.4 Airborne and Gaseous Wastes

A significant portion of human-generated waste ends up in the atmosphere. Industrial processes are major contributors to atmospheric waste. Power plants, manufacturing facilities, and chemical plants release greenhouse gases like carbon dioxide, methane, and nitrous oxide, contributing significantly to the approximately 53.0 gigatons of CO_2 equivalent emitted globally from all anthropogenic activities in 2023.[11] In addition to greenhouse gases, these industries also emit air pollutants such as sulfur dioxide, nitrogen oxides, and particulate matter. Industrial volatile organic compounds (VOCs) can persist in the atmosphere and travel long distances from their source.

Waste incineration facilities, while reducing solid waste volume, contribute their own atmospheric emissions. These include dioxins and furans, heavy metals, and particulate matter. Even with modern pollution controls, these facilities release significant quantities of CO_2 and other pollutants into the air.

Agricultural practices contribute substantially to atmospheric waste. Globally, small farmers burn approximately 10.5 billion metric tonnes of crop waste annually, releasing an estimated 16.6 billion tonnes of CO_2 into the atmosphere.[12] This practice also contributes to the emission of 9.8 billion tonnes of CO_2 equivalent, along with 1.1 billion tonnes of smog precursors and 66 million tonnes of particulate matter ($PM_{2.5}$), which have severe implications for air quality and public health.

Manure from livestock operations also contributes a substantial portion of agricultural waste. In 2022, global livestock production is estimated to have generated approximately 9.26 billion metric tonnes of manure.[13] While manure can be a valuable resource when properly managed, providing nutrients for crops and improving soil quality, improper management can lead to environmental issues, such as nutrient runoff, water contamination, and emissions of methane—a potent greenhouse gas. Methane emissions from livestock alone accounted for about 32% of global methane emissions in 2021.[14]

Urban areas generate significant atmospheric waste through transportation emissions, construction dust, and other activities. Vehicle exhaust contains a complex mixture of pollutants including nitrogen oxides, carbon monoxide, and fine particulates. Urban heat islands can trap these pollutants, creating zones of concentrated atmospheric waste.

The impact of atmospheric waste extends far beyond its point of origin. Air currents can transport pollutants thousands of kilometers, affecting regions far from emission sources. Atmospheric waste can return to Earth as acid rain, damaging ecosystems and infrastructure. Many air pollutants eventually settle into oceans and waterways, contributing to water pollution (see Section 3.3). The atmospheric waste burden significantly impacts human health, causing respiratory diseases, cardiovascular problems, and other health issues, particularly in urban areas and industrial regions.

Unlike solid waste in landfills or plastic pollution in oceans, atmospheric waste can't be easily contained or cleaned up once released. This makes prevention and emission reduction particularly crucial for managing this waste stream. The global nature of atmospheric waste transport means that local emissions can have worldwide consequences, particularly in the context of climate change.

3.5 The Waste Shadow Zone

A substantial amount of waste escapes conventional statistics altogether, existing in what can be described as the "waste shadow zone". This refers to waste generated but not accounted for in official reports due to mismanagement, illegal dumping, or other unmonitored disposal methods.

The statistical waste data cited above primarily covers municipal solid waste—waste generated by households, commercial establishments, and institutions, typically managed by local governments. This includes waste sent to landfills, incinerators, or recycling facilities. However, it does not account for waste that enters the environment through mismanagement, littering, or illegal dumping, particularly in developing countries, where waste management infrastructure is lacking.

One of the largest contributors to the waste shadow zone is marine waste. Plastic waste entering the world's oceans, for example, is estimated at 1.7 million metric tons annually, according to scientific studies and models (see Section 3.3). Yet, even this significant figure might underestimate the true scale of marine pollution. Various types of industrial and agricultural waste could also be making their way into oceans but remain unquantified due to poor tracking and misreporting.

Waste that escapes proper disposal channels and enters the oceans is particularly difficult to measure accurately. The plastic pollution problem is perhaps the most well-known example, but other categories, such as industrial waste, including hazardous materials, also evade official disposal statistics.

Additionally, airborne waste contributes to the shadow zone. Some of the waste that reaches oceans does not arrive *via* rivers or land runoff but through the air. PFAS (per- and polyfluoroalkyl substances), commonly known as "forever chemicals," are synthetic compounds that persist in the environment and are now found in rainwater globally.[15] These chemicals, which do not degrade over time, end up in oceans and waterways, further complicating waste tracking and management efforts. PFAS are notorious for their environmental persistence and harmful health impacts, including potential links to cancer, immune system damage, and developmental issues.

The concept of the waste shadow zone highlights the gap between official statistics and the real impact of waste on the environment. While current data provide a glimpse into the global waste crisis, they do not fully capture the extent of pollution caused by unregulated or improperly managed waste. Marine plastic pollution and the spread of airborne contaminants like PFAS demonstrate that waste continues to escape conventional tracking methods.

Comprehensive monitoring and data collection are necessary to fully understand the waste shadow zone. Enhanced collaboration between governments, researchers, and environmental organizations is essential for improving waste management practices and addressing the environmental consequences of unmonitored waste. Until then, it is crucial to recognize that official waste statistics likely underestimate the true scale of waste pollution in our environment, leaving us with an incomplete picture of the global waste crisis.

References

1. GDP (constant 2015 US$), World Bank Open Data, [cited 2024 Sep 18]. Available from: https://data.worldbank.org/indicator/NY.GDP.MKTP.KD.
2. G. Hondroyiannis, E. Sardianou, V. Nikou, K. Evangelinos and I. Nikolaou, Waste generation and macroeconomic drivers: a panel study for European countries and regions, *Manage. Environ. Qual.*, 2024, **35**(5), 1118–1136.
3. United Nations, Department of Economic and Social Affairs, Population Division (2019): World Population Prospects 2019: Highlights (ST/ESA/SER.A/423), [cited 2024 Jul 12]. Available from: https://population.un.org/wpp/Publications/Files/WPP2019_Highlights.pdf.
4. United Nations Environment Programme (2024). Global Waste Management Outlook 2024: Beyond an age of waste – Turning rubbish into a resource, Nairobi, [cited 2024 Jul 18]. Available from: https://wedocs.unep.org/20.500.11822/44939.
5. Oar, Oap, Basic Information about Landfill Gas, [cited 2024 Sep 18]. Available from: https://www.epa.gov/lmop/basic-information-about-landfill-gas.
6. P. W. Tait, J. Brew, A. Che, A. Costanzo, A. Danyluk and M. Davis, *et al.*, The health impacts of waste incineration: a systematic review, *Aust. N. Z. J. Public Health*, 2020, **44**(1), 40–48.

7. Baker-Brown D., *What a Waste! The Re-Use Atlas*, 2019, [cited 2024 Sep 18]. Available from: https://datatopics.worldbank.org/what-a-waste/.
8. Waste statistics, Europa.eu. [cited 2024 Jul 11]. Available from: https://ec.europa.eu/eurostat/statistics-explained/index.php?title=Waste_statistics.
9. OECD, *Global plastics outlook: Economic drivers, environmental impacts and policy options*, OECD, 2022.
10. M. Eriksen, W. Cowger, L. M. Erdle, S. Coffin, P. Villarrubia-Gómez and C. J. Moore, *et al.*, A growing plastic smog, now estimated to be over 170 trillion plastic particles afloat in the world's oceans—urgent solutions required, *PLoS One*, 2023, **18**(3), e0281596.
11. EDGAR – the emissions database for global atmospheric research, Europa.eu. [cited 2024 Oct 27]. Available from: https://edgar.jrc.ec.europa.eu/report_2024.
12. M. Shafer, Global crop waste burning – micro-biochar; how a small community development organization learned experientially to address a huge problem one tiny field at a time, *Sustainable Earth*, 2020, **3**, 18.
13. J. Qi, H. Yang, X. Wang, H. Zhu, Z. Wang and C. Zhao, *et al.*, State-of-the-art on animal manure pollution control and resource utilization, *J. Environ. Chem. Eng.*, 2023, **11**(5), 110462.
14. Reducing enteric methane emissions through market-based policy instruments, EntericMethane, FAO, 2023, [cited 2024 Aug 18]. Available from: https://www.fao.org/in-action/enteric-methane/countries/highlight-detail/reducing-enteric-methane-emissions-through-market-based-policy-instruments/en.
15. I. T. Cousins, J. H. Johansson, M. E. Salter, B. Sha and M. Scheringer, Outside the safe operating space of a new planetary boundary for per- and polyfluoroalkyl substances (PFAS), *Environ. Sci. Technol.*, 2022, **56**(16), 11172–11179.
16. W. Zhao, J. Xu, W. Fei, Z. Liu, W. He and G. Li, The reuse of electronic components from waste printed circuit boards: a critical review, *Environ. Sci. Adv.*, 2023, **2**, 196–214.

CHAPTER 4

Environmental, Societal, and Economic Impact of Waste

Waste generation has become one of the most pressing issues of modern society, impacting nearly every facet of our lives. The continued growth in population, coupled with increasing consumerism, has led to unprecedented levels of waste, placing immense pressure on our environment, public health, and economic systems. Current waste management practices, rooted in a linear economy model, often exacerbate these issues by focusing on short-term solutions that prioritize disposal over sustainability.

The environmental consequences of waste are profound, from pollution of air, water, and soil to the depletion of essential natural resources. Improper waste management contributes significantly to climate change, ecosystem degradation, and biodiversity loss, while also posing serious public health risks. Social justice concerns are also tightly intertwined with waste, as marginalized communities often bear the brunt of its negative impacts.

Economically, the costs of managing waste are staggering, with billions of dollars spent annually on waste collection, treatment, and disposal. At the same time, the failure to recover and repurpose valuable materials represents a substantial lost opportunity. However, there is hope. By transitioning to a circular economy—where resources are reused, repurposed, and recycled—society can reduce waste, conserve resources, and unlock new economic opportunities. Nanotechnology, in particular, offers innovative solutions that could revolutionize resource efficiency, providing a pathway to addressing these challenges.

Although nanotechnology is not the sole answer to the waste crisis, it stands out as a critical tool in the broader toolkit of sustainable solutions. Its unique properties can benefit a wide array of materials, products, and applications, from enhancing the recyclability of complex waste streams to

Waste Not! How Nanotechnologies Can Increase Efficiencies Throughout Society
By Michael Berger

Published by the Royal Society of Chemistry, www.rsc.org

improving the durability and efficiency of everyday products. By integrating nanotechnological innovations with other waste management strategies, society can significantly enhance its ability to reduce waste and mitigate environmental harm.

This chapter provides a comprehensive exploration of the waste crisis as a necessary foundation for understanding how nanotechnologies can address these systemic challenges. By laying out the environmental, societal, and economic impacts of waste, the chapter highlights the areas where nanotechnological innovations can make the most significant contributions. Nanotechnology's potential to enhance recyclability, resource efficiency, and product durability offers a transformative approach to minimizing waste. Therefore, understanding the full scope of the problem is essential for appreciating the solutions discussed in subsequent chapters.

4.1 Waste and the Linear Economy

The current waste crisis is largely a product of the linear "take-make-dispose" economic model, which has dominated global production and consumption patterns since the industrial revolution. In this system, raw materials are extracted, transformed into products, and, after a relatively short period of use, discarded as waste. This model relies on the assumption of infinite resources, with little consideration for the environmental and societal consequences of waste generation.[1]

The linear economy is characterized by its inefficiency in managing materials, driven by a lack of incentives for waste reduction, reuse, or recycling. Products are often designed for planned obsolescence—manufactured with short lifespans and limited repairability to encourage consumers to replace items frequently. This approach not only contributes to growing volumes of waste but also perpetuates the demand for virgin resources, further depleting finite natural reserves such as minerals, fossil fuels, and forests.

Several industries, such as electronics, textiles, and packaging, exemplify this "take-make-dispose" system. In these sectors, products are often designed with minimal consideration for their end-of-life phase, making them difficult to recycle or reuse. As a result, waste piles up in landfills, pollutes waterways, and contributes to greenhouse gas emissions, threatening both the environment and public health.

Moreover, the linear economy thrives on economic drivers such as cheap raw material extraction and consumer demand for novelty. Companies often find it more cost-effective to extract new resources rather than invest in sustainable production methods, while consumers are encouraged to purchase new products due to trends and marketing, rather than repairing or reusing existing goods.

In contrast, as illustrated in Figure 4.1, the circular economy model seeks to keep materials in use for as long as possible, extending product lifecycles and reducing waste. This approach involves rethinking the design, use, and disposal of products to create systems that minimize resource extraction and

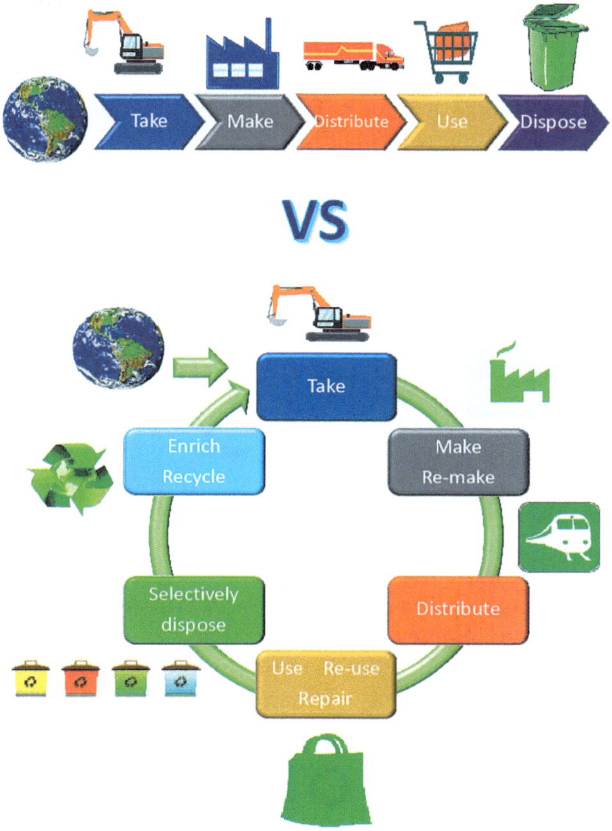

Figure 4.1 Linear economy model *versus* circular economy model. Reproduced from ref. 28, https://doi.org/10.1109/ACCESS.2019.2955008, under the terms of the CC BY 4.0 license, https://creativecommons.org/licenses/by/4.0/.

environmental harm.[2] Strategies within a circular economy include designing products for durability and recyclability, implementing extended producer responsibility, and adopting business models based on leasing, sharing, and remanufacturing. For example, rather than purchasing products outright, consumers might lease products that companies maintain, repair, and eventually recycle.

Transitioning from a linear to a circular economy represents a fundamental shift in how we think about resources and waste. It requires not only redesigning products and industrial processes but also fostering collaboration between sectors and stakeholders to create closed-loop systems that eliminate waste. However, this shift is not without challenges—there are technical barriers to redesigning complex products for recyclability, financial hurdles for industries accustomed to cheap raw materials, and regulatory gaps that fail to incentivize sustainable practices.

As we will discuss in the sections below, nanotechnology offers a critical solution in enabling this transition by providing innovative tools for waste reduction, resource efficiency, and sustainable product design. For instance, nanomaterials can make products more durable, extend their usable life, and improve recyclability. Nanotechnology is also being applied to develop advanced recycling techniques, enabling the breakdown and reuse of materials that are traditionally hard to process, such as electronics or plastics. This potential makes nanotechnology a key enabler of the circular economy, capable of transforming waste management and resource utilization in the years to come.

4.2 Growing Demand for Resources

The growing demands for resources, driven by population growth and consumerism, are leading to not only higher waste generation but also significant pressure on the environment. This results in the depletion of resources, deforestation, loss of biodiversity, and increased greenhouse gas emissions. As such, the first of the "3 Rs"—reducing—becomes a crucial component in any path towards sustainability.

Global energy consumption has grown by 54% from 1990 to 2020, with the International Energy Agency (IEA) reporting an increase from around 261 million terajoules (TJ) in 1990 to approximately 400 million TJ in 2020.[3]

The 4% increase in electricity demand projected for 2024, the highest since 2007 (excluding rebounds in 2010 and 2021), highlights the continued surge in resource consumption, as shown in Figure 4.2. This growth is fueled primarily by rising demand in China, India, and the United States. A similar trend is expected for 2025, with electricity demand outpacing the forecasted global GDP growth of 3.2%.[30]

Freshwater consumption has also grown substantially over the past century. The United Nations estimates that global water use is growing at about 1% annually, more or less in tune with the current population growth rate. However, already half of the global population (4 billion people) live under conditions of severe water scarcity at least 1 month of the year. Half a billion people in the world face severe water scarcity all year round.[4]

The extraction and use of natural resources have also increased. The International Resource Panel (IRP) reports that global material extraction has grown by 2.3% annually since 1970, from 30 billion metric tonnes to 106.6 billion tonnes in 2024.[5] That amounts to 13.2 tons of materials demand for every person on the planet. Some of the major materials contributing to this increase include:

- **Metal ores:** Metals accounted for 9% of global material extraction in 1970 (2.7 billion tonnes) and grew to 10% (9.6 billion tonnes) by 2020, averaging 2.6% annual growth. This increase highlights their importance in construction, energy, transport, manufacturing, and consumer goods. Iron ore saw the fastest growth due to rising steel demand in

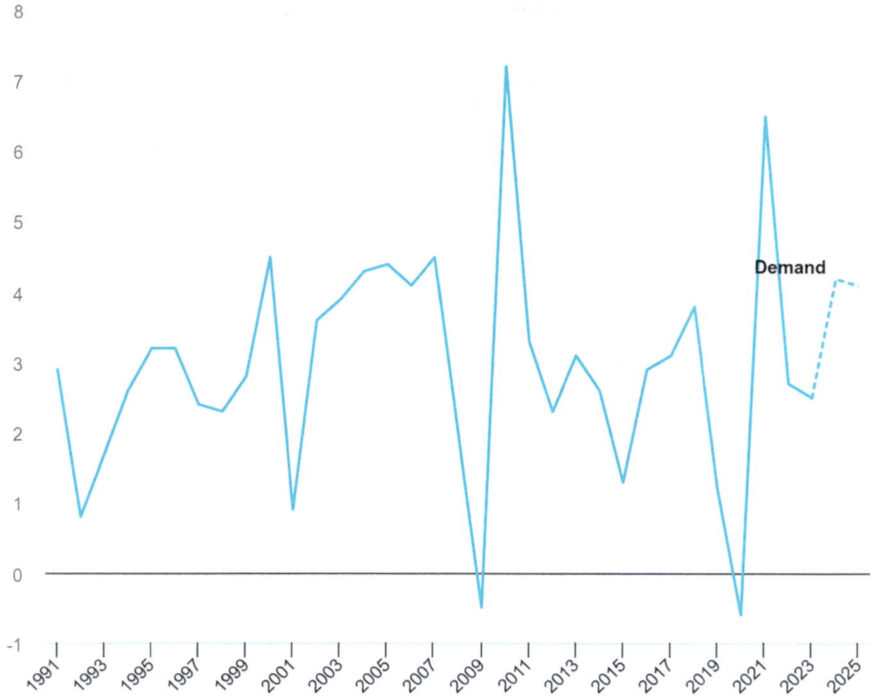

Figure 4.2 Year-on-year growth rate in global electricity demand, 1991–2025 (in %). Reproduced from ref. 29, https://www.iea.org/data-and-statistics/charts/year-on-year-growth-rate-in-global-electricity-demand-1991-2025, under the terms of the CC BY 4.0 license, https://creativecommons.org/licenses/by/4.0/deed.en.

construction and urbanization in the Global South. Metals are also crucial for the energy transition, requiring more aluminum, copper, cobalt, nickel, and lithium for energy infrastructure, storage, and transport electrification, along with expanded recycling capabilities.

- **Non-metallic minerals**: Sand, gravel, and clay for construction and industrial use saw the highest material use growth, averaging 3.2% per year. Extraction rose from 9.6 billion tonnes in 1970 to 45.3 billion tonnes in 2020, driven by global infrastructure expansion. The share of non-metallic minerals increased from 31% to nearly 50% of global material extraction, marking a shift from biomass to mineral-based resources.
- **Biomass**: Biomass, including crops, timber, and wild-caught fish, accounted for 41% of global material extraction in 1970 but fell to just over 25% by 2020. As countries industrialized, they shifted from biomass-based to mineral-based materials and energy systems. Despite the lower proportion, total biomass demand rose from 12.6 to 24.8 billion tonnes, averaging 1.3% annual growth. Crop harvests grew by 2.2%

annually, making up 40% of biomass in 2020, while grazed biomass increased at a similar rate, driven by rising animal-based diets. Biomass subcategories with alternatives, like wood and wild-caught fish, grew more slowly.

- **Fossil fuels**: Fossil fuel use, including coal, petroleum, and natural gas, grew from 6.1 billion tonnes in 1970 to 15.4 billion tonnes in 2020, though their share of global extraction fell from 20% to 16%. Fossil fuel extraction grew by 2.1% annually, with natural gas leading at 2.8%, coal at 2.1%, and petroleum at 1.3%, driven by increased coal- and gas-fired power generation. Recently, coal use has stagnated due to lower gas prices, renewable energy growth, and improved energy efficiency.

The growing demand for materials over the past five decades has been closely tied to economic activity, infrastructure development, and improving living standards. This rise has been driven by large infrastructure investments and increased material consumption in middle-income countries, especially in Asia. The growth in per capita demand for materials reflects a surge in consumption that has outpaced both GDP and population growth. This trend has contributed to stagnant material efficiency globally and heightened environmental pressures at every stage of the supply chain. As demand continues to rise, balancing economic growth with sustainable material use will be critical to mitigating the associated environmental impacts.

4.3 Environmental Impact

Waste generation leads to a cascade of environmental challenges that affect ecosystems and human health. Pollution, resource depletion, and biodiversity loss are the core environmental challenges linked to improper waste management.

The production of municipal solid waste alone is expected to rise from 2.1 billion tonnes in 2023 to 3.8 billion tonnes by 2050. In 2020, waste management worldwide had a direct cost of approximately USD 252 billion. When accounting for the indirect impacts such as pollution, health issues, and climate change due to inadequate waste disposal, this figure increases to USD 361 billion. If significant steps are not taken to address waste management, the annual global cost could nearly double by 2050, reaching an estimated USD 640.3 billion.[6]

Addressing these challenges requires understanding the systemic nature of the problem and its long-term consequences for both the planet and society.

4.3.1 Pollution and Ecosystem Degradation

Improper waste management contributes heavily to pollution affecting various ecosystems, affecting air, water, and soil. Landfills, one of the most prevalent waste disposal methods, release significant amounts of greenhouse gases like methane and carbon dioxide, intensifying global climate change. Methane is especially concerning due to its potency as a greenhouse gas. Additionally, leachate produced in landfills contains toxic chemicals that seep into groundwater, threatening both human health and biodiversity in surrounding ecosystems.

Incineration, another common disposal method, exacerbates the pollution problem by emitting harmful pollutants such as dioxins, furans, and heavy metals into the atmosphere.[7] These substances can lead to serious health issues for nearby populations, including respiratory problems and cancer. Furthermore, plastic waste is a severe environmental hazard, especially in marine environments, where it accumulates, leading to microplastic pollution. Marine life ingesting these particles faces severe consequences, which, in turn, disrupts the broader marine food chain.

Electronic waste, an increasingly significant part of the waste stream, poses specific environmental risks due to its toxic components, including lead, mercury, and cadmium. When improperly disposed of, e-waste contaminates soil and water sources, posing risks to both ecosystems and human populations. The improper management of e-waste highlights the need for specialized recycling efforts that can safely extract valuable materials while minimizing environmental damage.[8]

4.3.2 Resource Depletion

The continued reliance on the linear take-make-dispose economic model is contributing to the depletion of natural resources. As global demand for goods increases, so too does the extraction of raw materials like minerals, fossil fuels, and forests. The extraction process leads to environmental destruction, such as habitat loss and increased greenhouse gas emissions, while also straining finite resources.

E-waste, in particular, contributes to resource depletion as valuable materials like rare earth metals are discarded instead of recovered. With global material extraction increasing by 2.3% annually since 1970, the failure to implement widespread recycling programs exacerbates the need for virgin material extraction. This ongoing depletion not only drives environmental degradation but also threatens the availability of essential materials needed for future technological and industrial developments.

4.3.3 Biodiversity Loss

Waste generation and its disposal have profound effects on biodiversity.[9] Landfills and waste dumping sites encroach upon natural habitats, displacing wildlife and fragmenting ecosystems. Marine plastic pollution has devastating effects, with numerous species becoming entangled in or ingesting plastic debris, leading to injuries, malnutrition, and death. The ongoing destruction of these ecosystems threatens the survival of countless species, contributing to a global biodiversity crisis.

Electronic waste further accelerates biodiversity loss through its contamination of soil and water in ecosystems surrounding dumping grounds. Toxic chemicals from e-waste, like heavy metals, enter the food chain, disrupting not only wildlife but also human health. The damage to biodiversity also has long-term repercussions for ecosystem services, which provide critical benefits like climate regulation and clean water supplies.

4.3.4 Long-term Environmental Impacts and Interconnectedness

The environmental consequences of waste generation are persistent and multifaceted, with pollutants remaining in ecosystems for decades or even centuries. Plastic waste, for instance, can take hundreds of years to degrade, breaking down into microplastics that continuously harm marine and terrestrial ecosystems. These pollutants accumulate over time, disrupting ecosystems, diminishing biodiversity, and compromising essential ecosystem services like water purification and soil fertility. In particular, toxic chemicals from landfills, incineration, and e-waste can seep into soil and water sources, leading to long-term contamination that endangers both wildlife and human populations.

The long-term impacts of resource depletion are similarly significant. As finite resources become scarcer due to unsustainable extraction practices, ecosystems face further degradation, and industries reliant on these materials must cope with rising costs and supply instability. This depletion also undermines future technological advancements, particularly in sectors like renewable energy, which depend on critical materials found in electronic waste.

These environmental impacts are deeply interconnected. The depletion of resources, the destruction of habitats, and the accumulation of pollutants exacerbate each other, leading to a cascading effect.[10] As biodiversity is lost due to habitat destruction and pollution, ecosystems become less resilient to climate change and other stressors. This degradation of natural systems further intensifies the waste crisis, as weakened ecosystems are less capable of absorbing or mitigating environmental harm.

Ultimately, the systemic nature of waste generation requires holistic solutions. Addressing one issue in isolation, such as pollution or resource depletion, without considering its broader implications can lead to unintended consequences elsewhere in the ecosystem. Thus, tackling the waste problem demands a comprehensive approach that acknowledges the long-term and interlinked nature of environmental impacts.

4.4 Health and Social Justice Concerns

The environmental impacts of waste discussed in the previous sections have profound implications for human health and social equity. The waste crisis affects public health, exacerbates social inequalities, and disrupts crucial ecosystem services that support human well-being. These interconnected issues highlight the complex challenges society faces in addressing the growing problem of waste.

4.4.1 Public Health

Improper waste management poses significant risks to public health, both directly and indirectly.[11] Landfills and open dumping sites, which are still common in many parts of the world, serve as breeding grounds for disease vectors such as mosquitoes, flies, and rodents. These sites also release methane and other gases that can cause respiratory issues for nearby residents. The leachate produced by landfills contaminates soil and groundwater, potentially leading to waterborne diseases and long-term environmental degradation.

Incineration, another common waste disposal method, emits toxic pollutants such as dioxins, furans, and heavy metals into the atmosphere. These emissions can cause respiratory illnesses and cardiovascular diseases and increase cancer risks in populations living near incineration facilities. The health impacts of waste are often more severe in developing countries, where waste management infrastructure may be inadequate and informal waste picking is common.

Water contamination from industrial waste and landfill leachate poses a significant threat to public health, particularly in areas with inadequate water treatment facilities. This contamination can lead to waterborne diseases and chemical exposure, affecting both human populations and aquatic ecosystems. The burning of waste, especially plastics, contributes to air pollution, exacerbating respiratory conditions and contributing to the formation of smog and particulate matter.

The issue of food chain contamination is becoming increasingly concerning as microplastics and other pollutants enter the food web. The long-term health effects of consuming contaminated food and water are not yet fully understood, but early research suggests potential risks to human health.

Electronic waste (e-waste) presents a unique set of health challenges. Improper handling of e-waste exposes workers and communities to toxic substances like lead, mercury, and cadmium. These chemicals can cause neurological damage, kidney problems, and other severe health issues. As the volume of e-waste continues to grow with the rapid pace of technological advancement, addressing this specific waste stream becomes increasingly critical.

4.4.2 Environmental Justice

The burden of waste management is not distributed equally across society, leading to significant environmental justice concerns.[12] Waste management facilities, such as landfills and incinerators, are often disproportionately located in low-income and minority neighborhoods. These communities face higher exposure to pollution, odors, and associated health risks, creating a cycle of environmental injustice that perpetuates social and economic inequalities.[13]

The global waste trade exacerbates these inequalities on an international scale. Developed countries often export their waste to developing nations, creating environmental and health burdens for communities in receiving countries. This practice not only shifts the physical burden of waste but also the associated social and economic costs, further widening the gap between developed and developing nations.

The informal waste sector, particularly waste picking and recycling, employs millions globally, with an estimated 15–20 million informal waste pickers operating in developing countries. This sector fills critical gaps in waste management, especially where formal systems are lacking, but it also exposes workers to social inequalities. Informal waste pickers often work in hazardous conditions without proper protection or recognition, and they typically earn very low wages. These workers are predominantly part of marginalized communities, which exacerbates existing social and economic inequalities.[14]

Urban-rural disparities in waste management infrastructure further compound these issues. Rural areas often lack proper waste management facilities, leading to improper disposal methods that can contaminate local environments and water sources. This disparity not only affects public health but also contributes to the degradation of rural ecosystems and agricultural lands.

The intersection between environmental justice and long-term systemic challenges in waste management underscores the need for sustainable, inclusive solutions. Marginalized communities that disproportionately bear the negative impacts of waste often have fewer resources and political influence to demand better waste management practices. Without addressing these inequalities, society risks perpetuating a system that prioritizes short-term economic gains over long-term environmental sustainability.

4.4.3 Waste and Ecosystem Services

Waste not only pollutes the environment but also degrades the essential ecosystem services that support human well-being and economic development. Ecosystem services are the benefits that people derive from ecosystems, such as clean air, fresh water, fertile soil, and biodiversity. When waste is mismanaged, it can disrupt these services, leading to far-reaching consequences for both human and ecological health.

Plastic waste that enters marine environments can harm marine life through ingestion and entanglement, and can also accumulate in the food chain, potentially affecting human health. Landfills and open dumping sites can contaminate soil and groundwater, reducing the productivity of agricultural land and compromising the quality of drinking water sources.

Incineration of waste releases toxic pollutants into the air, contributing to air pollution and respiratory diseases.

The degradation of ecosystem services by waste also has significant economic costs. For instance, the loss of pollination services due to the decline of bee populations, partly attributed to pesticide use and habitat loss, can have severe impacts on agricultural productivity.[15] The contamination of water bodies by industrial waste can lead to increased water treatment costs and reduced revenues from fishing and tourism.

Protecting and restoring ecosystem services is, therefore, essential for addressing the waste crisis and promoting sustainable development. This requires a holistic approach that considers the entire lifecycle of products and materials, from extraction to disposal.

4.4.4 Socio-cultural Factors and Industry Practices

Societal and cultural dynamics play a significant role in shaping waste generation and management practices, influencing everything from individual consumer habits to large-scale industry operations. How societies view consumption, disposal, and environmental responsibility directly affects the scale of the waste problem. At the same time, many industries prioritize short-term profits over long-term sustainability, perpetuating wasteful practices that further strain global resources. Understanding these socio-cultural factors, alongside the profit-driven behaviors of key industries, is essential for developing more sustainable waste management solutions and shifting toward a circular economy. By fostering behavioral changes, increasing public awareness, and encouraging responsible industry practices, society can begin to address the root causes of waste generation and move towards a more sustainable future.

4.4.4.1 Consumer Behavior and Awareness

The growing culture of consumerism, characterized by a "throwaway" mentality, has been a key driver of waste generation. In many parts of the world, products are designed for disposability and rapid replacement, often at the expense of durability and repairability. This culture of consumption is fueled by marketing and the appeal of novelty, leading to increased demand for short-lived goods. However, shifts in consumer behavior, such as the adoption of minimalism, buying second-hand items, and prioritizing repair and reuse, can contribute significantly to waste reduction. Global movements like the zero-waste lifestyle and sustainable fashion highlight the potential for change when consumers become more mindful of their choices.

Public awareness and education also play a critical role in transforming waste management practices. Raising awareness about the environmental, social, and economic impacts of waste can inspire more responsible consumption patterns. Educational programs targeting all age groups and communities can foster a culture of waste reduction, recycling, and sustainability. Countries like Japan and Germany, which have embedded recycling and waste separation practices into daily life, demonstrate how public education can drive significant improvements in waste management. Awareness campaigns and community initiatives can empower individuals to take action, but they must be accompanied by accessible waste management systems to enable these behaviors.

4.4.4.2 *Profit-driven Industry Practices*

Many industries, driven by profit maximization, have historically prioritized short-term gains over long-term environmental sustainability, contributing to excessive waste generation. The fast fashion industry epitomizes this dynamic, producing cheap, trendy clothing at an accelerated pace, leading to vast amounts of textile waste. This industry alone is responsible for significant environmental degradation, from water pollution to the release of microplastics during laundering. Similarly, the food and beverage sector heavily relies on single-use plastics and non-recyclable packaging, contributing to the growing global plastic waste crisis.

Other industries, such as mining, electronics, pharmaceuticals, and construction, also generate substantial waste and pollutants. The electronics industry, in particular, is a significant contributor to e-waste, with the rapid pace of technological obsolescence leading to the disposal of large amounts of hazardous materials.

Some industries are beginning to embrace more sustainable practices. The rise of sustainable fashion brands demonstrates that it is possible to reduce textile waste by adopting circular models of production, reuse, and recycling. In the food and beverage industry, innovations like biodegradable packaging and reusable container programs are gaining traction, offering viable alternatives to single-use plastics.

4.4.4.3 *Technological Innovations and Industry Solutions*

In many cases, industries generate significant waste simply because sustainable alternatives have yet to be widely developed or adopted. For example, the healthcare industry generates a considerable amount of waste from disposable medical supplies and diagnostic tools. As industries seek to balance public health with environmental concerns, the development of sustainable technologies becomes critical. Innovations such as biodegradable medical equipment, nanotechnology-based waste treatment systems, and advanced recycling techniques can reduce the environmental footprint of these sectors.

As we will discuss in later sections, nanotechnologies, in particular, offer promising solutions for addressing industry-specific waste challenges. Nanomaterials can enhance the durability and recyclability of products, while advanced nanotechnology-based filters and purification systems can help mitigate the pollution caused by industrial waste streams. As these technologies evolve, they have the potential to transform industries that are currently significant contributors to waste generation.

4.5 The Economic Costs and Opportunities

Waste generation presents not only environmental and social challenges but also significant economic implications. These economic impacts are multifaceted, ranging from the direct costs of waste management to the broader, often hidden, costs that affect public health, ecosystems, and global trade. At the same time, the shift toward more sustainable waste practices presents vast economic opportunities, including the creation of new industries, the recovery of valuable resources, and the advancement of innovative technologies. To address the waste crisis, understanding both the financial burden and the untapped potential of waste is crucial.

4.5.1 Waste Management Costs

The growing volume and complexity of waste create substantial economic burdens for governments, businesses, and taxpayers. Municipalities and local governments are responsible for the costs of waste collection, transportation, treatment, and disposal. Building and maintaining infrastructure, such as landfills, recycling plants, and waste-to-energy facilities, requires continuous and significant investment. Inefficient or outdated waste management systems lead to higher operational costs, putting pressure on public budgets and constraining economic growth.

Globally, the direct cost of managing municipal solid waste (MSW)—which includes expenses for collection, transportation, processing, and disposal—reached \$252.3 billion in 2020. This figure only covers MSW, which represents waste generated by households and small businesses, and is just one facet of the broader waste management challenge. When factoring in external costs like environmental degradation and public health impacts from mismanaged MSW, the total rises to nearly \$496 billion. Data for industrial waste management costs are less readily available, but these costs are significant and vary depending on a country's economic structure and urbanization. Importantly, the projected cost of MSW alone could exceed \$640 billion by 2050 if current practices remain unchanged. However, a shift towards more sustainable practices, such as increased recycling and waste reduction, could stabilize MSW management costs. Adopting a circular economy approach, which focuses on minimizing waste and reusing resources, could keep MSW-related expenses near 2020 levels, underscoring the economic and environmental benefits of improved waste management strategies.[16]

4.5.2 Lost Economic Opportunities

Waste not only incurs costs but also represents missed economic opportunities. When materials that could be recycled or repurposed are sent to landfills or incinerators, society loses the economic value embedded in those resources. Implementing circular economy principles—where waste is seen as a resource rather than a problem—can unlock new business opportunities, stimulate innovation, and create jobs in industries focused on resource recovery and recycling.

The economic potential exists across multiple waste streams. For instance:

- Organic waste could be converted to compost or biogas.
- Construction and demolition waste contains valuable recoverable materials.
- Textile waste represents significant untapped value in fiber recovery.
- Plastic waste could be recycled into new products or chemical feedstocks.

The electronics industry provides one of the most compelling examples of this untapped potential. Currently, the global e-waste management system incurs a net annual loss of USD 37 billion due to inefficient recycling practices and externalized environmental and societal costs (2022 data).[17] This includes USD 78 billion in externalized costs such as pollution from hazardous substances like lead and mercury, plastic leakages, and contributions to global warming. However, the economic value of metals contained in global e-waste was estimated at USD 91 billion in 2022, yet only 22.3% of this e-waste was documented as formally collected and recycled. Proper recycling could not only generate significant revenue—currently estimated at USD 28 billion annually—but also prevent up to 93 billion kg of CO_2-equivalent emissions by reducing the need for virgin material extraction.[18]

By enhancing waste management infrastructure and embracing circular economy practices, society can move beyond merely mitigating costs to actively capturing value from discarded materials. This shift creates opportunities for innovative business models across various sectors. For instance, companies in the fashion industry are shifting toward "clothing-as-a-service" models, where consumers rent rather than purchase clothing, thereby extending garment lifespans and reducing textile waste. The rise of sharing economies, refurbishment markets, and product leasing demonstrates how businesses can profit by rethinking waste as a valuable resource.

Countries like the Netherlands exemplify this transition's potential. The Dutch government has set an ambitious goal to reduce raw material consumption by 50% by 2030, creating a thriving recycling industry and new opportunities for sustainable business models.[18] This approach highlights how waste reduction can serve as a driver of economic growth, rather than merely a cost to be managed.

4.5.3 Externalized Costs

A significant portion of the true cost of waste is externalized, meaning that the negative impacts of waste—such as environmental degradation, public health risks, and climate change—are not fully reflected in the price of goods and services. These externalized costs are instead borne by society at large, leading to hidden economic burdens that manifest over time in the form of health care expenses, ecosystem restoration, and lost agricultural productivity.

For example, the health impacts of air pollution from waste incineration are not factored into the operational costs of waste facilities. Incineration plants release toxic pollutants like dioxins and heavy metals, which contribute to respiratory diseases, cancers, and cardiovascular problems.[19] The healthcare costs associated with treating these diseases, as well as the loss of productivity due to illness, are often externalized, placing additional strain on national health systems.

Similarly, the contamination of water bodies by landfill leachate or industrial waste leads to the degradation of aquatic ecosystems and the loss of ecosystem services, such as clean drinking water and fertile soil for agriculture. These long-term costs are rarely considered in the pricing of waste management systems but can have devastating economic impacts on local communities.

Pollution is the leading environmental cause of disease and premature death, responsible for over 9 million deaths annually, with most attributed to air pollution. This toll exceeds deaths from AIDS, tuberculosis, and malaria combined. Air pollution alone causes 7 million premature deaths each year and cost the global economy \$8.1 trillion in 2019, equaling 6.1% of global GDP. In low- and middle-income countries, pollution-related mortality and morbidity contribute to economic losses of 5% to 14% of GDP.[20]

Internalizing these costs through policy instruments, such as extended producer responsibility, landfill taxes, and incentives for sustainable production, can help align economic incentives with environmental and social well-being. For instance, the European Union's EPR policies have led to more sustainable product designs and increased recycling rates, shifting some of the waste management burden away from taxpayers and onto producers, where it rightfully belongs.[21]

4.5.4 Global Waste Trade

The global waste trade is a complex and often controversial aspect of international waste management. Developed nations frequently export waste, particularly plastic and electronic waste, to developing countries where processing costs are lower, and environmental regulations are less stringent. This practice, while economically beneficial for exporting countries, has severe environmental and social consequences for receiving nations.

Much of the world's plastic waste used to be sent to China until the country implemented its 2018 ban on most imported plastic waste due to the environmental damage caused by improper recycling and disposal.[22] Since then, waste exports have shifted to other developing countries, such as Malaysia, Indonesia, and Vietnam, where waste infrastructure is often inadequate to handle the volume and complexity of the materials received. This has led to open dumping, burning of plastics, and severe pollution of rivers and oceans, exacerbating environmental degradation and threatening public health.[23]

At the same time, the global waste trade provides economic opportunities for some developing countries that have developed specialized recycling industries. For example, countries like India[24] and Ghana[25] have large informal sectors focused on recycling e-waste, providing jobs and economic activity. However, the lack of regulation in these sectors often leads to unsafe working conditions and environmental harm, illustrating the need for stronger international cooperation and enforcement of treaties like the Basel Convention.

The disruptions caused by China's waste import ban have forced many developed nations to rethink their waste management strategies. Countries like Japan[26] and Germany[27] are investing in domestic recycling infrastructure, reducing their reliance on global waste exports and creating jobs in green industries. This shift toward localized waste processing and recycling is a positive step toward a more sustainable global waste management system, but it will require continued investment, innovation, and international collaboration to address the growing waste crisis.

4.6 Circular Economy Potential

The circular economy presents a transformative model for addressing waste-related challenges while simultaneously offering significant economic and environmental benefits. By rethinking traditional linear systems that prioritize resource extraction, use, and disposal, the circular economy focuses on designing out waste and pollution, keeping products and materials in use for as long as possible, and regenerating natural systems. This shift has the potential to unlock substantial economic value and help mitigate some of the most pressing environmental crises of our time.

Figure 4.3 illustrates this process through four key strategies: Narrow, Slow, Regenerate, and Cycle. In practice, this means not only minimizing the material input in product design (Narrow) but also extending product life cycles (Slow) by encouraging repair, reuse, and refurbishment. Furthermore, the principle of Regenerate emphasizes the importance of clean, renewable inputs, reducing dependency on finite resources. Finally, materials are kept in circulation for as long as possible (Cycle), thus reducing waste and the need for virgin material extraction.

Regenerative agricultural practices or renewable energy sources (Regenerate) are becoming integral components of sustainable food systems and energy

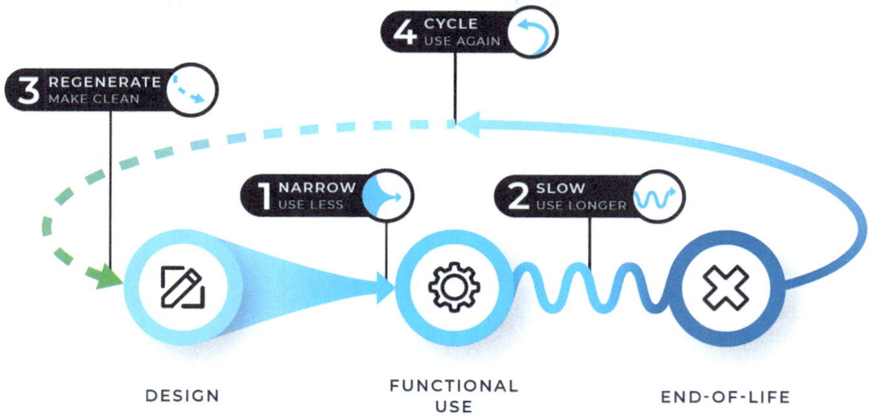

Figure 4.3 The four flows of the circular economy: a comprehensive framework for managing resource flows in a circular economy by using less (Narrow), using for longer (Slow), using again (Cycle), and using clean, regenerative materials and energy (Regenerate). Reproduced from ref. 30, https://circularity-gap.world/2024, under the terms of the CC BY-SA 4.0 license, https://creativecommons.org/licenses/by-sa/4.0/deed.en.

grids. At the same time, the manufacturing industry can benefit from strategies that keep machinery, electronics, and textiles in use for longer periods (Slow), further reducing waste and enhancing resource efficiency.

This holistic approach has been shown to yield both ecological and economic dividends, creating new business models centered on resource efficiency, reducing greenhouse gas emissions, and protecting natural ecosystems. By embracing these principles, industries and governments alike can transition toward more sustainable, resilient economies while safeguarding the planet's vital systems.

Implementing these circular strategies can significantly reduce material extraction needs—by up to one-third—while still meeting key societal demands such as housing, nutrition, and mobility. In addition to resource savings, the circular transition is expected to generate new employment opportunities across sectors like manufacturing and construction, boosting demand for skilled workers and fostering the creation of decent jobs. Furthermore, the development of local industries focused on recovering, processing, and recycling materials strengthens regional supply chains, supporting long-term sustainable growth. Achieving these economic benefits will depend on the right fiscal and financial incentives, including tax reforms, subsidies for circular businesses, and financial instruments like green bonds that make circular solutions more attractive and viable.

Technological innovation is a key enabler of the circular economy. Advances in materials science, such as the development of durable and recyclable nanomaterials, are allowing companies to design products that last longer and can be more easily disassembled and recycled. In the waste management sector, technologies like AI-driven waste sorting, advanced recycling processes, and blockchain-based tracking systems are improving the efficiency and transparency of recycling operations.

Additionally, digital platforms are playing a vital role in facilitating circular business models. Online platforms for sharing, renting, and reselling products—such as car-sharing services, peer-to-peer clothing rental, and electronics refurbishment marketplaces—are making it easier for consumers to participate in the circular economy and reduce their own environmental footprints.

References

1. M. Geissdoerfer, P. Savaget, N. M. P. Bocken and E. J. Hultink, The Circular Economy – A new sustainability paradigm?, *J. Clean. Prod.*, 2017, **143**, 757–768.
2. P. Lacy, J. Long and W. Spindler, *The circular economy handbook: Realizing the circular advantage*, Palgrave Macmillan, Basingstoke, England, 1st edn, 2021.
3. Source: IEA World Energy Balances 2022 [cited 2024 Jul 16]. Available from: https://www.iea.org/data-and-statistics/data-product/world-energy-statistics-and-balances.
4. M. M. Mekonnen and A. Y. Hoekstra, Four billion people facing severe water scarcity, *Sci. Adv.*, 2016, **2**(2), e1500323.
5. United Nations Environment Programme: Global Resources Outlook 2024: Bend the Trend – Pathways to a liveable planet as resource use spikes, International Resource Panel, Nairobi, [cited 2024 Sep 19]. Available from: https://www.unep.org/resources/Global-Resource-Outlook-2024.
6. Environment UN, *Global Waste Management Outlook 2024*, UNEP – UN Environment Programme, 2024, [cited 2024 Sep 18]. Available from: https://www.unep.org/resources/global-waste-management-outlook-2024.
7. EMEP/EEA air pollutant emission inventory guidebook 2023, European Environment Agency, 2023, [cited 2024 Sep 20]. Available from: https://www.eea.europa.eu/publications/emep-eea-guidebook-2023.
8. L. Andeobu, S. Wibowo and S. Grandhi, Environmental and health consequences of E-waste dumping and recycling carried out by selected countries in Asia and Latin America, *Sustainability.*, 2023, **15**(13), 10405.
9. A. S. Kolawole and A. O. Iyiola, Environmental pollution: Threats, impact on biodiversity, and protection strategies, in *Sustainable Development and Biodiversity*, Springer Nature Singapore, Singapore, 2023, pp. 377–409.
10. V. P. Singh, S. Yadav, K. K. Yadav and R. N. Yadava, An overview on environmental degradation and mitigation, in *Environmental Degradation: Challenges and Strategies for Mitigation*, Springer International Publishing, Cham, 2022, pp. 3–15.
11. A. David Raj, R. Padmapriya and A. David Raj, Climate crisis impact on ecosystem services and human well-being, in *Climate Crisis, Social Responses and Sustainability*, Springer Nature Switzerland, Cham, 2024, pp. 3–36.

12. J. Loos, F. Benra, M. Berbés-Blázquez, L. L. Bremer, K. M. A. Chan and B. Egoh, *et al.*, An environmental justice perspective on ecosystem services, *Ambio*, 2023, **52**(3), 477–488.
13. P. Vineis and A. Gambhir, Social inequalities and the environmental crisis: need for an intergenerational alliance: Tenth anniversary of Frontiers in Public Health, *Front. Public Health*, 2023, **11**, 1226961.
14. J. Gutberlet, Global plastic pollution and informal waste pickers, *Cambridge Prisms*, 2023, **1**, e9.
15. Y. Yang, Y. Wu, H. Long, X. Ma, K. Shariati and J. Webb, *et al.*, Global honeybee health decline factors and potential conservation techniques, *Food Secur.*, 2023, **15**(4), 855–875.
16. United Nations Environment Programme, International Solid Waste Association, *Global Waste Management Outlook 2024 – Beyond an age of waste: Turning rubbish into a resource*, United Nations Environment Programme, 2024.
17. The Global E-waste Monitor 2024, E-Waste Monitor, 2024 [cited 2024 Sep 20]. Available from: https://ewastemonitor.info/the-global-e-waste-monitor-2024/.
18. Circular Dutch economy by 2050, Government.nl. 2019 [cited 2024 Sep 20]. Available from: https://www.government.nl/topics/circular-economy/circular-dutch-economy-by-2050.
19. N. Tangri, Waste incinerators undermine clean energy goals, *PLoS Clim.*, 2023, **2**(6), e0000100.
20. Pollution, World Bank, [cited 2024 Sep 21]. Available from: https://www.worldbank.org/en/topic/pollution.
21. P. K. Mallick, K. B. Salling, D. C. A. Pigosso and T. C. McAloone, Designing and operationalising extended producer responsibility under the EU Green Deal, *Environ. Challenges*, 2024, **16**, 100977.
22. E. Brady, The effects of china's ban on imported scrap plastic on global recycling efforts, Earth.org., 2021 [cited 2024 Sep 21]. Available from: https://earth.org/china-ban-on-imported-scrap-plastic/.
23. A. Jain, Trash trade wars: Southeast Asia's problem with the world's waste, Council on Foreign Relations, 2020 [cited 2024 Sep 21]. Available from: https://www.cfr.org/in-brief/trash-trade-wars-southeast-asias-problem-worlds-waste.
24. T. S. Krishnan, Repositioning the "informal economy": The hidden potential for a new circular economy in India's electronic waste industry, in *Palgrave Advances in the Economics of Innovation and Technology*, Springer International Publishing, Cham, 2023, pp. 1–22.
25. M. O. Asibey, R. S. King, A. M. Lykke, D. K. B. Inkoom and R. S. Wireko-Gyebi, Informal electronic waste recycling in Ghanaian cities: environmental risks awareness and attitudes, *SN Soc. Sci.*, 2022, **2**, 115.
26. N. Kutty and N. Tochibayashi, *Why Japan sees regeneration as the key to a circular economy*, World Economic Forum, 2022, [cited 2024 Sep 21]. Available from: https://www.weforum.org/agenda/2022/11/is-regeneration-the-key-to-the-future-of-the-circular-economy/.

27. H. Engel, S. Helmcke, R. Heuss and S. Hieronimus, *Net-zero Germany: Chances and challenges on the path to climate neutrality by 2045*, McKinsey & Company, 2021.
28. A. Pont, A. Robles and J. A. Gil, E-WASTE: Everything an ICT scientist and developer should know, *IEEE Access*, 2019, 7, 169614–169635.
29. IEA, *Electricity Mid-Year Update - July 2024*, IEA, Paris, 2024, [cited 2024 Sep 18]. Available from: https://www.iea.org/reports/electricity-mid-year-update-july-2024.
30. CGR 2024, Circularity-gap.world, [cited 2024 Sep 21]. Available from: https://circularity-gap.world/2024.

Section II: An Essential Overview of Nanotechnology

CHAPTER 5

Novel Nanomaterials and Their Importance

Nanomaterials represent a frontier of scientific innovation, offering properties and functionalities distinct from their bulk counterparts. In this chapter, we explore novel types of nanomaterials and their significance across various applications.

Size becomes a defining characteristic when exploring the properties of nanomaterials, influencing behaviors at both the nano- and micro-scale. Figure 5.1 illustrates the relative sizes of structures, ranging from atoms and DNA to biological entities like eukaryotic cells and hair strands, placing nanomaterials into a broader scale context.

This visual representation helps us understand the unique position of nanomaterials within the size spectrum, which spans from roughly 1 to 100 nanometers. As materials shrink to these scales, phenomena such as quantum confinement and surface effects become increasingly significant. The figure underscores the vast range of dimensions that we consider when discussing nanoscale objects, setting the stage for how such minute sizes can lead to radically different properties compared to their bulk counterparts.

This book's examination of quantum confinement, surface interactions, and structural design reveals how nanomaterial innovations are transforming electronics, energy storage, medicine, and environmental science. This fundamental understanding underpins the development of efficient waste reduction techniques, energy-saving technologies, and sustainable solutions—practical applications explored throughout subsequent chapters. These core principles of nanoscience illuminate how theoretical insights drive real-world technological advances.

Waste Not! How Nanotechnologies Can Increase Efficiencies Throughout Society
By Michael Berger

Published by the Royal Society of Chemistry, www.rsc.org

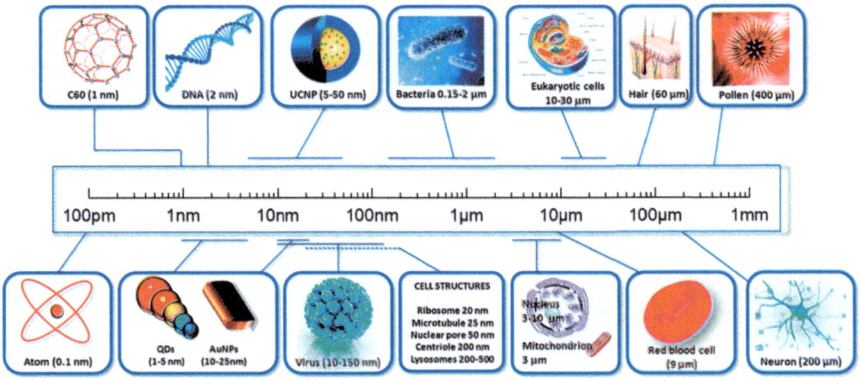

Figure 5.1 A comparison of sizes of nanomaterials. Reproduced from ref. 15 with permission from the Royal Society of Chemistry.

5.1 Quantum Size Effect

At the nanoscale, materials exhibit behavior different from their bulk counterparts due to the quantum size effect. This effect arises when the size of a material is reduced to a point where the movement of electrons and holes is restricted, resulting in distinct energy levels (see Figure 5.2). In these confined spaces, the material's electronic, optical, and magnetic properties change in significant ways. This phenomenon is rooted in quantum mechanics and becomes particularly noticeable when the dimensions of the material are comparable to the wavelength of electrons, which leads to a marked departure from classical behavior.[1]

In nanomaterials, the quantum size effect results in a range of unique properties. One key aspect is the restriction of electron movement, which forces electrons into discrete energy levels, akin to what happens in atoms. This confinement causes materials to absorb and emit light differently based on their size, a phenomenon often visible in quantum dots, where smaller dots emit blue light and larger ones emit red. These optical shifts make quantum dots valuable in applications such as display technologies and bioimaging. Similarly, metals like gold, which in bulk form are known for their shiny yellow appearance, can take on striking red or purple hues at the nanoscale due to this effect.

The quantum size effect also impacts electrical and magnetic properties. In nanomaterials, the confinement of electrons alters their conductivity, allowing for innovations like more efficient transistors and sensors. Magnetic nanoparticles exhibit superparamagnetism, where they can behave as single magnetic domains, making them highly useful in areas such as magnetic data storage and targeted drug delivery. The thermal properties of nanomaterials are similarly affected, as heat conduction at the nanoscale differs due to changes in the way vibrations (phonons) propagate, which can lead to more efficient thermoelectric materials for energy harvesting.

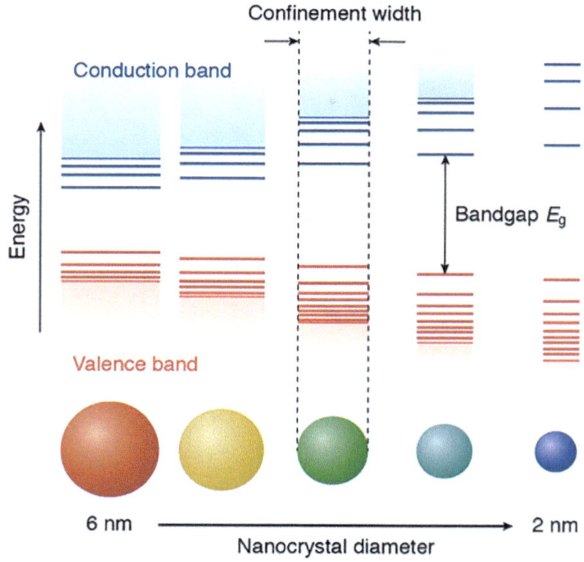

Figure 5.2 Illustration of the quantum confinement effect. The diagram shows how quantum dot size influences energy band structure and light emission. As quantum dots decrease in size, the bandgap widens, shifting emitted light from red to blue, demonstrating how size controls emitted light color. Adapted from ref. 16, https://doi.org/10.1021/acssensors.2c00149, under the terms of the CC BY 4.0 license, https://creativecommons.org/licenses/by/4.0/.

These remarkable properties emerging from the quantum size effect open up a wide range of applications in electronics, medicine, and energy, pushing the boundaries of what nanotechnology can achieve. While challenges remain in controlling the precise size, shape, and composition of nanomaterials, ongoing research and advancements in synthesis and characterization methods continue to unlock the potential of materials at the nanoscale. The quantum size effect is not just a scientific curiosity but a gateway to developing materials with custom-tailored properties, driving forward innovations in many technological fields.

5.2 Surface Effects

At the nanoscale, materials also exhibit unique behaviors due to the surface effect, which becomes increasingly prominent as the size of the material shrinks. When materials are reduced to nanometer dimensions, the surface area relative to the volume grows dramatically, as illustrated in Figure 5.3. This increased surface-to-volume ratio means that a significant proportion of the atoms or molecules in the material are located at or near the surface, making surface properties much more influential on the material's overall behavior.

Figure 5.3 Illustration of the effect of increased surface area. The image illustrates how dividing a cube, with an assumed 1 cm side length, exponentially increases surface area. The solid cube on the left has a surface area of 6 cm^2. Dividing it into 64 smaller cubes (right) increases the total surface area to 24 cm^2. Further splitting each of these 64 cubes into 64 more cubes for a total of 4096 cubes results in a total surface area of 96 cm^2. Repeating this split again results in 262 144 cubes with a total surface area of 384 cm^2. This exponential growth in surface area is crucial in distinguishing bulk from nanomaterials, as nanoscale materials have vastly larger surface areas relative to their volume, enhancing their reactivity and functionality in various applications like catalysis and drug delivery.

One of the most noticeable consequences of the surface effect is the enhanced reactivity of nanomaterials. With more atoms exposed on the surface, nanomaterials often display heightened chemical activity compared to their bulk counterparts. This is especially important in applications like catalysis, where materials such as titanium dioxide or gold nanoparticles can accelerate chemical reactions far more efficiently than their bulk forms. These surface atoms tend to have fewer neighboring atoms, leading to higher energy states that make them more reactive and, therefore, more capable of interacting with surrounding molecules.

In addition to chemical reactivity, the surface effect plays a crucial role in determining the electronic and optical properties of nanomaterials. The high proportion of surface atoms can alter the electronic structure, influencing how electrons are distributed across the material. This is particularly evident in nanostructures used in sensors or electronic devices, where the increased surface sensitivity to the surrounding environment allows for more precise detection of molecules or changes in conditions. The surface can be further modified through the introduction of functional groups or coatings, which can be engineered to enhance specific interactions, making nanomaterials ideal for targeted applications in medicine or environmental monitoring.

The surface effect also impacts mechanical and thermal properties. In nanomaterials, surface forces become dominant, often leading to increased strength and durability. For instance, nanostructures like carbon nanotubes or graphene sheets are exceptionally strong because of the influence of surface interactions, which helps maintain structural integrity even at small scales. In thermal applications, the surface effect can lead to reduced heat conductivity due to phonon scattering at the material's surface. This behavior is harnessed in nanomaterials designed for thermal insulation or energy-efficient devices, where controlling heat flow is essential.

While the quantum size effect becomes significant at sizes below 10 nanometers, the surface effect can become equally important as surface-to-volume ratios increase. These two effects often interplay, with quantum confinement influencing electronic behavior and surface interactions driving chemical reactivity and strength. Together, they offer a combined understanding of why nanomaterials exhibit such radically different properties from their bulk counterparts.

The importance of surface effects is not just in enhancing material properties but in enabling entirely new functionalities. Nanomaterials can be functionalized by attaching specific molecules or ligands to their surfaces, which can change how they interact with biological systems, chemicals, or other materials. For example, nanoparticles designed for drug delivery can have their surfaces tailored to bind specifically to certain types of cells, allowing for precise targeting of therapeutic agents in the body. Similarly, in environmental applications, surface-engineered nanomaterials can be used to selectively capture pollutants or toxins, offering a powerful tool for cleaning up contaminated water or air.

As with the quantum size effect, the surface effect is a key factor in the behavior and utility of nanomaterials. It opens up a range of possibilities across industries, from more efficient catalytic processes to advanced medical treatments. Understanding and controlling these surface interactions allow scientists to design materials with highly specific, enhanced properties, driving innovation in areas like energy storage, electronics, and environmental sustainability. Fully harnessing the surface effect requires precise control over nanomaterial synthesis and surface engineering, which remains an area of ongoing research and development.

5.3 Key Nanomaterials

Nanomaterials, characterized by their nanoscale dimensions, exhibit unique properties that significantly differ from their bulk counterparts. These properties are intricately linked to the dimensionality of the materials, which can range from zero-dimensional (0D) structures, where all dimensions are measured within nanometers, to one-dimensional (1D), two-dimensional (2D), and three-dimensional (3D) structures, each offering distinct physical and chemical characteristics. This section explores key nanomaterials within each dimensionality category, highlighting their properties, applications, and the innovative possibilities they present. By understanding the diverse range of nanomaterials and their unique properties, we can better appreciate their potential to revolutionize various fields, from electronics and energy to healthcare and environmental science.

5.3.1 Organic *vs.* Inorganic Nanomaterials

Before exploring nanomaterials by their dimensionality, it is essential to understand their fundamental classification into organic and inorganic categories. This distinction significantly influences their properties, behavior, and potential applications. As illustrated in Figure 5.4, nanomaterials can be broadly categorized into inorganic and organic nanoparticles, with some hybrid materials bridging these categories.

This fundamental classification provides a framework for understanding the diverse range of nanomaterials and their applications. As we explore specific dimensional categories, this organic–inorganic distinction will remain relevant to understanding material properties and behavior.

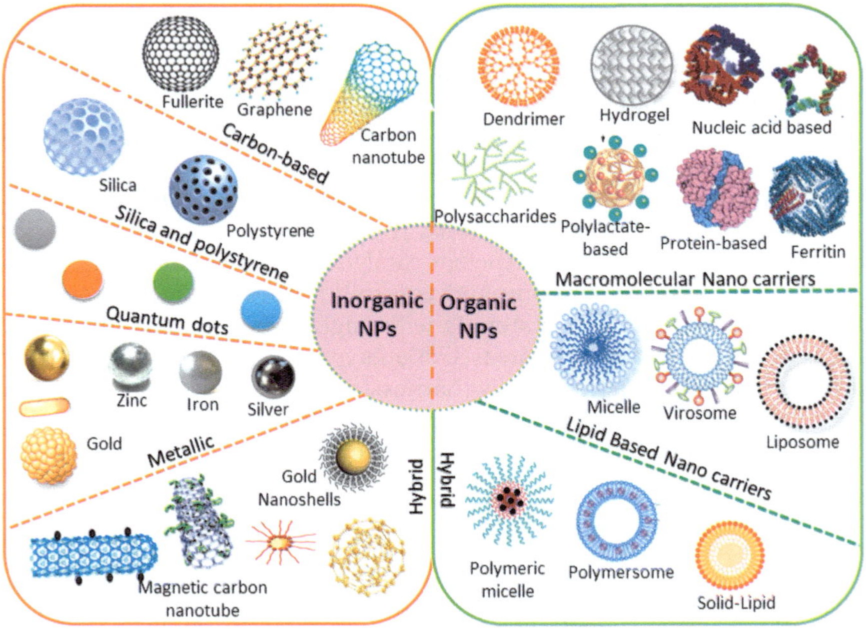

Figure 5.4 Classification of nanomaterials into organic and inorganic categories, showing representative examples and hybrid materials at their interface. Reproduced from ref. 17, https://doi.org/10.1039/D4NA00214H, under the terms of the CC BY 3.0 license, https://creativecommons.org/licenses/by/3.0/.

5.3.1.1 Inorganic Nanomaterials

Inorganic nanomaterials comprise non-carbon-based materials and certain carbon allotropes. These include:

- Metallic nanoparticles (gold, silver, zinc, and iron)
- Metal oxide nanoparticles
- Quantum dots
- Carbon-based materials (fullerites, graphene, and carbon nanotubes)
- Silica-based materials

Inorganic nanomaterials generally exhibit superior stability, durability, and well-defined electronic and optical properties. Their robust nature makes them particularly suitable for applications in electronics, catalysis, and sensing. For instance, metallic nanoparticles like gold and silver display unique optical properties through surface plasmon resonance, while quantum dots offer size-dependent luminescence properties valuable in imaging and display technologies.

5.3.1.2 Organic Nanomaterials

Organic nanomaterials are primarily carbon-based structures derived from biological or synthetic organic compounds. Key categories include:

- macromolecular nano carriers (dendrimers and protein-based structures),
- lipid-based nano carriers (liposomes, micelles, and virosomes),
- polymeric nanostructures,
- polysaccharide-based materials, and
- nucleic acid-based structures.

These materials often excel in biocompatibility and biodegradability, making them particularly valuable for biomedical applications. Their surface chemistry can be readily modified, allowing for targeted drug delivery, gene therapy, and tissue engineering applications. For example, liposomes and polymeric nanoparticles are widely used as drug delivery vehicles due to their ability to encapsulate therapeutic compounds and their favorable interaction with biological systems.

5.3.1.3 Hybrid Nanomaterials

At the interface of organic and inorganic materials lie hybrid nanomaterials, which combine properties from both classes. Examples include:

- gold nanoshells with organic coatings,
- magnetic nanoparticles with polymer shells,

- polymer-functionalized quantum dots, and
- core–shell structures combining organic and inorganic components.

These hybrid materials often achieve enhanced functionality through the synergistic combination of organic and inorganic properties. For instance, magnetic nanoparticles coated with biocompatible polymers can combine the magnetic properties of the inorganic core with the biological compatibility of the organic shell, enabling applications in targeted drug delivery and magnetic resonance imaging.

5.3.2 Zero-dimensional (0D) Nanomaterials

Zero-dimensional nanomaterials, such as quantum dots and nanoparticles, possess physical dimensions in all three spatial directions but are characterized by their nanoscale size in such a way that electrons are quantum confined in all three dimensions. This confinement leads to discrete energy levels and significantly alters their physical, chemical, and optical properties compared to their bulk counterparts.[2]

Specifically, the significance of quantum effects becomes pronounced when the nanomaterial size falls around or below a material-dependent threshold, known as the exciton Bohr radius, which typically ranges from 1–10 nanometers (see Figure 5.5).

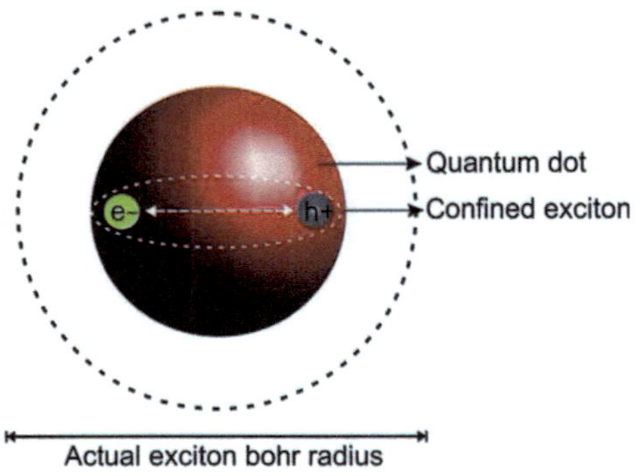

Figure 5.5 Quantum confinement and the Bohr radius. The exciton Bohr radius, represented by the dashed circle, marks the threshold below which quantum confinement effects dominate. In this scenario, the electron and hole are confined within the quantum dot, with their spatial separation limited by the dot's reduced size. As the size of the nanomaterial shrinks to this scale, its electronic properties shift, leading to discrete energy levels, similar to those observed in atoms. This phenomenon significantly alters the optical and electronic behavior of the material, allowing tunable properties based on particle size. Adapted from ref. 18, https://www.intechopen.com/chapters/70534, under the terms of the CC BY 3.0 license, https://creativecommons.org/licenses/by/3.0/.

For example, in materials like cadmium selenide for quantum dots and gold for nanoparticles, reaching sizes at or below their respective exciton Bohr radius results in pronounced quantum confinement effects. The electrons within these materials are highly restricted in all three dimensions, mimicking conditions as if they were trapped in an extremely tiny box—a scenario physicists describe as 'quantum confinement in all three dimensions' or 'zero-dimensional'.

5.3.2.1 Nanoparticles

In the case of nanoparticles, such as gold nanoparticles, this quantum confinement results in unique optical, electronic, and magnetic properties due to the discrete energy levels that electrons can occupy, contrasting sharply with the continuous energy levels found in bulk materials. The enhanced reactivity and unique optical properties of gold nanoparticles, driven by their sizes reaching the exciton Bohr radius, are extensively utilized in catalysis to accelerate chemical reactions and in biomedical applications for imaging and targeted drug delivery.

5.3.2.2 Quantum Dots

Quantum dots, particularly those made from semiconductor materials like cadmium selenide and indium phosphide, exhibit size-dependent emission wavelengths due to quantum confinement effects.

As the size of the quantum dot decreases, fitting within the material-specific exciton Bohr radius, the confinement of electrons becomes more pronounced, leading to a blue shift in the emission wavelength. This tunable property is invaluable in technologies requiring precise control over photon emission, including light-emitting diodes (LEDs), quantum computing, and biomedical imaging for tracking and labeling within cellular processes.

5.3.2.3 Fullerenes

Fullerenes are a class of carbon allotropes formed into spherical, ellipsoidal, or cylindrical shapes, with buckminsterfullerene (C_{60}) being a prime example. Their structure results in unique electronic, photophysical, and chemical properties. These materials are pivotal in two advanced application areas:

- **Organic photovoltaics**: Fullerenes serve as efficient electron acceptors in organic solar cells, improving solar energy conversion efficiency. The use of C_{60} in polymer solar cells exemplifies their contribution to enhancing renewable energy technologies.
- **Drug delivery**: Leveraging their cage-like structures, fullerenes can encapsulate and deliver therapeutic agents directly to targeted sites, such as tumor cells for cancer treatment. This capability, combined with their antioxidant properties, also shows promise in developing treatments for neurodegenerative diseases by neutralizing free radicals.

The exploration of fullerenes, including C_{70} and higher-order variants like C_{84}, continues to reveal their potential in both energy solutions and healthcare, demonstrating their significant role in nanotechnology's advancement.

5.3.3 One-dimensional (1D) Nanomaterials

Moving on to one-dimensional nanomaterials, we find structures with high aspect ratios, such as nanowires and carbon nanotubes, which exhibit distinct properties along their length.

5.3.3.1 Nanowires

With their length significantly greater than their diameter, nanowires exhibit anisotropic electrical, thermal, and mechanical properties. They are excellent conductors of heat and electricity along their length, while their confined structure enhances their quantum mechanical effects.

Nanowires are key components in the development of smaller, more efficient electronic devices, sensors, and are also being explored for their potential in next-generation computing technologies. Silicon, silver, and zinc oxide nanowires are among the most studied materials for their applications in electronics, photonics, and sensing.

5.3.3.2 Carbon Nanotubes (CNTs)

CNTs are cylindrical nanostructures composed of carbon atoms arranged in a hexagonal lattice. They come in two varieties: single-walled (SWCNTs) and multi-walled (MWCNTs).

With their remarkable strength, flexibility, and electrical conductivity, CNTs are exploited in a myriad of applications, from reinforcing materials in composites to serving as conductive elements in electronic devices and sensors. Their unique thermal and mechanical properties also make them attractive for use in heat dissipation systems and as reinforcing agents in high-performance composites.

5.3.4 Two-dimensional (2D) Nanomaterials

Two-dimensional nanomaterials, such as graphene and MXenes, have gained significant attention due to their exceptional properties and potential applications in various fields.

5.3.4.1 Graphene

Consisting of a single layer of carbon atoms arranged in a two-dimensional honeycomb lattice, graphene is renowned for its exceptional strength, electrical and thermal conductivity, and flexibility.

These properties make it a revolutionary material for flexible electronics, energy storage devices, and a host of other applications where high conductivity and mechanical strength are desirable. Graphene-based sensors, transparent conductive films, and high-frequency transistors are just a few examples of the potential applications of this remarkable material.

5.3.4.2 MXenes

MXenes are 2D materials made from transition metal carbides, nitrides, and carbonitrides. They are characterized by their hydrophilic surfaces and high electrical conductivity.

Their unique combination of properties makes MXenes suitable for energy storage, electromagnetic interference shielding, and as sensors, offering broad potential for technological innovation. $Ti_3C_2T_x$ and V_2CT_x are among the most studied MXenes for their applications in supercapacitors, batteries, and water purification.

5.3.5 Three-dimensional (3D) Nanomaterials

Three-dimensional nanomaterials, such as metal–organic frameworks and nanofibers, offer unique properties and applications due to their porous structures and high surface areas.

5.3.5.1 Metal–Organic Frameworks (MOFs)

MOFs are porous crystalline structures formed from metal ions or clusters coordinated to organic ligands. Their high surface area and tunable pore sizes offer exceptional gas storage and separation capabilities.

These materials find applications in gas storage, catalysis, and drug delivery, where their porosity can be exploited for selective adsorption and release. MOFs like MOF-5, HKUST-1, and ZIF-8 are widely investigated for their potential in hydrogen storage, carbon dioxide capture, and controlled drug release.

5.3.5.2 Nanofibers

Produced through electrospinning and other methods, nanofibers have diameters in the nanometer range but can be several micrometers long. They exhibit high surface area, excellent mechanical properties, and high porosity.

Their use spans across filtration, tissue engineering, drug delivery, and as substrates for energy storage, benefiting from their structural and surface characteristics. Polymeric nanofibers, such as those made from polycaprolactone (PCL) and polyvinyl alcohol (PVA), are extensively studied for their applications in wound dressings, scaffolds for tissue engineering, and drug delivery systems.

5.3.6 Hybrid or Special-class Nanomaterials

In addition to the nanomaterials categorized by dimensionality, several hybrid or special-class nanomaterials exhibit unique properties and applications due to their distinct structures and compositions.

5.3.6.1 Perovskites

Perovskites are a class of materials characterized by their distinct crystal structure, typically of the form ABX_3, where 'A' and 'B' are cations of different sizes, and 'X' is an anion. Their structure lends itself to a variety of compositions and configurations.

With outstanding photovoltaic efficiency, perovskites are revolutionizing solar cell technology. Their tunable optical properties and ease of fabrication also make them promising candidates for light-emitting diodes, lasers, and photodetectors. Methylammonium lead iodide ($CH_3NH_3PbI_3$) and formamidinium lead iodide ($CH(NH_2)_2PbI_3$) are among the most studied perovskite materials for their applications in solar cells and optoelectronics.

5.3.6.2 Nanocellulose

Derived from cellulose, the most abundant natural polymer, nanocellulose includes nano-sized fibers or crystals. It stands out due to its renewable nature, biodegradability, and impressive mechanical properties.

Its high strength, lightweight, and versatile chemical functionality make nanocellulose ideal for reinforcing materials in composites, as well as for use in flexible electronics, sustainable packaging, and biomedical applications such as wound dressings and drug delivery systems. Cellulose nanofibrils (CNFs) and cellulose nanocrystals (CNCs) are two main types of nanocellulose being investigated for their applications in various fields.

5.3.6.3 Core–Shell Nanoparticles

These nanoparticles consist of a core material that is encased by a shell made of a different material. The core and shell materials can be engineered to achieve specific properties.

Core–shell nanoparticles are used in various applications, including targeted drug delivery, where the core can carry therapeutic agents and the shell can provide biocompatibility or targeting capabilities; in catalysis, where the core–shell structure can enhance catalytic activity; and in optoelectronics, where their optical properties can be tuned for specific uses. Gold-silica and magnetic-polymeric core–shell nanoparticles are examples of these versatile nanostructures.

5.3.6.4 *Nanocomposites*

Nanocomposites are multiphase materials where one of the phases has at least one dimension in the nanometer range. They combine nanoparticles with bulk materials (like polymers, metals, or ceramics) to improve their properties.

The resulting materials can exhibit enhanced mechanical strength, electrical conductivity, thermal stability, and other improved properties. Applications range from stronger and lighter structural materials to conductive materials in electronics and improved barrier properties in packaging. Polymer-clay nanocomposites and carbon nanotube-reinforced composites are notable examples of nanocomposites with enhanced properties.

5.3.6.5 *Dendrimers*

Dendrimers are highly branched, tree-like molecules. Their structure allows for precise control over their composition and functionality.

Due to their well-defined structure and ability to encapsulate molecules, dendrimers are used in drug delivery systems, gene therapy, and as catalysts. Their ability to host guest molecules makes them suitable for medical imaging and diagnostics. Polyamidoamine (PAMAM) and polypropylenimine (PPI) dendrimers are widely studied for their applications in drug delivery and biomedical imaging.

5.3.6.6 *Biomimetic Nanomaterials*

These are nanomaterials inspired by biological systems. They mimic the structure or functionality of natural materials or biological processes.

Biomimetic nanomaterials find applications in regenerative medicine, such as designing materials that mimic the extracellular matrix for tissue engineering, in drug delivery systems mimicking biological carriers, and in creating surfaces with special wettability or adhesion properties inspired by plant leaves or gecko feet. Peptide-based nanofibers and synthetic nacre-like nanocomposites are examples of biomimetic nanomaterials.

5.3.6.7 *Aerogels*

Aerogels are distinguished by their macroscopic size but with an internal nanostructured network that makes them incredibly porous and lightweight. This structure consists of a mesh of interconnected nanoscale pores, which gives aerogels their unique properties. Although aerogels are not nanoscale in their external dimensions, the nanoscale features within their structure are critical to their functionality.

With their nanostructured internal makeup, aerogels exhibit extremely low thermal conductivity, making them some of the best thermal insulators available. This, combined with their light weight and high surface area, allows for their use in a wide range of applications, including thermal insulation for buildings and aerospace, lightweight structural components, and adsorbents for environmental cleanup efforts. Silica, carbon, and metal oxide aerogels are among the most studied types of aerogels for their diverse applications.

5.3.6.8 *Hydrogels*

Hydrogels consist of three-dimensional networks of hydrophilic polymers that can absorb and retain large volumes of water. While hydrogels themselves are macroscopic and their defining feature is the ability to swell with water, the nanoscale becomes relevant in their structure and functionality when considering the interactions at the molecular level and the potential for nanoscale modifications.

At the molecular level, the properties of hydrogels are influenced by the nanoscale arrangement of polymer chains and cross-links that define their porosity, mechanical strength, and swelling behavior. These nanostructures within hydrogels play a critical role in controlling the diffusion of molecules, making them highly effective for sustained release in drug delivery applications.

5.3.6.9 *Nanocomposite Hydrogels*

The incorporation of nanoparticles within hydrogels to create nanocomposite hydrogels introduces nanoscale functionality. These nanoparticles can enhance mechanical properties, introduce electrical conductivity, or impart specific biological functions, such as antimicrobial activity or targeted drug delivery, further expanding the application scope of hydrogels.

Hydrogels are extensively used in biomedical applications due to their biocompatibility and ability to closely mimic the natural tissue environment. This includes use in drug delivery systems for controlled release, wound dressings that provide a moist healing environment, and tissue engineering scaffolds that support cell growth and differentiation. The nanoscale modifications within hydrogels or the addition of nanoparticles enhance their functionality in these applications, offering precise control over physical properties and interactions with biological systems. Poly(ethylene glycol) (PEG) and poly(vinyl alcohol) (PVA) hydrogels are widely investigated for their applications in drug delivery and tissue engineering.

5.4 Structure–Property Relationships in Nanomaterials

The properties of nanomaterials are intricately linked to their structural features, including size, shape, composition, and surface chemistry, as illustrated in Figure 5.6. These relationships are fundamental to nanomaterial

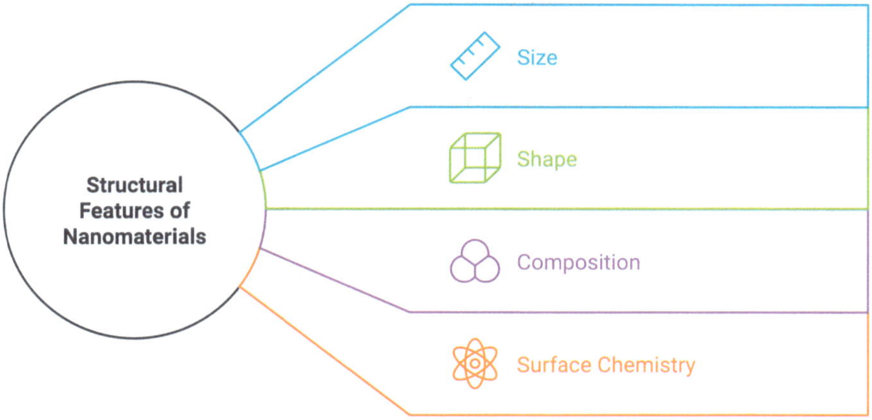

Figure 5.6 Key structural features of nanomaterials that determine their properties and behavior. These four fundamental aspects—size, shape, composition, and surface chemistry—form the basis for understanding structure–property relationships in nanomaterials.

design, as they allow for the customization of properties to meet specific needs in diverse applications. Understanding the structure–property relationships in nanomaterials is critical for tailoring their electronic, optical, mechanical, and chemical properties. This knowledge enables researchers to harness unique behaviors that emerge at the nanoscale, driving innovations in fields such as energy, electronics, catalysis, and medicine.[3]

5.4.1 Physical Size Effects

As materials are reduced to nanoscale dimensions, their properties undergo dramatic changes that extend beyond the quantum effects discussed earlier. While quantum confinement dominates at the smallest scales (typically below 10 nanometers), a broader range of size-dependent phenomena emerges across the nanoscale regime.

One of the most significant size-dependent transformations is the exponential increase in surface area-to-volume ratio. As particles become smaller, a greater proportion of their atoms reside at the surface rather than in the bulk. This fundamental change in atomic distribution leads to enhanced reactivity, altered melting points, and modified mechanical properties. For instance, gold nanoparticles exhibit melting temperatures significantly lower than bulk gold, with the melting point depression becoming more pronounced as particle size decreases.

Size effects also manifest in mechanical properties. Nanoscale materials often display enhanced strength and hardness compared to their bulk counterparts due to the limited volume available for defect propagation. This phenomenon, known as the Hall–Petch effect,[4] explains why nanocrystalline materials can be significantly stronger than their coarse-grained counterparts, though this relationship may break down at extremely small grain sizes.

The size of nanomaterials also influences their interaction with biological systems. Cellular uptake, biodistribution, and clearance mechanisms all depend strongly on particle dimensions. For example, nanoparticles between 20–50 nanometers typically show optimal cellular uptake, while larger or smaller particles may be less effectively internalized by cells.[5] This size dependence is crucial for designing drug delivery systems and therapeutic agents.

In catalysis, particle size plays a critical role beyond simple surface area effects. The density and arrangement of surface atoms change with particle size, leading to the emergence of specific catalytically active sites. This explains why certain particle sizes exhibit optimal catalytic activity for specific reactions, even when normalized for surface area.[6]

These physical size effects, distinct from quantum phenomena, provide additional tools for engineering nanomaterial properties. By precisely controlling dimensions at the nanoscale, researchers can tune material behavior for specific applications, from structural materials to biomedical devices.

5.4.2 Shape Effects

The shape of a nanomaterial is another critical factor that influences its properties. Nanomaterials can be synthesized in various shapes, such as spheres, rods, wires, cubes, and platelets, each exhibiting distinct characteristics. The anisotropic nature of non-spherical nanomaterials gives rise to shape-dependent properties, which can be harnessed for specific applications.

Gold nanorods provide a striking illustration of shape effects. Unlike spherical gold nanoparticles, which exhibit a single surface plasmon resonance (SPR) peak, gold nanorods display two SPR peaks: a transverse mode and a longitudinal mode. The longitudinal SPR peak is highly sensitive to the aspect ratio (length-to-width ratio) of the nanorods, with higher aspect ratios resulting in a red shift of the peak. This tunability of the optical properties based on shape is exploited in applications such as photothermal therapy, where gold nanorods are used to convert near-infrared light into localized heat for the targeted destruction of cancer cells.[7]

The shape of carbon nanomaterials, such as carbon nanotubes and graphene, also has a profound impact on their properties. The high aspect ratio and cylindrical structure of carbon nanotubes contribute to their exceptional mechanical strength and electrical conductivity. The unique electronic structure of carbon nanotubes, which can be either metallic or semiconducting depending on their chirality (the angle at which the graphene sheet is rolled), opens up possibilities for their use in nanoelectronics and sensors. Similarly, the two-dimensional structure of graphene gives rise to its extraordinary electronic, thermal, and mechanical properties, making it a promising material for a wide range of applications.[8]

5.4.3 Composition Effects

The composition of nanomaterials, including their chemical makeup and atomic arrangement, plays a significant role in determining their properties. By varying the composition of nanomaterials through alloying, doping, or functionalization, researchers can fine-tune their electronic, optical, magnetic, and chemical properties to suit specific applications.

Alloying is a powerful strategy for modifying the properties of metallic nanomaterials.[9] By combining two or more metals at the nanoscale, new materials with enhanced properties can be obtained. For example, alloying gold and silver nanoparticles results in a material with improved optical and catalytic properties compared to pure gold or silver nanoparticles. The composition of the alloy, as well as the atomic arrangement of the constituent elements, can be tailored to optimize the desired properties.

Doping is another effective method for modulating the properties of nanomaterials, particularly semiconductors.[10] By introducing small amounts of impurities (dopants) into the crystal structure of a semiconductor, its electronic properties can be altered. For instance, doping semiconductor nanoparticles with transition metal ions can lead to enhanced magnetic properties, making them suitable for applications in spintronics and magnetic resonance imaging (MRI). Doping can also be used to improve the performance of semiconductor nanomaterials in optoelectronic devices, such as solar cells and photodetectors, by controlling their charge carrier concentration and mobility.

Functionalization, which involves the attachment of functional groups or molecules to the surface of nanomaterials, is another powerful tool for tailoring their properties.[11] By selecting appropriate functional groups, the solubility, stability, and reactivity of nanomaterials can be modified. For example, functionalizing carbon nanotubes with hydrophilic groups can improve their dispersibility in aqueous media, enabling their use in biological applications. Functionalization can also be used to introduce new functionalities, such as targeting ligands for drug delivery or selective binding sites for chemical sensing.

5.4.4 Advanced Surface Engineering

While the fundamental effects of increased surface area were introduced earlier, the ability to engineer and control surface properties opens new possibilities for nanomaterial design. Surface engineering goes beyond simply exploiting the natural consequences of high surface area-to-volume ratios, enabling precise control over material behavior and functionality.

Two key approaches illustrate the power of surface engineering. Surface passivation involves forming a thin protective layer on the nanomaterial to reduce reactivity and improve stability. For example, coating semiconductor nanoparticles with zinc sulfide can passivate surface defects and enhance their photoluminescence.[12] Similarly, ligand capping, where organic molecules are attached to the surface, offers another versatile control method. When gold nanoparticles are capped with thiol ligands, their surface properties change dramatically, altering their optical behavior and preventing unwanted aggregation.[13]

Surface modification extends beyond these basic approaches to enable precise control over how nanomaterials interact with their environment. Through careful selection of surface chemistry, it becomes possible to control properties like solubility, reactivity, and biological interactions. For instance, a hydrophobic nanoparticle can be made hydrophilic through surface modification, dramatically changing its behavior in aqueous environments. Similarly, adding specific molecular recognition elements to surfaces enables targeted interactions, crucial for applications like drug delivery or sensing.

Interface engineering focuses on controlling the boundary between nanomaterials and their surroundings. The properties of these interfaces often determine how effectively nanomaterials perform in real applications. For example, in nanoelectronics, the nature of interfaces between different materials can significantly affect electron transport and device performance. Understanding and controlling these interfaces is essential for developing effective nanotechnology applications.[14]

Modern characterization techniques have revolutionized our understanding of surface properties. Advanced microscopy and spectroscopy methods allow us to observe and measure surface characteristics with unprecedented precision. This improved understanding enables more rational design of surface properties, moving beyond trial-and-error approaches to intentional engineering of desired characteristics.

Despite significant progress in surface engineering, challenges remain in creating stable and reproducible surface modifications at industrial scales. Ongoing research continues to develop new methods for controlling surface properties while maintaining the practical requirements of cost and scalability.

5.5 Synthesis and Characterization of Nanomaterials

The creation of nanomaterials with precisely controlled properties requires both sophisticated synthesis methods and advanced characterization techniques. While bulk materials can often be produced through relatively straightforward processes, engineering matter at the nanoscale demands exquisite control over growth and assembly. Similarly, conventional material characterization tools must be adapted or entirely new approaches developed to probe structures and behaviors at the nanometer scale. This section explores the primary methods for both creating nanomaterials and understanding their properties—essential knowledge for the practical applications discussed in later chapters.

5.5.1 Synthesis Methods

Nanomaterials can be synthesized through a wide range of methods, which can be broadly classified into two categories: top-down and bottom-up approaches. Each approach offers distinct advantages and limitations, and the choice of method depends on the desired material properties, scalability, and application requirements.

5.5.1.1 Top-down Methods

Top-down methods involve breaking down bulk materials into smaller, nanoscale structures. These methods rely on physical processes to sculpt or etch away material until the desired nanostructures are obtained. Some common top-down methods include:

- **Mechanical milling**: This method uses high-energy ball milling to grind bulk materials into nanoparticles. The process involves placing the material in a rotating drum along with milling balls, which collide with the material, causing it to fracture and reduce in size. Mechanical milling is a simple and scalable method but may introduce impurities, defects, and non-uniform particle sizes due to the high-energy collisions. Additionally, the process can generate heat, which may affect sensitive materials, and controlling particle size with high precision can be challenging, leading to a wide size distribution that may impact the material's final properties.
- **Laser ablation**: In this method, a high-energy laser beam is focused on a bulk material, causing it to vaporize and form a plasma. The vaporized material then condenses into nanoparticles. Laser ablation offers good control over particle size and can produce high-purity nanoparticles. However, this method can be limited in terms of production rates and is less scalable for industrial applications due to high energy costs.
- **Lithography**: Lithography techniques, such as electron beam lithography (EBL) and nanoimprint lithography (NIL), use a patterned mask or template to selectively remove material from a substrate, creating nanoscale structures. EBL uses a focused electron beam to write patterns directly onto a resist-coated substrate, while NIL uses a prepatterned mold to transfer patterns onto a substrate. Lithography methods offer excellent control over the size, shape, and arrangement of nanostructures, but they are often time-consuming, expensive, and difficult to scale for large-area applications.

5.5.1.2 Bottom-up Methods

Bottom-up methods involve building nanomaterials from individual atoms, molecules, or nanoparticles. These methods rely on the self-assembly of building blocks through chemical or physical interactions. Some common bottom-up methods include:

- **Chemical vapor deposition (CVD)**: In CVD, volatile precursors are vaporized and transported into a reaction chamber, where they decompose and react on a heated substrate to form a thin film or nanostructures. CVD is widely used for the synthesis of carbon nanotubes, graphene, and semiconductor nanowires. It offers good control over the composition and structure of the resulting nanomaterials but requires precise control of reaction conditions and precursor delivery, and scaling up can be challenging for large-area coatings.
- **Atomic layer deposition (ALD)**: ALD is a variation of CVD that enables the precise tuning of film thickness at the atomic level. In ALD, precursors are sequentially introduced into the reaction chamber, where they react with the substrate surface in a self-limiting manner. By repeating this process, ultrathin films with excellent conformality and uniformity can be obtained. ALD is particularly useful for coating complex 3D structures and creating nanolaminate materials. It is frequently employed in microelectronics, especially for high aspect ratio structures.
- **Sol–gel synthesis**: The sol–gel process involves the formation of a colloidal suspension (sol) from precursor molecules, followed by the gelation of the sol to form a continuous network (gel). The gel is then dried and heat-treated to obtain the final nanomaterial. Sol–gel synthesis is widely used for the preparation of metal oxide nanoparticles, thin films, and aerogels. It offers good control over the composition and morphology of the resulting materials, but care must be taken during the drying process to avoid cracking.
- **Colloidal synthesis**: Colloidal synthesis involves the formation of nanoparticles in a liquid medium through the reduction of metal salts or the decomposition of organometallic precursors. The size, shape, and surface properties of the nanoparticles can be controlled by adjusting reaction parameters such as temperature, pH, and surfactants. Surface functionalization is also important to control properties like dispersibility and biocompatibility, which is particularly useful in biological applications. Colloidal synthesis is widely used for the preparation of metal, metal oxide, and semiconductor nanoparticles with well-defined sizes and shapes.

5.5.2 Characterization Techniques

Characterization of nanomaterials is essential for understanding their properties, assessing their quality, and predicting their performance in various applications. A wide range of techniques are employed to probe the structural, chemical, optical, and electronic properties of nanomaterials. Some common characterization techniques include:

- **Scanning probe microscopy (SPM)**: SPM techniques, such as atomic force microscopy (AFM) and scanning tunneling microscopy (STM), use a sharp probe to scan the surface of a sample and generate high-resolution images. AFM measures the force between the probe and the sample surface, while STM measures the tunneling current between the probe and the sample. SPM techniques offer atomic-scale resolution and can provide information on surface topography, mechanical properties, and electronic structure. However, these methods are typically slower and more limited to surface analysis, which may not capture bulk properties.
- **Electron microscopy**: Electron microscopy techniques, such as scanning electron microscopy (SEM) and transmission electron microscopy (TEM), use a focused electron beam to image the sample. SEM provides information on the surface morphology and composition of the sample, while TEM offers higher resolution and can reveal internal structure and crystallinity. High-resolution TEM (HRTEM) is particularly useful in resolving atomic-scale features, which is critical for nanomaterial studies. Electron microscopy is widely used for the characterization of nanoparticles, nanowires, and thin films.
- **Spectroscopy**: Spectroscopic techniques, such as UV-visible spectroscopy, Raman spectroscopy, and X-ray photoelectron spectroscopy (XPS), use the interaction of electromagnetic radiation with the sample to obtain information on its chemical composition, electronic structure, and optical properties. UV-visible spectroscopy measures the absorption or reflectance of light by the sample, while Raman spectroscopy probes the vibrational modes of molecules. XPS measures the binding energies of electrons ejected from the sample surface, providing information on the chemical state and composition of the sample. Energy-dispersive X-ray spectroscopy (EDS), often used with SEM, provides elemental composition data for the sample.
- **X-Ray diffraction (XRD)**: XRD is a technique used to determine the crystalline structure and phase composition of a sample. It involves the scattering of X-rays by the atoms in the sample, which produces a diffraction pattern that can be analyzed to obtain information on lattice spacing, crystallite size, and strain. XRD is particularly useful for the characterization of crystalline nanomaterials, such as nanoparticles and thin films. However, it is less effective for analyzing amorphous materials, where no clear crystalline structure is present.
- **Thermal analysis**: Thermal analysis techniques, such as thermogravimetric analysis (TGA) and differential scanning calorimetry (DSC), measure the response of a sample to temperature changes. TGA measures the change in mass of the sample as a function of temperature, providing information on its thermal stability and composition. DSC measures the heat flow into or out of the sample as a function of temperature, revealing phase transitions, melting points, and crystallization behavior.

In addition to these techniques, many other characterization methods are employed depending on the specific properties and applications of the nanomaterials. These include magnetic measurements, electrical characterization, and surface area and porosity analysis.

The synthesis and characterization of nanomaterials are intertwined processes, as the properties of the materials depend on the synthesis method and conditions used. By carefully controlling the synthesis parameters and employing appropriate characterization techniques, researchers can tailor the properties of nanomaterials to suit specific applications. As nanotechnology continues to advance, new synthesis methods and characterization techniques are being developed to push the boundaries of what is possible with nanomaterials.

5.6 Advanced Fabrication Techniques for More Complex Structures and Devices

The transition from simple nanomaterials to complex nanosystems requires increasingly sophisticated fabrication techniques. While basic synthesis methods can produce individual nanoparticles or simple structures, creating functional devices demands precise control over multiple components and their integration. Modern nanofabrication combines traditional top-down approaches like lithography with novel bottom-up assembly methods, often guided by computational design. This section examines these advanced fabrication strategies and their role in creating the next generation of nanoscale devices.

5.6.1 Nanoscale 3D Printing

Advanced 3D printing techniques, such as two-photon polymerization (2PP), enable the fabrication of intricate three-dimensional nanostructures with sub-100 nm resolution. 2PP offers precise control over the polymerization process by using a focused laser to induce polymerization only at the focal point, allowing for the creation of complex geometries that were previously unattainable. This technique has applications in fields like photonic crystals, microfluidic devices, and biomimetic structures. However, despite its precision, 2PP faces challenges in terms of fabrication speed, limiting its scalability for large-scale industrial applications.

5.6.2 Nanoscale Lithography

While 3D printing allows for complex 3D structures, lithography techniques provide essential high-resolution patterning at the nanoscale. Advanced methods like extreme ultraviolet (EUV) lithography and nanoimprint lithography (NIL) represent the forefront of nanoscale patterning. EUV lithography uses short-wavelength light to achieve features below 20 nm, making it crucial for the continued miniaturization of semiconductor devices. However, EUV lithography faces challenges such as the high cost of equipment and the technical difficulty of generating intense EUV light sources. NIL, on the other hand, offers a high-throughput, cost-effective alternative for creating nanoscale patterns over large areas. It is especially useful for mass production of devices like optical components and biosensors, though ensuring uniformity across large surfaces can still be a challenge.

5.6.3 Nanorobotics and Nanomanipulation

The emergence of nanorobotics and nanomanipulation tools, such as AFM-based nanomanipulators and nanorobotic arms, has opened new possibilities for manipulating and assembling nanostructures with atomic or molecular precision. These tools allow for the precise placement, manipulation, and characterization of nanoscale components, enabling the construction of advanced nanodevices and systems. For example, nanorobotics is being used in the assembly of quantum dots for quantum computing, in nanoelectronic circuits, and in the investigation of single-molecule interactions. While these tools offer remarkable precision, they are often limited by slow operational speeds and the complexity of scaling such detailed manipulation for industrial production.

5.6.4 Hybrid Fabrication Techniques

The integration of multiple fabrication methods enables the creation of nanostructures with enhanced complexity and functionality. Hybrid approaches that combine top-down lithography with bottom-up self-assembly allow for the precise definition of nanostructures along with tailored chemical functionalities. This synergy is especially useful for creating nanostructures with specialized surfaces for catalysis or drug delivery, or for fabricating complex nanoelectronic devices. An example of such a hybrid approach is combining NIL with CVD to produce nanowires with both precise structural control and functionalized surfaces. These hybrid techniques offer versatility, though aligning multiple fabrication processes remains a technical challenge.

5.6.5 Computational Modeling and Simulation

In parallel with experimental techniques, computational methods provide powerful tools for guiding the design and optimization of nanostructures. By simulating nanoscale behavior, researchers can predict material properties and improve device performance before physical fabrication begins. Some key computational techniques include:

- **Density functional theory (DFT)**: A quantum mechanical method used to investigate the electronic structure of materials. DFT can predict properties such as band structure, optical properties, and chemical reactivity, making it crucial for understanding nanomaterials used in applications like quantum computing and catalysis.
- **Molecular dynamics (MD)**: This simulation method models the motion and interactions of atoms and molecules over time. MD is essential for studying the dynamic behavior of nanomaterials, including self-assembly processes and thermal or mechanical properties. For example, MD simulations can help optimize the design of nanomaterials for drug delivery by predicting how nanoparticles will behave in biological environments.
- **Finite element analysis (FEA)**: A numerical method used to simulate the mechanical, thermal, and electromagnetic behavior of nanoscale systems. FEA is particularly valuable in optimizing the design of nanoscale devices, such as nanomechanical resonators or heat-transfer devices, by modeling stress–strain responses or heat dissipation.

These computational techniques complement experimental work by reducing trial-and-error in the fabrication process, ultimately speeding up the development of new nanomaterials and devices.

These advanced nanofabrication and manipulation techniques represent the leading edge of nanotechnology, enabling researchers to explore new phenomena and engineer devices with unprecedented precision. As challenges in speed, scalability, and cost are overcome, these methods will continue to drive innovation across fields such as medicine, electronics, and materials science. Hybrid approaches and computational modeling will play an increasingly important role in enhancing the complexity and functionality of nanostructures, pushing the boundaries of what is possible at the nanoscale.

References

1. C. K. Ghosh, Quantum effect on properties of nanomaterials, in *Introduction to Nano*, Springer Berlin Heidelberg, Berlin, Heidelberg, 2015, pp. 73–111.
2. G. W. Bryant, Understanding quantum confinement in zero-dimensional nanostructures: Optical and transport properties, in *Science and Engineering of One- and Zero-Dimensional Semiconductors*, Springer US, Boston, MA, 1990, pp. 243–254.
3. A. A. Toropov, N. Sizochenko, A. P. Toropova and J. Leszczynski, Towards the development of global nano-quantitative structure–property relationship models: Zeta potentials of metal oxide nanoparticles, *Nanomaterials*, 2018, **8**(4), 243.

4. Y. Li, A. J. Bushby and D. J. Dunstan, The Hall-Petch effect as a manifestation of the general size effect, *Proc. Math. Phys. Eng. Sci.*, 2016, **472**(2190), 20150890.
5. P. Foroozandeh and A. A. Aziz, Insight into cellular uptake and intracellular trafficking of nanoparticles, *Nanoscale Res. Lett.*, 2018, **13**(1), 339.
6. D. Astruc, Introduction: Nanoparticles in catalysis, *Chem. Rev.*, 2020, **120**(2), 461–463.
7. R. Kumar, L. Binetti, T. H. Nguyen, L. S. Alwis, A. Agrawal and T. Sun, *et al.*, Determination of the aspect-ratio distribution of gold nanorods in a colloidal solution using UV-visible absorption spectroscopy, *Sci. Rep.*, 2019, **9**, 17469.
8. S. Rathinavel, K. Priyadharshini and D. Panda, A review on Carbon Nanotube: An overview of synthesis, properties, functionalization, characterization, and the application, *Mater. Sci. Eng., B*, 2021, **268**, 115095.
9. K.-H. Huynh, X.-H. Pham, J. Kim, S. H. Lee, H. Chang and W.-Y. Rho, *et al.*, Synthesis, properties, and biological applications of metallic alloy nanoparticles, *Int. J. Mol. Sci.*, 2020, **21**(14), 5174.
10. A. C. Silva, E. A. Alvin, F. R. dos Santos, S. L. de Matos, J. M. de Oliveira and A. S. Silva, *et al.*, Doped semiconductor nanocrystals: Development and applications, in *Materials at the Nanoscale*, ed. A. Mallik, 2021, DOI: 10.5772/intechopen.96753.
11. Y. Kumar, A. S. Sinha, K. D. Nigam, D. Dwivedi and J. S. Sangwai, Functionalized nanoparticles: Tailoring properties through Surface Energetics and Coordination Chemistry for Advanced Biomedical Applications, *Nanoscale.*, 2023, **15**(13), 6075–6104.
12. H.-Y. Cheng, O. Acar, W. Y. Shih and W.-H. Shih, Enhancing the photoluminescence of SNS quantum dots by ZnS treatment, *Chem. Phys. Lett.*, 2020, **754**, 137696.
13. K. Mahato, S. Nagpal, M. A. Shah, A. Srivastava, P. K. Maurya and S. Roy, *et al.*, Gold Nanoparticle Surface Engineering Strategies and their applications in Biomedicine and Diagnostics, *3 Biotech*, 2019, **9**, 57.
14. K. Borse, R. Sharma, D. Gupta and A. Yella, Interface engineering through electron transport layer modification for high efficiency organic solar cells, *RSC Adv.*, 2018, **8**(11), 5984–5991.
15. A. Gnach, T. Lipinski, A. Bednarkiewicz, J. Rybka and J. A. Capobianco, Upconverting nanoparticles: assessing the toxicity, *Chem. Soc. Rev.*, 2015, **44**(6), 1561–1584.
16. C. S. M. Martins, A. P. LaGrow and J. A. V. Prior, Quantum dots for cancer-related miRNA monitoring, *ACS Sens.*, 2022, 7(5), 1269–1299.
17. M. A. Darwish, W. Abd-Elaziem, A. Elsheikh and A. A. Zayed, Advancements in nanomaterials for nanosensors: a comprehensive review, *Nanoscale Adv*, 2024, **6**(16), 4015–4046.
18. G. Ramalingam, P. Kathirgamanathan, G. Ravi, T. Elangovan, B. Arjun Kumar, N. Manivannan *et al.* Quantum Confinement Effect of 2D Nanomaterials, in *Quantum Dots - Fundamental and Applications*, IntechOpen, 2020.

CHAPTER 6

Application Areas of Nanotechnology

Nanotechnology offers powerful new tools for addressing technical and environmental challenges. By engineering materials and devices at dimensions one-billionth of a meter, it becomes possible to create properties and capabilities that were previously impossible. These innovations are revolutionizing medicine through precise drug delivery, electronics through molecular-scale components, renewable energy through enhanced solar cells and batteries, and environmental protection through targeted pollutant removal.

The impact of nanotechnology extends far beyond simply making things smaller. Each reduction in scale reveals new material properties, enabling breakthroughs in energy storage, water purification, and sustainable manufacturing. This chapter explores the major application areas where nanotechnology is driving innovation, from advanced sensors and protective coatings to next-generation electronics and drug delivery systems. Throughout these applications, we'll examine how nanoengineering is simultaneously advancing technology and supporting global sustainability efforts through improved efficiency, reduced waste, and more precise control over matter itself.

6.1 Sensors

Sensors are devices that detect and respond to changes in their environment, converting the physical or chemical changes into electrical signals. They play a crucial role in a wide range of applications, including monitoring environmental conditions, ensuring safety and security, enabling precise control in industrial processes, and enhancing the functionality of consumer electronics (see Figure 6.1). Nanotechnology has significantly impacted the

Waste Not! How Nanotechnologies Can Increase Efficiencies Throughout Society
By Michael Berger

Published by the Royal Society of Chemistry, www.rsc.org

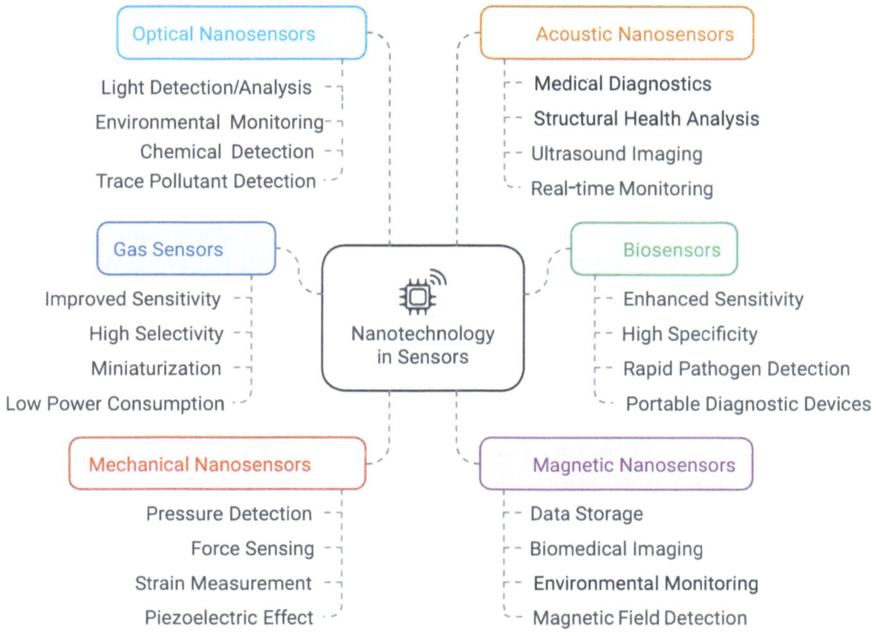

Figure 6.1 Overview of nanosensor types and their key applications and features.

field of sensors by improving their sensitivity, selectivity, and response time, while also reducing their size and power consumption.

6.1.1 Gas Sensors

Gas sensors are designed to detect and quantify the presence of gases in various environments by converting chemical reactions into measurable electrical signals. Nanotechnology has significantly enhanced gas sensors by incorporating materials like metal oxide semiconductors (*e.g.*, tin dioxide, zinc oxide), carbon nanotubes, and 2D materials such as graphene. These nanoscale materials exhibit high surface-to-volume ratios, improving sensitivity and selectivity, even for detecting trace amounts of gases.[1]

Nanotechnology-enabled gas sensors are critical in applications like air quality monitoring, industrial safety, and medical diagnostics. They can detect hazardous gases, volatile organic compounds, and other pollutants with exceptional precision. Furthermore, the miniaturization of gas sensors using nanomaterials allows for low power consumption and integration into portable devices, expanding their use in wearable technologies and environmental monitoring systems.

6.1.2 Biosensors

Biosensors utilize biological molecules, such as enzymes, antibodies, or DNA, to detect specific biological analytes, converting biochemical responses into electrical signals. Nanotechnology has revolutionized biosensors by incorporating nanomaterials like quantum dots, carbon nanotubes, and nanowires, which significantly enhance their sensitivity and specificity. These nanoscale materials enable the detection of extremely low concentrations of biomolecules, making biosensors invaluable in medical diagnostics, environmental monitoring, and drug discovery. Nanostructured biosensors can rapidly identify pathogens, pollutants, or disease biomarkers, offering real-time, highly accurate results.[2] Their small size and high performance also allow for integration into portable diagnostic devices, expanding their use in point-of-care testing and personal health monitoring.[3]

6.1.3 Mechanical Nanosensors

These sensors are designed to detect changes in mechanical properties such as pressure, force, and strain at the nanoscale. Often integrated into nanoelectromechanical systems (NEMS), these sensors are employed in fields such as medical diagnostics, industrial monitoring, and automotive systems. Their ability to detect subtle changes in mechanical forces makes them ideal for applications like vibration detection, force sensing, and even wearable devices that monitor physical exertion.

Piezoelectric nanosensors, which leverage the piezoelectric effect in nanomaterials, are particularly useful in converting mechanical stress into electrical signals. This property is advantageous in fields such as acoustics, where detecting sound vibrations with high sensitivity is crucial, or in medical implants where pressure changes within the body need to be closely monitored.

6.1.4 Magnetic Nanosensors

Magnetic nanosensors exploit the magnetic properties of nanoscale materials to detect changes in magnetic fields. These sensors have applications in data storage, biomedical imaging, and security systems. For instance, in the medical field, magnetic nanosensors are used in magnetic resonance imaging (MRI) to improve sensitivity, allowing for more precise imaging of biological tissues. They are also employed in environmental monitoring systems to detect magnetic pollutants or in high-precision navigation technologies that rely on magnetic field detection.[4]

6.1.5 Optical Nanosensors

This sensor category plays a pivotal role in industries where light-based detection and analysis are essential. These sensors work by interacting with light at the nanoscale, offering enhanced sensitivity for detecting changes in light intensity, wavelength, or polarization. Optical nanosensors are often used in applications such as environmental monitoring, where they can detect trace amounts of pollutants by analyzing light absorption or scattering patterns.[5]

Additionally, they find use in medical diagnostics, where their high sensitivity enables the detection of minute changes in biological samples, such as in blood or tissue analysis.[6]

6.1.6 Acoustic Nanosensors

Nanoacoustics is the study of sound waves at the nanoscale, where the interactions between mechanical vibrations and materials occur in ways that are distinct from those observed at larger scales.[7] This emerging field leverages the principles of acoustics in nanoscale systems to develop highly sensitive devices capable of detecting minute sound waves or mechanical vibrations. Acoustic nanosensors, built upon these principles, convert these vibrations into electrical signals with high precision. Utilizing piezoelectric nanomaterials, these sensors enhance sensitivity to detect even subtle acoustic changes, making them invaluable in applications such as medical diagnostics, environmental monitoring, and structural health analysis. For instance, in ultrasound imaging and sonar systems, acoustic nanosensors provide real-time monitoring of mechanical vibrations, while in industrial settings, they help detect early signs of structural failure through vibration analysis.

The diverse range of nanosensors, from gas and biosensors to mechanical, magnetic, and optical nanosensors, demonstrates the transformative impact of nanotechnology across various industries. These sensors not only enhance the precision, sensitivity, and efficiency of data collection but also open new possibilities for innovation in fields such as medicine, environmental monitoring, energy, and electronics. Advanced sensing systems will play an increasingly vital role in addressing the complex global sustainability challenges that we will address in this book.

6.2 Coatings

Coatings are critical components across various industries, providing protection, enhancing performance, and extending the lifespan of materials and devices. They serve various functions, including corrosion resistance, thermal insulation, wear resistance, and anti-fouling properties. With advancements in nanotechnology, coatings have become more efficient, thinner, and capable of offering multiple functionalities. This section explores several types of coatings, focusing on their applications and how nanotechnology is enhancing their effectiveness (see Figure 6.2).

6.2.1 Protective Coatings

Protective coatings play an essential role in industries where materials are exposed to harsh environmental conditions, such as corrosion, extreme

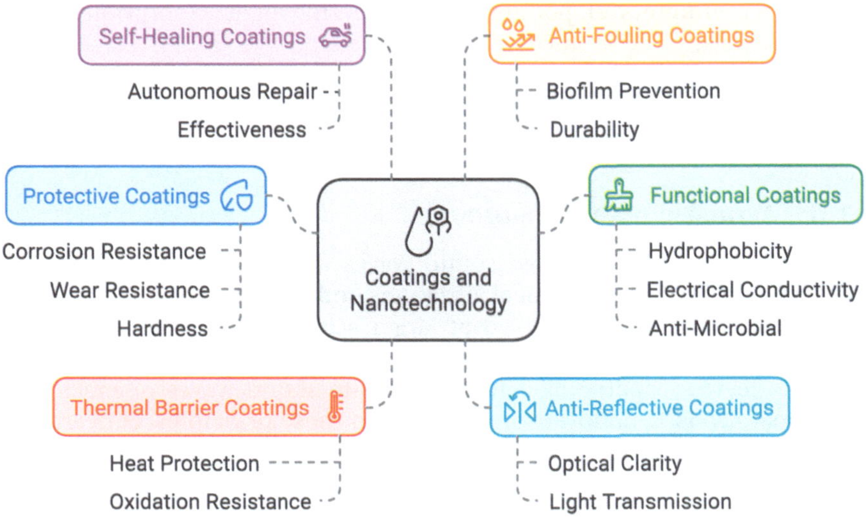

Figure 6.2 Overview of nanotechnology-enabled coating types and their key properties.

temperatures, and wear. Nanotechnology has significantly improved the durability and performance of protective coatings. Nanostructured coatings, such as those made from nanocomposites, offer enhanced hardness, wear resistance, and anti-corrosion properties.

For example, corrosion-resistant coatings prevent oxidation and corrosion in metals, which is crucial in sectors like marine, automotive, and aerospace.[8] By incorporating nanoparticles such as zinc oxide or silica, these coatings provide superior resistance to environmental degradation compared to traditional coatings, offering more uniform surface coverage and reducing micro-defects that typically initiate corrosion.

Additionally, wear-resistant coatings, commonly used in manufacturing, improve resistance to abrasion. Coatings that contain nanoparticles of ceramic materials or carbon-based nanostructures, such as carbon nanotubes, significantly reduce friction and wear, which prolongs the service life of mechanical parts.

6.2.2 Functional Coatings

These coatings go beyond basic protection by adding properties like hydrophobicity, electrical conductivity, or anti-microbial effects. They are widely applied in healthcare, electronics, and consumer products.

Hydrophobic coatings, for instance, repel water and keep surfaces dry and contamination-free. Nanostructured materials, such as silica nanoparticles, create surfaces with extreme water repellency, also known as superhydrophobic surfaces. These are commonly applied to self-cleaning windows and anti-icing surfaces in aviation and automotive industries.[9]

Anti-microbial coatings, which are embedded with silver or copper nanoparticles, have strong anti-microbial properties, making them essential in healthcare settings where surfaces need to remain free of pathogens.[10] Such coatings are applied to medical devices and hospital surfaces, and they have even been integrated into consumer products like mobile phones to reduce the spread of bacteria and viruses.

Another important type of functional coating is the conductive coating used in electronics to improve the conductivity of components. Nanomaterials such as graphene and carbon nanotubes are often used to enhance conductivity while maintaining flexibility and transparency.[11] These kinds of coatings are crucial in applications like touchscreens, flexible electronics, and solar panels.

6.2.3 Thermal Barrier Coatings

Thermal barrier coatings (TBCs) are essential in protecting components in high-temperature environments, such as those found in aerospace, automotive, and power generation sectors. These coatings insulate components like turbine blades, helping to prevent thermal degradation and extend operational lifetimes. Recent advances in nanotechnology have led to the development of more efficient and durable TBCs.[12]

Nanostructured materials, particularly ceramic-based nanocoatings like zirconia, offer superior thermal resistance, improved thermal shock tolerance, and a longer lifespan compared to traditional coatings. The nanoscale structure allows for better control over thickness and uniformity, enabling coatings that are thinner and lighter while still providing excellent protection.

Additionally, the use of nanomaterials can improve the coating's resistance to oxidation and wear, further enhancing the performance and reliability of systems operating under extreme conditions. These advancements are critical for improving fuel efficiency and reducing maintenance costs in industries where high-temperature performance is key.

6.2.4 Anti-reflective Coatings

Anti-reflective coatings (ARCs) are crucial for reducing unwanted reflections and improving the efficiency of devices like solar panels, lenses, and displays. Recent advancements in nanotechnology have significantly enhanced the performance of these coatings. For instance, nanostructured coatings inspired by the natural structures found in moth eyes have been shown to minimize reflections across a wide range of wavelengths, improving optical clarity and light transmission. This approach has been particularly effective in applications like solar panels and optical devices where reflections reduce efficiency.[13]

Atomic layer deposition (ALD) is another innovative method used to create highly effective ARCs. ALD allows for the precise layering of materials at the nanoscale, leading to coatings that are extremely thin but highly durable. This method has been employed to reduce reflections on complex 3D surfaces, which was previously difficult with conventional techniques. Such advancements are expanding the application of ARCs in areas ranging from optical lenses to next-generation solar cells.[14]

6.2.5 Self-healing Coatings

Self-healing coatings are an innovative type of coating designed to autonomously repair damage, such as scratches or microcracks, to extend the lifespan of the material they protect. These coatings incorporate healing agents like microcapsules or polymers that are released when the coating is damaged, effectively filling in the cracks or scratches and restoring the coating's functionality.

Nanotechnology has advanced this field by creating coatings that can self-repair more efficiently and under a wider range of conditions. For example, nanocapsules embedded in polymer matrices release healing agents when the surface is compromised, ensuring durability and reducing maintenance costs in industries such as aerospace, automotive, and marine.

Incorporating nanoparticles improves the speed and effectiveness of the self-healing process, enhancing both the mechanical properties and longevity of the coatings. Additionally, efforts to combine self-healing capabilities with other functions, like anti-corrosion or wear resistance, are creating multifunctional coatings that provide superior protection.[15]

6.2.6 Anti-fouling Coatings

Anti-fouling coatings are crucial in preventing the accumulation of unwanted organisms and contaminants on surfaces, particularly in marine environments. Traditional coatings often rely on toxic biocides like copper, which, while effective, pose environmental risks. Recent advances in nanotechnology offer more sustainable solutions by creating nanostructured coatings that repel organisms without relying on harmful chemicals.

Nanocomposites incorporating materials like zinc oxide (ZnO), titanium dioxide (TiO_2), and carbon nanotubes have shown promise in preventing biofilm formation and reducing microbial adhesion. These nanomaterials not only inhibit fouling but can also enhance durability, water resistance, and corrosion protection, extending the lifespan of the coated surfaces.[16]

The advancements in coatings, driven by nanotechnology, are transforming their capabilities across multiple industries. Whether for protective, functional, thermal, or self-healing purposes, nanostructured coatings offer enhanced performance and durability while reducing material usage and environmental impact. The ability to integrate multiple properties, such as wear resistance, corrosion protection, and even anti-fouling characteristics, makes modern coatings indispensable for applications ranging from aerospace to healthcare. As research progresses, coatings will continue to evolve, offering even greater efficiency and multifunctionality in diverse industrial environments.

6.3 Catalysts

Catalysts are substances that accelerate the rate of chemical reactions without being consumed in the process. They play an essential role in chemical manufacturing, energy conversion, and environmental applications by enabling more efficient and sustainable production processes. Reducing catalyst particles to the nanoscale has revolutionized the field of catalysis by dramatically increasing the surface area-to-volume ratio, thereby improving the activity, selectivity, and stability of catalytic processes. These enhanced properties can be tailored to specific reactions and applications, allowing for more precise control over chemical processes.[17]

6.3.1 Gold Nanoparticles

Gold, traditionally known for its chemical inertness, exhibits unique catalytic properties when reduced to the nanoscale.[18] Gold nanoparticles have been found to be highly effective catalysts for various chemical reactions, such as the oxidation of carbon monoxide, selective hydrogenation, and coupling reactions. Their high surface-to-volume ratio, coupled with the ability to fine-tune their size and shape, provides precise control over their catalytic activity and selectivity. Applications of gold nanoparticle catalysts include air purification, fine chemical synthesis, and pollution control. For example, gold nanoparticles can be used to catalyze the oxidation of toxic pollutants into less harmful substances, such as in the oxidation of carbon monoxide to carbon dioxide in air filtration systems. Gold nanoparticles are highly efficient in catalyzing this reaction at low temperatures, making them valuable in specific industrial processes like air filtration systems.

6.3.2 Platinum and Palladium-based Catalysts

Platinum and palladium are widely recognized for their catalytic efficiency, particularly in reactions related to energy production and environmental remediation. Nanotechnology has enhanced the catalytic performance of these metals by reducing their size to the nanoscale, which increases their active surface area.[19]

In fuel cells, platinum-based nanocatalysts are crucial for facilitating the oxygen reduction reaction and hydrogen oxidation reaction, both of which are key to improving the efficiency of energy conversion. Nanostructured palladium is commonly used in catalytic converters for the automotive industry, helping to reduce nitrogen oxide emissions. These catalysts have also found applications in organic synthesis, where their high selectivity ensures cleaner reactions with fewer by-products.

6.3.3 Nanostructured Electrocatalysts

Electrocatalysts are essential for driving chemical reactions involved in energy conversion processes, such as in fuel cells and water-splitting devices. Unlike traditional catalysts, nanostructured electrocatalysts are specifically designed to facilitate electron transfer at the electrode–electrolyte interface, enhancing reaction kinetics. The development of these nanocatalysts has led to significant improvements in efficiency and durability.

For example, platinum nanoparticles are used as electrocatalysts in proton exchange membrane fuel cells (PEMFCs), where they facilitate the conversion of hydrogen into electricity with minimal energy loss. Similarly, in water electrolysis, nanostructured electrocatalysts based on transition metals, such as cobalt or nickel, are employed to enhance the efficiency of hydrogen production, a critical component for sustainable energy systems.[20]

6.3.4 Bimetallic and Alloy Catalysts

Nanotechnology has enabled the creation of bimetallic and alloy catalysts, which combine the beneficial properties of two or more metals at the nanoscale. These catalysts offer enhanced catalytic performance due to synergistic effects between the metals.[21]

For instance, palladium-gold and platinum-nickel alloys have shown remarkable efficiency in hydrogenation reactions and fuel cell applications, respectively. The ability to engineer these materials at the nanoscale allows for fine-tuning their properties, such as enhancing their resistance to poisoning (a common issue in catalysis where the catalyst surface becomes inactive due to contaminants).[22]

6.3.5 Environmental Catalysis

Nanocatalysts are particularly effective in environmental applications, such as air and water purification.[23] Titanium dioxide nanoparticles, for example, act as photocatalysts in the degradation of organic pollutants in wastewater and the removal of volatile organic compounds from the air. The high surface area of these nanoparticles, combined with their ability to absorb ultraviolet light, makes them ideal for breaking down contaminants through photocatalytic oxidation. Furthermore, nanoscale iron catalysts are employed in environmental remediation to reduce toxic substances, such as chlorinated organic compounds, into harmless products through reduction reactions.

While nanocatalysts offer numerous advantages, they also present challenges. The potential environmental and health impacts of nanoparticles are still being studied, and their long-term effects remain uncertain. Additionally, the scalability and cost-effectiveness of producing nanocatalysts for large-scale industrial applications require further development.

As research progresses, the development of multifunctional nanocatalysts that can perform multiple reactions simultaneously is gaining momentum. These catalysts are designed to improve the overall efficiency of complex chemical processes, reducing energy consumption and waste production. The incorporation of nanocatalysts into renewable energy systems, such as solar-driven water-splitting devices, promises to revolutionize clean energy production and storage.

The economic impact of nanocatalysts is significant, with potential cost savings in industrial processes due to increased efficiency and reduced energy consumption. However, accurate characterization of nanocatalysts remains crucial for optimizing their performance. Techniques such as transmission electron microscopy, X-ray photoelectron spectroscopy, and atomic force microscopy are commonly used to analyze the structure and composition of these materials at the nanoscale.

6.4 Drug Delivery

Drug delivery systems are designed to transport therapeutic agents to specific target sites in the body, improving the efficacy and safety of treatments. While this topic may seem tangential to sustainability, nanotechnology-enabled drug delivery systems offer significant potential for waste reduction and improved resource efficiency in healthcare. These innovations contribute to sustainability by reducing the quantity of drugs needed for effective treatment, minimizing pharmaceutical waste and environmental contamination, extending drug shelf life, and potentially decreasing packaging requirements. Moreover, some nanocarriers can be produced using sustainable, bio-based materials, aligning with green chemistry principles. By enabling more efficient treatments, these systems may also indirectly reduce energy consumption in healthcare settings.

Traditional drug delivery methods often face challenges such as poor solubility, limited bioavailability, and off-target effects. Nanotechnology has revolutionized drug delivery by enabling the development of nanoscale carriers with enhanced stability, biocompatibility, and targeting capabilities (see Figure 6.3). These advancements not only improve patient outcomes but also contribute to more sustainable healthcare practices.

6.4.1 Liposomes

Liposomes are nanoscale vesicles composed of phospholipid bilayers that can encapsulate and transport both hydrophilic and hydrophobic drugs. Their biocompatible nature, tunable size, and surface modifications enable targeted delivery and controlled release of therapeutic agents, minimizing side effects and improving treatment outcomes. Liposomes have applications in cancer therapy, gene therapy, and vaccine delivery.[24]

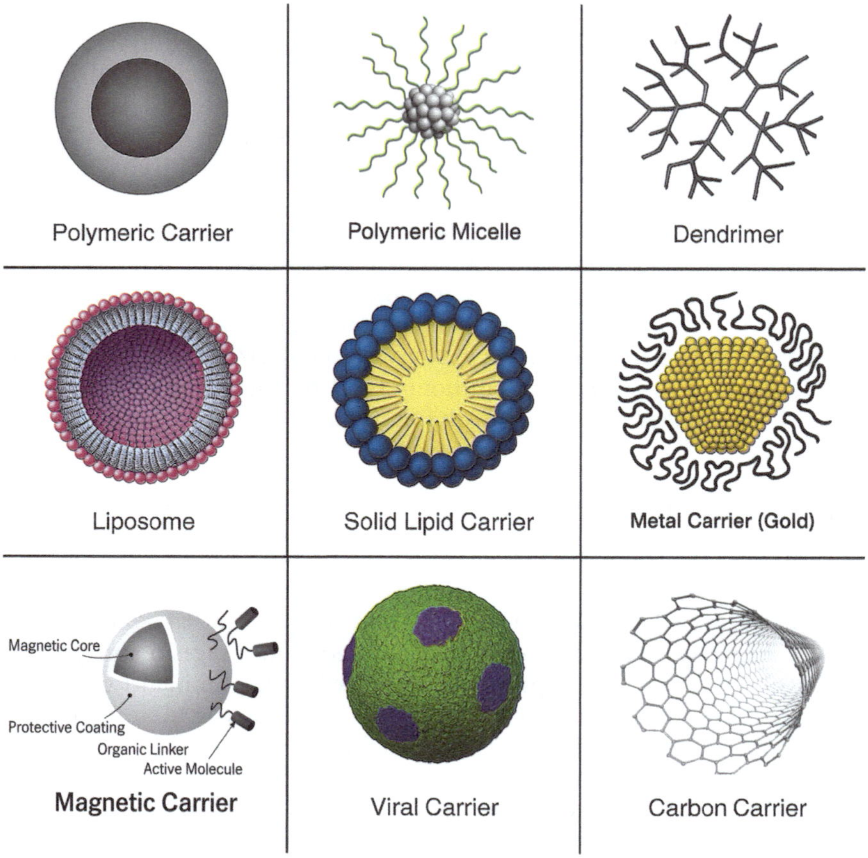

Figure 6.3 Different types of nanocarriers for drug delivery. Adapted from ref. 72 with permission from the author.

6.4.2 Dendrimers

Dendrimers are highly branched, three-dimensional macromolecules with a well-defined architecture and multiple functional groups on their surface. These nano-sized carriers can encapsulate or conjugate various therapeutic agents, such as drugs, genes, or imaging agents, facilitating targeted delivery and controlled release. Dendrimers have potential applications in cancer therapy, gene delivery, and medical imaging.[25]

6.4.3 Polymeric Nanoparticles

Polymeric nanoparticles are biodegradable carriers made from natural or synthetic polymers. They offer versatility in drug encapsulation, controlled release profiles, and surface functionalization for targeted delivery. These nanoparticles are particularly useful for delivering hydrophobic drugs and protecting sensitive therapeutics from degradation.[26]

6.4.4 Metal Nanoparticles

Metal nanoparticles, such as gold and silver nanoparticles, have unique optical and physical properties that make them suitable for both drug delivery and diagnostic applications. They can be easily functionalized with drugs and targeting ligands, and their photothermal properties can be exploited for controlled drug release or cancer therapy.[27]

6.4.5 Carbon Nanotubes

Carbon nanotubes are cylindrical nanostructures with high surface area and the ability to penetrate cell membranes. They can be functionalized with drugs, genes, and targeting moieties for efficient intracellular delivery. However, concerns about their long-term toxicity have limited their clinical applications.[28]

6.4.6 Magnetic Nanoparticles

Magnetic nanoparticles consist of a magnetic core surrounded by a biocompatible coating, which can be functionalized with drugs or targeting agents. These nanoparticles can be guided to specific sites in the body using an external magnetic field, allowing for targeted drug delivery and controlled release. They are also used in hyperthermia therapy, where they generate localized heat to destroy cancer cells.[29]

6.4.7 Polymeric Micelles

Polymeric micelles are nanoscale carriers formed by the self-assembly of amphiphilic block copolymers in an aqueous environment. They have a hydrophobic core, which can encapsulate poorly soluble drugs, and a hydrophilic shell that enhances circulation time. Polymeric micelles are particularly useful for delivering hydrophobic drugs and improving their bioavailability.[30]

6.4.8 Solid Lipid Nanoparticles

Solid lipid nanoparticles are composed of a solid lipid core that can encapsulate lipophilic drugs. These carriers offer advantages such as controlled drug release, good stability, and protection of sensitive drug molecules from degradation. SLNs are being explored for a variety of pharmaceutical and cosmetic applications.[31]

6.4.9 Viral Nanoparticles

Viral nanoparticles use viruses or virus-like particles as carriers to deliver therapeutic agents. These nanoparticles can be engineered to target specific cells and can encapsulate drugs, genes, or vaccines. Their natural ability to enter cells makes them highly efficient delivery vehicles, although their immunogenicity can present challenges in clinical applications.[32]

6.4.10 Stimuli-responsive Drug Delivery Systems

These advanced nanocarriers are designed to release their payload in response to specific stimuli such as pH, temperature, or light. This approach allows for precise control over drug release at the target site, further minimizing side effects and improving therapeutic efficacy.[33]

6.4.11 Challenges and Considerations

Despite their potential, nanoscale drug delivery systems face several challenges in development and implementation. Scalability and reproducibility in manufacturing remain significant hurdles, as producing consistent batches of nanocarriers at industrial scales can be complex. Ensuring stability during storage and in biological environments is crucial for maintaining the efficacy of these drug delivery systems.

There are also concerns about the potential toxicity and immunogenicity of nanocarriers, which necessitate thorough safety evaluations. Achieving efficient targeting and cellular uptake presents another challenge, as the behavior of nanoparticles in complex biological systems can be difficult to predict and control.

Additionally, regulatory hurdles and the need for comprehensive safety assessments can slow the progression of nanomedicines from laboratory to clinic. While several nanomedicine products have received regulatory approval, demonstrating the clinical potential of these technologies, rigorous safety evaluations and long-term studies remain necessary to ensure their safe and effective use in broader medical applications.

6.5 Energy Storage and Conversion

Efficient energy storage and conversion are crucial components of sustainable energy systems, playing a vital role in reducing waste and optimizing resource use. Nanotechnology has revolutionized this field by enabling the development of nanoscale materials and structures with improved electrochemical properties, leading to enhanced performance, increased energy density, and potentially reduced environmental impact.[34]

6.5.1 Lithium-ion Batteries

Nanostructured materials, such as silicon and metal oxides, can be used as anodes in lithium-ion batteries, providing increased energy storage capacity and faster charge/discharge rates compared to conventional materials.[35] This leads to improved performance and longer lifetimes for electric vehicles and portable electronics, potentially reducing electronic waste and the need for frequent battery replacements. Moreover, nanomaterials can enable the use of more abundant and less toxic elements in battery production, contributing to resource conservation and reduced environmental impact.

6.5.2 Solar Cells

Nanoscale materials, such as quantum dots, perovskites, and nanostructured semiconductors, can enhance light absorption and charge separation in solar cells, leading to higher conversion efficiencies. These materials can also enable the development of flexible, lightweight, and low-cost photovoltaic devices. By improving the efficiency and durability of solar cells, nanotechnology contributes to the broader adoption of renewable energy sources, reducing reliance on fossil fuels and associated waste products. Additionally, some nanomaterials used in solar cells can be synthesized using low-energy processes or from abundant, non-toxic precursors, aligning with sustainable manufacturing principles.[36]

6.5.3 Supercapacitors

Nanostructured carbon materials, such as graphene and carbon nanotubes, are being used to develop high-performance supercapacitors. These energy storage devices offer rapid charge/discharge capabilities and long cycle lives, complementing batteries in various applications. The use of carbon-based nanomaterials in supercapacitors can reduce reliance on rare or toxic elements, promoting more sustainable energy storage solutions.[37]

6.5.4 Thermoelectric Materials

Nanostructured thermoelectric materials can convert waste heat into electricity more efficiently than their bulk counterparts. By harvesting waste heat from industrial processes or vehicle exhaust, these materials can improve overall energy efficiency and reduce thermal pollution.[38] This application of nanotechnology directly contributes to waste reduction by converting what would otherwise be wasted energy into useful electrical power.

While nanotechnology offers significant potential for improving energy storage and conversion technologies, challenges remain in scaling up production, ensuring long-term stability, and managing potential environmental and health risks associated with nanomaterials. Ongoing research focuses on developing more sustainable synthesis methods, improving recycling processes for nano-enabled devices, and designing materials with reduced environmental impact throughout their lifecycle.

6.6 Structural Materials

Structural materials are critical in applications requiring strength, stiffness, and durability. Nanotechnology has significantly transformed this field by enabling the creation of nanocomposites and nanostructured materials with enhanced mechanical properties, making them lightweight while offering superior performance. These advanced materials find uses in industries such as aerospace, automotive, construction, and sports equipment.

6.6.1 Carbon Nanotube-reinforced Composites

Incorporating carbon nanotubes into polymer, metal, or ceramic matrices results in composites with remarkable improvements in strength, stiffness, and toughness, without adding significant weight. CNT-reinforced composites offer several advantages, including improved wear resistance, thermal stability, and electrical conductivity.[39] These composites are widely used in sectors such as aerospace, automotive, sports equipment, and infrastructure, where both high performance and lightweight materials are essential.

6.6.2 Nanostructured Metal Alloys

Nanotechnology has enabled the creation of metal alloys with nanostructured grains, which enhance their mechanical properties.[40] For instance, nanostructured aluminum alloys are significantly stronger and more durable than traditional aluminum while remaining lightweight. These materials are used in automotive and aerospace applications where both strength and weight reduction are critical. Additionally, nanostructured titanium alloys are used in biomedical implants due to their superior strength, biocompatibility, and corrosion resistance.

6.6.3 Aerogels

Aerogels are highly porous, ultra-light materials composed of nanostructured networks made from silica, carbon, or polymers. These materials exhibit exceptional thermal insulation, low density, and excellent mechanical properties. Aerogels are commonly used in applications such as thermal insulation for buildings, aerospace components, and even energy storage devices like batteries and supercapacitors due to their unique ability to retain structural integrity while being lightweight.[41]

6.6.4 Nanocellulose-based Materials

Nanocellulose, derived from plant fibers, is a renewable nanomaterial that offers excellent mechanical strength, flexibility, and low density. It is increasingly being used to reinforce composites in applications such as packaging materials, automotive components, and construction materials. Nanocellulose is biodegradable, making it an attractive alternative to traditional synthetic fibers and polymers in sustainable material design. It is also being explored for use in lightweight structural components, potentially reducing the environmental impact of industries that rely on plastics and metals.[42]

6.6.5 Nanoclay Composites

Nanoclay materials are layered silicate minerals that, when dispersed into a polymer matrix at the nanoscale, form nanocomposites with enhanced mechanical, thermal, and barrier properties.[43] These materials are used in packaging, automotive components, and construction. In packaging, nanoclay composites improve the barrier properties against gases like oxygen and moisture, extending the shelf life of food products. In automotive applications, these materials enhance the strength and heat resistance of components while reducing weight.

6.6.6 Nanostructured Ceramics

Ceramics, known for their high strength and resistance to heat, corrosion, and wear, can have their mechanical properties significantly enhanced at the nanoscale. Nanostructured ceramics offer increased toughness, reducing the brittleness that is typical in conventional ceramics.[44] This makes them highly suitable for applications in extreme environments, such as high-temperature components in engines and turbines, or protective coatings in industrial machinery. Additionally, nanoceramics are used in biomedical applications, such as dental implants and bone replacement materials, due to their biocompatibility and superior mechanical properties.

6.7 Water Purification

Access to clean water is a global challenge, and nanotechnology has made significant contributions to developing efficient and effective water purification technologies by utilizing nanoscale materials and structures with unique properties. These advanced methods offer enhanced removal of contaminants and provide scalable solutions for both large-scale and point-of-use applications.

6.7.1 Carbon Nanotube-based Membranes

Carbon nanotube-based membranes are highly effective in selectively filtering out contaminants, such as heavy metals, organic pollutants, and pathogens. Their size exclusion properties and large surface area enable precise filtration at the molecular level. These membranes are commonly used in applications like desalination and wastewater treatment, offering significant improvements in efficiency and durability compared to traditional filtration methods.[45]

6.7.2 Magnetic Nanoparticles

Magnetic nanoparticles, when functionalized with specific chemical groups or coatings, can selectively adsorb contaminants, such as heavy metals, organic pollutants, and pathogens. These particles can be easily separated from water using an external magnetic field, allowing for rapid and efficient purification without the need for complex filtration systems. This technology is well-suited for applications in environmental cleanup and industrial water treatment.[46]

6.7.3 Silver Nanoparticles

Silver nanoparticles are known for their strong antimicrobial properties, making them highly effective in eliminating bacteria, viruses, and other pathogens from water. They are used in disinfection systems, where they help to prevent microbial growth. Silver nanoparticles can be incorporated into membranes or point-of-use devices to provide an added layer of protection against waterborne diseases.[47]

6.7.4 Titanium Dioxide Nanoparticles for Photocatalysis

TiO_2 nanoparticles are widely used in photocatalytic water purification systems. When exposed to UV light, they generate reactive oxygen species that degrade organic pollutants and kill bacteria. This photocatalytic process is used in wastewater treatment and the removal of toxic organic compounds. An advantage of TiO_2-based systems is their ability to harness solar energy for water purification, making them an eco-friendly solution for both large-scale and off-grid applications.[48]

6.7.5 Zeolite Nanoparticles

Nanostructured zeolites have excellent ion exchange and adsorption properties, allowing them to remove heavy metals, radioactive elements, and other ions from contaminated water. These materials are used in applications such as water softening, desalination, and industrial wastewater treatment. Their high surface area and chemical stability make them particularly effective for long-term use in harsh environments.[49]

6.7.6 Graphene Oxide Membranes

Graphene oxide membranes are ultra-thin and offer exceptional performance in filtering out contaminants such as salts, heavy metals, and organic pollutants. Their high adsorption capacity and tunable pore size allow for efficient filtration at lower pressures, reducing energy consumption. GO membranes are used in a variety of applications, from industrial water treatment to next-generation desalination technologies, and they hold promise for portable and decentralized water purification systems.[50]

6.7.7 Nanofibers

Nanofiber-based membranes offer a flexible and efficient approach to water filtration. These membranes consist of extremely fine fibers with a high surface area, making them highly effective at capturing a wide range of contaminants, including bacteria, viruses, and particulate matter. Nanofiber filters can be customized for specific applications and are ideal for both large-scale water treatment plants and portable filtration devices.[51]

6.7.8 Zero-valent Iron Nanoparticles

Zero-valent iron (nZVI) nanoparticles are widely used for environmental remediation, particularly in treating contaminated groundwater. These nanoparticles promote chemical reactions that reduce toxic contaminants such as chlorinated solvents and heavy metals to less harmful forms. nZVI is highly effective in degrading organic pollutants and reducing the concentration of toxic metals, offering a promising solution for environmental clean-up and water restoration efforts.[52]

6.8 Electronics

Nanotechnology has revolutionized the field of electronics by enabling the development of nanoscale devices and components that exhibit enhanced performance, reduced power consumption, and entirely new functionalities. Electronics, which involve the generation, manipulation, and detection of electronic signals, are foundational to communication, computing, and sensing systems. The application of nanotechnology has resulted in smaller, faster, and more efficient electronic components, leading to a wide range of innovations across industries.

6.8.1 Nanoscale Transistors and Integrated Circuits

One of the most significant contributions of nanotechnology to electronics is the ability to scale down the size of transistors, the fundamental building blocks of integrated circuits (see Figure 6.4). Traditional semiconductor technologies have approached their physical limits, but by utilizing nanoscale materials such as carbon nanotubes and silicon nanowires, researchers have been able to further miniaturize transistors while maintaining or even improving their performance. Nanoscale transistors offer higher switching speeds, lower power consumption, and increased integration density, all of which are essential for the continued advancement of computing technology.

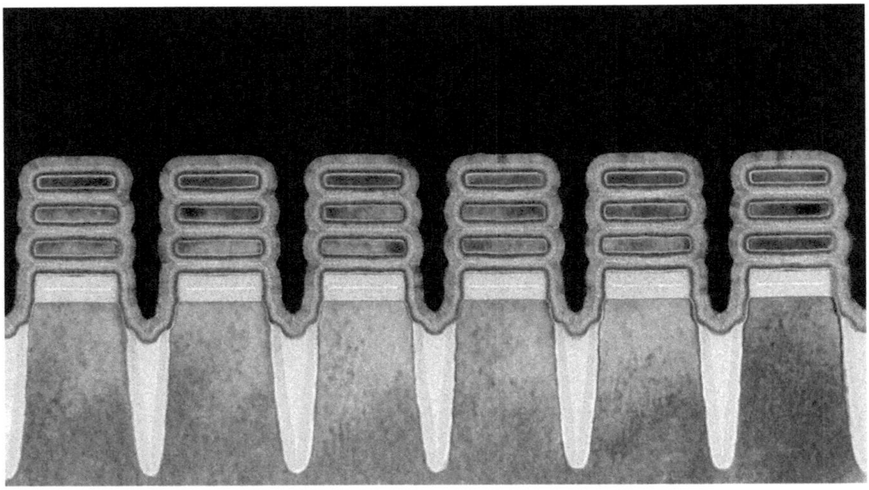

Figure 6.4 Row of 2 nm nanosheet devices as seen using transmission electron microscopy. 2 nm is smaller than the width of a single strand of human DNA. Reproduced from ref. 73. Reprint courtesy of IBM Corporation © 2024.

The development of complementary metal–oxide–semiconductor (CMOS) technology, driven by advances in nanoscale engineering and materials, has allowed for the production of more powerful and energy-efficient processors, thereby enhancing the performance of computers, smartphones, and other digital devices. These advancements contribute to the ongoing trend of miniaturization in consumer electronics, making devices more compact while increasing their computational power.

Several nanomaterials hold significant promise for overcoming the limitations of current CMOS technology. Carbon nanotubes are being extensively researched for their potential to replace silicon in transistors due to their exceptional electrical conductivity and ability to sustain high current densities. Graphene, a two-dimensional material composed of a single layer of carbon atoms, is also being investigated for its high electron mobility and flexibility, offering potential applications in next-generation transistors and flexible electronics.[53]

Additionally, semiconductor nanowires, such as silicon or germanium nanowires, are under exploration for their ability to improve performance in nanoscale transistors. Another emerging material is molybdenum disulfide, a type of transition metal dichalcogenide, which shows promise for ultra-thin, flexible, and transparent electronics. While these nanomaterials have not yet been fully integrated into commercial CMOS processes, their potential to revolutionize electronics is driving intense research and development efforts.

6.8.2 Advanced Memory Devices

In addition to improving processing power, nanotechnology has played a critical role in the development of next-generation memory devices.

Traditional memory technologies, such as dynamic random-access memory (DRAM) and flash memory, face challenges related to scaling, speed, and power efficiency.

Nanotechnology has enabled the creation of new memory architectures, such as resistive random-access memory (ReRAM), phase-change memory (PCM), and magnetic random-access memory (MRAM). These memory architectures are no longer confined to research and have made their way into specific commercial markets, though they haven't fully replaced traditional memory technologies yet.[54]

ReRAM utilizes nanoscale resistive materials to store data by altering the resistance of a material in response to an electrical signal. This memory type offers high-speed operation, low power consumption, and excellent scalability, making it ideal for applications requiring fast data processing, such as artificial intelligence (AI) and big data analytics.

PCM, another promising memory technology, exploits the ability of certain materials to switch between amorphous and crystalline states, with each state corresponding to a different electrical resistance. PCM enables fast data writing and erasing, long retention times, and high endurance.

MRAM, on the other hand, leverages the magnetic properties of nanoscale materials to store data by altering the magnetic orientation of a thin film. MRAM provides non-volatility, fast read/write speeds, and excellent durability, making it suitable for applications requiring reliable data storage.

The development of these nanoscale memory technologies has the potential to transform the performance and efficiency of electronic systems, particularly in high-performance computing, mobile devices, and data centers, where energy consumption and storage capacity are critical considerations.

6.8.3 Flexible and Wearable Electronics

Nanotechnology has also enabled the development of flexible and wearable electronics, which are transforming industries such as healthcare, fashion, and consumer electronics. By incorporating nanoscale materials like conductive polymers, carbon nanotubes, and graphene into electronic devices, researchers have been able to create flexible, lightweight, and durable components that can be integrated into non-traditional form factors (see Figure 6.5).[55]

Flexible electronics have numerous applications, including wearable health monitoring devices, e-textiles, foldable displays, and implantable medical devices. For example, flexible nanosensors embedded in wearable devices can continuously monitor vital signs such as heart rate, body temperature, and glucose levels, providing real-time health data that can be shared with medical professionals. Similarly, foldable displays made from nanomaterials allow for more compact and portable devices, such as smartphones and tablets, that can be folded or rolled without compromising their functionality.

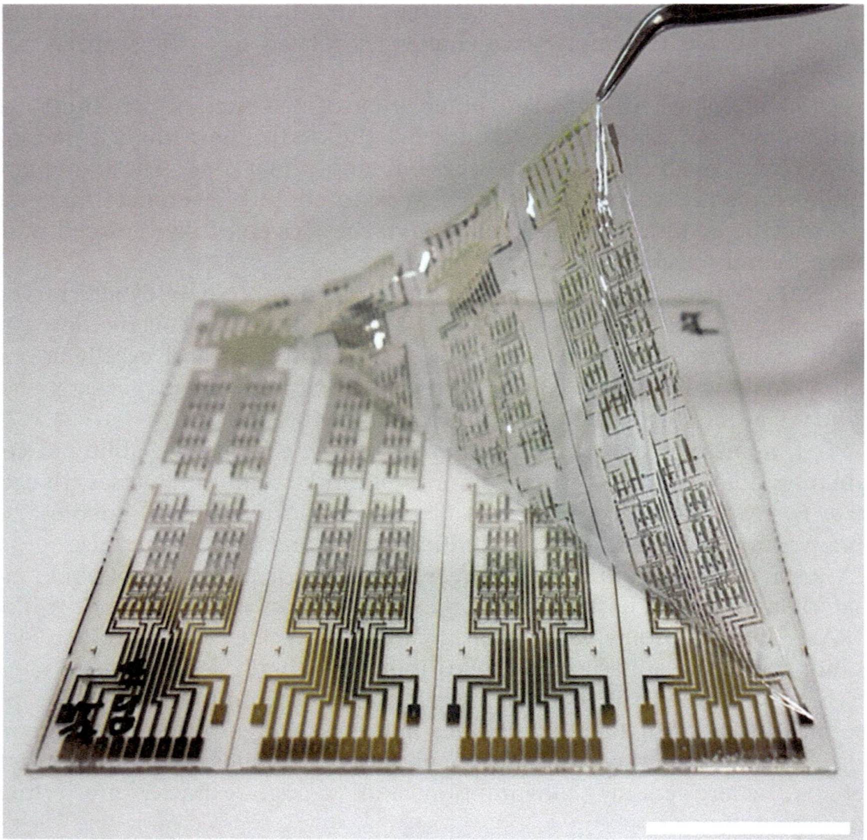

Figure 6.5 Photograph of organic CMOS logic circuit on a one-micron substrate. The total thickness is less than 3 microns. Scale bar: 25 mm. Adapted from ref. 73, https://doi.org/10.1038/srep25714, under the terms of the CC BY 4.0 license, https://creativecommons.org/licenses/by/4.0/.

These advancements in flexible electronics are made possible by the unique properties of nanomaterials. For instance, graphene, a two-dimensional sheet of carbon atoms, is highly conductive, strong, and flexible, making it an ideal material for creating bendable electronic components. Carbon nanotubes also offer excellent electrical conductivity and mechanical strength, making them suitable for use in flexible circuits, sensors, and displays.

6.8.4 Nanoscale Sensors and Actuators

Nanotechnology has significantly enhanced the sensitivity, speed, and precision of electronic sensors and actuators, which are critical components in numerous applications, including environmental monitoring, healthcare, and industrial automation. Nanosensors, which leverage the unique properties of nanomaterials, can detect minute changes in physical, chemical, or biological parameters, enabling high-precision measurements (see Sections 10.1 and 10.2 for a detailed discussion of nanosensors).

For example, nanoscale gas sensors, which incorporate materials like carbon nanotubes or metal oxide nanoparticles, can detect trace amounts of hazardous gases, pollutants, or volatile organic compounds with remarkable accuracy. These sensors are widely used in air quality monitoring, industrial safety systems, and environmental applications.

Similarly, nanotechnology has improved the performance of biosensors used in medical diagnostics. Nanomaterials such as quantum dots and gold nanoparticles have been integrated into biosensors to enhance their sensitivity and selectivity, enabling the detection of low concentrations of biomarkers for diseases such as cancer, diabetes, and cardiovascular conditions. These nanoscale biosensors are increasingly being incorporated into portable diagnostic devices, expanding their use in point-of-care testing and personal health monitoring.

6.8.5 Nanoscale Energy Harvesting and Storage

Energy harvesting and storage are essential for powering portable and wearable devices. Nanotechnology has contributed to the development of nanoscale energy harvesters and storage devices that are more efficient and compact than traditional systems.[56] For instance, piezoelectric nanogenerators can convert mechanical energy, such as body movements or vibrations, into electrical energy, providing a sustainable power source for wearable electronics.

Additionally, nanoscale batteries and supercapacitors offer improved energy storage capacity, faster charge/discharge rates, and longer lifetimes compared to conventional energy storage technologies. These advancements are crucial for extending the battery life of portable electronic devices and reducing the need for frequent recharging.

6.9 Optical Devices and Optoelectronics

Optical devices and optoelectronics involve the generation, manipulation, and detection of light, as well as the interaction between electrical and optical phenomena. These technologies play a crucial role in various applications, such as communication, sensing, and display technologies. Nanotechnology has significantly advanced this field by enabling the development of nanoscale optical and optoelectronic devices with improved performance and new functionalities.

Quantum dot light-emitting diodes (QLEDs) use quantum dots as the light-emitting material, resulting in higher brightness, improved color purity, and longer lifetimes compared to conventional organic light-emitting diodes (OLEDs). Applications include displays for televisions, smartphones, and other electronic devices.[57]

Nanophotonics is a branch of optics that deals with the interaction of light with nanoscale structures and materials, enabling novel devices and functionalities. Examples include photonic crystals for controlling the propagation of light, plasmonic devices for enhanced light-matter interaction and sensing, and nanoscale lasers for high-density optical data storage and communication.[58]

Optoelectronic devices combine electronic and optical components to convert between electrical and optical signals. Nanotechnology has enabled the development of high-performance optoelectronic devices, such as photodetectors with improved sensitivity and response times, and light-emitting diodes (LEDs) with higher efficiency and longer lifetimes. Applications include optical communication systems, solar cells, and various sensing technologies.

6.10 Metamaterials

Metamaterials are engineered materials with properties that do not exist in nature, made possible by designing their structures on the nanoscale rather than relying on chemical composition. These materials, along with their two-dimensional counterparts known as metasurfaces, manipulate waves—whether electromagnetic, acoustic, or mechanical—in ways that break the boundaries of traditional physics. Metasurfaces, in particular, are ultra-thin layers that can control wavefronts with subwavelength precision, offering a more compact and efficient means of wave manipulation. Nanotechnology has played a pivotal role in enabling the precise control that is required to craft these unique materials and surfaces. Together, metamaterials and metasurfaces offer groundbreaking applications across various fields, including optics, acoustics, electromagnetics, mechanics, and thermal management, each with its own set of innovative possibilities.[59]

6.10.1 Optical Metamaterials

Optical metamaterials are among the most studied and advanced in the field. They have the ability to control light in unconventional ways, particularly through the phenomenon of negative refraction, where light bends in the opposite direction compared to its behavior in natural materials (see Figure 6.6). This unique characteristic allows the creation of devices such as superlenses, which can capture images with resolutions far beyond the diffraction limit of conventional lenses, offering unprecedented clarity for imaging systems.

Additionally, optical metamaterials are instrumental in developing cloaking technologies that can bend light around an object, effectively rendering it invisible. These innovations have far-reaching applications in telecommunications, medical imaging, and next-generation display technologies, where manipulating light with precision is essential for improved performance and functionality. The fabrication of these materials relies

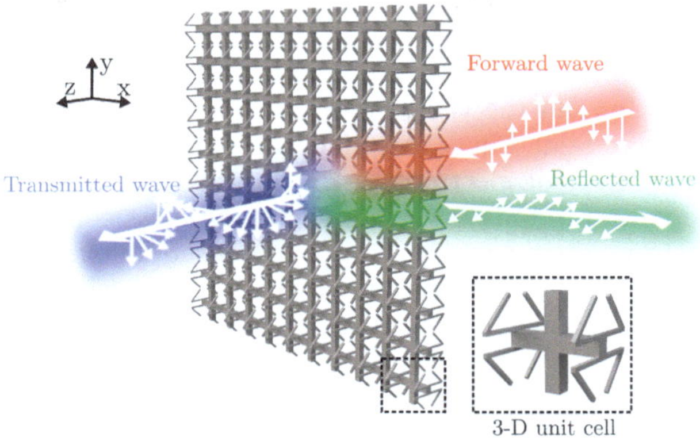

Figure 6.6 A conceptual illustration of a 3-D optical metamaterial controlling the flow of light. The structured lattice manipulates electromagnetic waves, such as visible light, by altering their transmission, reflection, and refraction paths in ways not possible with conventional materials. The transmitted wave (in blue) passes through the material, while the reflected (green) and forward (red) waves demonstrate the metamaterial's ability to direct light in unconventional ways. This structure, composed of repeating unit cells, enables phenomena like negative refraction and cloaking, with potential applications in optical devices such as invisibility cloaks and superlenses. Reproduced from ref. 74, https://doi.org/10.3390/electronics11030410, under the terms of the CC BY 4.0 license, https://creativecommons.org/licenses/by/4.0/.

heavily on nanotechnology, which enables the fine-tuning of structures at the nanoscale, thus optimizing their interaction with light.[60]

6.10.2 Acoustic Metamaterials

Acoustic metamaterials, on the other hand, manipulate sound waves in ways that challenge conventional acoustics. These materials are designed to control how sound is transmitted, absorbed, or reflected, enabling remarkable control over acoustic phenomena. Acoustic metamaterials can exhibit properties such as negative bulk modulus or negative mass density, allowing sound waves to bend, reflect, or even pass around objects in unusual ways. This leads to the development of devices that reduce noise pollution by selectively blocking certain frequencies while allowing others to pass through.

In addition to noise reduction, these materials are being used in the creation of acoustic cloaking devices, which make objects impervious to detection by sound waves. In medical diagnostics, acoustic metamaterials improve the precision of ultrasound imaging, enhancing the clarity of the images by manipulating the sound waves as they pass through different media. The ability to finely control the propagation of sound through nanoscale structures holds great promise for advancements in healthcare, environmental monitoring, and architectural acoustics.[61]

6.10.3 Electromagnetic Metamaterials

Electromagnetic metamaterials are engineered to manipulate electromagnetic waves beyond the visible spectrum, including microwaves, radio waves, and infrared light. These materials are highly valuable in applications requiring the control of electromagnetic interference (EMI), which is increasingly important in today's world of dense wireless communication. For example, electromagnetic metamaterials are used in stealth technology, where they can absorb or redirect radar waves, rendering objects less detectable to radar systems. This technology has significant implications for defense and aerospace industries.

Additionally, electromagnetic metamaterials are enabling the development of advanced antennas with improved directionality and signal strength, critical for the next generation of telecommunications and satellite communications. Microwave imaging systems are also benefiting from the use of these materials, allowing for non-invasive medical and security screening with enhanced resolution and accuracy. By leveraging nanotechnology to create intricate patterns that interact with electromagnetic waves at various frequencies, these materials can enhance the efficiency and performance of devices across multiple industries.[62]

6.10.4 Mechanical Metamaterials

Mechanical metamaterials are distinct from others in that they manipulate mechanical forces, such as tension or compression, rather than waves of energy. These materials can exhibit highly unusual mechanical properties, such as negative Poisson's ratio, where the material expands laterally when stretched, instead of contracting. This property is particularly useful in creating impact-resistant materials, which are lightweight yet incredibly strong, making them ideal for use in protective gear and structural reinforcements in aerospace and automotive industries.

In addition to protective materials, mechanical metamaterials are being explored for use in soft robotics, where their flexibility and adaptability allow for the development of robots capable of more precise, intricate movements. These materials also show promise in energy-absorbing applications, where they can efficiently dissipate energy during impacts, enhancing safety and durability.[63]

6.10.5 Thermal Metamaterials

Thermal metamaterials provide a new frontier in controlling heat and thermal energy. These materials can manipulate the flow of heat in ways that traditional materials cannot, such as by directing heat around an object rather than allowing it to pass through. This is particularly useful in thermal cloaking technologies, which could prevent thermal detection by guiding heat around objects. Moreover, thermal metamaterials are being developed for use in passive cooling systems that reduce heat absorption in buildings and electronic devices, enhancing energy efficiency without the need for external power sources.[64]

Nanotechnology enables the creation of these materials with precisely tuned thermal properties, offering new opportunities for advanced thermal management in electronics, energy systems, and even construction, where controlling heat flow is essential for improved performance and sustainability.

6.10.6 Negative-index Metamaterials

This is a subset of both optical and acoustic metamaterials that stand out for their ability to exhibit a negative refractive index. This means that waves, whether they are light or sound, bend in the opposite direction compared to conventional materials. Negative-index metamaterials are particularly exciting because they open the door to phenomena such as superlensing, which allows for imaging at resolutions previously thought impossible due to the diffraction limit. Additionally, these materials are the basis for cloaking devices that can render objects invisible to both light and sound. In sensing technology, negative-index materials enable the development of ultra-sensitive sensors capable of detecting even the smallest changes in environmental conditions or biological processes.

6.11 Nanorobotics

Nanorobotics is the study and application of nanoscale devices that are capable of performing precise tasks and interacting with their environment on the atomic and molecular levels. These devices have garnered significant research interest due to their ability to self-propel, be guided by external fields, and integrate with biological and non-biological systems for a wide variety of applications.

Recent advancements have allowed for the creation of nanorobots that leverage various propulsion mechanisms such as self-electrophoresis, diffusiophoresis, and microbubble recoil, enabling autonomous motion in different environments. These nanorobots exhibit ultrafast speeds, reaching up to 350 body lengths per second, and can follow complex trajectories. Fabrication techniques such as microfluidics, self-assembly, and chemical synthesis are utilized to develop nanorobots with specific functions.

Nanorobots are increasingly being explored for environmental remediation, offering innovative solutions for sustainability. They can be designed to remove pollutants from water, such as heavy metals and microplastics, through targeted adsorption and self-propelling mechanisms. Biohybrid nanorobots, for instance, have been used to remove heavy metal ions from contaminated water, and other types of nanorobots are capable of gathering microplastics or absorbing oil in oil spill cleanups. These devices contribute to more efficient, sustainable environmental management by enhancing the speed and efficacy of pollutant removal and offering possibilities for reuse in ongoing remediation efforts.[65]

6.11.1 Medical Applications of Nanorobots

One of the most promising areas for nanorobots is in healthcare. These devices can be programmed for highly targeted drug delivery, ensuring that therapeutic agents reach the exact site of disease, such as cancer cells, while minimizing damage to healthy tissues. This precise targeting reduces side effects and enhances treatment efficacy. Nanorobots can also be designed for minimally invasive surgeries, traveling through the body to perform tasks such as removing blockages in blood vessels, repairing tissues at the cellular level, or delivering stem cells to damaged areas.

Recent advancements include the development of magnetic nanorobots that can be guided externally to specific areas within the body. For example, magnetic nanorobots have shown potential in targeted cancer therapies, where they can be directed to tumor sites to deliver drugs or generate localized heat for hyperthermia treatment (see Figure 6.7). Similarly, photoresponsive nanorobots can be activated by light to release drugs at the precise location of a tumor, further reducing systemic side effects.

6.11.2 Environmental and Industrial Applications

Nanorobots are not limited to medical applications. In environmental monitoring, nanorobots can detect and neutralize pollutants at the source. For instance, nanorobots equipped with chemical sensors could be deployed to contaminated water sources, where they would identify toxic substances and either neutralize them or provide real-time data for remediation efforts. Nanorobots could also play a crucial role in detecting harmful gases or pollutants in the atmosphere, contributing to cleaner air and water.

In manufacturing, nanorobots are envisioned to revolutionize precision assembly at the molecular scale, enabling the construction of advanced materials and devices with unprecedented accuracy. This could lead to more efficient production processes, reduced material waste, and the creation of novel materials with enhanced properties.

6.11.3 Nanomanipulators and Fabrication

Nanorobots also include nanomanipulators, which are tools used to manipulate individual atoms or molecules with extreme precision. These manipulators are essential in the fabrication of nanoscale devices and systems, such as nanoelectronics and quantum computing components. By arranging atoms or molecules in precise configurations, nanomanipulators enable the creation of structures with highly controlled properties, which are critical for advanced technologies.

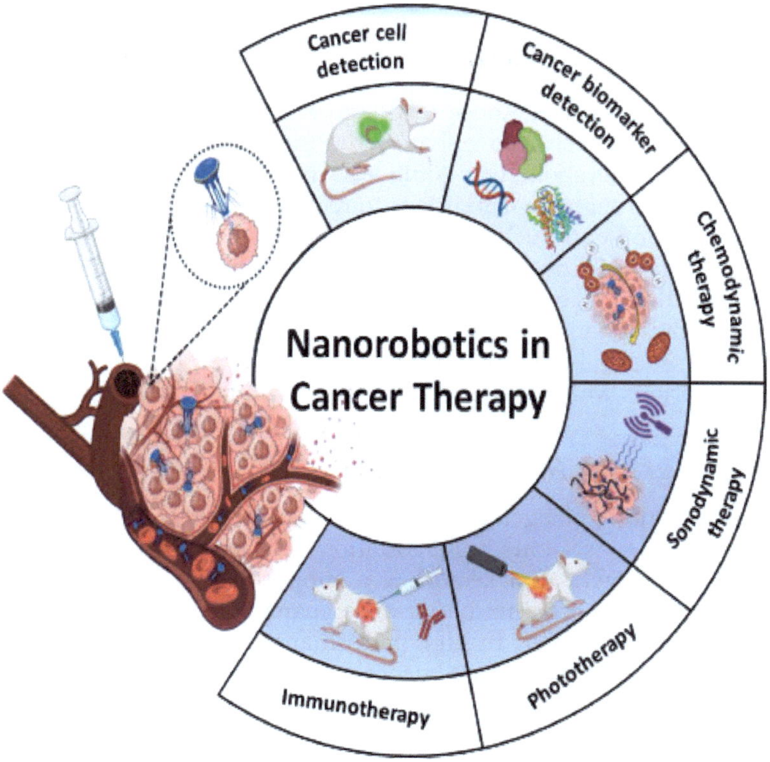

Figure 6.7 Nanorobots in cancer therapy represent a groundbreaking approach that allows targeted drug delivery and enhanced treatment options. Illustrated here are the multiple therapeutic applications of nanorobots, including cancer cell detection, biomarker tracking, and treatments such as chemodynamic, sonodynamic, phototherapy, and immunotherapy. Nanorobots use precise propulsion mechanisms to access difficult-to-reach areas, enabling advanced diagnostic and therapeutic interventions. Reproduced from ref. 75 with permission from Elsevier, Copyright 2024.

Nanomanipulation has been crucial in the development of carbon nanotube-based electronics, quantum dots for imaging, and molecular-scale sensors. As fabrication techniques continue to evolve, nanorobots may play an even greater role in the next generation of technology.

6.11.4 Challenges and Future Directions in Nanorobotics

Despite their potential, nanorobotics faces several significant challenges. Fabricating nanorobots with the precision and functionality needed for real-world applications is a complex process. Scalability is another issue, as the mass production of nanorobots with consistent quality remains difficult. Additionally, ensuring the safe and reliable operation of nanorobots within biological systems or harsh environmental conditions presents a unique set of challenges, including potential immune responses in medical applications and durability in industrial environments.

As research progresses, the integration of artificial intelligence (AI) with nanorobotics is expected to enhance the autonomy and decision-making capabilities of nanorobots. This could lead to the development of nanorobots that can adapt to changing conditions in real-time, improving their efficacy in complex environments such as the human body or industrial systems.

Furthermore, advancements in energy storage and harvesting at the nanoscale will be critical to enabling the prolonged operation of nanorobots in various applications. Future nanorobots may be equipped with nanoscale batteries or energy harvesting mechanisms, allowing them to operate for extended periods without the need for external power sources.

6.12 Thermal Management

Efficient thermal management is pivotal in numerous fields, including electronics, energy systems, transportation, and advanced materials. While thermal management may seem like a supporting function rather than a distinct application area, its importance spans across many of the sectors discussed in this chapter. In high-performance electronics, nanoscale materials are necessary to dissipate heat effectively, preventing device failure. In energy systems, thermal control plays a vital role in improving efficiency and reliability, particularly in energy storage and conversion technologies. Similarly, in coatings, aerospace, and automotive industries, maintaining thermal stability is key to ensuring the durability and performance of components.

Nanotechnology enables the development of advanced thermal management materials and systems that address these challenges in ways previously impossible with traditional materials. By reducing the size of materials to the nanoscale, researchers have developed innovative thermal interface materials, nanofluids, and coatings that offer superior heat transfer and insulation properties. These advances are essential for pushing the boundaries of what is possible in other application areas, from wearable electronics to energy-efficient buildings.

As such, thermal management serves as an integral component across various nanotechnology applications, providing the critical heat control needed to sustain the performance and longevity of next-generation technologies (see Figure 6.8). This section explores how nanotechnology is transforming thermal management through nanoscale innovations that impact multiple fields.

6.12.1 Nanoscale Thermal Interface Materials

Thermal interface materials (TIMs) are used to enhance heat transfer between two surfaces, such as between a microprocessor and a heat sink in

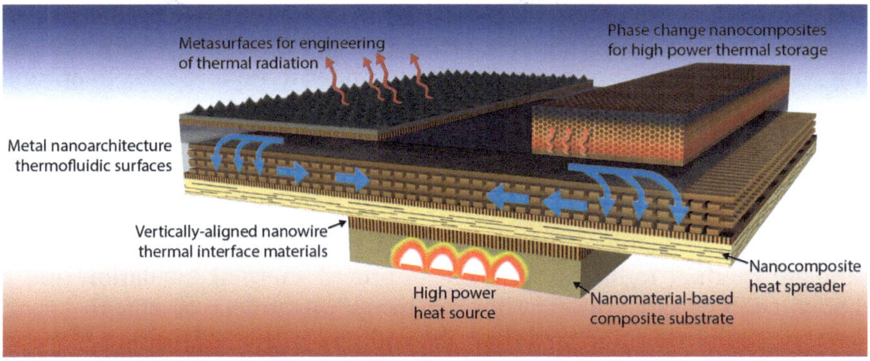

Figure 6.8 Conceptual representation of a holistic integration of nanomaterials into a complete thermal architecture, featuring multiple advanced layers designed for efficient heat management. The top metasurface layer is engineered to control thermal radiation, optimizing heat emission. Below it, phase change nanocomposites store thermal energy by utilizing materials that undergo phase transitions, offering high power thermal storage capacity. Metal nanoarchitectures enhance thermofluidic heat transfer, while vertically aligned nanowire thermal interface materials facilitate efficient heat conduction between layers. The structure is supported by a nanomaterial-based composite substrate that provides both mechanical stability and thermal performance. A nanocomposite heat spreader ensures even distribution of heat from the high-power heat source at the base, effectively dissipating thermal loads throughout the system. Reproduced from ref. 76 with permission from IOP Publishing Ltd, Copyright 2018.

electronics. Conventional TIMs often suffer from high thermal resistance, limiting their effectiveness. Nanotechnology has facilitated the development of nanoscale TIMs, which improve heat transfer efficiency by reducing the interfacial resistance.

Carbon nanotube-based TIMs, for instance, exhibit exceptional thermal conductivity, allowing for rapid dissipation of heat in electronic devices. Additionally, graphene-based TIMs are increasingly being explored due to their high thermal conductivity and mechanical stability, offering potential for integration into high-performance computing and advanced microelectronics. These nanomaterials help minimize overheating issues, extending the lifespan of devices while reducing energy consumption.[66]

6.12.2 Nanofluids for Enhanced Heat Transfer

Nanofluids, which are suspensions of nanoparticles in a base fluid, have emerged as highly effective materials for improving heat transfer in various systems. The addition of nanoparticles, such as aluminum oxide, copper oxide, or titanium dioxide, to fluids like water or oil enhances the thermal conductivity of the fluid, improving heat dissipation in applications such as industrial cooling, automotive radiators, and solar collectors. The small size and large surface area of the nanoparticles allow for more efficient heat exchange, making nanofluids an attractive option for energy-efficient cooling systems. Research into nanofluids continues to explore ways to optimize their stability, thermal performance, and environmental impact.[67]

6.12.3 Nanostructured Coatings for Thermal Insulation

Nanostructured coatings are being increasingly applied to surfaces to manipulate thermal radiation, offering improved thermal insulation and radiative cooling capabilities. These coatings can reflect or absorb specific wavelengths of thermal radiation, depending on the application.[68] For example, nanostructured ceramics are used as thermal barrier coatings (TBCs) in aerospace and automotive engines to protect components from extreme heat while maintaining structural integrity. In passive radiative cooling systems, nanostructured coatings that reflect solar radiation while emitting infrared radiation help reduce the heat load on buildings and electronic devices, improving energy efficiency without the need for additional power consumption.[69]

6.12.4 Phase-change Materials Enhanced with Nanoparticles

Phase-change materials (PCMs) store and release thermal energy during phase transitions, such as from solid to liquid. Nanotechnology has improved the thermal performance of PCMs by incorporating nanoparticles, which enhance the heat storage capacity, thermal conductivity, and phase-change kinetics. Nanoparticles, such as silver or carbon nanotubes, embedded in PCMs improve heat dissipation, making them more effective in applications like thermal energy storage, battery cooling, and temperature regulation in buildings. By efficiently storing and releasing heat, nanotechnology-enhanced PCMs can contribute to more sustainable energy systems by reducing reliance on active cooling methods.[70]

6.12.5 Thermal Management in Electronics

The increasing miniaturization of electronic components and the growing demand for more powerful devices have made thermal management a critical concern in the electronics industry. Nanotechnology offers innovative solutions, such as using nanostructured materials for heat dissipation in integrated circuits and semiconductors. For instance, graphene and carbon nanotubes are being explored as potential materials for next-generation heat spreaders in microelectronics, due to their high thermal conductivity and ability to manage heat at the nanoscale. Additionally, the development of nanoscale thermoelectric materials, which can convert waste heat into electricity, provides an opportunity to recover energy from heat-intensive processes, improving the overall energy efficiency of electronic devices.[71]

6.12.6 Challenges and Future Directions

While nanotechnology has advanced thermal management significantly, challenges remain in terms of scalability, cost, and long-term stability. Manufacturing nanoscale TIMs, nanofluids, and coatings in a consistent, cost-effective manner requires further research. Additionally, the long-term effects of nanoparticles on human health and the environment need to be thoroughly evaluated. Future developments in nanotechnology will likely focus on integrating thermal management solutions with energy-harvesting systems, such as using thermoelectric nanomaterials to capture and reuse waste heat in electronics and industrial processes. Continued innovations in this area are essential for addressing the growing demands of high-performance technologies while improving sustainability and energy efficiency.

References

1. H. Nazemi, A. Joseph, J. Park and A. Emadi, Advanced micro- and nano-gas sensor technology: A review, *Sensors*, 2019, **19**(6), 1285.
2. V. Naresh and N. Lee, A review on biosensors and recent development of nanostructured materials-enabled biosensors, *Sensors*, 2021, **21**(4), 1109.
3. D. Han, H. Hosamo, C. Ying and R. Nie, A comprehensive review and analysis of nanosensors for structural health monitoring in bridge maintenance: Innovations, challenges, and future perspectives, *Appl. Sci.*, 2023, **13**(20), 11149.
4. Y. Slimani and E. Hannachi, Magnetic nanosensors and their potential applications, in *Nanosensors for Smart Cities*, Elsevier, 2020, pp. 143–155.
5. N. H. Anh, M. Q. Doan, N. X. Dinh, T. Q. Huy, D. Q. Tri and L. T. Ngoc Loan, *et al.*, Gold nanoparticle-based optical nanosensors for food and health safety monitoring: recent advances and future perspectives, *RSC Adv.*, 2022, **12**(18), 10950–10988.
6. J. T. Metternich, B. Hill, J. A. C. Wartmann, C. Ma, R. M. Kruskop and K. Neutsch, *et al.*, Signal amplification and near-infrared translation of enzymatic reactions by nanosensors, *Angew. Chem., Int. Ed.*, 2024, **63**(9), e202316965.
7. C. Peng, M. Chen, J. B. Spicer and X. Jiang, Acoustics at the nanoscale (nanoacoustics): A comprehensive literature review.: Part I: Materials, devices and selected applications, *Sens. Actuators, A*, 2021, **332**(Pt 2), 112719.
8. L. T. Nhiem, D. T. Y. Oanh and N. H. Hieu, Nanocoating toward anti-corrosion: A review, *Vietnam J. Chem.*, 2023, **61**(3), 284–293.
9. S. Das, S. Kumar, S. K. Samal, S. Mohanty and S. K. Nayak, A review on superhydrophobic polymer nanocoatings: Recent development and applications, *Ind. Eng. Chem. Res.*, 2018, **57**(8), 2727–2745.
10. A. Jose, M. Gizdavic-Nikolaidis and S. Swift, Antimicrobial coatings: Reviewing options for healthcare applications, *Appl. Microbiol.*, 2023, **3**(1), 145–174.

11. Y. Li, Q. Ai, L. Mao, J. Guo, T. Gong and Y. Lin, *et al.*, Hybrid strategy of graphene/carbon nanotube hierarchical networks for highly sensitive, flexible wearable strain sensors, *Sci. Rep.*, 2021, **11**(1), 21006.
12. A. Mehta, H. Vasudev, S. Singh, C. Prakash, K. K. Saxena and E. Linul, *et al.*, Processing and advancements in the development of thermal barrier coatings: A review, *Coatings*, 2022, **12**(9), 1318.
13. A. Gärtner, A. Sabbagh, U. Schulz, F. Rickelt, A. Bingel and S. Wolleb, *et al.*, Combined antifogging and antireflective double nanostructured coatings for LiDAR applications, *Appl. Opt.*, 2023, **62**(7), B112–B116.
14. D. Astrauskytė, K. Galvanauskas, D. Gailevičius, M. Drazdys, M. Malinauskas and L. Grineviciute, Anti-reflective coatings produced *via* atomic layer deposition for hybrid polymer 3D micro-optics, *Nanomaterials*, 2023, **13**(16), 2281.
15. S. Sanyal, S. Park, R. Chelliah, S.-J. Yeon, K. Barathikannan and S. Vijayalakshmi, *et al.*, Emerging trends in smart self-healing coatings: A focus on micro/nanocontainer technologies for enhanced corrosion protection, *Coatings*, 2024, **14**(3), 324.
16. S. Kumar, F. Ye, S. Dobretsov and J. Dutta, Nanocoating is a New Way for biofouling prevention, *Front. Nanotechnol.*, 2021, **3**, 771098.
17. S. B. Somwanshi, S. B. Somvanshi and P. B. Kharat, Nanocatalyst: A brief review on synthesis to applications, *J. Phys.: Conf. Ser.*, 2020, **1644**(1), 012046.
18. H. Hassan, P. Sharma, M. R. Hasan, S. Singh, D. Thakur and J. Narang, Gold nanomaterials – The golden approach from synthesis to applications, *Mater. Sci. Energy Technol.*, 2022, **5**, 375–390.
19. M. Alaqarbeh, S. F. Adil, T. Ghrear, M. Khan, M. Bouachrine and A. Al-Warthan, Recent progress in the application of palladium nanoparticles: A review, *Catalysts*, 2023, **13**(10), 1343.
20. L. A. Omeiza, A. M. Abdalla, B. Wei, A. Dhanasekaran, Y. Subramanian and S. Afroze, *et al.*, Nanostructured electrocatalysts for advanced applications in fuel cells, *Energies*, 2023, **16**(4), 1876.
21. L. Liu and A. Corma, Bimetallic sites for catalysis: From binuclear metal sites to bimetallic nanoclusters and nanoparticles, *Chem. Rev.*, 2023, **123**(8), 4855–4933.
22. T. Wang, A. Chutia, D. J. L. Brett, P. R. Shearing, G. He and G. Chai, *et al.*, Palladium alloys used as electrocatalysts for the oxygen reduction reaction, *Energy Environ. Sci.*, 2021, **14**(5), 2639–2669.
23. R. Ningthoujam, Y. D. Singh, P. J. Babu, A. Tirkey, S. Pradhan and M. Sarma, Nanocatalyst in remediating environmental pollutants, *Chem. Phys. Impact*, 2022, **4**, 100064.
24. F. Rommasi and N. Esfandiari, Liposomal nanomedicine: Applications for drug delivery in cancer therapy, *Nanoscale Res. Lett.*, 2021, **16**(1), 95.
25. A. P. Dias, S. da Silva Santos, J. V. da Silva, R. Parise-Filho, E. Igne Ferreira and O. E. Seoud, *et al.*, Dendrimers in the context of nanomedicine, *Int. J. Pharm.*, 2020, **573**, 118814.

26. M. A. Beach, U. Nayanathara, Y. Gao, C. Zhang, Y. Xiong and Y. Wang, *et al.*, Polymeric nanoparticles for drug delivery, *Chem. Rev.*, 2024, **124**(9), 5505–5616.
27. T. K. Sau, A. Biswas and P. Ray, Metal nanoparticles in nanomedicine: Advantages and scope, in *Metal Nanoparticles*, Wiley-VCH Verlag GmbH & Co. KgaA, Weinheim, Germany, 2017. pp. 121–168.
28. B. O. Murjani, P. S. Kadu, M. Bansod, S. S. Vaidya and M. D. Yadav, Carbon nanotubes in biomedical applications: current status, promises, and challenges, *Carbon Lett.*, 2022, **32**(5), 1207–1226.
29. C. Comanescu, Magnetic nanoparticles: Current advances in nanomedicine, drug delivery and MRI, *Chemistry*, 2022, **4**(3), 872–930.
30. A. Mustafai, M. Zubair, A. Hussain and A. Ullah, Recent progress in proteins-based micelles as drug delivery carriers, *Polymers*, 2023, **15**(4), 836.
31. D. Sivadasan, K. Ramakrishnan, J. Mahendran, H. Ranganathan, A. Karuppaiah and H. Rahman, Solid lipid nanoparticles: Applications and prospects in cancer treatment, *Int. J. Mol. Sci.*, 2023, **24**(7), 6199.
32. S. Iravani and R. S. Varma, Vault, viral, and virus-like nanoparticles for targeted cancer therapy, *Mater. Adv.*, 2023, **4**(14), 2909–2917.
33. M. A. Rahim, N. Jan, S. Khan, H. Shah, A. Madni and A. Khan, *et al.*, Recent advancements in stimuli responsive drug delivery platforms for active and passive cancer targeting, *Cancers*, 2021, **13**(4), 670.
34. S. Zhao, H. Wang, A. Ruediger, F. Gu, D. Yan and J. Lu, Editorial: Micro/nano materials for energy storage and conversion, *Front. Chem.*, 2023, **11**, 1150885.
35. Z. Bitew, M. Tesemma, Y. Beyene and M. Amare, Nano-structured silicon and silicon based composites as anode materials for lithium ion batteries: recent progress and perspectives, *Sustainable Energy Fuels*, 2022, **6**(4), 1014–1050.
36. L. Wang, M. P. R. Teles, A. Arabkoohsar, H. Yu, K. A. R. Ismail and O. Mahian, *et al.*, A holistic and state-of-the-art review of nanotechnology in solar cells, *Sustainable Energy Technol. Assess.*, 2022, **54**, 102864.
37. A. Nayak, B. Bhushan, S. Kotnala, N. Kukretee, P. Chaudhary and A. R. Tripathy, *et al.*, Nanomaterials for supercapacitors as energy storage application: Focus on its characteristics and limitations, *Mater. Today*, 2023, **73**, 227–232.
38. M. M. R. Al-Fartoos, A. Roy, T. K. Mallick and A. A. Tahir, Advancing thermoelectric materials: A comprehensive review exploring the significance of one-dimensional nano structuring, *Nanomaterials*, 2023, **13**(13), 2011.
39. S. Dinesh Kumar, M. Ravichandran, S. V. Alagarsamy, C. Chanakyan, M. Meignanamoorthy and S. Sakthivelu, Processing and properties of carbon nanotube reinforced composites: A review, *Mater. Today*, 2020, **27**, 1152–1156.
40. X. Li, L. Lu, J. Li, X. Zhang and H. Gao, Mechanical properties and deformation mechanisms of gradient nanostructured metals and alloys, *Nat. Rev. Mater.*, 2020, **5**(9), 706–723.

41. N. R. Khan, T. Sharmin and A. Bin Rashid, Exploring the versatility of aerogels: Broad applications in biomedical engineering, astronautics, energy storage, biosensing, and current progress, *Heliyon*, 2024, **10**(1), e23102.
42. M. Y. Khalid, A. Al Rashid, Z. U. Arif, W. Ahmed and H. Arshad, Recent advances in nanocellulose-based different biomaterials: types, properties, and emerging applications, *J. Mater. Res. Technol.*, 2021, **14**, 2601–2623.
43. M. N. Uddin, M. T. Hossain, N. Mahmud, S. Alam, M. Jobaer and S. I. Mahedi, *et al.*, Research and applications of nanoclays: A review, *SPE Polym.*, 2024, 507–535.
44. N. B. Singh, M. Guin and R. Tomar, Nanostructured ceramics, in *Ceramic Science and Engineering*, Elsevier, 2022. pp. 545–573.
45. J. Lee, Carbon nanotube-based membranes for water purification, in *Nanoscale Materials in Water Purification*, Elsevier, 2019, pp. 309–331.
46. M. Pinto, P. S. F. Ramalho, N. F. F. Moreira, A. G. Gonçalves, O. C. Nunes and M. F. R. Pereira, *et al.*, Application of magnetic nanoparticles for water purification, *Environ. Adv.*, 2020, **2**, 100010.
47. Y. Yu, Z. Zhou, G. Huang, H. Cheng, L. Han and S. Zhao, *et al.*, Purifying water with silver nanoparticles (AgNPs)-incorporated membranes: Recent advancements and critical challenges, *Water Res.*, 2022, **222**, 118901.
48. S. J. Armaković, M. M. Savanović and S. Armaković, Titanium dioxide as the most used photocatalyst for water purification: An overview, *Catalysts*, 2022, **13**(1), 26.
49. Á. J. de Ruíz-Baltazar, Advancements in nanoparticle-modified zeolites for sustainable water treatment: An interdisciplinary review, *Sci. Total Environ.*, 2024, **946**, 174373.
50. L. Zhu, X. Guo, Y. Chen, Z. Chen, Y. Lan and Y. Hong, *et al.*, Graphene oxide composite membranes for water purification, *ACS Appl. Nano Mater.*, 2022, **5**(3), 3643–3653.
51. Y. Tang, Z. Cai, X. Sun, C. Chong, X. Yan and M. Li, *et al.*, Electrospun nanofiber-based membranes for water treatment, *Polymers*, 2022, **14**(10), 2004.
52. K. Plessl, A. Russ and D. Vollprecht, Application and development of zero-valent iron (ZVI) for groundwater and wastewater treatment, *Int. J. Environ. Sci. Technol.*, 2023, **20**(6), 6913–6928.
53. J. Hecht, Nanomaterials pave the way for the next computing generation, *Nature*, 2022, **608**(7922), S2–S3.
54. M. Si, H.-Y. Cheng, T. Ando, G. Hu and P. D. Ye, Overview and outlook of emerging non-volatile memories, *MRS Bull.*, 2021, **46**(10), 946–958.
55. G. Bhatti, Y. Agrawal, V. Palaparthy, M. Kavicharan and M. Agrawal, Flexible electronics: A critical review, in *Springer Tracts in Electrical and Electronics Engineering*, Springer Nature Singapore, Singapore, 2024. pp. 221–248.
56. N. Farooq, Z. Rehman, M. I. Khan, S. Asghar, M. Saleem and R. Irshad, *et al.*, Nanomaterial-based energy conversion and energy storage devices: a comprehensive review, *New J. Chem.*, 2024, **48**(19), 8933–8962.

57. D. Tian, H. Ma, G. Huang, M. Gao, F. Cai and Y. Fang, *et al.*, A review on quantum dot light-emitting diodes: From materials to applications, *Adv. Opt. Mater.*, 2023, **11**(2), 2201965.
58. M. A. Iqbal, M. Malik, N. Anwar, S. Bakhsh, S. Javeed and S. Sarah Maidin, *et al.*, Basic concepts, advances and emerging applications of nanophotonics, *Arabian J. Chem.*, 2023, **16**(9), 105040.
59. A. Ali, A. Mitra and B. Aïssa, Metamaterials and metasurfaces: A review from the perspectives of materials, mechanisms and advanced meta-devices, *Nanomaterials*, 2022, **12**(6), 1027.
60. Y. Shi, Q. Song, I. Toftul, T. Zhu, Y. Yu and W. Zhu, *et al.*, Optical manipulation with metamaterial structures, *Appl. Phys. Rev.*, 2022, **9**(3), 031303.
61. G. Aydın and S. E. San, Breaking the limits of acoustic science: A review of acoustic metamaterials, *Mater. Sci. Eng., B*, 2024, **305**, 117384.
62. J. You, Q. Ma, L. Zhang, C. Liu, J. Zhang and S. Liu, *et al.*, Electromagnetic metamaterials: From classical to quantum, *Electromagn. Sci.*, 2023, **1**(1), 1–33.
63. P. Jiao, J. Mueller, J. R. Raney, X. R. Zheng and A. H. Alavi, Mechanical metamaterials and beyond, *Nat. Commun.*, 2023, **14**(1), 6004.
64. Y. Li, W. Li, T. Han, X. Zheng, J. Li and B. Li, *et al.*, Transforming heat transfer with thermal metamaterials and devices, *Nat. Rev. Mater.*, 2021, **6**(6), 488–507.
65. G. Chen, F. Zhu, A. S. J. Gan, B. Mohan, K. K. Dey and K. Xu, *et al.*, Towards the next generation nanorobots, *Next Nanotechnol.*, 2023, **2**, 100019.
66. X. Guo, S. Cheng, W. Cai, Y. Zhang and X.-A. Zhang, A review of carbon-based thermal interface materials: Mechanism, thermal measurements and thermal properties, *Mater. Des.*, 2021, **209**, 109936.
67. A. H. Alami, M. Ramadan, M. Tawalbeh, S. Haridy, S. Al Abdulla and H. Aljaghoub, *et al.*, A critical insight on nanofluids for heat transfer enhancement, *Sci. Rep.*, 2023, **13**(1), 15303.
68. C. Lin, K. Li, M. Li, B. Dopphoopha, J. Zheng and J. Wang, *et al.*, Pushing radiative cooling technology to real applications, *Adv. Mater.*, 2024, e2409738.
69. L. Liu, S. Wang, B. Zhang, G. Jiang, H. Liu and J. Yang, *et al.*, Present status and prospects of nanostructured thermal barrier coatings and their performance improvement strategies: A review, *J. Manuf. Process.*, 2023, **97**, 12–34.
70. Z. Said, A. K. Pandey, A. K. Tiwari, B. Kalidasan, F. Jamil and A. K. Thakur, *et al.*, Nano-enhanced phase change materials: Fundamentals and applications, *Prog. Energy Combust. Sci.*, 2024, **104**, 101162.
71. D. Thesiya, H. Patel and G. S. Patange, A comprehensive review electronic cooling: A nanomaterial perspective, *Int. J. Thermofluids*, 2023, **19**, 100382.
72. H. Khodabandehloo, H. Zahednasab and A. Ashrafi Hafez, Nanocarriers usage for drug delivery in cancer therapy, *Iran. J. Cancer Prev.*, 2016, **9**(2), e3966.

73. Y. Takeda, K. Hayasaka, R. Shiwaku, K. Yokosawa, T. Shiba and M. Mamada, *et al.*, Fabrication of ultra-thin printed organic TFT CMOS logic circuits optimized for low-voltage wearable sensor applications, *Sci. Rep.*, 2016, **6**, 25714.
74. A. Alex-Amor, Á. Palomares-Caballero and C. Molero, 3-D metamaterials: Trends on applied designs, computational methods and fabrication techniques, *Electronics*, 2022, **11**(3), 410.
75. A. Shinde, S. B. Prasad, D. A. Srinivasarao, S. Shah, P. Famta and P. Khairnar, *et al.*, Nano voyagers: Pioneering a new frontier in cancer treatment with nanorobots as drug transporters, *Appl. Mater. Today*, 2024, **38**, 102162.
76. M. T. Barako, V. Gambin and J. Tice, Integrated nanomaterials for extreme thermal management: a perspective for aerospace applications, *Nanotechnology*, 2018, **29**(15), 154003.

CHAPTER 7

Enabling Technologies, Interdisciplinary Collaboration, and Societal Implications of Nanotechnology

Nanotechnology has emerged as a transformative field with far-reaching implications across scientific disciplines and industries. As an enabling technology, nanotechnology provides a powerful toolset for investigating, fabricating, and manipulating materials at the most fundamental levels, driving advancements in fields as diverse as electronics, energy, healthcare, and environmental science.

However, the true potential of nanotechnology lies not only in its technical capabilities but also in its ability to foster interdisciplinary collaboration and convergence among traditionally distinct fields. As nanotechnology continues to evolve and mature, it is crucial to consider the societal and environmental implications of this rapidly advancing field to ensure its responsible development and deployment. This chapter explores the role of nanotechnology as an enabling technology, its impact on interdisciplinary research, and the broader societal and environmental considerations that must be addressed for advancing the field as a driver of sustainability.

7.1 Nanotechnology as an Enabling Toolset

At its core, nanotechnology serves as a comprehensive set of tools and methods that empower researchers and engineers from a wide range of disciplines to probe, understand, and manipulate matter at the nanoscale. The unique properties and behaviors that emerge at this scale, such as quantum

Waste Not! How Nanotechnologies Can Increase Efficiencies Throughout Society
By Michael Berger

Published by the Royal Society of Chemistry, www.rsc.org

confinement, high surface-to-volume ratios, and enhanced surface reactivity, form the basis for nanotechnology's enabling capabilities. By harnessing these nanoscale phenomena, researchers can develop novel materials, devices, and systems with unprecedented performance and functionality.

As outlined in Section 5.3, nanotechnology encompasses a wide range of tools and techniques that enable the characterization, fabrication, and manipulation of nanomaterials and nanostructures. These include advanced imaging and analysis techniques, such as scanning probe microscopy, electron microscopy, and X-ray characterization methods, which allow researchers to visualize and study nanoscale structures and properties with atomic precision. Nanofabrication techniques, such as nanolithography, self-assembly, and additive manufacturing, enable the precise control and patterning of materials at the nanoscale, opening up new possibilities for creating complex nanodevices and systems.

One of the key strengths of nanotechnology as an enabling toolset is its versatility and applicability across a wide range of scientific disciplines and industries. For example, in the field of electronics, nanotechnology has enabled the development of nanoscale transistors, memory devices, and sensors that offer improved performance, reduced power consumption, and new functionalities compared to their bulk counterparts. In the energy sector, nanotechnology has led to the development of high-efficiency solar cells, advanced battery materials, and novel catalysts for clean energy production and storage. In healthcare, nanotechnology is revolutionizing drug delivery, diagnostic imaging, and regenerative medicine, enabling targeted therapies, early disease detection, and tissue engineering.

The enabling capabilities of nanotechnology have also led to groundbreaking scientific discoveries and technological advancements. For instance, the development of nanofluidic devices has allowed researchers to study single-molecule dynamics and interactions, providing new insights into fundamental biological processes. The use of nanostructured materials has enabled the creation of ultra-strong and lightweight composite materials, flexible electronics, and high-performance sensors. The application of nanotechnology in environmental remediation has led to the development of nanomaterial-based membranes and catalysts for water purification, air pollution control, and oil spill cleanup.

As an enabling technology, nanotechnology not only empowers advancements within individual disciplines but also serves as a platform for convergence and integration among different fields. The tools and methods of nanotechnology are increasingly being adopted and adapted by researchers across the scientific spectrum, fostering interdisciplinary collaboration and knowledge sharing. This convergence is essential for tackling complex global challenges, such as sustainable energy and environmental sustainability, which require expertise and insights from multiple domains.

The development and deployment of nanotechnology also raises important considerations regarding safety, ethics, and societal impact. As the use of nanomaterials and nanotechnology-enabled products becomes more widespread, it is crucial to assess and manage potential risks to human health and the environment. This requires ongoing research into the toxicology and fate of nanomaterials, as well as the development of appropriate safety guidelines and regulations. Additionally, the responsible advancement of nanotechnology necessitates public engagement and dialogue to ensure that the benefits of this transformative field are distributed equitably and that societal concerns are adequately addressed.

7.2 Fostering Interdisciplinary Collaboration

Nanotechnology is inherently interdisciplinary, drawing from and integrating knowledge from diverse fields such as physics, chemistry, biology, materials science, and engineering. This interdisciplinary nature is one of the key strengths of nanotechnology, as it enables the convergence of different expertise, perspectives, and methodologies to address complex scientific and technological challenges. By fostering collaboration among researchers, engineers, and industry partners from various disciplines, nanotechnology can accelerate innovation, drive breakthrough discoveries, and create new opportunities for societal and economic impact.

7.2.1 The Role of Nanotechnology in Converging Different Fields

Nanotechnology serves as a unifying platform that brings together different scientific and engineering disciplines around the common goal of understanding and manipulating matter at the nanoscale. The tools, techniques, and principles of nanotechnology are applicable across a wide range of fields, from materials science and electronics to healthcare and environmental science. This cross-cutting nature of nanotechnology facilitates the integration of knowledge and expertise from diverse domains, enabling researchers and engineers to approach problems from multiple angles and develop innovative solutions that transcend traditional disciplinary boundaries.

For example, the integration of nanotechnology with environmental research leverages the expertise of chemists, physicists, biologists, and engineers to develop nanomaterials for pollution control, water purification, and renewable energy sources. Nanotechnology can enhance the efficiency of photovoltaic cells for solar energy and improve the capacity and charging times of batteries, contributing significantly to sustainable energy solutions.

Similarly, the integration of nanotechnology with energy research brings together scientists and engineers from fields such as materials science, chemistry, physics, and electrical engineering to develop advanced nanomaterials and devices for energy conversion, storage, and conservation.

Effective interdisciplinary collaboration is essential for realizing the full potential of nanotechnology. Collaboration among researchers, engineers, and industry partners from different disciplines enables the pooling of knowledge, skills, and resources, leading to synergistic outcomes that would be difficult to achieve by individual efforts alone. By working together, collaborators can leverage their complementary expertise to tackle complex problems, generate new ideas, and accelerate the translation of nanotechnology research into practical applications. Interdisciplinary collaboration also promotes creativity and innovation by exposing researchers and engineers to diverse perspectives and approaches. When experts from different fields come together, they can challenge existing paradigms, identify new research directions, and develop novel solutions that bridge disciplinary gaps. This cross-pollination of ideas and methodologies can lead to breakthrough discoveries and transformative technologies that have a lasting impact on society and the economy.

A structured framework for purposeful nanotechnology development, such as the one proposed in Figure 7.1, illustrates how collaboration and common purpose can drive responsible innovation and interdisciplinary success in the nanotechnology field.[24] This framework emphasizes not only the need for collaboration but also the integration of six key principles—funding, research agendas, openness, measurement, governance, and regulation. By aligning these principles with shared goals, stakeholders can ensure that innovations

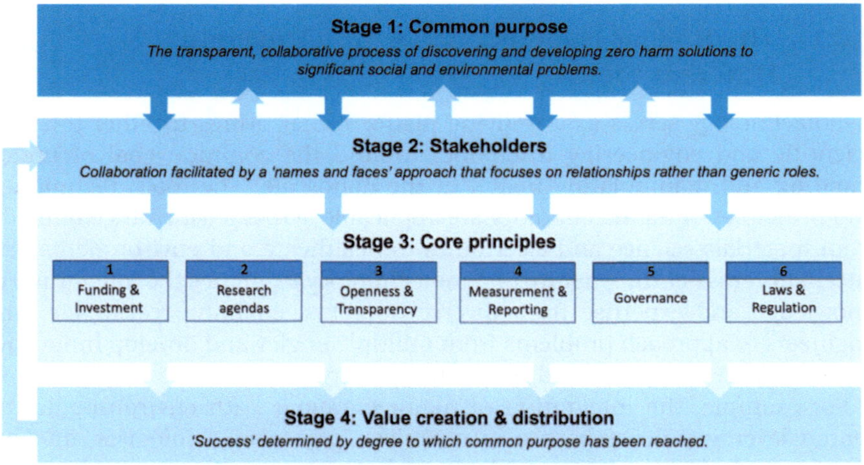

Figure 7.1 Framework for purposeful nanotechnology development: this framework illustrates a four-stage process that puts 'common purpose' at the center of nanotechnology innovation. The stages include defining a shared language and goals, fostering interdisciplinary collaboration, and ensuring the integration of ethical, environmental, and social considerations throughout the innovation cycle. Six core principles—funding, research agendas, openness, measurement, governance, and regulation—are prioritized to guide responsible and sustainable nanotechnology development. The framework emphasizes the creation of value for all stakeholders, with a focus on addressing significant societal and environmental challenges. Reproduced from ref. 24, https://doi.org/10.1016/j.impact.2024.100511, under the terms of the CC BY 4.0 license, https://creativecommons.org/licenses/by/4.0/.

are developed sustainably and ethically, while also addressing global challenges such as environmental degradation and health risks.

Moreover, collaboration between academia and industry is crucial for the successful commercialization and deployment of nanotechnology-based products and services. Industry partners bring valuable insights into market needs, scalability, and manufacturing challenges, while academic researchers contribute cutting-edge knowledge and expertise in nanotechnology fundamentals and applications. By working together, academia and industry can accelerate the transfer of nanotechnology innovations from the lab to the marketplace, create new business opportunities, and drive economic growth. The framework's emphasis on open communication and transparent stakeholder engagement ensures that these collaborations are not only effective but also aligned with societal expectations for safety and sustainability. This structured approach provides a roadmap for advancing nanotechnology in a way that balances innovation with responsibility, ensuring long-term positive impact.

7.2.2 Technological and Methodological Innovations from Interdisciplinary Collaborations

Interdisciplinary collaborations in nanotechnology have not only led to groundbreaking scientific discoveries but have also driven the development of new technologies and methodologies that are transforming the field. These innovations span across various aspects of nanotechnology, from nanofabrication techniques and characterization methods to computational models and simulation tools. These collaborations foster the integration of core disciplines such as materials science, chemistry, and physics, along with emerging fields like nanobiotechnology, enabling unprecedented control in the design and synthesis of nanomaterials.[1]

For example, the development of quantum dots, which originated from collaborations between materials science, physics, and biology, has revolutionized medical imaging and quantum computing. Similarly, lab-on-a-chip devices, born from the intersection of nanotechnology, microfluidics, and biotechnology, have transformed diagnostics and drug discovery processes. In nanobiotechnology, cross-disciplinary work has led to innovations such as nanoparticle-based drug delivery systems, improving the precision and effectiveness of treatments. Another notable example is the discovery and application of graphene, which, through the collaboration of physicists, chemists, and engineers, is now being used in flexible electronics and high-performance batteries.

The convergence of diverse fields, including biotechnology, cognitive science, and information technology, enhances the development of new approaches in nanotechnology. The use of cutting-edge tools, such as atomic force microscopes and advanced simulation technologies, plays a key role in enabling the precision and control needed for nanoscale innovations. The integration of diverse disciplines is not only driving innovation but is also being tracked through advanced bibliometric tools and science mapping techniques, reflecting the complex networks of knowledge exchange in nanotechnology.

The impact of interdisciplinary collaboration on innovation acceleration and technological breakthroughs in nanotechnology can be demonstrated through various studies and analyses. For example, a report by the National Science Foundation[2] highlights the exponential growth of nanotechnology-related scientific publications since 2000, with an average annual growth rate of 25%. The report also emphasizes the increasing interdisciplinary nature of nanotechnology research, with a significant proportion of publications involving authors from multiple disciplines.

Similarly, a study[3] that analyzed patent citations and author-inventor links in Chinese nanotechnology patents found evidence of growing interdisciplinary collaboration. The percentage of nanotechnology patents with interdisciplinary citations increased from 36.8% in the period 1991–2000 to 50.7% in the period 2001–2008, indicating a trend towards more interdisciplinary knowledge flows in nanotechnology innovation.

A bibliometric analysis[4] examined how nanotechnology research has become more interdisciplinary over time. The researchers analyzed a large dataset of 1 675 367 nanotechnology-related research papers published between 1980 and 2019. They used several statistical measures to assess the level of interdisciplinarity in these publications.

The results showed that nanotechnology research has become increasingly interdisciplinary over the past four decades. In other words, nanotechnology researchers are increasingly collaborating with experts from other scientific fields and incorporating knowledge from a wide range of disciplines into their work.

The study identified the top 10 scientific fields that contribute to nanotechnology research, including Physics, Chemistry, Materials Science, Engineering, and Biotechnology. Moreover, the authors found that nanotechnology is increasingly collaborating with fields such as Environmental Sciences, Toxicology, and Food Science, highlighting the expanding applications of nanotechnology in various areas. Overall, this study demonstrates that nanotechnology is a highly interdisciplinary field, with researchers from diverse scientific backgrounds working together to advance our understanding of the nanoscale world and develop new applications for nanomaterials and nanodevices.

These studies collectively demonstrate the growing trend of interdisciplinary collaboration in nanotechnology research and innovation, which has led to accelerated knowledge generation and technological breakthroughs in the field.

7.2.2.1 Nanofabrication Techniques

Nanofabrication, the process of creating structures and devices at the nanoscale, has been a key area of innovation driven by interdisciplinary collaborations. One notable example is the development of nanoimprint lithography (NIL), a high-throughput, low-cost nanopatterning technique that has emerged from the convergence of nanotechnology, materials science, and mechanical engineering. NIL involves the use of a nanopatterned mold to imprint nanoscale features onto a substrate, enabling the fabrication of complex nanostructures with high resolution and accuracy (see Figure 7.2). This technique has found applications in a wide range of fields, from electronics and photonics to biomedical devices and sensors.

Another example of an innovative nanofabrication approach that has been advanced through interdisciplinary collaborations is the directed self-assembly (DSA) of block copolymers. DSA leverages the self-organizing properties of block copolymers, which are composed of two or more chemically distinct polymer segments, to create ordered nanostructures with well-defined shapes and sizes. This technique has been developed through collaborations among researchers from chemistry, materials science, and chemical engineering, and has the potential to enable the scalable fabrication of sub-10 nm features for next-generation semiconductor manufacturing.

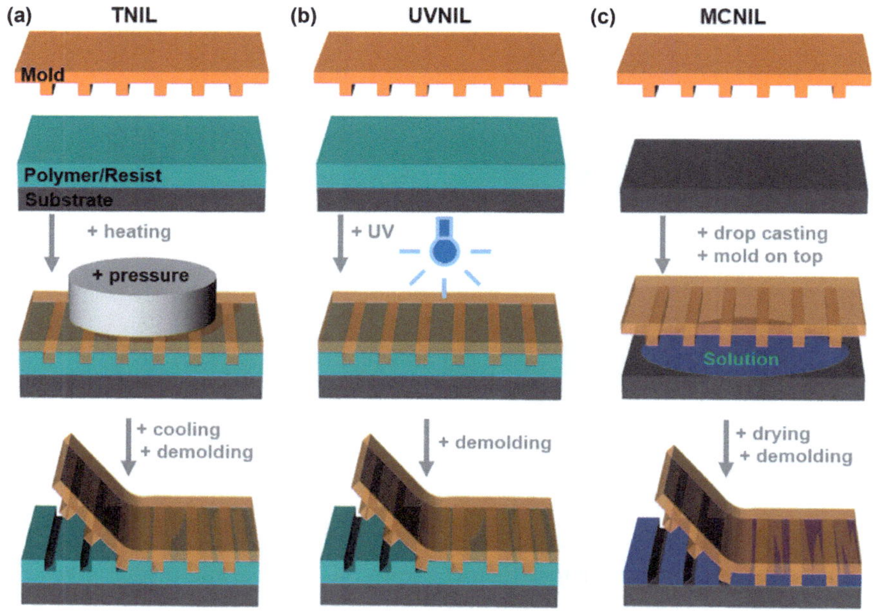

Figure 7.2 (a)–(c) Illustrative representation of various types of nanoimprint lithography (NIL): thermal nanoimprint lithography (TNIL) (a), UV nanoimprint lithography (UVNIL) (b), and molding in capillaries NIL (MCNIL) (c). Reproduced from ref. 25, https://doi.org/10.3390/polym13030445, under the terms of the CC BY 4.0 license, https://creativecommons.org/licenses/by/4.0/.

7.2.2.2 Characterization Methods

Interdisciplinary collaborations have also driven the development of advanced characterization methods that are providing new insights into the structure, properties, and behavior of nanomaterials. One prominent example is *in situ* transmission electron microscopy, a technique that allows researchers to observe dynamic processes in nanomaterials in real-time, under controlled environmental conditions. This technique has been developed through collaborations among researchers from nanotechnology, materials science, and physics, and has enabled the study of nanomaterial growth, phase transformations, and mechanical behavior with unprecedented spatial and temporal resolution.

Another innovative characterization method that has emerged from interdisciplinary collaborations is tip-enhanced Raman spectroscopy (TERS). TERS combines the high spatial resolution of scanning probe microscopy with the chemical sensitivity of Raman spectroscopy, enabling the characterization of nanomaterials and biological systems with nanoscale resolution and molecular specificity (see Figure 7.3). This technique has been developed through collaborations among researchers from nanotechnology, chemistry, and optics, and has found applications in fields ranging from materials science and catalysis to biomedicine and environmental monitoring.

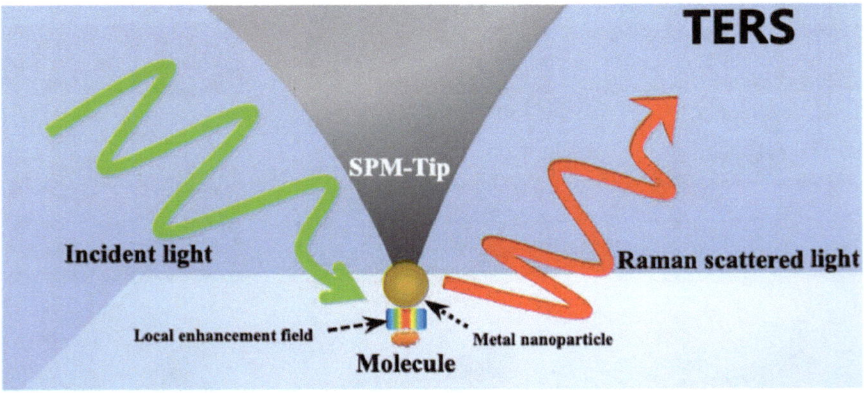

Figure 7.3 Schematic diagram of TERS principle. It illustrates the working principle of tip-enhanced Raman scattering. A scanning probe microscope (SPM) tip brings metal nanoparticles close to the molecule of interest. Incident light (green) irradiates the tip, creating a localized electromagnetic field. This field enhances the Raman signal from the molecule, which is then scattered (red light). The enhancement occurs in the gap between the metal tip and the sample surface, allowing high-resolution spectral and spatial imaging. Reproduced from ref. 26, https://doi.org/10.1016/j.revip.2022.100067, under the terms of the CC BY 4.0 license, https://creativecommons.org/licenses/by/4.0/.

7.2.2.3 Computational Models

Interdisciplinary collaborations have also driven the development of advanced computational models and simulation tools that are enabling the design and optimization of nanomaterials and nanosystems. One important example is multiscale modeling, which involves the integration of computational models across different length and time scales to predict the properties and behavior of nanomaterials and devices.[5] This approach has been developed through collaborations among researchers from nanotechnology, materials science, physics, and computer science, and has enabled the rational design of nanomaterials with tailored properties for specific applications.

Another area of innovation in computational modeling driven by interdisciplinary collaborations is the application of machine learning techniques to nanomaterials discovery and design.[6] Machine learning algorithms, such as deep learning and generative models, are being used to predict the properties of nanomaterials based on their composition and structure, and to identify new nanomaterials with desired functionalities. These techniques have been developed through collaborations among researchers from nanotechnology, materials science, computer science, and data science, and have the potential to accelerate the development of advanced nanomaterials for a wide range of applications, from energy storage and conversion to drug delivery and biomedical imaging.

The technological and methodological innovations emerging from interdisciplinary collaborations in nanotechnology are not only advancing the state-of-the-art in the field but are also creating new opportunities for scientific discovery and technological innovation. By bringing together diverse expertise and perspectives, these collaborations are driving the development of powerful new tools and approaches that are enabling the design, synthesis, and characterization of nanomaterials and nanosystems with unprecedented control and precision.

7.2.3 The Role of Policy and Governance

As nanotechnology continues to advance through interdisciplinary collaboration, it is essential to consider the role of policy and governance in guiding these efforts. Regulatory frameworks, ethical guidelines, and governance structures play a critical role in ensuring that nanotechnology research and development is conducted responsibly, with due consideration for potential risks and societal implications.

7.2.3.1 Regulatory Frameworks

The rapid pace of innovation in nanotechnology has necessitated the development of regulatory frameworks that can keep up with the evolving landscape. These frameworks aim to ensure the safety and responsible development of nanotechnology-based products and applications. For example, US agencies such as the Environmental Protection Agency (EPA)[7] and the Food and Drug Administration (FDA)[8] have been actively involved in developing guidelines for the assessment and regulation of nanomaterials in various contexts, from consumer products to medical applications.

The EPA's engagement with nanomaterials encompasses a comprehensive strategy under the Toxic Substances Control Act (TSCA). This includes developing a regulatory framework that mandates an information gathering rule on new and existing nanomaterials and premanufacture notifications for new nanomaterials. The agency is also involved in a wide array of research aimed at understanding, predicting, and managing the impacts of nanomaterials. This research is essential for characterizing nanomaterials' occurrence, studying their environmental transport and fate, and estimating human exposure levels. Furthermore, the EPA issued a final regulation that requires one-time reporting and recordkeeping of existing exposure and health and safety information on nanoscale chemical substances in commerce. This rule is part of the agency's effort to secure a more comprehensive understanding of nanoscale materials in commerce.

Nanotechnology holds potential across a broad array of FDA-regulated products, including medical products, foods, and cosmetics. The FDA is actively monitoring the science surrounding nanotechnology and has a robust research agenda to assess the safety and effectiveness of nanotech products. A key priority is understanding how the unique properties of nanomaterials might affect product safety, effectiveness, or other attributes.

The FDA has issued several guidance documents on the application of nanotechnology in regulated products. These documents, which represent the FDA's current thinking, cover areas such as manufacturing process changes, the application of nanotechnology, liposome drug products, and the safety of nanomaterials in cosmetics and food for animals. For example, in evaluating FDA-regulated products involving nanotechnology, considerations include whether a material or product is engineered to have nanoscale dimensions or to exhibit properties attributable to those dimensions, which may affect its regulatory status or the need for additional safety data.

In cosmetics, the FDA provides guidance on the safety substantiation of products containing nanomaterials, highlighting that traditional testing methods may not fully apply due to the distinctive properties of nanomaterials. The guidance underscores the importance of characterizing the nanomaterial and evaluating its physical and chemical properties, toxicity, and potential for systemic exposure.

In the European Union (EU), the regulation of nanomaterials[9] is integrated within its comprehensive chemical regulatory frameworks, notably REACH (Registration, Evaluation, Authorisation, and Restriction of Chemicals) and CLP (Classification, Labelling, and Packaging). These frameworks are designed to ensure the safe use of all chemicals, including those in nanoform, assessing their hazardous properties and ensuring that any risks they may pose are adequately controlled. Nanomaterials in the EU must be registered, and depending on the market volume, manufacturers or importers are required to submit detailed information regarding both human health and environmental effects. This includes data on hazardous nanoforms and an estimation of exposure throughout their lifecycle.

The European Commission has also provided a recommendation for defining nanomaterials based on the size of constituent particles, aiming for a consistent approach across EU and national legislation, policy, or research programs. This definition applies to natural, incidental, or manufactured materials and supports the implementation of regulatory provisions for nanomaterials across various sectors, including biocides, cosmetics, and food.

ECHA collaborates with Member State authorities, the European Commission, NGOs, industry associations, and international organizations like the OECD to implement EU chemicals legislation for nanomaterials.[10] As of January 1, 2020, explicit legal requirements under REACH apply to companies manufacturing or importing nanoforms. These include detailed reporting obligations covering the characterization of nanoforms, chemical safety assessments, registration information requirements, and downstream user obligations.

This regulatory approach reflects the EU's commitment to balancing the technological and commercial opportunities of nanomaterials with the need to address potential health and environmental risks.

Collaborative efforts among regulatory bodies, researchers, and industry partners have been crucial in shaping these frameworks. By engaging in dialogue and knowledge-sharing, these stakeholders can work together to develop evidence-based regulations that promote innovation while protecting public health and the environment. Interdisciplinary collaboration has also been important in addressing the challenges associated with regulating nanomaterials, such as the need for standardized testing methods and the consideration of nanomaterials' unique properties and behaviors.

7.2.3.2 Ethical Guidelines

The development and application of nanotechnology raise a range of ethical considerations, from the potential risks and benefits to society to questions of equity and access. To address these concerns, there has been a growing emphasis on the development of ethical guidelines for nanotechnology research and innovation.

Interdisciplinary collaboration has played a key role in shaping these guidelines, bringing together perspectives from ethics, social science, law, and policy, along with insights from the scientific and engineering communities. This collaborative approach has led to the development of frameworks such as "responsible research and innovation" (RRI), which aims to align nanotechnology research and development with societal values and needs.[11]

RRI emphasizes the importance of anticipating and addressing potential ethical and societal implications throughout the research and innovation process, rather than as an afterthought. This involves engaging with diverse stakeholders, including the public, to understand their concerns and aspirations, and to co-create solutions that are socially acceptable and desirable. Interdisciplinary collaboration is essential for operationalizing RRI, as it requires the integration of technical, social, and ethical considerations in the design and governance of nanotechnology.

7.2.3.3 Impact on Fostering Responsible Research and Innovation

The evolution of regulatory frameworks and ethical guidelines in response to nanotechnology advancements has had a significant impact on fostering responsible research and innovation. By providing a shared set of principles and practices, these frameworks have helped to guide interdisciplinary collaboration in nanotechnology towards socially responsible and beneficial outcomes.

The adoption of RRI principles has encouraged researchers and innovators to consider the wider societal implications of their work, leading to the development of nanotechnology applications that address pressing global challenges, such as clean energy, sustainable food production, and affordable healthcare. It has also promoted greater transparency and public engagement in nanotechnology research and development, helping to build trust and support for the field.

The collaborative development of regulatory frameworks and ethical guidelines has helped to create a more level playing field for nanotechnology research and innovation, promoting international cooperation and knowledge-sharing. This has been particularly important given the global nature of nanotechnology research and the need for coordinated governance approaches to address transboundary risks and challenges.

As nanotechnology continues to evolve and mature, ongoing collaboration among policymakers, researchers, industry partners, and other stakeholders will be critical for ensuring that the field remains aligned with societal values and priorities.

7.2.4 Strategies for Promoting Interdisciplinary Knowledge Sharing and Tech Transfer

To foster interdisciplinary collaboration in nanotechnology, establishing effective strategies for knowledge sharing and technology transfer is essential. These strategies can be grouped into several key areas, each contributing to the broader goal of advancing collaboration across disciplines.

Interdisciplinary education and training play a crucial role in building a workforce with the skills and mindset needed for collaboration. Cross-disciplinary courses, internships, workshops, and mentorship opportunities provide students and researchers exposure to different disciplines, fostering a culture of cooperation. This exposure equips them with the tools to navigate complex interdisciplinary challenges.

Additionally, establishing dedicated research centers and consortia offers a platform for collaboration, where experts from various fields can work together, exchange knowledge, and share resources. These centers act as hubs for research, education, and innovation, promoting a synergistic environment for interdisciplinary advancements in nanotechnology.

Another vital element is the development of knowledge-sharing platforms and networks. Online platforms, open-access repositories, and expert networks facilitate the exchange of ideas and best practices among researchers and engineers from different disciplines. These platforms make it easier to connect, communicate, and collaborate, thus accelerating technological progress.

Technology transfer offices and partnerships between academia and industry are also critical in bridging the gap between research and commercialization. Intellectual property support, licensing opportunities, entrepreneurship programs, and joint research and development initiatives help ensure that innovations move efficiently from the laboratory to the marketplace.

Lastly, collaborative funding mechanisms can incentivize interdisciplinary teamwork by providing resources for researchers to share and develop integrated solutions. Funding programs that prioritize collaboration encourage researchers to combine their expertise, fostering the development of innovative, cross-disciplinary approaches to complex problems.

Figure 7.4 illustrates these five interconnected strategies that support interdisciplinary collaboration in nanotechnology:

Implementing these approaches can create collaborative ecosystems that propels innovation and strengthens the advancement of nanotechnology across diverse fields.

7.2.5 Examples of Successful Interdisciplinary Collaborations

This section highlights notable examples of successful interdisciplinary collaborations that have led to significant advancements across various fields. By combining expertise from multiple disciplines, these collaborations have addressed complex challenges, ultimately resulting in innovations that benefit society. The examples cover the development of nanomedicine for cancer treatment, nanomaterials for energy storage, and nanosensors for environmental monitoring, each demonstrating the value of integrating knowledge from fields such as nanotechnology, chemistry, materials science, and engineering. These collaborations not only push scientific boundaries but also ensure practical, real-world applications of cutting-edge research.

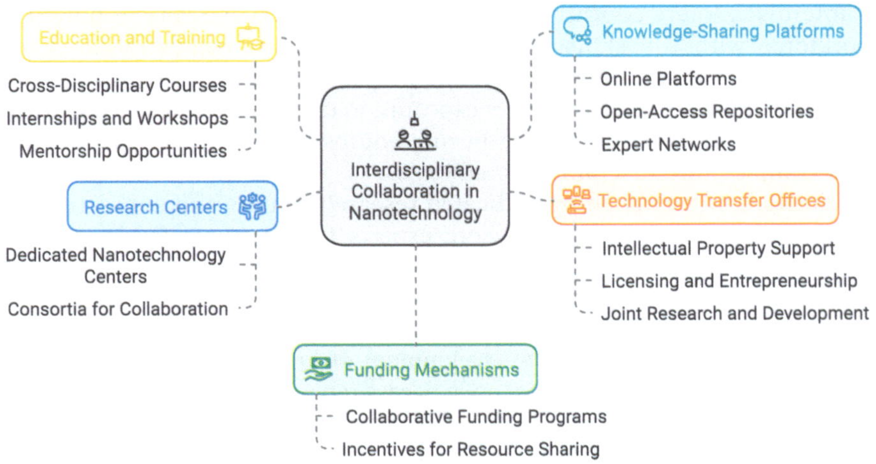

Figure 7.4 Key strategies for interdisciplinary collaboration in nanotechnology. The diagram highlights five key strategies—education and training, research centers, knowledge-sharing platforms, technology transfer offices, and funding mechanisms—that foster interdisciplinary collaboration and innovation in nanotechnology.

7.2.5.1 Nanomedicine for Cancer Treatment

The development of nanomedicine for cancer treatment exemplifies how interdisciplinary collaboration can lead to significant scientific and clinical breakthroughs. By bringing together expertise from nanotechnology, oncology, chemistry, and biomedical engineering, researchers have developed highly targeted, effective treatments that continue to evolve and improve through ongoing collaborative efforts.

One of the most significant breakthroughs has been the development of tumor-targeted nanocarriers. Chemists, oncologists, and nanotechnologists have worked closely to design these nanoparticles, which can precisely target cancer cells while minimizing damage to healthy tissues. Through this collaborative effort, researchers have developed hierarchical targeting systems that enable nanocarriers to overcome multiple biological barriers, reaching tumor cells and even penetrating organelles. This targeted approach has significantly improved therapeutic efficacy and reduced off-target toxicity.[12]

Liposomal drug delivery systems, such as Doxil® and Abraxane®, stand as prime examples of successful interdisciplinary innovations. The development of these FDA-approved treatments required the combined efforts of biomedical engineers, clinical oncologists, and chemists. Liposomes enhance drug stability and bioavailability, while reducing adverse effects, representing a major advancement in cancer nanomedicine. The transition of these therapies from the lab to clinical practice underscores the importance of collaboration across disciplines to ensure that scientific innovations can be safely and effectively applied in patient care.[13]

Progress in nanocarrier design has further demonstrated the value of interdisciplinary efforts. The development of second- and third-generation nanomedicines, which can penetrate deeply into tumors and target specific subcellular locations, has been driven by the combined expertise of molecular biologists, pharmacologists, and materials scientists. These advanced nanocarriers address complex challenges, such as overcoming multidrug resistance, and offer new possibilities for enhancing the precision of cancer treatments. This kind of innovation would not be possible without the seamless integration of knowledge from diverse fields.

The extensive number of clinical trials focused on nanoparticle-based therapies for cancer is another clear indicator of the power of interdisciplinary collaboration. With over 486 clinical trials conducted, these efforts require close cooperation between clinicians, pharmacologists, and nanotechnology experts to optimize nanoparticle formulations for therapeutic use. Cancer treatment remains a primary focus, with these collaborations ensuring that promising research is translated into viable clinical treatments. The success of these trials reflects the ability of interdisciplinary teams to bridge the gap between basic research and clinical application, ultimately improving patient outcomes.

7.2.5.2 Nanomaterials for Energy Storage

The development of advanced nanomaterials for energy storage is an outstanding example of how interdisciplinary collaboration can push the boundaries of technology. Researchers from materials science, chemistry, physics, and electrical engineering have come together to design and create nanostructured materials that deliver high energy density, fast charge-discharge rates, and extended cycle life. These efforts have driven the development of advanced energy storage devices such as lithium-ion batteries, supercapacitors, and fuel cells, which are essential for transforming the energy landscape and supporting the transition to more efficient energy storage systems.[14]

In lithium-ion batteries, a critical area of focus has been the creation of nanostructured electrodes. Materials scientists and chemists have worked together to increase the surface area of electrode materials, which allows for faster ion diffusion and more efficient charge transfer. For instance, nanostructured silicon, graphene, and metal oxides are being studied as alternatives to traditional graphite anodes. These materials offer significantly higher energy densities, which is essential for applications ranging from electric vehicles to portable electronics. The multidisciplinary collaboration involved in improving these materials ensures that they not only perform well but are also scalable and reliable for commercial use.[15]

Supercapacitors are another domain where interdisciplinary research has led to notable advancements. Chemists and electrical engineers have designed nanostructured carbon materials like graphene and carbon nanotubes, which offer high electrical conductivity and surface area. These properties are essential for boosting both the energy storage capacity and the power density of supercapacitors. The synergy between materials design, chemical synthesis, and electrochemical engineering has enabled the development of supercapacitors that are suited for rapid energy delivery in applications such as backup power systems and regenerative braking in electric vehicles.[16]

Fuel cells, an important technology for clean energy, have also benefited from nanomaterials research. By incorporating nanostructured catalysts, researchers from physics, chemical engineering, and materials science have significantly improved the efficiency and durability of fuel cells. Innovations in nanocatalysts, particularly in reducing the reliance on costly platinum, have helped make fuel cells more commercially viable. Nanomaterials such as platinum nanoparticles and alternative transition metal oxides have proven to enhance catalytic activity, reduce costs, and improve the long-term stability of fuel cells, making them more attractive for widespread use in energy storage.[17]

These interdisciplinary efforts have not only advanced scientific knowledge but also bridged the gap between fundamental research and industrial application. The development of nanomaterials for energy storage has directly influenced the performance of electric vehicles, large-scale energy storage systems for the electrical grid, and the integration of renewable energy sources. The collaboration between various fields ensures that innovations are not only optimized for technical performance but are also cost-effective, scalable, and suitable for real-world use.

7.2.5.3 Nanosensors for Environmental Monitoring

The development of nanosensors for environmental monitoring exemplifies the power of interdisciplinary collaboration. Scientists from fields such as nanotechnology, environmental science, chemistry, and electrical engineering have joined forces to create highly sensitive and specific nanoscale sensors capable of detecting a wide range of pollutants, toxins, and pathogens in air, water, and soil. These collaborations have produced innovative sensing technologies that can monitor environmental conditions in real-time, providing critical data to inform policy decisions, improve environmental health, and protect public safety.[18,19]

One of the key advancements in this area is the development of nanosensors based on carbon nanomaterials. These materials possess unique electrical and chemical properties that make them highly effective for detecting even trace amounts of harmful substances. For example, CNT-based sensors can be functionalized with chemical groups that bind to specific pollutants, enhancing the sensitivity and selectivity of detection. Collaborations between chemists and electrical engineers have optimized these sensors for environmental applications by improving their conductivity and response time, which is crucial for rapid and accurate monitoring of pollutants like heavy metals, volatile organic compounds, and greenhouse gases.

Another significant breakthrough is in the use of metallic nanoparticles as the active component in nanosensors. Researchers have incorporated gold, silver, and palladium nanoparticles into sensor platforms due to their exceptional surface reactivity and plasmonic properties, which allow for highly sensitive detection of environmental contaminants. These sensors can detect minute changes in the environment, such as variations in pollutant concentrations, by measuring changes in the optical or electrical properties of the nanoparticles. This interdisciplinary work, involving materials scientists and environmental engineers, has led to the development of highly efficient monitoring systems for detecting pollutants in water sources, such as heavy metals and organic pollutants.

Nanosensors based on biological elements have also emerged as an important innovation in environmental monitoring. Through collaboration between biochemists, microbiologists, and nanotechnologists, researchers have designed biosensors that use enzymes, antibodies, or DNA sequences as recognition elements. These biosensors can detect pathogens and toxins in the environment with high specificity. For instance, enzyme-linked nanosensors have been developed to monitor water quality by detecting biological contaminants, such as bacteria and viruses, providing critical data for maintaining safe water supplies.[20]

7.3 Engagement with Society in Research and Development

As nanotechnology continues to advance through interdisciplinary collaboration, it is increasingly important to engage with broader societal stakeholders in the research and development process. This engagement is crucial for ensuring that nanotechnology develops in a way that is socially responsible, ethically sound, and responsive to the needs and concerns of diverse communities.

7.3.1 Public Engagement Strategies

Effective public engagement is essential for building trust and support for nanotechnology research and innovation. This involves moving beyond one-way communication of scientific findings to more participatory and inclusive forms of engagement that allow for meaningful dialogue and exchange.

One key strategy for public engagement in nanotechnology is the use of deliberative forums, such as citizen panels or consensus conferences. These forums bring together diverse groups of citizens to learn about and discuss nanotechnology issues, provide input on research and policy priorities, and develop recommendations for responsible development. By creating a space for informed and inclusive dialogue, deliberative forums can help to surface important societal perspectives and values, and to build mutual understanding between scientific experts and the public.[21]

Another important strategy for public engagement is the use of science communication and outreach activities that are accessible, engaging, and responsive to public interests and concerns. This can involve the use of interactive exhibits, social media, and other digital platforms to share information about nanotechnology research and its potential applications. It can also involve partnerships with science museums, schools, and community organizations to develop educational programs and resources that foster public understanding and engagement with nanotechnology.

7.3.2 Participatory Design Processes

Participatory design is an approach to technology development that actively involves users and other stakeholders in the design process, with the goal of creating solutions that are more responsive to their needs and values.[22] In the context of nanotechnology, participatory design can help to ensure that research and innovation are guided by societal priorities and concerns.

One example of participatory design in nanotechnology is the involvement of patients and healthcare providers in the development of nanomedicine applications. By engaging these stakeholders early in the design process, researchers can gain valuable insights into the real-world needs and challenges of healthcare delivery and can work collaboratively to develop nanotechnology solutions that are more effective, accessible, and patient-centered.

Another example is the engagement of local communities in the development of nanotechnology applications for environmental remediation or sustainable agriculture. By involving community members in the design and testing of these applications, researchers can ensure that they are responsive to local knowledge, practices, and concerns, and can help to build local capacity and ownership for sustainable nanotechnology solutions.

7.3.3 Citizen Science in Nanotechnology

Citizen science is an approach to scientific research that involves the participation of non-professional scientists in the research process, often through the collection or analysis of data. In the context of nanotechnology, citizen science can help to broaden participation in the research enterprise and to tap into the diverse knowledge and perspectives of societal stakeholders.

One example of citizen science in nanotechnology is illustrated in a study that involved citizens in the development of nanotechnology for health applications. Fifty citizens participated in focus groups where they provided feedback on future nanotechnologies, such as diagnostic and monitoring devices. Through this engagement, participants were able to suggest how these technologies should be used and embedded into the healthcare system, expressing concerns about data privacy, accessibility, and the overall impact on well-being. This process allowed for the collection of valuable societal input, enabling researchers to align technological advancements with public values and preferences.[23]

Another example is the involvement of students and teachers in nanotechnology education and research through classroom-based projects and online learning communities. By engaging these stakeholders in hands-on, inquiry-based learning about nanotechnology, researchers can help to build scientific literacy and interest in the field and can inspire the next generation of nanotechnology researchers and innovators.

The engagement of broader societal stakeholders in nanotechnology research and development is essential for ensuring that the field advances in a way that is socially responsible, ethically sound, and responsive to the needs and concerns of diverse communities. By using public engagement strategies, participatory design processes, and citizen science approaches, the nanotechnology community can help to build trust, understanding, and collaboration between researchers and the public. This, in turn, can help to ensure that the benefits of nanotechnology are realized in a way that is equitable, sustainable, and aligned with societal values and priorities.

References

1. A. L. Porter and J. Youtie, How interdisciplinary is nanotechnology?, *J. Nanopart. Res.*, 2009, **11**(5), 1023–1041.
2. M. C. Roco, W. S. Bainbridge, B. E. Tonn and G. M. Whitesides, *Convergence of Knowledge, technology and society: Beyond convergence of nano-bio-info-cognitive technologies*, Springer, Cham, 2014.
3. G. Wang and J. Guan, Measuring science–technology interactions using patent citations and author-inventor links: An exploration analysis from Chinese nanotechnology, *J. Nanopart. Res.*, 2011, **13**(12), 6245–6262.
4. U. Finardi and C. Lamberti, Diversity and Interdisciplinarity in nanoscience and nanotechnology: A time-related analysis of the subject category, *J. Nanopart. Res.*, 2021, **23**, 7.
5. G. Fitzgerald, J. DeJoannis and M. Meunier, Multiscale modeling of nanomaterials, in *Modeling, Characterization, and Production of Nanomaterials*, Elsevier, 2015, pp. 3–53.
6. M. Wang, T. Wang, P. Cai and X. Chen, Nanomaterials discovery and design through machine learning, *Small Methods*, 2019, 3(5), 1900025.
7. US EPA, Research on nanomaterials, [cited 2024 Sep 30]. Available from: https://www.epa.gov/chemical-research/research-nanomaterials.
8. FDA.gov, Nanotechnology Fact Sheet [cited 2024 Sep 30]. Available from: https://www.fda.gov/science-research/nanotechnology-programs-fda/nanotechnology-fact-sheet.

9. Nanomaterials in EU legislation, Nanomaterials in EU Legislation, [cited 2024 Sep 30]. Available from: https://academy.europa.eu/courses/nanomaterials-in-eu-legislation.
10. Nanomaterials – ECHA, Europa.eu. [cited 2024 Sep 30]. Available from: https://www.echa.europa.eu/regulations/nanomaterials.
11. R. Von Schomberg, Introduction: Understanding Public Debate on Nanotechnologies Options for Framing Public Policy (August 1, 2010), in *Understanding Public Debate on Nanotechnologies*, ed. R. Von Schomberg and S. Davies, Publication Office of the European Union, Luxembourg, 2010. Available at SSRN: https://ssrn.com/abstract=2436346.
12. D. Fan, Y. Cao, M. Cao, Y. Wang, Y. Cao and T. Gong, Nanomedicine in cancer therapy, *Signal Transduction Targeted Ther.*, 2023, **8**(1), 293.
13. E. D. Namiot, A. V. Sokolov, V. N. Chubarev, V. V. Tarasov and H. B. Schiöth, Nanoparticles in clinical trials: Analysis of clinical trials, FDA approvals and use for COVID-19 vaccines, *Int. J. Mol. Sci.*, 2023, **24**(1), 787.
14. N. Farooq, Z. Rehman, M. I. Khan, S. Asghar, M. Saleem and R. Irshad, *et al.*, Nanomaterial-based energy conversion and energy storage devices: a comprehensive review, *New J. Chem.*, 2024, **48**(19), 8933–8962.
15. M. Adnan, The future of energy storage: Advancements and roadmaps for lithium-ion batteries, *Int. J. Mol. Sci.*, 2023, **24**(8), 7457.
16. K. O. Oyedotun, J. O. Ighalo, J. F. Amaku, C. Olisah, A. O. Adeola and K. O. Iwuozor, *et al.*, Advances in supercapacitor development: Materials, processes, and applications, *J. Electron. Mater.*, 2023, **52**(1), 96–129.
17. L. M. Shaker, A. A. Al-Amiery and W. K. Al-Azzawi, Nanomaterials: paving the way for the hydrogen energy frontier, *Discover Nano*, 2024, **19**(1), 3.
18. M. A. Darwish, W. Abd-Elaziem, A. Elsheikh and A. A. Zayed, Advancements in nanomaterials for nanosensors: a comprehensive review, *Nanoscale Adv.*, 2024, **6**(16), 4015–4046.
19. M. R. Willner and P. J. Vikesland, Nanomaterial enabled sensors for environmental contaminants, *J Nanobiotechnol.*, 2018, **16**(1), 95.
20. F. J. Tovar-Lopez, Recent progress in micro- and nanotechnology-enabled sensors for biomedical and environmental challenges, *Sensors*, 2023, **23**(12), 5406.
21. A. R. Jones, A. A. Anderson, S. K. Yeo, A. E. Greenberg, D. Brossard and J. W. Moore, Using a deliberative exercise to foster public engagement in nanotechnology, *J. Chem. Educ.*, 2014, **91**(2), 179–187.
22. M. van der Velden and C. Mörtberg, Participatory design and design for values, in *Handbook of Ethics, Values, and Technological Design*, Springer Netherlands, Dordrecht, 2015, pp. 41–66.
23. S. R. Jansma, A. M. Dijkstra and M. D. T. de Jong, How can I contribute? Citizen engagement in the development of nanotechnology for health, *Nanoethics*, 2021, **15**(3), 211–227.

24. P. Jeffcoat, C. Di Lernia, C. Hardy, E. J. New and W. Chrzanowski, (Re)imagining purpose: A framework for sustainable nanotechnology innovation, *NanoImpact*, 2024, **35**, 100511.
25. M. Handrea-Dragan and I. Botiz, Multifunctional structured platforms: From patterning of polymer-based films to their subsequent filling with various nanomaterials, *Polymers*, 2021, **13**(3), 445.
26. Y. Cao and M. Sun, Tip-enhanced Raman spectroscopy, *Rev. Phys.*, 2022, **8**, 100067.

Section III: A Role for Nanotechnology in Waste Solutions

CHAPTER 8

Nanotechnology for Waste Reduction and Prevention

Nanotechnology is increasingly recognized for its potential to revolutionize waste reduction and prevention strategies across numerous industries, offering the possibility of more efficient, durable, and sustainable materials and processes. By manipulating matter at the nanoscale, researchers have been able to enhance properties such as strength, electrical conductivity, thermal stability, and chemical reactivity in ways that traditional materials cannot achieve. These advancements not only optimize resource usage but also reduce emissions and material waste, contributing to sustainability goals in sectors as diverse as aerospace, automotive, energy, and packaging.

Carbon-based nanomaterials represent a significant area of development in nanotechnology. Beginning with carbon nanotubes in the 1990s and later expanded by the groundbreaking isolation of graphene in 2004, these materials have attracted intense research interest for their extraordinary properties. They offer exceptional mechanical strength, electrical conductivity, and thermal stability while maintaining very low density. While they show great promise for improving structural materials and energy efficiency across multiple industries, their widespread practical application has been slowed by challenges like processing complexity, scalability, and cost-effectiveness. However, carbon nanomaterials represent just one facet of the broader impact nanotechnology can have on waste reduction.

Nanocatalysis represents a key approach to waste prevention in industrial processes. By providing larger surface areas and higher reactivity, nanoscale catalysts significantly reduce energy consumption and by-product formation in chemical reactions. Advances in single-atom catalysts are driving this field forward, offering improvements in energy conversion, carbon dioxide reduction, and sustainable chemistry. Beyond catalysis, nanotechnology

Waste Not! How Nanotechnologies Can Increase Efficiencies Throughout Society
By Michael Berger

Published by the Royal Society of Chemistry, www.rsc.org

enables precision monitoring and control through smart manufacturing systems and agricultural applications, transforming how industries optimize resource use.

Nanomaterials are revolutionizing structural applications through enhanced strength, durability, and reduced weight, particularly in automotive and aerospace sectors. By incorporating materials like carbon-based nanostructures, manufacturers can create components that are lighter and stronger, leading to reductions in fuel use and raw material consumption. Beyond structural improvements, these materials enable functional advances such as embedded sensing capabilities and self-healing properties, creating a new generation of smart materials.

In packaging applications, biodegradable nanomaterials are providing sustainable solutions. Innovations such as nanocellulose and chitosan nanoparticles are being developed into eco-friendly alternatives to conventional plastics, with improved barrier properties, antimicrobial functionality, and reduced environmental impact. These materials not only extend the shelf life of products but also help address the growing global issue of plastic waste.

Nanotechnology plays a crucial role in improving energy efficiency across multiple sectors. Nanomaterials enhance energy storage capacity and conversion efficiency while enabling better building insulation and more efficient industrial processes. Through innovations in areas such as battery technology and smart windows, these materials are reducing energy waste throughout the built environment. Their enhanced durability in applications like coatings and construction materials further contributes to sustainability by extending product lifespans.

Together, these innovations demonstrate how nanotechnology has the potential to address some of the most pressing challenges related to waste and resource management in modern industries. This chapter deliberately presents a large number of detailed examples and case studies across multiple sectors—from industrial catalysis to smart packaging, precision agriculture to energy storage—to demonstrate both the scope and depth of ongoing research into how these diverse nanoscale innovations collectively offer a powerful toolset for advancing circular economy principles. While individual applications may seem small in scale and many are still in development or early stages of commercialization, their combined potential for improving resource efficiency, reducing waste, and enabling more sustainable practices represents a significant opportunity to address global environmental challenges. Through careful examination of both current applications and emerging possibilities, even when only in minute detail, we can better understand how nanotechnology can contribute to a more sustainable and efficient future."

8.1 Catalysis

Catalysis, a technology that accelerates the rate of a chemical reaction, is a critical component in contemporary chemical and energy production pathways. As an essential interdisciplinary technology within the chemical industry, catalysis has far-reaching societal consequences, even though it is often underestimated or overlooked. The production of fertilizers, enabled by catalysis, sustains a global population of eight billion individuals. Moreover, catalysis is crucial for producing fuels from oil at the necessary quality and quantity for modern energy consumption. More than 85% of all chemical products are manufactured using catalytic processes, which also enable the modern refinement of fuels.[1]

Catalytic processes are indispensable for maintaining air quality amid the pervasive use of automobiles, and environmental catalysis in exhaust gas after-treatment for vehicles, ships, and trains, as well as other mobile applications, warrants heightened attention. Catalysis and related processes account for 20–30% of the global GDP, either directly or indirectly. Of the 50 largest volume chemicals produced, 30 are synthesized through catalytic pathways, with the top 50 processes collectively emitting over 20 billion tons of CO_2 annually. The production of 18 key products represents 80% of the chemical industry's energy demand and 75% of its greenhouse gas emissions.[2]

As catalysis continues to play a foundational role in sectors ranging from healthcare to energy production, the need for more efficient and sustainable solutions has become increasingly urgent. Recent breakthroughs, particularly in single-atom catalysts (SACs), offer promising pathways to enhance catalytic efficiency while addressing critical environmental challenges.

SACs are a nanotechnology innovation, as they manipulate materials at the atomic scale by dispersing individual metal atoms on nanostructured supports (see Figure 8.1). This configuration maximizes the exposure of active sites, drastically improving catalytic efficiency while minimizing

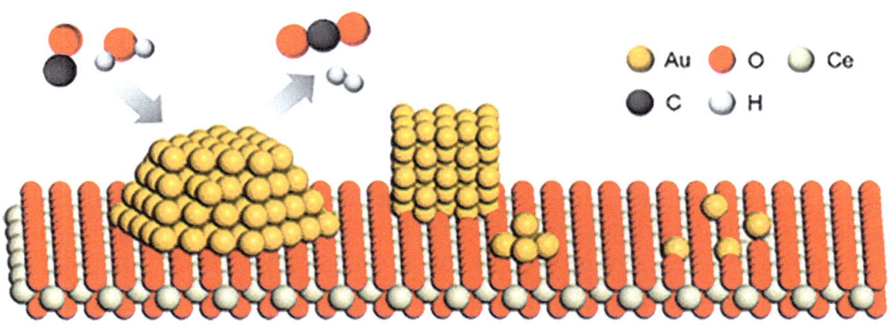

Figure 8.1 Representation of catalytic processes on a nanostructured surface: the image illustrates gold (Au) nanoparticles and single atoms on a cerium oxide (CeO_2) support, with the interaction between reactants (CO and H_2O) and catalyst sites leading to product formation. This showcases the role of size, shape, and metal-support interactions in heterogeneous catalysis, emphasizing the importance of well-defined nanomaterials for advanced catalytic reactions. Reproduced from ref. 249 with permission from American Chemical Society, Copyright 2020.

material usage. These catalysts are especially impactful in energy conversion processes, such as hydrogen evolution and CO_2 reduction, due to their high selectivity and activity.

SACs have shown potential to improve energy efficiency by 20–30% by 2050 in energy conversion and storage systems. In absolute terms, they could save up to 10 exajoules of energy annually and mitigate around 0.9 gigatons of CO_2-equivalent emissions.[3,4]

Building on the transformative potential of SACs, the field of catalysis is also witnessing significant innovations with nanostructured materials. These advanced materials, including metal–organic frameworks and carbon-based nanomaterials, are redefining what is possible by offering unique properties that further enhance catalytic performance. These nanomaterials can enhance catalytic processes due to their unique properties, such as high surface area, tunable pore sizes, and optimized active sites. By leveraging the capabilities of nanomaterials, researchers and industry professionals can develop more efficient, selective and stable catalysts, ultimately contributing to a more sustainable future.[5]

For example, MOFs offer vast surface areas and tunable pore structures, making them ideal for gas storage and catalysis. Similarly, graphene and CNTs provide exceptional conductivity and mechanical strength, enhancing the performance of energy storage devices and catalysts. These materials are being actively explored for applications in CO_2 reduction, hydrogen production, and environmental remediation.

However, adjusting the chemical reactivity of nanomaterials at the atomic scale is a critical task in catalysis research. To accomplish this difficult objective, it is essential to gain a deep understanding of the atomic-level geometric and electronic structures of these intricate systems.[6]

8.1.1 Enhanced Catalytic Activity

The significance of nanostructured materials in various applications, particularly in catalysis, is largely attributed to their high surface area-to-volume ratio. This property ensures an increased number of active sites available for chemical reactions, as a larger proportion of the atoms or molecules are exposed on the surface rather than concealed within the bulk material. In turn, this heightened surface area provides more active sites for catalytic reactions, where reactant molecules can adsorb, interact, and transform into products.[7]

Consequently, a greater number of reactant molecules can simultaneously engage with the catalyst, resulting in an overall increase in reaction rate, enhanced catalytic efficiency, and greater productivity due to the generation of more products within a shorter timeframe.

Furthermore, the high surface area of nanostructured materials can contribute to the selectivity of a catalytic process. Owing to their diminutive size and tunable morphology, nanostructured materials can be designed with specific shapes, sizes, and crystal facets. This allows for the selective stabilization of certain reaction intermediates, promoting desirable reaction pathways over less favorable ones. Such improved selectivity leads to the production of fewer byproducts and a diminished need for energy-intensive separation processes, ultimately rendering the overall process more sustainable and environmentally friendly.

Several crucial processes in the chemical industry, which involve the use of nanocatalysts or have the potential to benefit from their application, include the Haber–Bosch process, the production of hydrogen through water splitting, and olefin polymerization. The Haber–Bosch process, a key industrial method for producing ammonia from nitrogen and hydrogen gases, requires high temperatures and pressures, consuming over 1% of the world's power production. Additionally, this process is responsible for approximately 1.2% of global anthropogenic CO_2 emissions. Given ammonia's critical role in fertilizer production and its subsequent impact on global food supply, ongoing research into nanostructured catalysts aims to improve the efficiency and sustainability of ammonia production, reducing both energy consumption and greenhouse gas emissions.[8]

The production of hydrogen through water splitting is a promising avenue for generating clean and renewable energy. This process typically employs electrocatalysts to facilitate the oxygen evolution reaction (OER) and hydrogen evolution reaction (HER). Nanostructured materials, such as transition metal oxides, sulfides, and phosphides, have been extensively researched as electrocatalysts for both OER and HER, owing to their high surface area, unique electronic properties, and improved catalytic performance.[9,10]

Lastly, olefin polymerization is a process employed to produce various types of plastics, including polyethylene and polypropylene, which are prevalent in packaging, consumer goods, and automotive applications. Catalysts based on nanoscale transition metal complexes, such as zirconium or titanium, have demonstrated high activity and selectivity for olefin polymerization, resulting in more efficient and controlled production of polymer materials.[11]

Researchers also have shown that metal–organic framework-supported catalysts exhibit high activity and selectivity for the polymerization of ethylene and propylene, underscoring the potential advantages of employing nanostructured materials in olefin polymerization processes.

8.1.2 Improved Selectivity

Nanostructured materials are revolutionizing the field of catalysis, offering unparalleled opportunities to enhance the selectivity of various processes. By tailoring their size, shape, and composition, researchers can exert precise control over reaction pathways, leading to improved efficiency and specificity across a wide range of applications. A prominent example is shape-controlled platinum nanoparticles, which showcase the potential for driving more sustainable and efficient catalytic processes.

Noble metal nanocrystals often adopt conventional shapes like cubes, tetrahedra, and octahedra, which are primarily bound by low-index facets such as {111} and {100}. Facets are the flat surfaces of a crystal, and in nanocrystals, they play a crucial role in how well the material interacts with its surroundings. The numbers like {111} and {100} refer to specific orientations of the crystal's atoms on these flat surfaces. Some facets, especially those with more exposed atoms at their edges and corners, are more "active" and can significantly speed up chemical reactions.

Researchers found that when platinum nanocrystals are synthesized with unconventional shapes, such as tetrahexahedra (THH), they exhibit much higher catalytic activity. The THH structure, featuring high-index facets like {730}, {210}, and {520}, refers to more complex atomic arrangements that expose a greater number of active sites, making it more effective in catalysis (see Figure 8.2). These high-energy surfaces not only improve catalytic efficiency, surpassing traditional platinum catalysts by up to 400%, but also demonstrate remarkable stability, maintaining their shape even at extreme temperatures of up to 800 °C.[12] This stability is likely due to the adsorption of oxides or hydroxides during the synthesis process, making the THH shape energetically favorable. Advances in controlling the shape of nanocrystals like these open up new possibilities for creating more efficient catalysts.

This enhanced performance has the potential to contribute to the development of more efficient catalysts for a variety of applications, including fuel cells and chemical reactions involving small organic fuels like formic acid and ethanol. In the long run, this could lead to advancements in energy conversion technologies and more sustainable chemical processes.[13]

Another compelling example is the use of nanostructured metal–organic frameworks in catalysis. By adjusting the structure and composition of

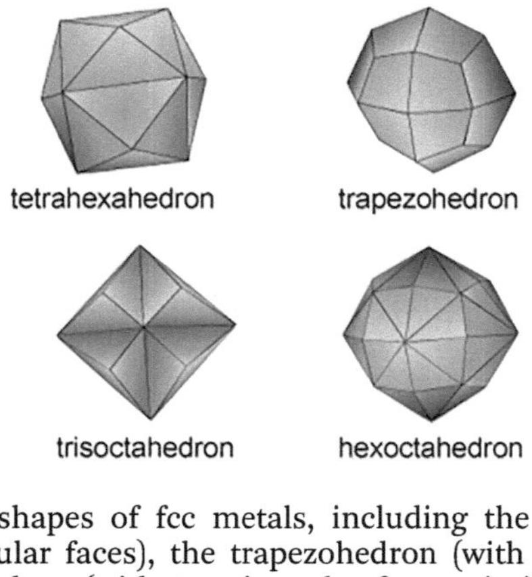

Figure 8.2 Unusual shapes of fcc metals, including the tetrahexahedron (with 24 triangular faces), the trapezohedron (with kite-shaped faces), the trisoctahedron (with 24 triangular faces, similar to an inflated octahedron), and the hexoctahedron (with 48 triangular faces). Each of these shapes has surfaces made up of high-index crystal faces. Reproduced from ref. 12 with permission from John Wiley & Sons, Copyright © 2007 Wiley-VCH Verlag GmbH & Co. KGaA, Weinheim.

MOFs, researchers can engineer materials with unique active sites and pore environments that are optimized for a particular reaction, thus improving selectivity.[14]

These advancements in nanostructured materials highlight their versatility, adaptability, and the economic and environmental benefits they bring to catalysis, emphasizing their promising potential in various industries. By increasing the selectivity and efficiency of catalytic processes, these materials can help minimize the consumption of raw materials and energy, as well as reduce the generation of unwanted by-products. This, in turn, can lead to more sustainable industrial practices and a reduced carbon footprint.

8.1.3 Increased Stability

The stability of catalysts is vital for the efficiency and longevity of catalytic processes in various industries, such as petrochemicals, pharmaceuticals, and environmental remediation. Incorporating nanostructured materials, like MOFs and zeolites, has proven to be a promising approach for enhancing catalyst stability. These porous materials protect catalytic active sites from deactivation while allowing reactants and products to diffuse through their pores. This approach has the potential to increase catalyst lifetimes and reduce the environmental impact of industrial processes.

Zeolites, in particular, have long been recognized as highly effective catalyst supports due to their unique properties.[15] These crystalline aluminosilicates feature a well-defined pore structure and an extraordinary surface area, making them ideal for hosting active catalytic species. Their ability to accommodate a variety of metal ions within their framework, along with the tunability of their pore sizes, allows for precise control over catalytic activity. In industrial applications such as hydrocracking and fluid catalytic cracking, zeolites have demonstrated remarkable stability under harsh conditions, maintaining their structural integrity even at elevated temperatures. This resilience, combined with their ability to promote selective reactions, positions zeolites as a cornerstone in modern catalytic systems.

While zeolites offer robustness and stability, MOFs provide a more customizable platform for catalyst design. MOFs introduce greater flexibility in terms of structure and functionality, thanks to their organic–inorganic hybrid composition. Like zeolites, they have a high surface area and adjustable pore sizes, but MOFs also allow for more sophisticated structural engineering. For example, researchers designed a nickel-based MOF with specific crystal facets to enhance the conversion of CO_2 into CO. By exposing the {100} facets, two closely positioned nickel atoms worked together, facilitating easier CO detachment during the reaction and improving overall catalytic efficiency. This illustrates how the modular nature of MOFs enables precise tuning of active sites for optimized performance in specific reactions.[16]

The complementary strengths of zeolites and MOFs open the door for synergistic hybrid systems. Zeolites provide the durability needed for industrial processes, while MOFs offer the flexibility to fine-tune catalyst performance. Together, these materials can be combined to create advanced catalytic systems that enhance both stability and efficiency, pushing the boundaries of catalytic technology across various industries.

8.1.4 Photocatalysis

Nanomaterials, specifically TiO_2 nanoparticles, have garnered significant attention for their potential use in photocatalytic applications. These applications span a wide range of environmentally beneficial processes, such as water splitting for hydrogen production, pollutant degradation for water purification, and CO_2 reduction to mitigate climate change.

Titanium dioxide's high surface area is a crucial factor in enhancing its photocatalytic performance. The increased surface area provides more active sites for the photocatalytic reactions to take place, boosting the overall efficiency of the process. For example, TiO_2 nanoparticles with a high surface area have demonstrated superior photocatalytic activity in water splitting, leading to an increased rate of hydrogen production.[17]

The unique electronic properties of TiO_2 nanoparticles also contribute to their improved light absorption and charge separation. When TiO_2 absorbs light, it generates electron–hole pairs, which are essential for driving photocatalytic reactions. The unique band structure of these nanoparticles allows for efficient separation and transfer of these charge carriers. As a result, more electron–hole pairs are available to participate in photocatalytic reactions, improving the overall efficiency of the process.[18]

In addition to TiO_2 nanoparticles, several other nanoparticles are widely used in photocatalysis. Zinc oxide (ZnO) exhibits strong photocatalytic activity similar to TiO_2, particularly in pollutant degradation and water splitting. Graphene-based materials enhance charge separation and light absorption, improving overall photocatalytic performance. Cadmium sulfide (CdS) nanoparticles are responsive to visible light, making them efficient for solar-driven applications like hydrogen production. Silver (Ag) nanoparticles, through localized surface plasmon resonance, increase light harvesting, while bismuth oxyhalide (BiO_X) and cobalt oxide (Co_3O_4) nanoparticles excel in utilizing visible light for pollutant degradation and water splitting. Copper oxide (CuO) is also effective in visible-light-driven photocatalysis, particularly for hydrogen generation and CO_2 reduction. These materials, often combined or modified, offer diverse solutions for enhancing photocatalytic efficiency compared to bulk material catalysts.

The development and application of nanocatalysts are still an active area of research. As new nanostructured materials and catalytic systems continue to be discovered and optimized, their impact on various chemical processes is likely to grow, potentially leading to more sustainable and efficient production methods across the industry.

8.1.5 Case Study: Tailored Nanograins Enable Greener Ammonia Production

Ammonia is indispensable for producing the fertilizers and chemicals that sustain food and material production for billions globally. However, conventional industrial scale ammonia synthesis comes at great environmental costs. The predominant century-old Haber–Bosch method accounts for over 2% of energy use worldwide, while generating substantial carbon emissions.

As global ammonia demand continues rising, there is heightened impetus to develop alternative green production routes with improved energy efficiency and sustainability.

Previous attempts at electrochemically extracting ammonia from air have confronted major roadblocks like poor yields, low efficiency, and lack of stability. Molecular nitrogen (N_2) bonds require tremendous energy input to break and activate. Most catalysts fail to adsorb or crack N_2 molecules effectively before competing side reactions dominate. Transition metals demonstrate potential, but their conductivity often limits performance.

Seeking to overcome these hurdles, a team of researchers at Nanyang Technological University, Singapore, tailored an electrocatalyst consisting of ultrafine molybdenum oxide (MoO_2) nanograins anchored throughout nitrogen-doped carbon fiber supports.[19]

Revolutionizing ammonia production, this research pioneers an eco-friendly alternative to the conventional Haber–Bosch process, mitigating negative environmental impacts and embracing the potential for a greener industrial and agricultural future.

Their optimization process focused on manipulating MoO_2 crystal boundaries to generate abundant oxygen vacancies within the structures. These vacancy sites can strongly trap N_2 molecules and provide reactive interfaces to cleave their bonds with less energy, drastically improving reaction kinetics. Meanwhile, the surrounding conductive carbon matrix enhances electron transport to molybdenum centers and prevents aggregation.

According to the researchers, careful heating of the MoO_2/carbon catalyst to 700 °C (MoO_2/C_{700}) during preparation proved optimal for producing tiny, interfaced grains with maximized vacancies (Figure 8.3 depicts the schematic representation of the MoO_2/C_x catalyst obtained through electrospinning). When they tested it for electrolytic ammonia synthesis under alkaline

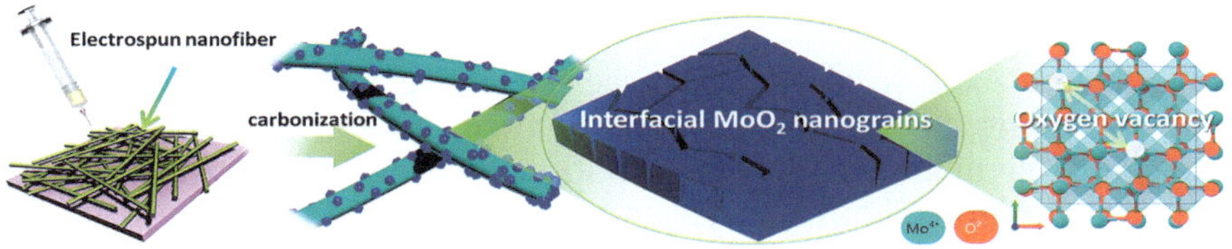

Figure 8.3 Schematic representation of the synthesis of MoO_2/C_x. Adapted from ref. 250 with permission from the National Academy of Sciences.

conditions, this thermally activated material achieved record-breaking rates over four times higher than recently reported metal-based counterparts, converting over a quarter of passed charge into NH_3. The robust catalyst sustained this high-efficiency ammonia production steadily for 60 hours.

Insight into the underlying mechanisms comes from spectroscopic analyses before and after reacting the optimized MoO_2/C_{700} nanostructures with nitrogen.

The data implies that N_2 molecules initially bind strongly with oxygen vacancy sites, facilitating the crucial transfer of electrons from adjacent molybdenum and carbon to weaken and break the nitrogen-nitrogen triple bond. Further incoming protons then attach to the dissociated atomic nitrogen, ultimately forming ammonia products that remain adhered until increasing buildup.

Advanced simulations corroborate this proposed reaction pathway, revealing over two times stronger nitrogen absorption on oxygen vacancy-laden molybdenum oxide than on regular oxide surfaces.

Subsequent energetics also become more favorable for hydrogenation steps, specifically *via* an alternate route that proceeds through a key N_2H_2 intermediate.

When the team compared their MoO_2/C_{700} catalyst under the same electrochemical conditions, the combination of specially designed gaps in the material and the carbon-supported nanostructure showed a significant improvement in ammonia production, especially when measured against pure carbon fibers and standard commercial molybdenum oxide.

While requiring further optimization for industrial implementation, these findings provide proof-of-concept for a more sustainable electrocatalytic production technique to obtain essential ammonia chemicals. The facile and inexpensive nature of the high-performance catalyst synthesis developed here could allow scaling up environmentally benign processes to displace conventional energy-hungry and polluting industrial procedures.

Looking beyond this initial demonstration, the conceptual framework of deliberately engineering oxygen defects in nanostructured composites could aid in designing improved catalysts for other challenging reactions like nitrogen reduction.

8.2 Advanced Packaging and Biodegradable Materials

Nanomaterials are revolutionizing the packaging industry, particularly in food packaging, by enabling the development of lightweight, durable, and biodegradable materials. These innovations not only reduce waste but also enhance the protection of packaged goods, ensuring superior quality and extended shelf life. As a result, nanotechnology-based packaging materials contribute to a more sustainable and environmentally friendly packaging sector.

Additionally, nanotechnology can significantly improve food quality and safety through enhanced vapor and gas barrier properties, antimicrobial features of nanocomposite films, and nanosensor-based smart packaging. However, alongside these benefits, it is crucial to address potential health risks associated with nanoparticles, such as migration into food and consumer exposure, due to their unique properties compared to bulk materials.[20]

To address these concerns, multilayer packaging materials, often designed as sandwich structures, have been developed. These materials consist of three distinct layers: barrier, active, and control layers, which together offer superior or extended functional performance compared to single-layer materials (see Figure 8.4).[21]

The barrier layer, in direct contact with the external environment, acts as a shield against degrading substances, ensuring the retention and protection of active ingredients within the packaging. The active layer, located in the center, contains active ingredients such as antioxidants and antimicrobials that remain stable until needed. These ingredients then diffuse through the control layer at a regulated rate to interact with the packaged item.

Beyond food packaging, the principles of multilayer technology and the integration of nanomaterials also have significant applications in pharmaceutical and cosmetic packaging, where enhanced barrier properties and controlled release mechanisms are critical.

Nanomaterials can enhance barrier properties, provide superior protection against moisture, and deliver antimicrobial properties in pharmaceutical packaging. For instance, nanoclay and silver nanoparticles can be employed to create high barrier and antimicrobial films, which help preserve the efficacy of pharmaceutical products and prolong their shelf life.

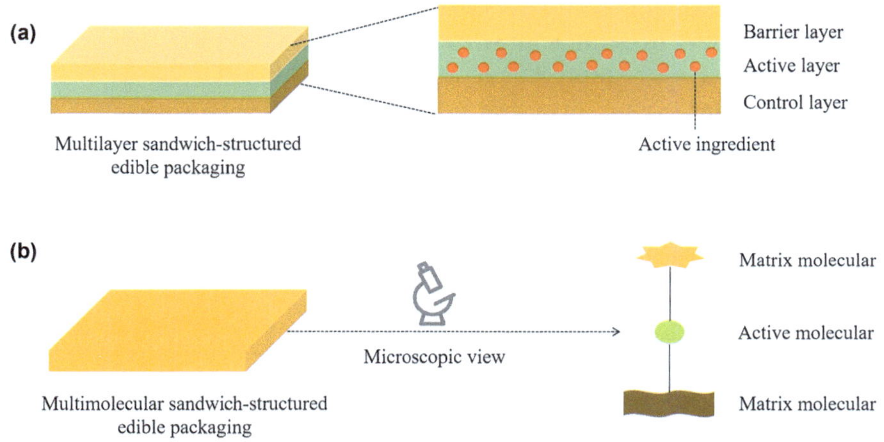

Figure 8.4 Schematic representation of biodegradable packaging materials with sandwich structure. Reproduced from ref. 251, https://doi.org/10.1038/s41538-022-00132-8, under the terms of the CC BY 4.0 license, https://creativecommons.org/licenses/by/4.0/.

8.2.1 Nanocomposites

Nanocomposites, often created by embedding nanoparticles such as clay, carbon nanotubes, or graphene into a polymer matrix, exhibit unique properties like enhanced strength, thermal stability, and improved barrier capabilities. These materials can also be formed by incorporating nanoparticles into other types of matrices, such as metals or ceramics, allowing for tailored enhancements in mechanical, electrical, and thermal properties across a wide range of applications.

For instance, incorporating nanoclay into polymer films significantly improves their barrier properties against oxygen, carbon dioxide, and water vapor, thus preserving the freshness of food products. Moreover, silver nanoparticles in packaging materials provide antimicrobial properties that inhibit bacterial growth and prevent spoilage.

While nanocomposites offer remarkable benefits, the quest for sustainable packaging solutions also drives the development of biodegradable materials. These materials, enhanced with nanomaterials, provide an eco-friendly alternative to traditional plastics by improving the properties of natural polymers like cellulose, chitosan, and starch.

Using nanoclay in plastics results in more effective packaging materials that protect goods from moisture, gases, and other contaminants, thereby reducing waste by extending perishable goods' shelf life and minimizing the need for additional packaging layers. Naturally occurring clay minerals such as kaolin, talc, or montmorillonite are often used in polymer nanocomposites production, significantly enhancing plastics' mechanical properties, chemical resistance, and gas barrier properties.[22]

In applications like polypropylene or polylactic acid packaging films, nanoclays prevent oxygen or flavoring diffusion, consequently prolonging the shelf life of food products. Layered silicates, abundant in nature and producible at low cost, contribute to waste reduction by minimizing spoilage and the need for extra protective layers.

8.2.2 Biodegradable Materials

Bio-based packaging uses natural polymers instead of conventional plastics. Nanomaterials are playing a role in the development of biodegradable packaging materials by enhancing the properties of natural polymers such as cellulose, chitosan, and starch. These solutions offer an environmentally friendly alternative to conventional petroleum-based plastics, which are notorious for their adverse impact on ecosystems and waste volume.[23]

One example is the use of nanocellulose derived from plant sources, like wood fibers, to create transparent, flexible, and biodegradable films. These films have excellent barrier properties against oxygen and water vapor, making them suitable for food packaging applications. They can protect perishable goods while reducing the environmental footprint associated with traditional packaging materials.

Another example is the incorporation of chitosan nanoparticles into packaging materials. Chitosan, derived from chitin found in the exoskeletons of crustaceans and insects, is a biodegradable polymer with antimicrobial properties. By incorporating chitosan nanoparticles into films, the resulting material not only prevents microbial growth and spoilage but also decomposes naturally, contributing to waste reduction and a more sustainable packaging approach.

Starch-based nanocomposites are another promising avenue for biodegradable packaging. By reinforcing starch films with nanoparticles, the resulting material exhibits enhanced mechanical properties and barrier performance. This development allows starch-based packaging to compete with petroleum-based plastics in terms of functionality while maintaining their biodegradable nature.

8.2.3 Active Packaging

Active packaging involves using materials that interact with the packaged product to maintain or enhance its quality. This can include releasing or absorbing substances to extend the shelf life of food products, control moisture, or indicate freshness. Nanoparticles play a crucial role in advancing active packaging systems by providing unique properties. For instance, silver nanoparticles embedded in packaging materials offer antimicrobial properties that inhibit the growth of bacteria, fungi, and other microorganisms causing food spoilage.[24]

Nanotechnology has also enabled the creation of active packaging systems that can release natural antioxidants or antimicrobial agents to further extend food shelf life. For instance, incorporating nanocapsules containing essential oils, such as oregano or thyme oil, into packaging materials can slowly release these oils over time, providing a steady antimicrobial effect to protect the food against spoilage.[25]

8.2.4 Intelligent Packaging

Building on the innovations in active packaging, intelligent packaging systems go a step further by incorporating sensors and indicators to monitor the condition of packaged products in real-time (see Figure 8.5). These systems utilize nanosensors to detect changes in temperature, humidity, or the presence of harmful substances, ensuring the safety and quality of the packaged goods. As a result, the shelf life of perishable items can be extended, reducing food waste and offering consumers fresher products for longer periods.[26]

For example, nanosensors can detect temperature fluctuations in perishable goods like dairy products, meats, and seafood. As technology continues to advance, another innovative approach in packaging is the development of edible packaging materials. These materials are designed to be consumed along with the food product, thereby minimizing waste.

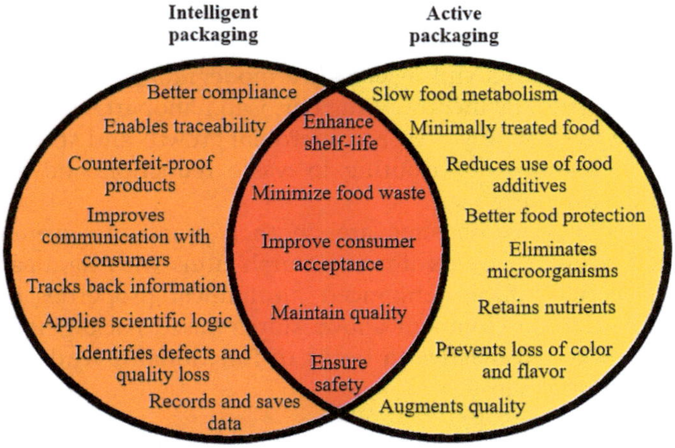

Figure 8.5 Venn diagram identifying similarities and differences between intelligent and active packaging. Reproduced from ref. 252, https://doi.org/10.1002/fbe2.12038, under the terms of the CC BY 4.0 license, https://creativecommons.org/licenses/by/4.0/.

Another example is the integration of humidity sensors into packaging materials for products sensitive to moisture, such as electronics or pharmaceuticals. These nanosensors can indicate if moisture levels have exceeded safe limits, signaling possible damage or reduced product efficacy.

Nanosensors can also identify the presence of harmful substances or pathogens in food packaging. For instance, a nanosensor might change color when it detects food spoilage indicators like ammonia or biogenic amines. This visual alert enables consumers to make informed decisions about the safety and quality of the food product before consumption.

In addition to detecting spoilage, intelligent packaging systems can monitor the presence of allergens, pesticides, or other contaminants. For example, nanosensors embedded in packaging could detect traces of peanuts or gluten, providing valuable information to individuals with allergies or specific dietary requirements.

Lastly, nanosensors in intelligent packaging systems can be employed to detect and monitor gases that indicate product freshness, such as oxygen or carbon dioxide levels. This is particularly useful for modified atmosphere packaging (MAP), where the gas composition inside the package is controlled to prolong the shelf life of perishable goods like fruits, vegetables, and meats.

8.2.5 Edible Packaging

Edible packaging materials can be consumed along with the food product, further minimizing waste. These innovative packaging solutions are designed to be safe for human consumption while protecting and preserving the contents within.

Researchers have developed edible films made from proteins such as gelatin, whey protein, or soy protein. These protein-based films can be fortified with nano-encapsulated nutrients like vitamins, minerals, or antioxidants, effectively turning the packaging into an additional source of nutrition. Moreover, the films can be enhanced with flavors or antimicrobial agents to improve taste and safety.[27]

Another example involves polysaccharides, such as chitosan, alginate, or pectin, which can be used to create biodegradable and edible packaging materials. These materials can be tailored to provide specific properties, such as moisture resistance, by altering their composition. Furthermore, they can be combined with lipids, like beeswax or vegetable oils, to improve their water vapor barrier properties and overall performance.

Additionally, lipid-based films made from natural sources like milk proteins or plant-based oils can provide a suitable barrier against oxygen and other gases, ensuring the food product's freshness. These films can also be fortified with nano-encapsulated nutrients or antimicrobial agents to enhance their functional properties.

8.2.6 Antioxidant Packaging

Oxidation is a chemical process that occurs when a substance loses electrons, typically to oxygen molecules. In the context of food quality, oxidation can lead to the degradation of essential nutrients, the development of off-flavors, and changes in the texture, color, and overall appearance of food products. This degradation is particularly pronounced in products containing fats and oils, as they are susceptible to oxidative reactions that produce rancidity and reduce shelf life.

By incorporating antioxidants into packaging materials, it is possible to actively protect food items from the harmful effects of oxidation and maintain their quality for extended periods.

One example of this approach is the use of nano-encapsulated vitamin E or other natural antioxidants, which are integrated into packaging films. These antioxidants are typically encapsulated within nanoscale structures, such as liposomes, biopolymer-based nanospheres, or nanoparticles, which are designed to control the release of the antioxidant over time. The encapsulation process not only improves the stability of the antioxidants but also allows for a more precise and targeted delivery.[28]

As the packaging film comes into contact with the food product, the nano-encapsulated antioxidants are gradually released onto the food surface. This controlled release helps to neutralize reactive oxygen species, preventing the oxidative reactions that cause food degradation. By slowing down the oxidative process, the antioxidants effectively extend the shelf life of the food product and maintain its nutritional value, taste, and overall quality.

In addition to nano-encapsulated vitamin E, other natural antioxidants such as carotenoids, flavonoids, and polyphenols can also be incorporated into nanotechnology-based packaging systems. These antioxidants can be sourced from various plant materials, providing an eco-friendly and sustainable alternative to synthetic antioxidants.

8.2.7 Antimicrobial Packaging

Antimicrobial packaging materials are an innovative approach to addressing the issue of food spoilage and contamination caused by microorganisms like bacteria, fungi, and mold. These microorganisms are responsible for degrading the quality, taste, and nutritional value of food products, leading to significant food waste and potential health risks for consumers. Incorporating antimicrobial agents into packaging materials can enhance the safety and shelf life of food products, while also supporting sustainability efforts and reducing food waste.

One of the key techniques used in antimicrobial packaging is the incorporation of silver or zinc nanoparticles. Both silver and zinc have proven antimicrobial properties that effectively inhibit the growth of various microorganisms. Nanoparticles, due to their small size and high surface area, can efficiently release antimicrobial ions and interact with microbial cells, disrupting their cellular processes and leading to their inactivation.[29] This helps maintain the freshness and quality of food products while reducing the risk of contamination.

Another antimicrobial agent used in packaging materials is chitosan, a natural biopolymer derived from chitin, which is found in the exoskeletons of crustaceans like shrimp and crabs. Chitosan exhibits strong antimicrobial activity against a broad spectrum of microorganisms, including bacteria, fungi, and mold. When incorporated into packaging materials, chitosan can form a protective barrier that prevents microbial contamination and enhances the shelf life of food products.

Additionally, other bioactive compounds can be used in antimicrobial packaging materials to further improve their effectiveness. These compounds, such as essential oils, plant extracts, and bacteriocins, possess inherent antimicrobial properties that can target specific microorganisms or work synergistically with other antimicrobial agents. By incorporating these bioactive nanoencapsulated compounds into packaging materials, food products can be protected from spoilage, and their safety can be enhanced.[30]

8.2.8 Self-healing Materials

Self-healing packaging materials are primarily composed of polymers that have been infused with microcapsules or nanoparticles filled with a healing agent. The primary goal of these materials is to maintain the integrity of the packaging and ensure the protection of the packaged goods, even when subjected to physical damage.

Self-healing packaging materials work on the principle of autonomous healing, which mimics natural biological systems that possess the ability to self-repair when damaged. The nanocapsules embedded within the polymers remain dormant until the material experiences damage, such as cracks, tears, or punctures. Upon damage, the embedded healing agent is released and interacts with the polymer matrix, filling the gaps and restoring the material's original properties.[31]

Some common healing agents used in self-healing packaging materials include epoxy resins, polyurethane, and silicone-based materials. These agents are chosen for their ability to form strong bonds with the polymer matrix and effectively repair the damaged areas, ensuring the longevity of the packaging.

Apart from enhanced product protection, self-healing packaging materials can extend the life of packaging, leading to less waste and reduced environmental impact.

8.2.9 Case Study: Cellulose Nanocrystals as an Alternative to Petroleum-based Plastics

Researchers from The Institute of Scientific and Industrial Research at Osaka University and collaborating partners have made strides in developing eco-friendly alternatives to conventional, single-use plastics. Their study introduces highly adaptable hydrogels and moldings that could potentially replace traditional plastics.[32]

Addressing the worldwide plastic waste issue demands a multifaceted approach and the study's authors emphasize the need for "inexpensive, high-performance, plastic-like alternatives that don't persist in the environment." Despite being an active research area, no proposed alternatives have yet fully met society's needs.

The team focused on cellulose nanofibers, which provide plants with rigidity and lightness. These nanofibers make wood, pound-for-pound, stronger than steel by some measures. Their customizable hierarchical structure has garnered interest in synthetic tissue and bioengineering applications.

While several techniques exist for molding nanofibers into controlled orientations or anisotropy, a simple method for molding cellulose nanofibers across multiple spatial axes has remained elusive. To address this gap, Kasuga and his colleagues employed electrophoretic deposition to fabricate anisotropic cellulose-nanofiber-based materials.

The study yielded several impressive outcomes. Researchers could orient cellulose nanofibers horizontally, randomly, and vertically by merely adjusting the applied voltage. They also created a multilayer hydrogel with alternating nanofiber orientations, mimicking biological tissue (see Figure 8.6). Furthermore, they easily crafted intricate structures like microneedles and mouthpiece molds, which demonstrated a smooth surface and minimal hydrogel cracking due to the uniform nanofiber orientation.

This technique extends beyond cellulose nanofibers, as the researchers also successfully applied it to sodium alginate and nanoclay. This enables

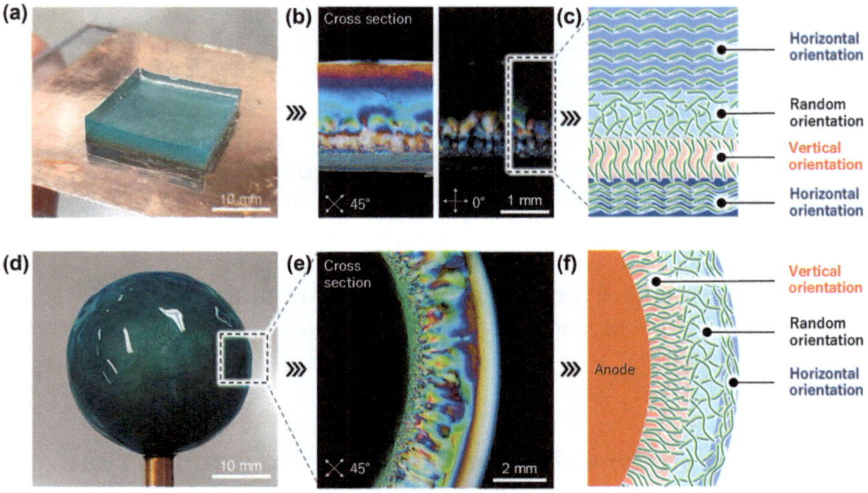

Figure 8.6 (a) Multilayered CNF hydrogel. (b) Cross-section of the multilayered CNF hydrogel observed through a cross-polarizer. (c) By controlling the applied voltage, CNFs were deposited in the order horizontal, vertical, random, and horizontal with respect to the anode. (d) Cartilage-like CNF hydrogel formed around a spherical electrode. (e) Cross-section observed through a cross-polarizer. (f) Corresponding schematic diagram. Adapted from ref. 253, https://doi.org/10.1021/acsnano.2c06392, under the terms of the CC BY 4.0 license, https://creativecommons.org/licenses/by/4.0/.

the easy preparation of multicomponent materials with controlled nanoscale orientations. The study's immediate application involves the simple manufacturing of complex, hierarchical hydrogels and moldings across various spatial scales. These eco-friendly materials have potential uses in healthcare, biotech, and other sectors, ultimately reducing reliance on petroleum-based plastics.

The development of nanomaterial composites that mimic biological structures could play a vital role in addressing the global plastic waste problem. By harnessing the simplicity and flexibility of electrophoretic and electrochemical deposition-induced orientational control, researchers can continue to optimize and refine these materials, potentially offering a sustainable alternative to petroleum-based plastics.

8.3 Lightweight Materials with Enhanced Durability

This section examines the significant impact of nanomaterials, particularly in the automotive and aerospace industries. Advanced materials, like carbon nanotubes, are already proving effective at enhancing strength-to-weight ratios, which is essential for reducing material use, fuel consumption, and emissions. Beyond these sectors, lightweight and durable materials also play a crucial role in other industries.

In construction, they reduce raw material use and improve energy efficiency in buildings, while in renewable energy, such materials enhance the performance and longevity of wind turbines and solar panels. The packaging industry benefits from lighter materials that decrease transportation emissions and promote better recycling, while sports equipment manufacturers rely on these materials for stronger, more durable products that lower environmental impact through longer life cycles. The marine industry, too, utilizes lightweight, corrosion-resistant materials to improve fuel efficiency and reduce maintenance, contributing to more sustainable operations.

These advances demonstrate how the use of nanomaterials can lead to significant sustainability improvements, reducing resource consumption, emissions, and waste generation.

8.3.1 Impact of Nanomaterials in the Automotive and Aerospace Industries

The car and aerospace manufacturing industries are sectors where nanomaterials could have a significant impact. Research shows that nanocomposite automotive components—polymeric, elastomeric and electro-conductive nanocomposites—could achieve up to 25% weight savings compared to conventional plastics and up to 80% over steel.[33]

This is highly relevant if you consider that the body shell of a car on average makes up about 40% of its total weight. A 10% reduction in vehicle weight can result in a 2.7–3.6% reduction in CO_2 emissions, emphasizing the benefits of reducing car weight. The use of lightweight nanocomposites also enables optimized car body designs, potentially improving aerodynamic flow by 10%, reducing rolling resistance of tires by 20%, and increasing powertrain efficiency by 7.5%.[34]

Lighter structural materials play a significant role in the aerospace industry as well.[35] Lighter aircraft offer numerous benefits that contribute to waste reduction and increased sustainability, apart from significant fuel savings. A generally accepted finding is that a 1% reduction in aircraft weight can result in a 1.5% improvement in fuel efficiency.[36] To illustrate the impact of this efficiency improvement, considering the total annual global aviation fuel consumption of 375 billion liters (2024 estimate),[37] every 1% weight reduction in all aircraft results in annual fuel savings of about 5.6 billion liters. This savings directly translates to lower CO_2 emissions.

In addition to fuel savings, lighter aircraft offer other advantages that help reduce or even prevent waste. Firstly, lighter aircraft designs require fewer raw materials for their construction, which helps conserve natural resources and reduce waste generated during the manufacturing process. Furthermore, lighter aircraft often incorporate advanced materials with enhanced durability and fatigue resistance, resulting in a longer service life. This leads to a reduced frequency of maintenance, repairs, and eventual replacement, ultimately reducing waste generation throughout the aircraft's lifecycle.

8.3.2 Innovative Advances in Durable, Lightweight Materials

Several noteworthy examples of pioneering research focus on developing lighter, more durable materials.

Researchers at the City University of Hong Kong have developed a low-cost method to transform 3D printable polymers into lightweight, ultra-tough, biocompatible hybrid carbon microlattices using nano- and microstructuring combined with 3D printing (see Figure 8.7). These materials are 100 times stronger and more ductile than the original polymers. This innovation enables the creation of sophisticated 3D parts with tailored mechanical properties for various applications, including coronary stents, bio-implants, micro-robots, and energy harvesting and storage devices.[38]

The manufacturing sector, which is increasingly focused on its environmental impact and seeking sustainable and long-lasting materials, also stands to benefit from nanomaterial advancements. For instance, galvanized steel is widely used in the automotive industry due to its excellent mechanical properties and corrosion resistance, but its weight has led to increased use of polymer-metal composites as alternatives. Traditional techniques for bonding polymers to galvanized steel are unsuitable for mass production due to harsh chemicals or specialized equipment requirements.

Developing lightweight materials that are both strong and highly ductile has been a long-desired goal in the field of structural materials, although these properties were generally considered to be mutually exclusive for a long time.

MIT chemical engineers have developed a new material using a unique polymerization process.[39] This material is stronger than steel yet as light as

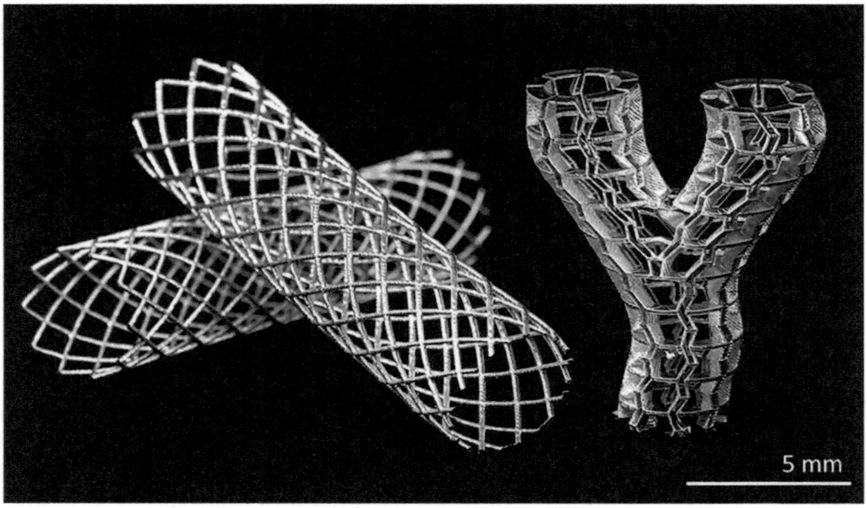

Figure 8.7 Examples of functionally coated core–shell coronary stents with partially carbonized core. Adapted from ref. 254 with permission from Elsevier, Copyright 2022.

plastic and can be easily produced in large quantities. This two-dimensional polymer self-assembles into sheets, which could be used as a lightweight, durable coating for various applications such as car parts, cell phones, or even building materials for bridges and other structures. The elastic modulus of the material is 4–6 times greater than bulletproof glass, and its yield strength is twice that of steel, despite having only one-sixth the density. Additionally, the material is impermeable to gases, making it ideal for protective coatings.

In another example of steel modification, researchers from The Institute of Industrial Science at The University of Tokyo developed a cost-effective and straightforward method to bond polymers to galvanized steel, creating a lightweight, durable material suitable for industrial-scale production.[40] The team developed a method involving pre-treating the steel with an acid wash and dipping it in hot water, creating nanoscale needle structures on the surface for the polymer to adhere to. This process, called injection-molded direct joining, results in strong mechanical linkages. By optimizing the hot water temperature and treatment time, the team significantly increased the tensile-shear strength compared to untreated metal. The method, which avoids harsh chemicals and complicated procedures, can be adapted for various hybrid joining applications and is suited for industrial-scale production.

8.3.3 Case Study: 10× Stronger than Steel with only 5 Percent of the Density

At MIT, a group of scientists has developed an exceptionally strong, lightweight material by compressing and fusing graphene flakes. The sponge-like structure has a density of only 5 percent yet boasts a strength ten times greater than steel.[41] Until now, transforming the 2D strength of graphene into practical 3D materials has been challenging.

The key to these new 3D structures lies in their unique geometric configuration rather than the material itself. This suggests that similar strong, lightweight materials can be created using various materials with analogous geometric features. Researchers examined the material's behavior at the atomic level, developing a mathematical framework that closely aligns with experimental observations.

They compressed graphene flakes using heat and pressure, resulting in a stable, strong structure reminiscent of certain corals and microscopic organisms called diatoms. These shapes exhibited remarkable strength. Computational simulations showed that one sample had 5 percent the density of steel but ten times the strength.

Using a high-resolution, multi-material 3D printer, the team produced these new configurations, which were then tested for tensile and compressive properties. Numerous potential applications requiring a combination of extreme strength and light weight may be feasible. The geometry discovered with graphene could be applied to other materials, such as polymers or metals, to achieve similar strength benefits while also offering cost, processing method, or other material property advantages.

There are numerous potential applications for this material in industries such as construction, transportation, and filtration, where a combination of extreme strength and light weight is required. For example, the same geometry could be applied to large-scale structural materials like concrete for a bridge, offering comparable strength with a fraction of the weight and providing excellent insulation. The porous geometry might also find use in filtration systems for water or chemical processing. The mathematical models developed could expedite the development of a wide range of applications for these novel materials.

The dominant factor for this new material is the geometry, which can be applied to numerous materials. Under heat and pressure, graphene naturally forms intricate geometric shapes resembling hole-filled balls called gyroids. These structures are so complex that conventional manufacturing methods would likely be unable to produce them. The team tested 3D-printed models of the structure, scaled up thousands of times from their natural size.

For actual synthesis, one possibility is using polymer or metal particles as templates, coating them with graphene through chemical vapor deposition before heat and pressure treatments, and then chemically or physically removing the polymer or metal phases to leave the 3D graphene in the gyroid form. The computational model provided in the current study serves as a guideline for evaluating the mechanical quality of the synthesis output.

8.4 Incorporation of Nanomaterials for Enhanced Functional Properties

The incorporation of nanomaterials into traditional materials can lead to significant improvements in their properties, resulting in more sustainable and efficient products that contribute to waste prevention and reduction. Nanomaterials, such as nanoparticles, nanofibers, and nanosheets, can be introduced into polymers, metals, ceramics, and other materials to create nanocomposites with superior characteristics. The unique properties of nanomaterials, including their high surface-to-volume ratio, mechanical strength, and electrical conductivity, can impart new functionalities and improved performance to conventional materials.

The incorporation of nanomaterials can also lead to the development of multifunctional materials with a combination of desirable properties. For instance, introducing nanoscale fillers into a polymer matrix can simultaneously enhance its mechanical strength, thermal stability, and electrical conductivity, allowing for the creation of versatile materials with a wide range of applications.

8.4.1 Enhanced Functional Properties of Nanomaterials

Nanomaterials stand out for their superior functional properties, largely due to their high surface-to-volume ratio. This unique characteristic boosts interactions within composite matrices, leading to stronger and more durable materials. Their ability to enhance mechanical strength, electrical conductivity, and thermal stability makes them ideal for applications in industries such as aerospace, automotive, and flexible electronics. Additionally, nanomaterials contribute to sustainability by improving product lifespans and reducing waste, positioning them as key players in the development of next-generation multifunctional materials.

8.4.1.1 High Surface-to-volume Ratio

Nanomaterials possess a significantly higher surface-to-volume ratio compared to their bulk counterparts. This increased surface area enhances the interaction between the nanomaterials and the matrix, leading to improved mechanical properties and functionality. For instance, nanoparticles can provide a larger interface for stress transfer in composites, resulting in enhanced strength and durability.

8.4.1.2 Mechanical Strength

Carbon nanotubes and graphene are known for their remarkable mechanical properties. When incorporated into polymers, metals, or ceramics, they can significantly enhance the material's tensile strength, stiffness, and toughness. This improvement in mechanical properties is crucial for applications requiring robust and durable materials, such as in aerospace and automotive industries.

8.4.1.3 Electrical Conductivity

The high electrical conductivity of certain nanomaterials, like graphene and metallic nanowires, makes them ideal for applications in flexible electronics and sensors. Incorporating these nanomaterials into polymers can create conductive composites suitable for use in wearable electronics, thereby extending the lifespan and reducing electronic waste.

8.4.1.4 Thermal Stability

Nanomaterials can also improve the thermal stability of composites. For example, adding nanoscale fillers to polymers can enhance their heat resistance, making them suitable for high-temperature applications. This is particularly important in industries like aerospace, where materials must withstand extreme thermal conditions.

8.4.1.5 Multifunctionality

The unique combination of properties offered by nanomaterials enables the development of multifunctional materials. For instance, a polymer matrix infused with nanoscale fillers can simultaneously exhibit improved mechanical strength, thermal stability, and electrical conductivity. Such multifunctional materials find applications across various fields, including electronics, automotive, and aerospace.

8.4.1.6 Sustainability and Waste Reduction

The enhanced durability and performance of nanocomposites lead to longer product lifespans and reduced material consumption. This not only improves sustainability but also contributes to waste reduction by decreasing the frequency of material replacement. Additionally, nanocomposites can be designed to be recyclable, further promoting environmental sustainability.

By leveraging these benefits, the incorporation of nanomaterials into traditional materials presents a promising pathway to developing advanced composites that meet the evolving demands of various industries while promoting sustainability and efficiency.

8.4.2 Applications of Plastic Nanocomposites

The research field of plastic additives has developed rapidly through the use of nanomaterials. In a nanocomposite material, the nanoscale additives can be in the form of platelets, fibers, or particles. Above all, they serve to improve tensile strength, heat resistance, flame retardancy, optical and electrical properties, and the barrier properties of the plastic.

8.4.2.1 Electrical Conductivity Enhancements

Incorporating metal nanowires into polymers can enhance their electrical conductivity, making them suitable for applications in flexible electronics and sensors. Researchers have demonstrated this by combining polyurethane nanofibers with silver nanowires to produce a stretchable and conductive material that is suitable for use as wearable electronics. These advanced materials can help reduce electronic waste by improving the durability and longevity of electronic devices.[42]

Additionally, studies have shown that the integration of silver nanowires into different polymer matrices, such as polydimethylsiloxane (PDMS), can create conductive networks that are both flexible and robust. These materials can be utilized in a variety of applications, including wearable electronics, soft robotics, and flexible sensors. The high aspect ratio of silver nanowires facilitates the formation of percolative networks, which are essential for maintaining conductivity under mechanical stress.[43]

The use of silver nanowires also extends to applications requiring electromagnetic interference (EMI) shielding. For instance, flexible and robust bacterial cellulose/silver nanowire papers have been developed that demonstrate high EMI shielding effectiveness. This highlights the potential of silver nanowire-based composites in advanced electronic applications that require both flexibility and high conductivity.[44]

By enhancing the electrical conductivity of polymers through the incorporation of silver nanowires, these materials not only improve the performance and durability of electronic devices but also contribute to the reduction of electronic waste by extending the lifespan of these devices.

8.4.2.2 Mechanical Property Improvements

Carbon nanotubes are well-known for their ability to impart electrical conductivity to thermosetting and thermoplastic polymers and to enhance the mechanical properties of thermoplastic composites. This has led to the development of multifunctional thermoplastic polymer composites that combine electrical conductivity with robust mechanical properties that have broad applications in various engineering fields, such as static charge dissipation, electrostatic painting processes, and electromagnetic shielding. These enhanced properties can contribute to waste prevention by increasing the lifetime of products and reducing material consumption.

Multi-walled carbon nanotubes (MWCNTs) have been shown to enhance the mechanical performance of polymer composites through several mechanisms. Their high aspect ratio and surface area provide efficient load transfer between the strong, stiff nanotubes and the polymer matrix. Well-dispersed MWCNTs act as nanofillers that carry applied stresses, restricting deformation of the more compliant polymer. The MWCNT–polymer interfacial interactions, such as van der Waals forces and chemical bonding, further promote effective stress transfer and reinforce the composite.[45]

Experimental results indicate significant improvements in mechanical properties with the addition of CNTs. For example, adding just 1 wt% MWCNTs to epoxy can enhance the tensile strength by 36% and modulus by 28%. These reinforcing effects are related to the high aspect ratio and surface area of CNTs, which provide an efficient load transfer mechanism between the stiff, strong nanotubes and the polymer chains.[46]

These studies illustrate the potential of CNTs to significantly improve the mechanical properties of thermoplastic composites, making them suitable for a wide range of applications in engineering fields where enhanced durability and performance are critical.

8.4.2.3 Thermal Stability Enhancements

Nanomaterials improve composite thermal stability by enhancing heat resistance in polymers, making them suitable for high-temperature uses like aerospace. Incorporating boron nitride (BN) nanomaterials significantly boosts thermal conductivity and stability in polymer composites, crucial for extreme thermal conditions. Research confirmed that polymer/BN composites offer substantial improvements in thermal performance, especially for aerospace components exposed to high temperatures.[47]

Additionally, integrating polyimide-based nanocomposites with various nanofillers has shown remarkable enhancements in thermal stability and mechanical properties. Polyimide composites are widely used in aerospace and electronics due to their ability to maintain performance at high temperatures. The incorporation of nanofillers like carbon nanotubes and graphene further improves their thermal stability and expands their application range.[48]

Moreover, research on Pd–Au–Si thin films has revealed that adding multi-walled carbon nanotubes and carbon black can increase the thermal stability and conductivity of these composites, making them suitable for high-temperature applications such as thermal batteries. This study showed that these composites could maintain structural integrity and performance at elevated temperatures, highlighting their potential for use in aerospace and defense applications.[49]

These advancements illustrate how the incorporation of nanomaterials into composites enhances their thermal stability, making them ideal for high-temperature applications in demanding industries.

8.4.2.4 Optical and Barrier Property Enhancements

Graphene, along with other 2D materials like hexagonal boron-nitride (h-BN), numerous transition metal dichalcogenides, such as molybdenum disulfide, and MXenes, have become sought-after fillers for nanocomposite materials due to their combined distinctive optical, electromagnetic, and conductive properties contributing to flexibility and mechanical resistance. These enhanced properties can contribute to waste prevention by increasing product lifetimes and reducing material consumption.

Graphene and h-BN are known for their exceptional barrier properties. Graphene's impermeability to gases and liquids, combined with its high mechanical strength, makes it an excellent candidate for protective coatings. Similarly, h-BN's high thermal stability and electrical insulation properties complement graphene, providing a synergistic effect when used together in nanocomposites.[50]

MXenes have shown promise in enhancing the barrier properties of nanocomposites due to their layered structure and high aspect ratio. MXenes can improve the gas and moisture barrier performance of polymer films, which is crucial for packaging applications.[51]

By leveraging these properties, the incorporation of these 2D materials into nanocomposites not only enhances their optical and barrier properties but also extends the lifespan and performance of products, contributing to sustainability and waste reduction.

8.4.2.5 Flame Retardancy Improvements

Nanomaterials can serve as effective flame retardants when incorporated into polymers. This enhancement is particularly beneficial in applications where fire resistance is critical, such as in building materials and electronics.

Researchers have modified carbon nanotubes with tannic acid and then wrapped them on the surface of ammonium polyphosphate to create intumescent flame retardants. This structure significantly reduces the peak heat release rate and total heat release, enhancing the flame-retardant properties without compromising mechanical properties.[52]

2D nanomaterials like MXenes have been demonstrated to enhance the thermal stability and flame resistance of polymers by forming protective barriers that slow down heat transfer and combustion processes.[53]

By integrating these nanomaterials, the flame retardancy of polymers is significantly enhanced, making them safer for use in critical applications while maintaining their structural integrity.

8.4.2.6 Durability and Weight Reduction

Carbon nanotubes and graphene can be embedded into polymer matrices as nanoscale reinforcements to create lightweight and strong composites. These make them suitable for structural applications in the automotive, aerospace, and construction industries. The enhanced durability and reduced weight of these materials can lead to improved fuel efficiency and longer service life, further contributing to waste reduction.

8.4.2.6.1 Carbon Nanotubes in Polymers. When incorporated into polymer matrices, CNTs significantly enhance the mechanical strength, thermal stability, and durability of the composites (see Figure 8.8). This makes them ideal for use in high-performance structural applications in the automotive and aerospace industries.

For instance, CNT-reinforced polymer nanocomposites have been shown to improve the structural integrity of automotive body panels, leading to lighter and more fuel-efficient vehicles.[54]

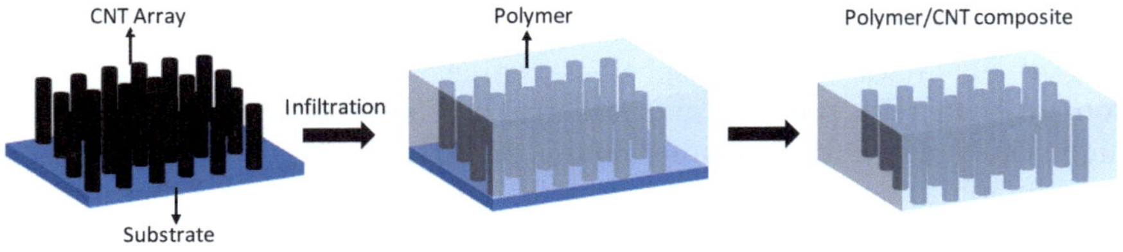

Figure 8.8 Schematic illustration of the fabrication of vertically aligned carbon nanotubes/polymer composites. Adapted from ref. 55, https://doi.org/10.3390/polym10050542, under the terms of the CC BY 4.0 license, https://creativecommons.org/licenses/by/4.0/.

8.4.2.6.2 Graphene in Polymers. Graphene, with its high surface area and extraordinary mechanical properties, is another effective reinforcement for polymer composites. Graphene nanoplatelets can be dispersed within polymers to produce composites with enhanced strength, flexibility, and thermal conductivity.

These graphene-based composites are particularly valuable in the aerospace industry, where reducing weight without compromising strength is critical. Studies have demonstrated that incorporating graphene into polymer matrices can lead to substantial weight reductions while maintaining or even improving the mechanical properties of the material.[55]

The integration of graphene into polymers improves the mechanical properties while also enhancing the thermal and electrical conductivity of the composites. This multi-functional enhancement allows for the development of advanced materials that are lighter and stronger, leading to improved fuel efficiency and reduced emissions in automotive applications. In aerospace, the use of these nanocomposites contributes to longer-lasting, more reliable components that can withstand harsh environmental conditions.

8.4.2.7 Challenges in Processing 2D Materials

Transferring the remarkable properties of 2D materials to polymeric matrices using traditional processing methods, such as directly mixing nanofillers with polymers in the melt, remains a challenge. Two primary limitations exist:

1. Producing 2D materials with controlled layer numbers, lateral dimensions, and surface functionalities.
2. Achieving uniform dispersions with minimal agglomeration while also managing the final morphology, including relative orientations, distribution, and content within the polymer matrix.

Efforts to optimize the final nanocomposites using 2D materials for specific applications, which require multiple properties, is still an ongoing active research area.[56]

To address these issues, researchers are exploring surface functionalization and chemical modifications of 2D materials. Functionalizing the surface of 2D materials with specific chemical groups can improve their compatibility with polymer matrices and facilitate better dispersion. Techniques such as grafting polymer chains onto the surface of 2D materials or introducing functional groups that promote strong interfacial bonding are being developed to enhance the integration of 2D materials into polymers.

Advanced processing techniques, such as solution processing, *in situ* polymerization, and electrospinning, are being investigated to improve the dispersion and alignment of 2D materials within polymer matrices. These methods can help achieve better control over the morphology and

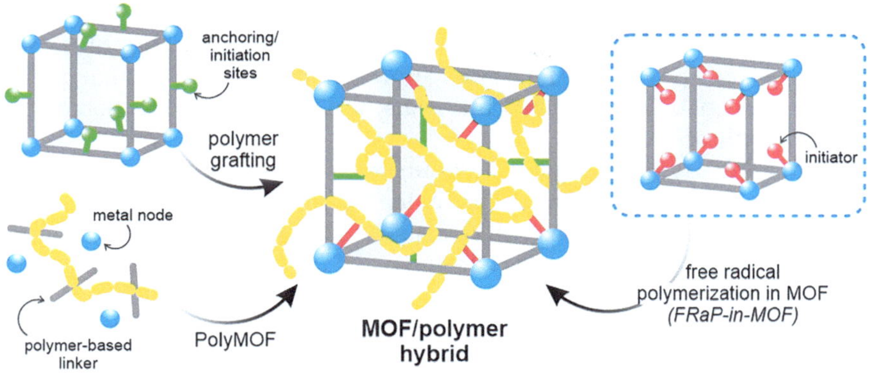

Figure 8.9 Synthetic strategies for the preparation of MOF/polymer hybrids. Adapted from ref. 57, https://doi.org/10.1039/D2MH01202B, under the terms of the CC BY 3.0 license, https://creativecommons.org/licenses/by/3.0/.

distribution of nanomaterials, leading to composites with superior properties. *In situ* free radical polymerization within metal–organic frameworks has been used to create highly uniform MOF/polymer hybrids (see Figure 8.9).[57] These hybrids enhance the mechanical strength and thermal stability of the composites due to improved dispersion and stronger adhesion between the polymer and MOF components. The polymerization process ensures better integration of the materials, leading to more robust composites with enhanced durability under mechanical stress and elevated temperatures.

One significant challenge to commercialization is the scalability and cost-effectiveness of producing high-quality 2D material-based nanocomposites. Developing cost-efficient methods for large-scale production while maintaining the quality and consistency of the 2D materials is crucial for their commercial viability. Research is ongoing to develop scalable manufacturing processes that can produce 2D materials with consistent quality and at a lower cost.

8.4.3 Applications of Ceramic Nanocomposites

Ceramics are a class of materials renowned for their corrosion resistance, chemical inertness, high strength, and thermal stability, making them ideal for applications in harsh environments or those requiring high-temperature exposure.

Despite their exceptional properties and well-established synthesis techniques, monolithic ceramic materials have not become the primary choice for bulk structural components due to their limited thermal transport, electrical properties, and brittleness. To address these challenges, engineers have developed ceramic matrix composites (CMCs) by incorporating nanoscale reinforcements such as nanorods, nanotubes, and particulates into bulk glass-ceramics and polycrystalline ceramic matrices.

The incorporation of nanomaterials as a reinforcing agent presents an elegant solution to these issues, as it enables the fine-tuning of ceramic structures at the nanoscale. This approach significantly enhances the functional capabilities of CMC materials, unlocking their full potential in various applications.[58]

Compared to their monolithic counterparts, these reinforced ceramic materials exhibit significant improvements in thermal transport and electrical conduction properties, while also limiting crack propagation, which increases their overall strength. The enhanced characteristics of CMCs make them attractive materials for a wide range of structural applications, bridging the gap between traditional ceramic materials and the evolving needs of various industries.

CMCs are versatile materials employed across commercial, laboratory, and industrial settings, with applications ranging from glass-ceramic-based LEDs to commercial cookware, high-temperature lab instruments, industrial catalytic reactors, and engine turbine blades. The use of CMCs can contribute to waste reduction by increasing the lifespan of components and reducing the frequency of material replacement.

8.4.4 Applications of Metal Nanocomposites

Nanocomposites in metals offer a range of enhanced properties that significantly improve their performance and sustainability. By integrating nanomaterials into metal matrices, it is possible to develop advanced materials that exhibit superior wear and corrosion resistance, enhanced fatigue resistance, improved thermal conductivity, and other beneficial properties. These improvements lead to longer service life, reduced maintenance, and lower material consumption, contributing to waste reduction and environmental sustainability. The following subsections explore the various applications and advantages of incorporating nanomaterials into metals.

8.4.4.1 Improved Wear and Corrosion Resistance

The incorporation of nanomaterials, such as nanoparticles or nanocoatings, significantly enhances the wear and corrosion resistance of metal components. This improvement leads to longer service life and reduced maintenance,[59] thereby contributing to waste reduction by decreasing the frequency of component replacements. The following examples illustrate the effectiveness of nanocomposites in improving wear and corrosion resistance in various applications:

- Bearings: The performance of bearings can be significantly improved by incorporating nanomaterials into the bearing materials or applying nanocoatings on the bearing surfaces. For instance, researchers demonstrated that implanting diamond nanoparticles on the surface of low-carbon steel reduces its wear by 50% compared to untreated samples.[60]

- Lubricants: Adding nanoparticles to lubricants enhances their performance, particularly in reducing friction and wear. This improvement results in smoother operations, longer life of cutting tools, and reduced fuel consumption. Nanoparticles fill the cavities of the molecular surface, forming an insulating nanofilm which reduces friction-induced heat emissions and prevents gradual deformation of material at solid surfaces.[61]
- Anti-corrosion coatings: Various research efforts have demonstrated the effectiveness of different types of nanocoatings in preventing corrosion. These ultra-thin layers of protective material applied on a surface are more effective than conventional coatings due to their lack of flaws and ability to create a barrier against corrosive substances. Self-healing nanocoatings offer superior corrosion protection and can repair themselves, which is achieved through the use of tiny, mechanized containers that release corrosion inhibitors when exposed to a corrosive solution.[62]

8.4.4.2 Enhanced Fatigue Resistance

Metal matrix nanocomposites (MMNCs) reinforced with nanomaterials, such as carbon nanotubes, exhibit improved fatigue resistance compared to conventional metal alloys. This improvement results in extended lifetimes for components used in industries that involve cyclic loading, such as automotive and aerospace, reducing waste generation from component failure and replacement.

For instance, multi-walled CNTs have been shown to enhance the fatigue strength of aluminum alloy 2024 by increasing the number of fatigue cycles. This illustrates the effectiveness of reinforcing MMNCs with nanomaterials in enhancing the fatigue resistance of metal components.[63]

8.4.4.3 Improved Thermal Conductivity

The incorporation of nanomaterials with high thermal conductivity, such as CNTs and in particular graphene, into metal matrices can result in materials with enhanced thermal conductivity.[64] These materials can be used in heat sink applications or cooling systems, improving their efficiency and reliability, and contributing to waste reduction by minimizing the need for material replacement.

A study that investigated the thermal conductivity of copper matrix composites reinforced with graphene nanoplatelets found that the thermal conductivity of the composites initially improves substantially by approximately 30% with a slight loading of 0.25 vol% graphene platelets.[65]

8.4.4.4 Additive Manufacturing

Incorporating nanomaterials into metal powders used in additive manufacturing processes, such as selective laser melting or electron beam melting, can lead to the production of components with enhanced properties and reduced material waste. Additive manufacturing processes enable the production of complex components with minimal material waste compared to traditional manufacturing methods, which often involve subtractive processes and generate significant waste.

Examples include:

- Titanium matrix composites reinforced with multi-walled CNTs show improved mechanical properties, including tensile strength and fatigue resistance, with reduced material waste.[66]
- Aluminum matrix composites reinforced with graphene exhibit enhanced mechanical properties, such as hardness and wear resistance.[67]

8.4.4.5 Lightweight Materials

The development of metal matrix nanocomposites with improved strength-to-weight ratios can lead to the production of lightweight materials, which can reduce overall material consumption and waste. Lightweight materials are beneficial in applications such as automotive and aerospace industries, where reduced weight can contribute to improved fuel efficiency and lower emissions.

Examples include:

- Aluminum matrix composites reinforced with CNTs have demonstrated significant improvements in the strength-to-weight ratio, making them attractive options for lightweight applications in automotive and aerospace industries.[68]
- Magnesium matrix composites with silicon carbide nanoparticles exhibit a higher strength-to-weight ratio compared to the unreinforced magnesium alloy, demonstrating its potential for use in lightweight applications, such as automotive components.[69]
- Aluminum matrix composite with graphene nanoplatelets exhibits an enhanced strength-to-weight ratio, making it a promising material for lightweight applications in the automotive and aerospace industries.[70]

8.4.4.6 Controlled Degradation

Metal matrix nanocomposites can be designed to exhibit controlled degradation rates, which can be beneficial in applications where temporary support is needed, such as in the medical field (*e.g.*, temporary implants, orthopedic devices, *etc.*). Once the support is no longer needed, the material safely degrades, reducing the need for additional surgeries or interventions and minimizing waste generation.

The following examples illustrate the potential of using nanomaterials to create metal matrix composites with controlled degradation rates:

- Magnesium-based composites reinforced with biodegradable polymers and bioactive nanoparticles are useful for orthopedic implants. These composites exhibit controlled degradation rates, providing temporary support during the healing process and then safely degrading, eliminating the need for additional surgeries or interventions.[71]
- Zinc-based composites reinforced with biodegradable polymers and bioactive glass nanoparticles for bone implant applications exhibit controlled degradation rates, making them suitable for temporary support in bone healing and regeneration, while minimizing waste generation.[72]
- Iron attracts less attention than magnesium and zinc but is one of the best candidates for biodegradable metal stents thanks its biocompatibility, great elastic moduli and high strength. Iron-based composites reinforced with biodegradable polymers and bioceramic nanoparticles exhibit controlled degradation rates, providing temporary support to the blood vessels and then gradually degrading, reducing the need for further intervention.[73]

These examples demonstrate the potential of using nanomaterials to create metal matrix composites with controlled degradation rates for applications where temporary support is needed. By safely degrading once their purpose is fulfilled, these materials can reduce the need for additional surgeries or interventions and minimize waste generation, contributing to overall sustainability and resource conservation in the medical field.

8.4.5 Applications of Nanocomposites in Smart and Functional Materials

Nanocomposites are also revolutionizing material science by enabling the development of smart and highly functional materials. This section briefly looks at four key applications where nanocomposites significantly enhance material properties and performance: self-healing materials, self-cleaning surfaces, advanced sensors, and high-efficiency membranes. By incorporating nanoscale components such as nanoparticles, nanotubes, and nanoplatelets, these materials exhibit remarkable capabilities in autonomous damage repair and cleaning, sensitive detection, and selective filtration. These innovations not only improve the longevity and functionality of products but also contribute to waste reduction and environmental sustainability across various industries.

8.4.5.1 Self-healing Materials

Incorporating nanomaterials into self-healing and self-cleaning materials can contribute to waste reduction by increasing the lifetime of products and reducing the need for maintenance and replacement. Self-healing materials containing nanocapsules filled with healing agents can repair damage, such as cracks and scratches, when the capsules rupture and release the healing agents. These materials can be employed in various applications, including coatings, paints, and structural components, prolonging their service life and reducing waste generation.

Examples of self-healing metal matrix composites:

- A self-healing aluminum-based metal matrix composite (MMC) created by embedding low melting point particles (LMPPs) within an Al 5083 matrix. These LMPPs, created through a two-step micro-encapsulation process, act as healing agents. The MMC, fabricated using spark plasma sintering, heals fractures through transient liquid phase bonding when external compressive force is applied during heat treatment, effectively filling cracks in areas with high Zn concentration.[74]
- The incorporation of CNTs into an epoxy matrix significantly improves the crack-closing and healing capabilities of the composite material. This enhancement is attributed to the unique properties of CNTs, which facilitate better structural integrity and durability. Such advancements suggest potential applications in various industries, including aerospace and construction, due to the improved resilience and self-repairing features of these composites.[75]
- A copper matrix composite reinforced with graphene nanoplatelets demonstrates that incorporating graphene significantly enhances the self-healing behavior of the composite. The material autonomously recovers from defects or damages, which is attributed to the unique properties of graphene that improve structural integrity and resilience. This advancement holds potential for various industrial applications, including electronics and mechanical manufacturing.[76]

8.4.5.2 Self-cleaning Surfaces

Self-cleaning surfaces maintain their cleanliness by incorporating photocatalytic nanoparticles, such as titanium dioxide, which break down dirt and grime upon exposure to light. These materials can be used in various applications, including building facades, windows, and solar panels. This innovation significantly reduces the need for manual cleaning and maintenance, thereby minimizing the generation of waste from cleaning agents and materials.[77]

TiO_2, when exposed to ultraviolet light, generates reactive oxygen species like hydroxyl radicals. These ROS have strong oxidative properties, enabling them to decompose organic matter, bacteria, and pollutants on the surface. The photocatalytic activity of TiO_2 makes it particularly effective in maintaining a clean surface by continuously breaking down contaminants upon light exposure.

- Building facades and windows: Self-cleaning coatings on building exteriors reduce the frequency of cleaning required, saving both time AND labor costs. This is particularly beneficial for high-rise buildings where cleaning is challenging and expensive. These coatings help maintain the pristine appearance of buildings, ensuring they look new for longer periods.
- Solar panels: Dust and dirt accumulation on solar panels can significantly reduce their efficiency. Self-cleaning coatings ensure that solar panels remain clean, maintaining optimal performance and reducing the frequency of manual cleaning. By reducing the need for cleaning, maintenance costs are lowered, making solar energy more cost-effective in the long run.
- Automotive industry: Self-cleaning coatings on car windshields and mirrors improve visibility by preventing the buildup of dirt and water. This contributes to safer driving conditions. The exterior of cars can remain cleaner for longer periods, enhancing the vehicle's appearance and reducing the frequency of washes.

The use of self-cleaning materials aligns with the growing emphasis on sustainability and resource efficiency across multiple industries. By reducing the need for frequent cleaning and maintenance, these materials help conserve water, reduce the use of chemical cleaning agents, and minimize waste production. This contributes to a more sustainable approach in product design and maintenance, supporting environmental conservation efforts and reducing the ecological footprint of various industries.

8.4.6 Nanocomposite Sensors

Nanocomposite sensors represent a significant advancement in sensing technology, offering enhanced sensitivity, selectivity, and reliability compared to traditional sensors. These sensors incorporate nanomaterials into a matrix, resulting in unique properties that enable superior detection capabilities across various applications. Nanocomposite sensors are particularly useful in structural health monitoring, environmental detection, industrial safety, and wearable electronics, contributing to waste reduction and resource efficiency by enabling early detection of issues, optimizing processes, and extending product life.

8.4.6.1 Mechanism and Advantages

Nanocomposite sensors leverage the exceptional properties of nanomaterials, such as high surface area-to-volume ratio, quantum confinement effects, and unique electrical, optical, and mechanical characteristics. When integrated into a composite structure, these nanomaterials can:

- Amplify signal transduction
- Improve selectivity towards specific analytes
- Enhance durability and stability
- Enable multifunctional sensing capabilities

8.4.6.2 *Strain and Pressure Sensors*

The implementation of nanocomposite sensors in load-bearing structures significantly contributes to waste reduction by enabling early detection of structural issues, facilitating timely repairs, and preventing catastrophic failures. By continuously monitoring structures, these sensors extend the lifespan of buildings, bridges, and other infrastructure, reducing the need for premature replacements.

For instance, flexible and stretchable strain sensors using a combination of carbon nanotubes and graphene embedded in a polymer matrix have demonstrated remarkable sensitivity, with a gauge factor of up to 38.5. These sensors provide precise strain measurements in concrete structures such as beams and columns. By facilitating early detection of structural issues, they help extend the lifespan of buildings and bridges, significantly reducing waste in the construction industry.[78]

8.4.6.3 *Gas and Chemical Sensors*

Nanocomposite sensors also play a crucial role in waste reduction across various sectors by detecting minute concentrations of volatile organic compounds and other hazardous chemicals. These advanced devices enable the rapid identification of leaks or contamination in industrial settings, preventing chemical wastage and reducing environmental contamination risks.

For example, a highly sensitive nanocomposite sensor based on MoS_2–CuO nanocomposites demonstrates impressive capabilities in detecting acetone vapor, a common VOC. With a remarkably low detection limit of 93 parts per billion, the sensor maintains high sensitivity over a linear range of 0.5 to 2 parts per million (ppm) acetone concentration. Moreover, it showcases rapid detection capabilities, with a response time of just 61 seconds for 10 ppm acetone concentration.[79] This precise detection can significantly contribute to waste reduction in industries where acetone is commonly used or produced.

8.4.6.4 *Biosensors*

Nanocomposite biosensors are particularly valuable in environmental protection, enabling early detection of water pollutants and allowing for rapid intervention to prevent widespread contamination. Their high sensitivity allows them to detect low concentrations of pollutants, facilitating proactive measures. In industrial settings, these biosensors provide real-time wastewater monitoring, ensuring effective treatment and minimizing pollutant release. This capability optimizes chemical use and reduces waste.

For example, electrochemical sensors enhanced with nanocomposite materials are highly effective for detecting a wide range of water contaminants, such as pesticides, nitrates, and other hazardous chemicals.[80] These sensors are noted for their sensitivity, portability, and rapid response times, making them ideal for real-time environmental monitoring. By facilitating proactive environmental management, nanocomposite biosensors play a crucial role in reducing water waste, chemical usage, and environmental impact.

8.4.6.5 Wearable Electronics and Health Monitoring

Nanocomposite sensors are also integral to the development of wearable and flexible electronics. By incorporating nanomaterials like silver nanowires and polyurethane nanofibers into stretchable and conductive composites, these sensors can monitor physiological parameters in real time, including heart rate and body temperature. This innovation contributes to the creation of durable and long-lasting wearable devices, reducing electronic waste by extending device lifespans.

Wearable sensors incorporating nanocomposite materials have been shown to offer both flexibility and high conductivity, making them suitable for use in applications like personal health monitoring and smart textiles. These sensors provide continuous data, enabling users to track their health metrics more accurately and reliably, while the durable materials reduce the frequency of device replacement, contributing to sustainability.[81]

8.4.6.6 Temperature Sensors

Nanocomposite temperature sensors are under active research for use in industrial applications where precise temperature control is critical. By integrating nanomaterials into sensor technology, the potential benefits include optimizing heating and cooling systems, reducing energy consumption, and preventing equipment damage. While these sensors show promise for enhancing performance in applications such as HVAC systems and localized temperature control in smart buildings, their widespread commercial use remains limited as most advancements are still in the research and development phase.

For instance, polymer-based nanocomposite temperature sensors provide high sensitivity and accurate temperature measurements, which are crucial for optimizing industrial processes and reducing energy consumption.[82]

CNTs are often used in nanocomposites to create temperature sensors with high thermal conductivity and sensitivity. These sensors are being researched for applications in energy-intensive environments like power plants, where monitoring equipment temperature is critical to prevent overheating and optimize energy efficiency.[83]

Metal oxide nanoparticles like ZnO and TiO_2 are being explored for their use in high-precision temperature sensors. These sensors offer enhanced durability and performance in extreme environments, making them promising for use in industries such as aerospace and automotive, where accurate temperature readings are essential for both safety and energy efficiency.[84]

In machining and manufacturing processes, where high temperatures can affect both tool life and product quality, thin-film nanocomposite sensors are being tested for real-time temperature monitoring. This technology has potential applications in optimizing industrial processes by minimizing energy waste and improving tool longevity.[85]

8.4.6.7 Contributions to Waste Reduction and Sustainability

Nanocomposite sensors contribute to waste reduction by extending product life, optimizing industrial processes, and reducing the need for maintenance or replacements. Their role in real-time monitoring, whether in infrastructure, industrial safety, or personal health, allows for proactive management, reducing waste generated by failures, inefficiencies, or premature equipment replacement. Additionally, the durability and multifunctionality of nanocomposite-based sensors reduce the frequency of sensor replacement, further contributing to sustainability goals across multiple industries.

8.4.7 Nanocomposite Membranes

Nanocomposite membranes represent a cutting-edge development in materials science, combining the benefits of traditional membrane technology with the unique properties of nanomaterials.[86] These advanced membranes incorporate various nanoparticles, nanofibers, or nanosheets into a polymer matrix, resulting in enhanced performance characteristics that surpass those of conventional membranes. The integration of nanomaterials can significantly improve membrane properties such as permeability, selectivity, mechanical strength, and fouling resistance.

This synergistic combination of materials at the nanoscale opens up new possibilities for addressing complex separation challenges in water treatment, gas separation, and resource recovery applications. As global concerns about water scarcity, environmental pollution, and resource depletion continue to grow, nanocomposite membranes emerge as a promising solution for more efficient and sustainable industrial processes.

8.4.7.1 Applications in Water Treatment

One of the most prominent applications of nanocomposite membranes is in water and wastewater treatment.[87] By incorporating nanomaterials like graphene oxide, carbon nanotubes, or metal–organic frameworks, these membranes demonstrate significantly enhanced permeability and selectivity for removing contaminants from water. The high surface area and unique chemical properties of these nanomaterials allow for the efficient filtration of heavy metals, organic pollutants, and microorganisms.[88]

The incorporation of MXene nanosheets into polymer-based membranes was shown to improve water flux and contaminant rejection. These membranes are particularly effective in desalination and wastewater recovery, where their enhanced antifouling properties extend operational lifetimes and reduce the need for chemical cleaning agents.[89] By reducing the frequency of membrane replacement and minimizing the use of harsh chemicals, nanocomposite membranes contribute to more sustainable water management practices.

8.4.7.2 Gas Separation and Resource Recovery

Nanocomposite membranes are also widely used in gas separation processes, where the integration of nanomaterials like zeolites or metal nanoparticles enhances gas selectivity and permeability.[90] These membranes are employed in industrial applications such as carbon dioxide capture, hydrogen purification, and natural gas processing. The use of nanocomposite membranes in gas separation processes not only improves efficiency but also helps reduce greenhouse gas emissions and energy consumption.

Graphene-based nanocomposite membranes have proven effective in separating CO_2 from methane, a critical process in natural gas purification. The addition of graphene enhanced the mechanical strength of the membrane while maintaining high gas selectivity, making it suitable for high-pressure applications in the energy industry.[91]

8.4.7.3 Fouling Resistance and Durability

Fouling, the accumulation of unwanted materials on the surface of membranes during filtration, is a major challenge that reduces the efficiency and lifespan of traditional membranes. Nanocomposite membranes, however, exhibit enhanced antifouling properties due to the presence of nanomaterials like TiO_2 or silver nanoparticles, which prevent the adhesion of foulants such as proteins, oils, and microorganisms.

The antifouling properties of nanocomposite membranes not only improve their longevity but also reduce the need for frequent cleaning or replacement, lowering operational costs and minimizing waste generation. This is particularly beneficial in industries such as food processing, pharmaceuticals, and petrochemicals, where membrane fouling is a common issue.[92]

8.4.8 Case Study: Nanocomposite Membrane Enhances Water Purification and Antimicrobial Performance

Researchers have presented a compelling case for the application of MXenes in next-generation water purification membranes.[89] They focused on $Ti_3C_2T_x$ MXene, a promising 2D material known for its hydrophilicity, mechanical strength, and rich surface chemistry. However, pure MXene membranes tend to suffer from swelling and aggregation in water, compromising their long-term performance. To address these challenges, the researchers developed a novel nanocomposite by incorporating a polycation, polydiallyldimethylammonium chloride (PDDA), into the MXene structure (see Figure 8.10).

The fabrication method represents a key innovation in this study. The researchers used a simple vacuum filtration process to create a layered membrane structure with PDDA intercalated between the MXene sheets. This approach leverages electrostatic interactions to achieve a stable and functional nanocomposite. The positively charged PDDA molecules serve multiple roles: they act as pillars to maintain spacing between MXene layers, crosslink the structure to enhance stability, and increase surface hydrophilicity.

Extensive characterization confirmed the successful integration of PDDA into the MXene structure. X-ray diffraction showed an increase in interlayer spacing from 14.7 Å to 16.4 Å, while electron microscopy revealed changes in surface morphology. Importantly, the nanocomposite maintained the desirable properties of MXene while addressing its key weaknesses.

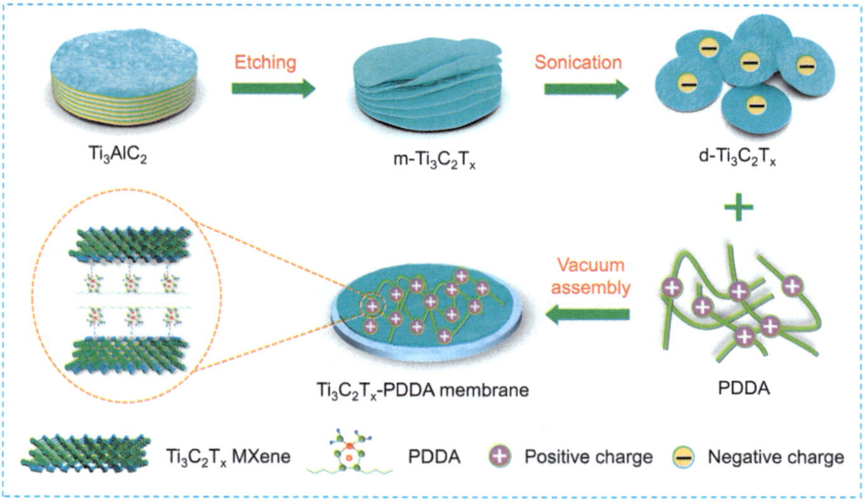

Figure 8.10 Schematic diagram of the fabrication of $Ti_3C_2T_x$/PDDA hybrid membranes. Reproduced from ref. 89, https://doi.org/10.3390/nano13212885, under the terms of the CC BY 4.0 license, https://creativecommons.org/licenses/by/4.0/.

The enhanced properties of the MXene-PDDA membrane are evident in its performance metrics. In filtration experiments, the nanocomposite achieved a water permeance of 73.4 liters per square meter per hour at 0.6 MPa—a significant improvement over pure MXene membranes. More critically, it demonstrated high selectivity, rejecting over 94% of magnesium chloride from a 2000 ppm solution. This combination of high flux and selectivity is crucial for efficient water purification processes.

The membrane's ability to remove antibiotics further showcases its versatility. It achieved a 95.8% rejection rate for erythromycin, addressing the growing concern of pharmaceutical contamination in water sources. Notably, the membrane maintained stable performance over extended operation, indicating good resistance to swelling and fouling—common issues that plague many advanced membranes.

Perhaps the most striking feature of the MXene-PDDA membrane is its multifunctionality. Beyond its impressive separation performance, the nanocomposite exhibits strong antimicrobial properties. Tests with both Gram-negative (*Escherichia coli*) and Gram-positive (*Staphylococcus aureus*) bacteria showed inhibition rates of 90% and 95%, respectively. Microscopy analysis revealed minimal bacterial adhesion to the membrane surface, suggesting excellent resistance to biofouling.

This dual functionality—enhanced separation and antimicrobial activity—demonstrates the power of nanocomposite design in achieving multiple property improvements simultaneously. The positively charged, highly hydrophilic surface created by the MXene-PDDA combination not only improves water transport but also actively resists bacterial colonization. This multi-pronged approach to performance enhancement could significantly extend membrane lifetimes and reduce maintenance requirements in real-world applications.

The environmental impact of this technology is potentially significant. By improving both water flux and contaminant rejection, the MXene-PDDA membrane could lead to more energy-efficient water treatment processes. Higher permeability means less pressure is needed to force water through the membrane, potentially reducing pump energy consumption. Meanwhile, improved selectivity and antimicrobial properties could decrease the need for additional treatment steps or chemical additives, further lowering the environmental footprint of water purification systems.

To evaluate the practical potential of their innovation, the researchers tested the membrane with real wastewater samples from a polluted river. The MXene-PDDA membrane effectively inhibited the growth of environmental microorganisms, achieving a 95% reduction compared to control samples. This real-world performance demonstrates the membrane's ability to handle complex water sources beyond idealized laboratory conditions—a crucial consideration for practical applications.

This case study effectively bridges the gap between cutting-edge nanomaterial research and practical water treatment applications. By incorporating MXene, an advanced 2D nanomaterial, into a functional membrane structure, the researchers have created a nanocomposite with enhanced properties directly relevant to water purification challenges. The simultaneous improvement in permselectivity and antimicrobial activity showcases how nanomaterial engineering can address multiple performance aspects in a single material system.

While further research is needed to scale up production and evaluate long-term stability under diverse conditions, this work provides a valuable proof-of-concept for next-generation water purification materials. The MXene-PDDA nanocomposite membrane exemplifies how advanced nanomaterials can be engineered to create multifunctional solutions for complex environmental challenges. As water scarcity and contamination continue to pose global threats, innovations like this offer promising pathways towards more efficient, effective, and sustainable water treatment technologies.

8.5 Waste Prevention

By enabling smart packaging, precision agriculture, and energy-efficient production processes, nanotechnology is fundamentally changing our approach to resource utilization and waste management. These nano-enabled solutions not only reduce material and energy waste but also enhance product longevity, optimize resource use, and improve overall system efficiency. From extending food shelf life through advanced sensing technologies to minimizing agrochemical runoff in farming, nanotechnology is proving to be a powerful tool in our quest for a more sustainable future.

As we take a closer look at the applications, emerging trends, and challenges of nano-enabled waste prevention, it becomes clear that this field holds immense potential to transform our production and consumption patterns. By preventing waste before it occurs, these technologies offer a proactive approach to environmental stewardship, potentially reducing the burden on waste management systems and conserving valuable resources. As with any emerging technology, careful consideration of potential risks and ethical implications is essential to ensure responsible development and deployment.

8.5.1 Smart Packaging

Smart packaging represents a revolutionary approach to waste prevention, particularly in the food and pharmaceutical industries. By integrating nanotechnology-based sensors and indicators, packaging can now actively monitor product quality, freshness, and safety.[93] This section explores how smart packaging systems leverage nanomaterials to provide real-time information about a product's condition, enabling consumers and businesses to make informed decisions that minimize waste. From color-changing labels that indicate spoilage to RFID tags enhanced with nanoparticles for improved tracking, we'll examine the cutting-edge technologies that are transforming packaging from a passive container into an active waste prevention tool.

8.5.1.1 Nanosensor-enabled Packaging

Smart packaging has advanced significantly with the integration of nanosensor technology, transforming traditional packaging into an active system capable of monitoring and responding to changes in product quality. These nanosensors, embedded within the packaging material, provide real-time data on freshness, safety, and environmental conditions like temperature and humidity. By detecting chemical, biological, and physical changes, they ensure product quality and longevity while minimizing waste.

Chemical nanosensors are particularly useful for monitoring spoilage by detecting gases such as ethylene or volatile organic compounds, which signal the ripening or decay of products.[94,95] Biological nanosensors, on the other hand, play a crucial role in ensuring food safety by identifying harmful pathogens such as *E. coli* or *Salmonella*.[96] Physical nanosensors offer another layer of protection by tracking environmental factors like temperature and humidity, which are critical for maintaining the integrity of perishable goods.[97]

The real-time monitoring provided by nanosensors brings substantial benefits in managing product freshness and reducing contamination risks. Color-changing labels or smart indicators can visually display freshness, alerting consumers or supply chain managers to potential spoilage. Additionally, nanosensors are essential for monitoring transport conditions, tracking temperature fluctuations or physical damage to ensure products remain within safe parameters.[98]

These capabilities not only enhance product safety but also optimize supply chain management. Continuous data on environmental conditions and location help reduce spoilage during transit. Wireless data transmission from nanosensor-enabled packaging improves transparency and efficiency, allowing for real-time adjustments to storage or transportation conditions, ensuring optimal handling.[99]

Smart packaging also enhances consumer engagement through digital platforms. By scanning a QR code on the packaging, consumers can access real-time information about product freshness, safety, and storage recommendations, fostering more informed and sustainable consumption choices.

One of the most advanced applications of nanosensors involves their integration with wireless communication systems. These nanosensors can transmit data to external devices, enabling remote monitoring and data logging, which aligns with the Internet of Things (IoT) framework. This significantly improves traceability and transparency, allowing supply chain managers to receive real-time alerts if temperature or humidity deviates from safe ranges during transport. This capability enables immediate corrective actions, preventing spoilage and ensuring product quality.

In addition to enhancing supply chain efficiency, nanosensor-enabled packaging contributes to waste reduction by providing early detection of spoilage or contamination. Continuous monitoring of storage conditions allows for timely interventions, significantly extending the shelf life of perishable goods and minimizing food waste.

The integration of nanomaterials into packaging also improves performance by enhancing barrier properties, mechanical strength, and antimicrobial capabilities. This combination of nanosensors with IoT technology represents a promising future for food packaging, although challenges related to cost, regulatory approval, and scalability need to be addressed.

The cost of integrating nanosensors into packaging remains high, necessitating innovations in manufacturing processes to reduce expenses and make the technology more accessible. Regulatory hurdles also pose a challenge, particularly regarding the safety of nanomaterials in contact with food. Clear guidelines and rigorous testing protocols are essential to ensure compliance and safety.[100]

Scalability is another hurdle. Developing efficient production methods for nanosensor-enabled packaging is crucial for widespread adoption. Additionally, managing the large volumes of data generated by these sensors will require secure and efficient data management systems to fully realize the technology's potential.

Despite these challenges, nanosensor-enabled packaging is poised to transform the industry by enhancing product safety, quality, and waste prevention. This technology empowers both consumers and businesses with real-time, accurate information, leading to more informed decisions and reduced waste. As advancements continue, nanosensor-enabled packaging will play a crucial role in sustainable consumption, with the potential to revolutionize the packaging industry by making products safer, reducing waste, and prioritizing sustainability.

8.5.1.2 Real-time Monitoring Capabilities

The real-time monitoring capabilities of smart packaging systems represent a major advancement in ensuring product quality and preventing waste. Building on the nanosensor technologies previously discussed, these systems offer a comprehensive solution for monitoring the condition of products throughout their entire lifecycle, providing real-time insights that enable proactive interventions.

Nanosensors play a crucial role in detecting various changes in product quality. For freshness monitoring, sensors can detect specific chemical markers that indicate spoilage. For example, colorimetric nanosensors change color in response to biogenic amines produced during the spoilage of meat, providing a visual cue for consumers or supply chain managers.[101] Similarly, biological nanosensors are designed to detect harmful contaminants like bacteria. Research has shown that aptamer-based nanosensors can rapidly detect pathogens in food products, significantly improving safety measures.[102]

These technologies are not limited to detecting contamination or spoilage. Physical nanosensors can also enhance product security by identifying if packaging has been tampered with. Recent innovations include nanocomposite films that change color irreversibly when tampered with, providing an immediate and clear indication of any breach in package integrity.[103]

Beyond detection, real-time tracking features offer valuable insights into the handling and condition of products across the supply chain. Nanosensors can record temperature variations, which is critical for maintaining the cold chain for perishable goods. For instance, nanoparticle-based time-temperature indicators have demonstrated effectiveness in monitoring products that are sensitive to temperature changes.[104] Additionally, these sensors enable dynamic shelf-life predictions. By tracking environmental conditions and product status, smart packaging systems can use machine learning algorithms to provide accurate predictions of a product's remaining shelf life, allowing for more efficient stock management.[105]

Geolocation tracking is another important feature that enhances logistics. When integrated with RFID technology, nanosensor-enabled packaging can provide real-time location data, reducing waste caused by lost or mishandled products. Nano-enhanced RFID tags have been shown to improve read ranges and data storage capabilities, further enhancing visibility within the supply chain.[106]

These detection and tracking features create a comprehensive monitoring system that ensures both product safety and quality. By providing early warnings of potential spoilage or contamination, these systems enable timely interventions that prevent the distribution of unsafe products, helping to reduce food waste.

The real-time data generated by these systems can be integrated with the IoT and smartphone applications, enhancing consumer interaction. Consumers can access up-to-date information about a product's freshness and safety by simply scanning a QR code on the packaging, empowering them to make more informed decisions.[107]

The vast amount of data generated by these monitoring systems presents challenges in terms of data management and interpretation. Efficient data analysis algorithms, as well as robust data privacy and security measures, will be essential to fully capitalize on the benefits of real-time monitoring in smart packaging.[108]

8.5.1.3 Consumer Interfaces and Information Delivery

Smart packaging builds on the integration of nanosensor technologies and real-time monitoring systems, focusing heavily on how this information is delivered to consumers. The success of these systems in reducing waste

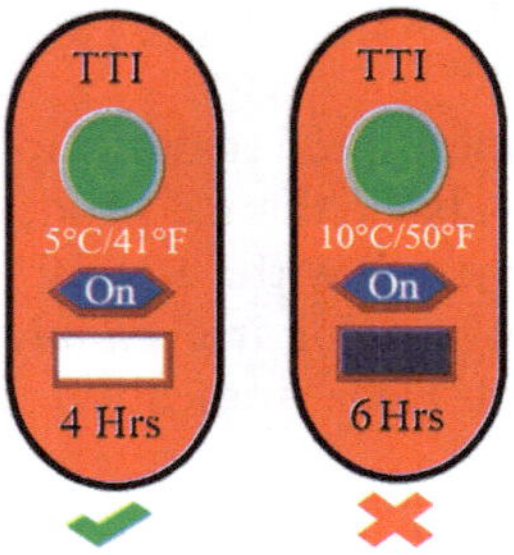

Figure 8.11 Mechanism of time–temperature indicator. Reproduced from ref. 252, https://doi.org/10.1002/fbe2.12038, under the terms of the CC BY 4.0 license, https://creativecommons.org/licenses/by/4.0/.

depends largely on the clarity and accessibility of the information provided, empowering consumers to make better-informed decisions regarding product use and disposal.

One key aspect of smart packaging is the combination of visual and digital systems. Packaging often incorporates both physical indicators and digital technologies to create an engaging, comprehensive information delivery system. For instance, color-changing labels or time-temperature indicators often act as the first point of interaction, alerting consumers to potential issues such as temperature changes or spoilage. When these visual cues signal a problem, users can turn to digital platforms, such as QR codes, to access more detailed product information. Research shows that when visual indicators, like shown in Figure 8.11, alert consumers to potential issues, they are more likely to engage with digital tools, enhancing their ability to make informed decisions about product quality and safety.

Augmented reality (AR) technology is another innovative tool advancing consumer engagement. AR applications overlay virtual information on physical packaging, providing dynamic data such as nutritional information, product freshness, and personalized usage recommendations. This interactivity creates a more engaging and informative experience, influencing consumer behavior in ways that encourage sustainable consumption and waste reduction.

The way consumers interact with smart packaging plays a crucial role in its effectiveness in reducing waste. For instance, advanced smartphone applications can interpret data from packaging sensors and provide personalized recommendations for storage or usage. Research indicates that these tailored suggestions may reduce household food waste more effectively than static labeling, as they adapt to the specific conditions and needs of the product and consumer.[109]

Additionally, the use of digital platforms enhances trust and transparency, particularly in industries such as pharmaceuticals. Packaging that provides real-time authenticated information about product efficacy and expiration could increase consumer confidence and adherence to proper usage guidelines. Blockchain technology is also being explored as a way to secure and verify the supply chain, further enhancing trust in the authenticity and handling of products.[110]

While smart packaging holds considerable promise, several challenges must be addressed. For example, it is essential to prevent information overload through user-centered design that presents data in an accessible and understandable format. Not all consumers are equally comfortable with or have access to digital technologies, and future innovations must consider this digital divide to ensure widespread adoption.

Privacy concerns also arise as smart packaging collects data on consumer behavior. Robust data protection measures are critical, and advanced encryption methods are being explored to ensure that the data collected through these systems is managed securely. Future developments could also see smart packaging systems integrated into broader smart home systems, such as refrigerators and pantry management tools, allowing for automated inventory management and further reducing household waste.

8.5.1.4 Challenges and Future Directions

Smart packaging, leveraging nanotechnology and IoT, offers significant potential for waste prevention through real-time product monitoring, yet several challenges still hinder widespread adoption: High production costs necessitate innovations in cost-effective manufacturing, while regulatory concerns regarding the safety of nanomaterials in food packaging require clear international guidelines.

Consumer education is vital for maximizing the benefits of smart packaging, and managing the large volumes of data generated presents additional challenges in terms of security and interpretation. The sustainability of smart packaging materials themselves must be ensured to avoid creating new waste streams.

Future developments may include biodegradable nanosensors, blockchain integration for improved traceability, and self-healing materials to extend product shelf life. Continued research is focusing on reducing costs, improving consumer interfaces, and enhancing the overall sustainability of packaging systems.

8.5.2 Precision Agriculture

Precision agriculture represents a paradigm shift in farming practices, offering a data-driven approach to crop and livestock management that significantly contributes to waste reduction in the agricultural sector. This innovative methodology leverages advanced technologies to optimize resource use, enhance crop yields, and minimize environmental impact. At its core, precision agriculture aims to apply the right treatment in the right place at the right time, ensuring that every input—be it water, fertilizer, or pesticides—is used with maximum efficiency.[111]

The concept of precision agriculture has evolved dramatically with the advent of nanotechnology. Nanomaterials and nanodevices have opened up new possibilities for ultra-precise monitoring and targeted interventions in agricultural systems. By operating at molecular and atomic scales, nanotechnology enables the development of highly sensitive sensors, smart delivery systems, and novel materials that can revolutionize farming practices.[112]

In the context of waste reduction, precision agriculture powered by nanotechnology offers several key advantages. First, it allows for the minimal use of agricultural inputs by providing real-time, localized data on crop and soil conditions. This precision helps prevent over-application of fertilizers and pesticides, reducing both economic waste and environmental pollution. Second, by optimizing water usage through advanced sensing and irrigation systems, it addresses one of the most critical issues in agriculture—water waste. Third, the ability to detect plant stresses and diseases early can prevent crop losses, thereby reducing food waste at the production stage.

Precision agriculture contributes to waste reduction indirectly by improving overall agricultural efficiency. By maximizing crop yields on existing farmland, it can potentially reduce the need for agricultural expansion, thus conserving natural habitats. The precise application of inputs also means fewer residual chemicals in the soil and water systems, contributing to long-term environmental sustainability.

Nanotechnology plays a crucial role in enabling these benefits. Nanoscale sensors can detect changes in plants and soil at a molecular level, providing unprecedented insights into crop health and environmental conditions. Nanoformulations of fertilizers and pesticides allow for controlled release and targeted delivery, significantly reducing the amount of chemicals needed. However, there are concerns about nanoparticle toxicity and their impact on human health and the environment. For instance, nanoparticles can accumulate in the soil and plants, which raises questions about their long-term effects on soil health, water systems, and food safety.

Nanomaterials can also enhance the efficiency of water treatment and desalination processes, making more water available for agriculture while reducing waste.

8.5.2.1 Nanosensor-enabled Precision Agriculture

The foundation of precision agriculture lies in accurate, real-time data collection, and nanosensors are at the forefront of this technological revolution (see Figure 8.12). These miniature devices, operating at the nanoscale, offer unprecedented sensitivity and specificity in detecting and measuring various agricultural parameters. The integration of nanosensors into farming

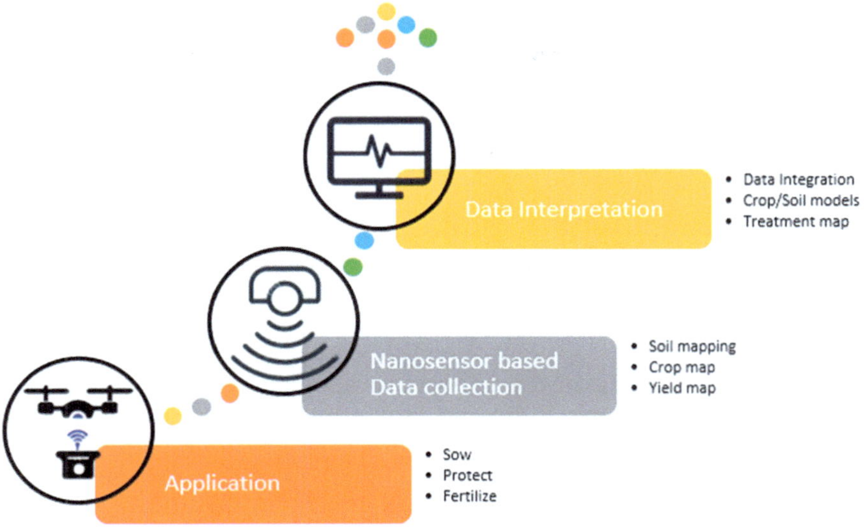

Figure 8.12 Illustration of nanosensor-driven precision agriculture: From real-time data collection and analysis to targeted field applications, optimizing resources and improving crop management through soil mapping, crop monitoring, and variable-rate input systems. Reproduced from ref. 112, https://doi.org/10.3390/agrochemicals2020016, under the terms of the CC BY 4.0 license, https://creativecommons.org/licenses/by/4.0/.

systems has opened up new possibilities for monitoring and managing crops with extraordinary precision.[113]

Types of nanosensors used in precision agriculture:

8.5.2.1.1 Soil Sensors. Nanosensors for soil analysis provide detailed, localized information about soil conditions, enabling farmers to make informed decisions about crop management.

- Moisture sensors: Nanostructured materials, such as carbon nanotubes and graphene oxide, are being used to create highly sensitive soil moisture sensors.[114] These sensors can detect minute changes in soil water content, allowing for precise irrigation scheduling and water conservation and thereby significantly improving water use efficiency in experimental crop fields.
- Nutrient sensors: Nanomaterial-based sensors can detect and quantify various soil nutrients, including nitrogen, phosphorus, and potassium.[115] In particular, ion-selective field-effect transistors (ISFETs) incorporating nanoparticles have shown promise in rapid, *in situ* nutrient analysis.[116]
- pH sensors: Nanocomposite materials are being utilized to create highly sensitive and stable pH sensors for soil analysis. These sensors can

provide real-time monitoring of soil acidity, crucial for maintaining optimal growing conditions for crops. Nanocomposite-based conductometric microsensor platforms are capable of measuring changes in soil pH. These sensors have shown good sensitivity across a wide pH range and can function effectively in different soil types. They exhibit fast response times, typically within seconds, and offer excellent long-term stability and repeatability.[117]

8.5.2.1.2 Plant Sensors.

Nanosensors attached to or inserted into plants can provide valuable data about plant health and physiological status, enabling early detection of stresses and diseases.

- Stress sensors: Advanced nanosensors are being developed to detect various plant stress indicators, including signaling molecules and plant hormones. These sensors can potentially alert farmers to different types of stress, such as drought, heat, or pathogen infection, before visible symptoms appear. For instance, multiplexed nanosensors capable of simultaneously monitoring multiple stress-related compounds in plants can detect distinct patterns and dynamics of stress signaling molecules under various environmental conditions.[118] Such technology could enable more timely and targeted interventions in crop management, potentially improving plant resilience and productivity in the face of changing environmental conditions.
- Disease detection sensors: Nanoparticle-based immunosensors can detect plant pathogens at very low concentrations, allowing for early intervention and reduced use of pesticides.[119] For instance, the detection of *Pseudomonas syringae* with a nanoparticle electrochemical biosensor was more sensitive than conventional PCR, enabling diagnosis before any disease symptoms appeared.[120] Additionally, an electrochemical impedance spectroscopy-based biosensor for *Pseudomonas syringae pv. lachrymans* exhibited a detection limit of just 2.6 $pg\,ml^{-1}$.[121] To put this exceptional sensitivity in context, a detection limit of 2.6 picograms per milliliter is roughly equivalent to detecting one grain of salt dissolved in more than 22 000 liters of water.
- Nutrient uptake sensors: Electrochemical nanosensors have been effectively utilized for the detection of various soil contents, including essential nutrient ions, to ensure optimal plant growth. Recent advances,[122] particularly involving metallic nanoparticles such as gold and nanocarbon materials like carbon nanotubes and graphene, have enhanced their detection capabilities. For instance, electrochemical nanosensors can precisely monitor nutrient ions such as H^+, K^+, and Na^+ in soil, providing critical data for maintaining optimal nutrient levels for plant growth.

8.5.2.1.3 Environmental Sensors. These sensors monitor the micro-climate around crops, providing data crucial for optimizing growing conditions and predicting potential issues.

- Temperature sensors: Nanostructured thermocouples and resistance temperature detectors offer high-precision temperature monitoring. For instance, a flexible, graphene/alginate composite temperature sensor with a resolution of 0.1 °C operates within a temperature range of 25 °C to 45 °C.[123] This sensor is potentially capable of mapping temperature variations across crop canopies, providing valuable data for optimizing agricultural practices.
- Humidity sensors: Cellulose nanofiber-based humidity sensors, fabricated using ionic liquid/cellulose nanofibers, offer high sensitivity and rapid response characteristics, essential for dynamic agricultural environments. With a high response of 27.95 pF/% RH and a fast recovery time of approximately 1 second, these sensors can continuously and accurately track humidity levels. This capability is crucial for optimizing irrigation and climate control within greenhouses, ensuring optimal conditions for crop growth and health.[124]
- Light sensors: Quantum dot-based light sensors can measure light intensity and spectral quality with high accuracy. This information is crucial for optimizing plant growth, especially in greenhouse environments. Quantum dot-based sensor arrays are capable of measuring both light intensity and spectral composition, allowing for precise control of supplemental lighting in greenhouse crops.[125]

8.5.2.1.4 Integration Methods for Nanosensors in Agricultural Systems. The effective deployment of nanosensors in agricultural settings requires careful consideration of integration methods. Several approaches are being explored:[126]

- Direct soil incorporation: Nanosensors can be directly mixed into the soil, allowing for widespread monitoring across fields. However, this method requires sensors that are robust enough to withstand soil conditions and have long operational lifespans.
- Plant attachment or insertion: Some nanosensors can be attached to plant leaves or inserted into plant tissues. This method allows for direct monitoring of plant physiological parameters but requires careful design to avoid damaging the plants.
- Drone-based deployment: Nanosensors can be incorporated into small, biodegradable capsules that are distributed across fields using drones. This method allows for flexible, targeted deployment of sensors.
- Integration with existing agricultural equipment: Nanosensors can be incorporated into farming equipment such as tractors or irrigation systems, allowing for continuous monitoring during regular farming operations.

- Wireless sensor networks: Multiple nanosensors can be linked into wireless networks, enabling comprehensive monitoring across large agricultural areas. These networks can integrate different types of sensors to provide a holistic view of crop and field conditions.

The integration of nanosensors into precision agriculture systems represents a significant advancement in our ability to monitor and manage agricultural processes. By providing highly accurate, real-time data at an unprecedented scale and resolution, these technologies are enabling farmers to make more informed decisions, optimize resource use, and significantly reduce waste in agricultural practices. As research in this field continues to advance, we can expect even more sophisticated and effective nanosensor systems to emerge, further revolutionizing precision agriculture.

8.5.2.2 Precision Application of Inputs

The integration of nanotechnology in precision agriculture has revolutionized the application of agricultural inputs, allowing for unprecedented efficiency and reduced waste. This section explores how nanomaterials and smart delivery systems are transforming the application of fertilizers, pesticides, and water in modern farming practices (see Figure 8.13).

8.5.2.2.1 Nanofertilizers and Smart Delivery Systems. Nanofertilizers represent a significant advancement in nutrient management, offering enhanced efficiency and reduced environmental impact compared to conventional fertilizers. These innovative formulations leverage nanotechnology to improve nutrient uptake and utilization by plants.[127]

Types of nanofertilizers:

- Nanoscale fertilizers: Nutrients are formulated into nanoparticles, typically ranging from 1–100 nm in size. This small size allows for increased surface area and reactivity, enhancing nutrient availability to plants.
- Nanoscale additives: Conventional fertilizers are modified with nanomaterials to improve their performance.
- Nanoencapsulated fertilizers: Nutrients are enclosed within nanoscale protective coatings, allowing for controlled release over time.

Key benefits of nanofertilizers:

- Enhanced nutrient use efficiency: Nanofertilizers can significantly improve the uptake and utilization of nutrients by plants. For instance, researchers developed zinc-complexed chitosan nanoparticles (Zn-CNP) for wheat ferti-fortification, achieving a comparable 36% grain zinc enrichment with Zn-CNP using 90% less zinc than traditional methods. Four-year field trials showed consistent results without affecting wheat

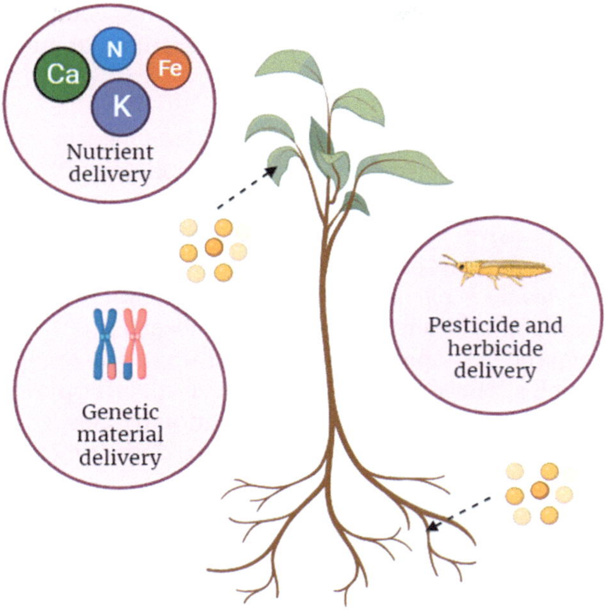

Figure 8.13 Nanoparticles as carriers for nutrient, genetic material, and pesticide delivery into plants. Nanoparticles can be introduced into plants through roots, leaves, or stems and are subsequently transported *via* the plant's vascular system (xylem and phloem). This method enables precise delivery of agricultural inputs, enhancing nutrient uptake, targeted pest control, and gene delivery, while minimizing environmental impact. Reproduced from ref. 131, https://doi.org/10.3390/nano14020131, under the terms of the CC BY 4.0 license, https://creativecommons.org/licenses/by/4.0/.

quality, proving Zn-CNP as an efficient, environmentally friendly nanofertilizer.[128]

- Controlled release: Nanoencapsulation allows for the gradual release of nutrients in response to environmental triggers such as soil pH, temperature, or moisture.[129] This synchronization with plant needs reduces nutrient losses through leaching or volatilization.
- Reduced environmental impact: The precision of nanofertilizers minimizes excess nutrient runoff, which can lead to eutrophication of water bodies. A field study on lettuce fields found that adding carbon nanoparticles (CNPs) to fertilizer reduced nitrate leaching by 57% when fertilizer was reduced to 70% of the recommended dose, without affecting lettuce yield. This method suggests that CNPs can lower fertilizer doses and minimize nitrate infiltration into groundwater while maintaining crop productivity.[130]
- Targeted delivery: Nanofertilizers can be designed to target specific plant tissues or cellular structures, maximizing nutrient absorption. Nanoparticle-based delivery systems have shown the ability to penetrate

plant cell walls and deliver nutrients directly to specific organelles, such as chloroplasts, thereby enhancing nutrient uptake and utilization by plants.[131]

Advanced delivery mechanisms are being developed to further enhance the precision and efficiency of nanofertilizer application:[132]

- Stimuli-responsive nanocarriers: These systems release nutrients in response to specific environmental cues. For instance, pH-sensitive nanoparticles can release phosphorus in acidic soil conditions, ensuring availability when plants can best absorb it.
- Bioconjugated nanoparticles: By attaching biological molecules to nanoparticles, these systems can target specific plant receptors or tissues. Research has shown that nanoparticles conjugated with plant hormones can enhance nutrient delivery to specific plant organs.
- Nanohybrid systems: Combining different nanomaterials can create multifunctional delivery systems. For example, a nanohybrid of graphene oxide and iron oxide nanoparticles has demonstrated enhanced delivery of both micronutrients and macronutrients in hydroponic systems.

8.5.2.2.2 Targeted Pest Control Using Nanoformulations. Nanotechnology is enabling the development of more effective and environmentally friendly pest control strategies. Nanoformulations of pesticides offer improved efficacy, reduced environmental persistence, and minimized non-target effects.[133]

Key advantages of nanopesticides:

- Enhanced efficacy: Nanoformulations can improve the solubility, stability, and dispersion of active ingredients, leading to better pest control with lower doses. Field trials have shown increases in efficiency against target organisms by up to 31.5% compared to conventional pesticides.
- Controlled release: Nanoencapsulated pesticides can release active ingredients gradually or in response to specific triggers. This selective, targeted, and long-term controlled release prolongs the effectiveness of treatments and reduces the need for frequent applications.
- Reduced environmental impact: The precision of nanoformulations minimizes pesticide drift and non-target exposure. Studies have shown up to 43.1% lower toxicity to non-target organisms compared to conventional pesticides, highlighting insufficient collateral environmental damage.
- Overcoming resistance: Nanocarriers can enhance the penetration and bioavailability of pesticides to target organisms, potentially overcoming resistance mechanisms.

- Improved stability and bioavailability: Nano-carriers enhance the stability of pesticides and facilitate their delivery to target organisms, increasing bioavailability.
- Reduced chemical input: Nanotechnology can diminish the amount of active compounds used, preserving the environment and human health while reducing the cost of crop production.

These advancements make nanopesticides a promising tool for sustainable agriculture, offering a balance between effective pest control and environmental preservation. However, the review also notes that more research is needed to improve their development, evaluation, and regulation to address potential risks and uncertainties.

Overall, the integration of these nanotechnology-enabled precision application systems is transforming agricultural input management. By allowing for the ultra-precise delivery of fertilizers, pesticides, and water, these technologies are significantly reducing waste, minimizing environmental impact, and optimizing resource use efficiency. As research in this field continues to advance, we can expect even more sophisticated and effective precision application systems to emerge, further enhancing the sustainability and productivity of modern agriculture.

8.5.2.3 Challenges and Future Directions

While the potential benefits of nanotechnology in agriculture are significant, several challenges and concerns need to be addressed:

- Cost and scalability: Currently, many nanomaterials and nanoformulations are expensive to produce at scale. Making these technologies economically viable for widespread agricultural use remains a challenge.
- Regulatory concerns and safety assessments: The unique properties of nanomaterials raise new regulatory questions. Comprehensive safety assessments are needed to understand potential long-term effects on human health and the environment. More research is required to improve the development, evaluation, and regulation of nanopesticides.
- Environmental persistence and bioaccumulation: While nanomaterials can reduce the immediate environmental impact of agrochemicals, their long-term fate in the environment is not fully understood. There are concerns about potential bioaccumulation in soil, water systems, and food chains that need to be addressed through ongoing research.
- Data management and privacy: Precision agriculture often involves collecting and analyzing large amounts of data. Ensuring data security and addressing privacy concerns for farmers will be crucial.
- Integration with existing practices: Adopting nanotechnology may require significant changes to existing agricultural practices. This could pose challenges for farmers, particularly in developing regions.

- Ethical considerations: The use of nanotechnology in food production raises ethical questions about food safety, labeling, and consumer choice.
- Unintended ecological impacts: While nanomaterials can reduce non-target effects, their novel properties might lead to unforeseen impacts on beneficial insects, soil microorganisms, or other ecological components.

Future developments in agricultural nanotechnology could have significant impacts on global food security, potentially increasing crop yields and resilience to climate change. However, realizing these benefits will require addressing the challenges outlined above.

A balanced approach to implementing nanotechnology in agriculture is crucial. This includes continued research into both benefits and risks, development of robust regulatory frameworks, and engagement with stakeholders including farmers, consumers, and environmental groups. By carefully navigating these challenges, nanotechnology has the potential to contribute significantly to sustainable and efficient agricultural practices.

8.5.3 Energy-efficient Production

Nanotechnology is revolutionizing industrial processes, paving the way for more energy-efficient and less wasteful production methods. This section examines how nanomaterials are enhancing chemical reactions, improving manufacturing efficiency, and contributing to more sustainable production across various industries.

According to data from the International Energy Agency, the industrial sector accounted for 37% (166 EJ) of global energy use in 2022, compared to 34% in 2002 (see Figure 8.14).[134] The sector was directly responsible for emitting 9.0 Gt of CO_2, accounting for a quarter of global energy system CO_2 emissions. Growth in energy consumption over the past decade has been driven largely by continued rising production in energy-intensive industry subsectors.

The unique properties of nanomaterials, such as their high surface area-to-volume ratio and quantum effects, enable unprecedented control over chemical reactions and material properties. This control translates into significant improvements in energy efficiency and waste reduction. From reducing reaction temperatures and pressures to increasing product yield and selectivity, nano-enabled solutions are minimizing energy consumption and waste generation at the source.

This section will investigate specific applications in industries such as petrochemicals, pharmaceuticals, and fine chemicals, exploring how nanotechnology is driving process intensification and contributing to more sustainable production methods. We will also examine the role of advanced nanomaterials in improving energy efficiency, the integration of

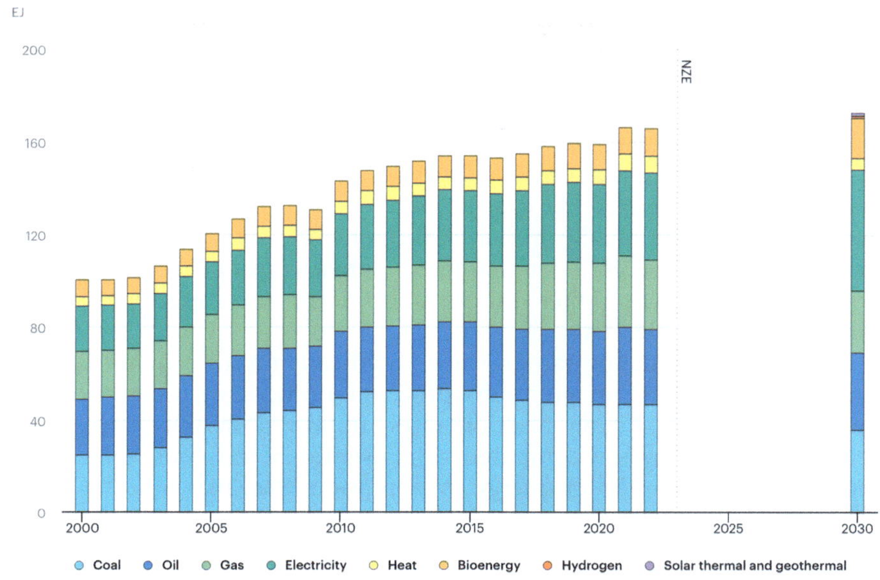

Figure 8.14 Industrial energy consumption (in exajoules, EJ) by fuel in the Net Zero Emissions (NZE) Scenario, 2000–2030. Reproduced from ref. 255, https://www.iea.org/data-and-statistics/charts/industrial-energy-consumption-by-fuel-in-the-net-zero-scenario-2000-2030, under the terms of the CC BY 4.0 license, https://creativecommons.org/licenses/by/4.0/deed.en.

nanotechnology in smart manufacturing systems, and the challenges and future directions of this rapidly evolving field.

8.5.3.1 Nanocatalysts in Chemical Reactions

Nanocatalysts play a crucial role in improving chemical reactions across various industries, leading to more energy-efficient and less wasteful production methods. Their unique properties, such as high surface area-to-volume ratios and precisely controlled structures, offer unprecedented enhancements in reaction efficiency, selectivity, and yield. This section explores how nanocatalysts are transforming processes in the petrochemical, pharmaceutical, and fine chemicals industries.

8.5.3.1.1 Petrochemical Industry. In the petrochemical sector, nanocatalysts are revolutionizing cracking and reforming processes. Zeolite-based nanocatalysts with precisely controlled pore sizes and acidities have significantly improved fluid catalytic cracking (FCC) processes, increasing gasoline yields by up to 15% while reducing coke formation, a common waste product in refineries.[135]

Nanostructured bimetallic catalysts, such as platinum-rhenium nanoparticles supported on alumina, have demonstrated enhanced activity and selectivity in naphtha reforming. These catalysts operate at temperatures 20–30 °C lower than conventional ones, resulting in substantial energy savings and longer catalyst lifetimes.[136]

A key application of nanocatalysts is in the production of propylene, a vital feedstock for materials like polypropylene, acrolein, acetone, and polyacrylonitrile. Traditionally produced through the cracking of petroleum-derived naphtha or as a by-product of ethylene production, increasing demand for propylene and environmental concerns have spurred the search for more sustainable methods.

Platinum-based nanocatalysts show significant promise in propane dehydrogenation (PDH), an alternative process for producing propylene. These nanocatalysts offer higher reaction rates, improved stability, and greater selectivity for propylene compared to conventional catalysts, resulting in increased yields and lower energy consumption. Research has shown that platinum-based nanocatalysts can enhance propylene production efficiency by up to 30%, significantly cutting energy use and greenhouse gas emissions.[137]

8.5.3.1.2 Pharmaceutical Industry. The pharmaceutical industry is benefiting from nanocatalysts that facilitate more efficient and environmentally friendly synthesis routes. Palladium nanoparticles supported on graphene oxide exhibit exceptional activity in Suzuki–Miyaura cross-coupling reactions, a key process in synthesizing many drug molecules. These nanocatalysts enable reactions to proceed in aqueous media at room temperature, reducing the need for organic solvents and energy-intensive heating.[138]

Gold nanoparticle catalysts are also finding use in the selective oxidation of alcohols, an essential step in synthesizing many pharmaceutical intermediates. These catalysts offer high selectivity and can operate under mild conditions, reducing the formation of by-products and the energy required for subsequent purification steps.[139]

8.5.3.1.3 Fine Chemicals Industry. In the fine chemicals industry, nanocatalysts are enabling more precise and efficient syntheses. Titanium dioxide nanoparticles, doped with noble metals such as platinum or palladium, have shown remarkable activity in photocatalytic organic transformations. These catalysts harness light energy to drive reactions, reducing the need for high temperatures and pressures.[140]

Recent advancements in single-atom catalysts (SACs) are pushing the boundaries of catalytic efficiency.[141] For example, single platinum atoms dispersed on titanium nitride have shown exceptional activity in the hydrogenation of nitroaromatics, a key reaction in producing aniline derivatives. These SACs achieve nearly 100% atom efficiency, maximizing the use of precious metals and minimizing waste.[142]

The development of these nanocatalysts represents a significant step towards more sustainable chemical processes. By enabling reactions to proceed under milder conditions with higher selectivity and reduced energy input, nanocatalysts are driving a shift towards greener industrial chemistry and more energy-efficient production methods.

8.5.3.2 Process Intensification Through Nanotechnology

Nanotechnology significantly contributes to process intensification by integrating multiple steps of chemical production into fewer, more efficient processes. This integration enhances productivity and addresses environmental concerns by minimizing waste generation and energy consumption.

8.5.3.2.1 Membrane Reactors. Membrane reactors represent a significant advance in process intensification by integrating reaction and separation processes into a single unit operation. These reactors employ selective membranes to continuously remove products or introduce reactants during the reaction, thereby shifting equilibrium and enhancing overall efficiency. It's important to note that membrane reactors encompass a wide range of applications, including both chemical and biological processes.[143]

In chemical membrane reactors, nanoporous membranes made from materials such as silica, zeolites, and metal–organic frameworks enhance selectivity and throughput. The precise control over pore size at the nanoscale allows for molecular-level separation, which can be tailored for specific reactions. For instance, in the production of hydrogen through steam reforming of methane, palladium-based nanomembranes can selectively remove hydrogen as it forms, shifting the equilibrium to increase conversion and yield.[144] Similarly, in esterification reactions, hydrophilic nanomembranes can selectively remove water byproducts, driving the reaction to completion and improving product purity.

Catalytic membrane reactors further enhance this concept by incorporating catalysts either on the membrane surface or within the reaction chamber. This configuration is particularly beneficial for equilibrium-limited reactions, as the continuous removal of products can significantly increase conversion rates beyond what would be possible in traditional batch reactors.

In biotechnology, membrane bioreactors combine the principles of membrane separation with biological processes. These systems often involve microorganisms or enzymes and are used in wastewater treatment, fermentation processes, and the production of certain pharmaceuticals. The membranes in these bioreactors can serve multiple functions, such as retaining biomass, selectively removing inhibitory products, or controlling the introduction of substrates.[145]

Across all types of membrane reactors, the integration of reaction and separation processes reduces the need for additional downstream processing steps, thereby minimizing waste generation and energy consumption. This aligns perfectly with the goals of process intensification, offering more compact, efficient, and sustainable production methods for a wide range of industries, from petrochemicals to pharmaceuticals and environmental engineering.

The ongoing development of novel nanomaterials and membrane technologies continues to expand the potential applications of membrane reactors, promising even greater advancements in process intensification and sustainable chemical production in the future.

8.5.3.2.2 **Microreactors.** Microreactors, characterized by their small internal volumes typically in the microliter to milliliter range, represent a significant advancement in process intensification. These reactors, often incorporating nanostructured surfaces, provide enhanced heat and mass transfer properties due to their high surface area-to-volume ratio. This unique characteristic allows them to operate under more controlled conditions, enabling precise temperature and pressure regulation.[146]

Microreactors can be broadly categorized into several types, each with specific advantages:

- Continuous flow microreactors are the most common type, where reactants are continuously pumped through small channels. These are particularly useful for multi-step syntheses and can be easily scaled up by running multiple reactors in parallel.
- Droplet-based microreactors generate discrete droplets of reactants, each acting as an individual microreactor. This approach is beneficial for reactions requiring precise control over reagent ratios and reaction times.
- Falling film microreactors use gravity to create thin liquid films over catalytic surfaces, maximizing gas-liquid interactions. These are particularly useful for gas-liquid reactions such as hydrogenations.
- Packed bed microreactors contain small particles of catalyst or other reactive materials. They are effective for heterogeneous catalysis and can provide high catalyst surface area per reactor volume.

The unique properties of microreactors make them powerful tools for process intensification across various industries. Their ability to precisely control reaction conditions, improve mixing and heat transfer, and enhance safety has led to significant improvements in reaction efficiency and product quality.

In the pharmaceutical industry, for instance, microreactors have been used to optimize the synthesis of active pharmaceutical ingredients, resulting in reduced solvent use, increased yields, and improved purity. As research in this field progresses, microreactors are expected to play an increasingly important role in the development of more sustainable and efficient chemical processes, contributing to the broader goals of green chemistry and process intensification.

8.5.3.2.3 Energy Storage and Conversion. In batteries, nanostructured materials such as silicon nanowires and graphene can be incorporated into electrode materials, markedly increasing the capacity and efficiency of lithium-ion batteries.[147] These nanomaterials can better accommodate the volume changes during charge–discharge cycles, resulting in improved cycle life and higher energy density.

For instance, silicon nanowires can expand and contract without fracturing, a common issue with bulk silicon, thereby enhancing the longevity and performance of the batteries.[148] Additionally, the development of solid-state batteries is being propelled by nanotechnology, where nanoscale electrolytes improve ionic conductivity and safety compared to traditional liquid electrolytes.[149]

Supercapacitors also benefit from nanotechnology, particularly through the use of carbon nanotubes and graphene to create electrodes with extremely high surface areas.[150] These materials facilitate rapid charge and discharge cycles with high power density, making them ideal for applications requiring quick energy bursts, such as regenerative braking systems in vehicles. The high conductivity and stability of these nanomaterials further enhance the performance and durability of supercapacitors. Hybrid supercapacitors, which combine battery-like energy storage mechanisms with capacitive properties, leverage nanomaterials to deliver both high energy and power densities, bridging the gap between batteries and conventional capacitors.

Fuel cells are another area where nanotechnology makes a significant impact. Nanocatalysts enhance fuel cell efficiency by providing a larger active surface area for electrochemical reactions. This not only reduces the amount of precious metal catalysts required, lowering costs, but also improves the overall sustainability of fuel cells.

Ongoing research into alternative nanocatalysts, including palladium or non-precious metal catalysts, aims to further decrease costs and enhance efficiency. Additionally, nanostructured membrane electrode assemblies in fuel cells improve proton conductivity and durability, contributing to better overall performance.

The advancements brought by nanotechnology in these areas not only improve energy density and efficiency but also extend the lifespan of energy storage devices, reducing waste associated with frequent replacements. In the context of chemical processing, more efficient energy storage and conversion enable better utilization of intermittent renewable energy sources, thereby reducing the carbon footprint of chemical manufacturing. Enhanced energy management facilitated by nanotechnology allows for smoother integration of renewable energy into industrial processes, contributing to more sustainable and efficient chemical production.

8.5.3.3 Advanced Materials for Energy Efficiency

Nanostructured materials and nanocoatings are contributing significantly to energy efficiency by improving insulation and reducing energy loss in industrial settings. These advanced materials leverage the unique properties that emerge at the nanoscale to create more effective solutions for thermal management and energy conservation.

8.5.3.3.1 Nanostructured Materials for Improved Insulation. Nanostructured materials such as aerogels[151] and nanoporous insulators provide superior thermal insulation properties. These materials reduce energy loss in buildings and industrial processes, leading to significant energy savings and reduced carbon footprints.

Aerogels, often called "frozen smoke" due to their appearance, are ultralight materials composed of up to 99.98% air. Their nanoporous structure severely restricts heat transfer through the material, resulting in thermal conductivity values as low as 0.013 $W\,m^{-1}\,K^{-1}$ (at room temperature and atmospheric pressure), which is significantly lower than traditional insulation materials.[152] Silica aerogels are among the most common and have been used in various applications, from building insulation to space exploration.[153]

Another example of nanostructured insulation is vacuum insulated panels with nanoporous cores.[154] These panels use materials like fumed silica or aerogels as the core, which is then evacuated and sealed. The combination of the nanoporous structure and the vacuum results in thermal conductivity values as low as 0.0045 $W\,m^{-1}\,K^{-1}$, making them up to 10 times more effective than conventional insulation materials of the same thickness.

In industrial settings, nanostructured insulation materials are being applied to high-temperature processes. For instance, in oil refineries and petrochemical plants, nanoporous ceramic insulation is used to reduce heat loss from pipes and vessels operating at temperatures exceeding 1000 °C. This not only saves energy but also improves process efficiency and worker safety by reducing surface temperatures.

8.5.3.3.2 Heat-reflective Nanocoatings for Industrial Equipment. Heat-reflective nanocoatings applied to industrial equipment minimize heat absorption and improve thermal management. These coatings reduce the need for cooling and air conditioning, thereby conserving energy and reducing waste heat emissions.

One type of heat-reflective nanocoating uses yttrium oxide (Y_2O_3) nanofiber membranes. When applied to surfaces, these coatings can reflect up to 95% of infrared radiation.[155] Nanomaterials like these could potentially be used in various industries, including oil and gas, where they could be applied to storage tanks to reduce evaporation losses and maintain product quality.

Another example is the use of titanium dioxide nanoparticles in coatings. When properly engineered at the nanoscale, TiO_2 can provide both UV protection and infrared reflection. These coatings have been applied to building facades and roofs, reducing the urban heat island effect and lowering cooling costs.[156]

8.5.3.4 Smart Manufacturing Systems

The integration of sensors into manufacturing processes enables real-time monitoring and optimization, leading to more efficient and less wasteful production. This paradigm shift towards smart manufacturing is revolutionizing industrial processes across various sectors, from chemical production to electronics manufacturing.

8.5.3.4.1 (Nano)Sensors for Real-time Process Monitoring and Optimization. Sensors embedded in manufacturing systems provide real-time data on process conditions such as temperature, pressure, and chemical composition. This data enables precise control and optimization, reducing material waste and enhancing process efficiency.

One example of this technology is the use of traditional gas sensors in chemical reactors. These sensors can detect changes in gas composition, allowing for real-time adjustment of reaction conditions. In pharmaceutical manufacturing, this capability has led to improved yield and purity of active pharmaceutical ingredients, reducing waste and energy consumption.

In the semiconductor industry, optical emission sensors are used for *in situ* monitoring of plasma etching processes. These sensors can detect changes in plasma composition with high sensitivity, allowing for more precise control of the etching process.[157] This results in improved yield and reduced material waste in the production of microchips.

Colorimetric sensors are another innovation in this field. These sensors change color in response to specific chemical or physical stimuli, providing a visual indication of process conditions. For instance, in food processing, gold nanoparticle sensors can be used to monitor pH levels or detect the presence of contaminants, ensuring product quality and safety while minimizing waste due to spoilage or contamination.[158]

Once nanosensors move from the lab into real applications, they offer potential improvements in sensitivity and specificity. Nanosensors such as carbon nanotube-based sensors, quantum dot sensors, and nanoparticle-based colorimetric sensors could provide even more precise and sensitive monitoring capabilities. This could further enhance process control, reduce material waste, and improve product quality across various industries.

8.5.3.4.2 Nanoelectronics in Energy-efficient Control Systems. Nanoelectronics has the potential to enhance the performance of control systems used in manufacturing, allowing for more precise and efficient control of production processes. This leads to reduced energy consumption and minimizes waste.

One significant application is in the development of neuromorphic computing systems for process control. These systems, inspired by the structure and function of biological neural networks, use nanoelectronic devices like memristors to create highly efficient, low-power computing architectures.[159] In complex manufacturing processes, such as those in the automotive or aerospace industries, these systems potentially could process vast amounts of sensor data in real-time, making rapid adjustments to optimize production efficiency and quality.

Another example is the use of carbon nanotube field-effect transistors (CNTFETs) in control systems. CNTFETs are advancing towards commercial viability, with researchers demonstrating their swift production in commercial facilities using existing silicon transistor manufacturing equipment.[160] This type of transistor is significantly more energy-efficient than its silicon counterparts and has the potential to enable innovative three-dimensional microprocessors. Historically, CNFETs have been produced in limited quantities within academic settings, but the new manufacturing methods pave the way for broader industrial adoption.

8.5.3.4.3 Integration of Artificial Intelligence and Machine Learning. The true power of smart manufacturing systems is evident when sensors and nanoelectronics are coupled with artificial intelligence (AI) and machine learning (ML) algorithms. Currently, regular sensors and conventional electronics are widely used in these systems to enable predictive maintenance, adaptive process control, and continuous optimization. These sensors and electronics analyze data on vibration, temperature, and other parameters to predict equipment failures and schedule maintenance activities, reducing downtime and extending equipment life.[161]

In additive manufacturing (3D printing), ML algorithms process data from these sensors to monitor factors like material deposition rate, temperature gradients, and cooling rates. This allows for real-time adjustments to printing parameters, ensuring consistent part quality and reducing the need for post-processing or part rejection.

Moving to nanoscale sensors, although not yet widely adopted, could offer significant benefits. Nanosensors provide higher sensitivity and precision, detecting minute changes that regular sensors might miss. They can deliver enhanced data quality, leading to more reliable maintenance predictions and more precise control in manufacturing processes. Additionally, their smaller size allows for integration into more components, offering comprehensive monitoring and lower power consumption.

The integration of nanosensors and AI in smart manufacturing systems will represent a significant leap forward in industrial efficiency and sustainability. However, the path to widespread adoption is not without obstacles. Key challenges include the standardization of nanosensor interfaces, ensuring the long-term stability of nanoelectronic components in harsh industrial environments, and developing robust data analytics platforms capable of processing vast amounts of data.

As these technologies mature and become more widely adopted, they will play a pivotal role in shaping the future of industry. Smart manufacturing systems leveraging nanotechnology are set to transform industrial processes, leading to unprecedented levels of efficiency, quality, and sustainability. This technological revolution will be instrumental in realizing the vision of Industry 4.0 and beyond, paving the way for a more productive and environmentally conscious manufacturing landscape.

8.5.3.5 Challenges and Future Directions

While the benefits of nanotechnology in energy-efficient production are significant, several challenges need to be addressed to realize its full potential.

8.5.3.5.1 Scaling Up Nano-enabled Technologies. One of the main challenges is scaling up the production and application of nano-enabled technologies to industrial levels. Research and development efforts are needed to develop cost-effective manufacturing processes that can produce nanomaterials in large quantities.

The transition from laboratory-scale production to industrial-scale manufacturing of nanomaterials presents several hurdles:

- Maintaining quality and consistency: Ensuring uniform properties and performance of nanomaterials when produced in large quantities is crucial. Variations in size, shape, or composition can significantly affect their performance in applications.
- Cost-effectiveness: Developing economically viable production methods that can compete with existing technologies is essential for widespread adoption.
- Equipment and process adaptation: Existing industrial equipment and processes may need significant modifications to handle nanomaterials effectively, requiring substantial investment.
- Supply chain integration: Establishing reliable supply chains for raw materials and developing logistics for handling and transporting nanomaterials safely are necessary considerations.

8.5.3.5.2 Safety and Environmental Considerations. The safety and environmental impact of nanomaterials must be carefully evaluated. This includes understanding the potential risks associated with the production, use, and disposal of nanomaterials, and developing guidelines to ensure their safe and sustainable use.

Key areas of focus include:

- Occupational safety: Developing protocols to protect workers involved in the production and handling of nanomaterials from potential exposure risks.
- Environmental impact assessment: Studying the long-term effects of nanomaterials on ecosystems and developing strategies to mitigate any negative impacts.
- Life cycle analysis: Conducting comprehensive life cycle assessments of nano-enabled products to ensure that their overall environmental impact is indeed positive.
- Regulatory frameworks: Establishing and refining regulations and standards for the safe production, use, and disposal of nanomaterials across different industries and applications.
- Public perception and education: Addressing public concerns about nanotechnology through transparent communication and education initiatives.

8.5.4 Case Study: Nanosensors Enable Real-time Detection of Plant Stress Signaling Molecules

Plant stress detection has long been a critical challenge in agriculture, with the ability to identify early signs of stress potentially saving crops and improving yields. Traditional methods of stress detection often rely on visual symptoms, which appear only after significant damage has already occurred. Over the years, researchers have explored various technologies to detect plant stress earlier, including chlorophyll fluorescence spectroscopy and hyperspectral imaging. However, these approaches primarily detect metabolic changes that occur after the initial stress perception and signaling within plants.

A groundbreaking study showcases an innovative approach to monitoring plant stress responses in real-time using advanced nanosensors.[162] The research team developed a novel salicylic acid (SA) nanosensor and combined it with an existing hydrogen peroxide (H_2O_2) sensor, enabling simultaneous detection of these crucial signaling molecules in living plants under various stress conditions.

The centerpiece of this research is the SA nanosensor, created through a meticulous screening process using corona phase molecular recognition (CoPhMoRe). By testing different cationic polymer-wrapped single-walled carbon nanotubes, the team identified a sensor (S3) with high specificity for SA. This sensor exhibited a dissociation constant of 32 μM and could detect SA at concentrations as low as 4.4 μM, outperforming its affinity for other plant hormones and signaling compounds.

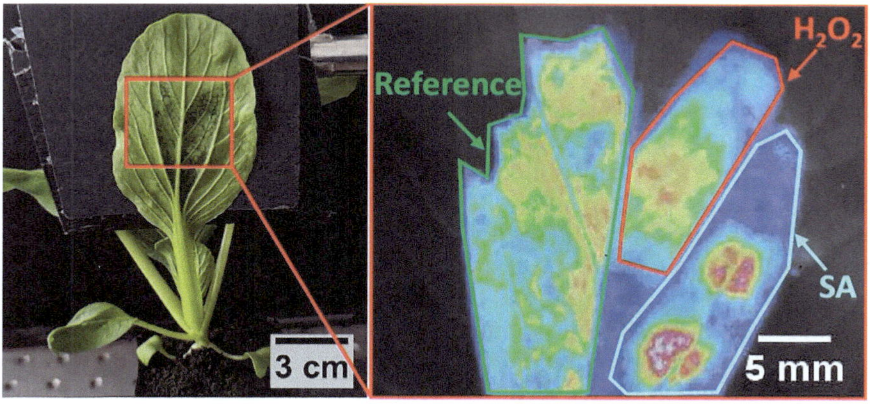

Figure 8.15 Schematic demonstrating the multiplexed sensor platform for the early decoding of different plant stresses using an in planta SA nanosensor (blue) paired with an H_2O_2 nanosensor (red) and a reference sensor (green). Adapted from ref. 256, https://doi.org/10.1038/s41467-024-47082-1, under the terms of the CC BY 4.0 license, https://creativecommons.org/licenses/by/4.0/.

Rigorous testing demonstrated the SA nanosensor's effectiveness and safety in plant systems (see Figure 8.15). Using confocal microscopy, the researchers observed that the sensor primarily accumulated in the cytoplasm and chloroplasts of plant cells—ideal locations for monitoring SA production.

The team validated the sensor's performance using genetically modified *Arabidopsis thaliana* plants with controllable SA production, showing a strong correlation between the sensor's output and SA levels measured *via* traditional LC–MS methods.

The true potential of this nanosensor technology shines through in its application to real-world plant stress monitoring. By introducing both SA and H_2O_2 sensors into pak choi leaf tissue, the researchers uncovered distinct temporal patterns of these signaling molecules in response to various stressors:

- Mechanical damage triggered a swift H_2O_2 surge but minimal SA production within the 4-hour observation window.
- Bacterial infection with Xanthomonas campestris pv. campestris initiated an H_2O_2 wave followed by SA production approximately 87 minutes later.
- Exposure to high light intensity prompted both H_2O_2 and SA production, with SA levels rising around 22.5 minutes after stress onset.
- Heat stress elicited a response similar to high light, with SA production beginning roughly 35.5 minutes post-stress.

These unique stress signatures highlight the potential of multiplexed nanosensors for rapid, non-invasive identification and differentiation of various plant stressors. Furthermore, the research team leveraged this experimental data to construct a mathematical model describing the interplay between H_2O_2 and SA signaling cascades, offering new insights into stress-specific response mechanisms.

A key advantage of this nanosensor platform is its versatility across plant species. The sensors functioned effectively in both the model plant Arabidopsis and the commercially relevant pak choi, underscoring their broad applicability. Moreover, the ability to detect stress responses within hours, before visible symptoms manifest, could revolutionize agricultural practices by enabling timely interventions to mitigate crop losses.

This research represents a significant step forward in bridging the gap between cutting-edge nanomaterials and practical agricultural solutions. By harnessing the unique properties of SWNTs, the team has created powerful tools for probing complex plant biology. The simultaneous monitoring of multiple signaling molecules demonstrates how engineered nanomaterials can address intricate biological questions within a single system.

While further work is needed to expand the range of detectable molecules and assess long-term sensor stability under diverse environmental conditions, this study provides compelling evidence for the potential of nanosensor technology in next-generation plant monitoring. As global agriculture faces mounting challenges from climate change, innovations like these multiplexed nanosensors offer promising avenues for developing more resilient and sustainable crop production systems.

8.6 Energy Efficiency

Nanotechnology can contribute to reducing energy waste by improving the efficiency of energy generation, storage, transmission and utilization (see Figure 8.16). This section explores how nanomaterials and nanostructures are revolutionizing energy efficiency, contributing to sustainable development and waste reduction.

8.6.1 Nanomaterials for Enhanced Energy Generation

The pursuit of efficient, sustainable energy sources has led researchers to harness the unique properties of nanomaterials, potentially revolutionizing renewable energy technologies. Solar energy conversion and thermoelectric materials showcase the transformative potential of nanotechnology in addressing global energy challenges.

8.6.1.1 Solar Energy Conversion

Quantum dots have emerged as promising materials in photovoltaics due to their tunable optical and electronic properties. These semiconductor nanocrystals, exhibiting quantum confinement effects, can theoretically be

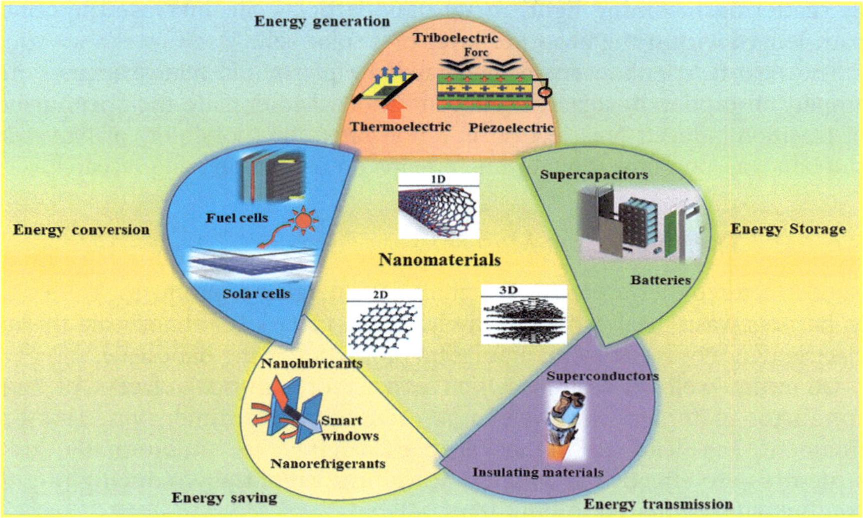

Figure 8.16 Nanomaterials play a crucial role across various energy sectors, including conversion, storage, saving, and transmission. Key applications involve technologies such as fuel cells, solar cells, supercapacitors, and batteries for energy conversion and storage. They also contribute to energy saving with innovations like nanolubricants, smart windows, and nanorefrigerants. In energy transmission, superconductors and insulating materials are essential components. Additionally, energy generation methods, including thermoelectric, piezoelectric, and triboelectric technologies, benefit from the integration of 1D, 2D, and 3D nanomaterials. Reproduced from ref. 257 with permission from the Royal Society of Chemistry.

engineered to absorb light across a broader spectrum of solar radiation than conventional solar cells.[163]

The phenomenon of multiple exciton generation (MEG) in QDs further pushes the boundaries of potential solar cell efficiency. By generating multiple electron–hole pairs from a single high-energy photon, MEG offers a path to surpass traditional efficiency limits in solar energy conversion. Notably, devices incorporating PbTe QDs have demonstrated external quantum efficiencies exceeding 100%, underscoring the practical enhancement of photocurrent achievable through MEG.[164]

Perovskite materials, particularly organic–inorganic hybrids, have garnered significant attention in the solar energy community due to their rapid improvement in efficiency over the past decade. The ability to control perovskite crystalline structure at the nanoscale has proven crucial in optimizing light absorption and charge transport.[165]

Plasmonic nanostructures offer yet another avenue for potentially enhancing solar cell performance. These noble metal nanoparticles, typically gold or silver, employ various mechanisms to boost light absorption.

By scattering incoming light, metal nanoparticles can increase the optical path length within the absorber layer of a solar cell.[166]

The near-field enhancement generated by plasmonic nanostructures may amplify absorption in adjacent semiconductors. Additionally, the phenomenon of plasmon-induced hot electron injection offers the possibility of harvesting sub-bandgap photons, though this remains an area of ongoing research.

8.6.1.2 Thermoelectric Materials

Nanostructured thermoelectric materials are emerging as promising candidates to harness waste heat more effectively. These materials enhance the thermoelectric figure of merit by manipulating properties at the nanoscale.[167]

Quantum well structures, which consist of nanoscale layers of semiconductor materials, confine charge carriers in one dimension. This confinement can lead to higher Seebeck coefficients. Additionally, these structures are effective at phonon scattering, potentially reducing thermal conductivity compared to bulk materials.

Nanocomposite thermoelectric materials take a different approach by incorporating nanoscale inclusions or precipitates into a bulk matrix to enhance phonon scattering. The concept of all-scale hierarchical architecture, which integrates nano-, meso-, and microscale structures, has shown potential to scatter phonons across a wide range of mean free paths. Recent studies have demonstrated that these nanocomposites significantly improve thermoelectric performance.[168]

For instance, PbS–Ag nanocomposites have achieved a figure of merit as high as 1.7 at 850 K, a 30% improvement over previous PbS records.[169] This increase is primarily attributed to enhanced electrical conductivity and effective phonon scattering. Such advancements have practical significance for the development of more efficient and cost-effective thermoelectric devices used in waste heat recovery and energy harvesting applications.

While nanoscale engineering offers promising avenues, many of these techniques are still in the research phase. Practical implementation in commercial energy systems will require ongoing efforts to address scalability, cost-effectiveness, and long-term stability.

8.6.2 Energy Storage

The efficient storage and distribution of energy are crucial components in the transition to a sustainable energy future. Nanotechnology offers promising avenues for enhancing the performance of energy storage devices and improving the efficiency of power distribution systems.

8.6.2.1 Advanced Battery Technologies

Nanostructured materials are at the forefront of battery technology advancements, addressing key challenges in energy density, charging speed, and cycle life. High surface area nanostructures, such as graphene and carbon nanotubes, show promise for increasing the electrode–electrolyte interface in batteries.[170] This enhanced interface could improve charge and discharge rates, as well as overall capacity. The high conductivity of these materials may also contribute to improved battery performance.

Researchers are actively exploring a range of nanostructured designs. One such approach involves the development of 3D porous electrodes that aim to maximize surface area while maintaining efficient electron and ion transport pathways. Another technique focuses on integrating nanoparticles of active materials into conductive matrices, which could enhance both capacity and conductivity. However, challenges remain regarding scalable manufacturing and ensuring the long-term stability of these nanostructured electrodes.

8.6.2.2 Solid-state Electrolytes

Nanocrystalline ceramic electrolytes are being investigated for their potential to enable safer, higher-capacity solid-state batteries. These materials offer several advantages over conventional liquid electrolytes, including enhanced safety due to their non-flammable nature, the potential for higher energy density by enabling the use of lithium metal anodes, and improved thermal stability. However, significant challenges remain, such as improving ionic conductivity at room temperature and ensuring good interfacial contact between the electrolyte and electrodes.[171]

8.6.2.3 Silicon Anodes

Nanostructured silicon is being explored as a promising anode material due to its high theoretical capacity. The nanostructuring approach aims to address the significant volume changes that silicon undergoes during charging and discharging, which typically lead to rapid degradation in bulk form. Various strategies are being investigated to mitigate this issue, including the use of silicon nanowires, porous silicon structures, and silicon-carbon nanocomposites. While laboratory progress has been encouraging, there is still considerable work needed to scale up production and ensure long-term cycle stability.[172]

8.6.2.4 Supercapacitors

Supercapacitors, also known as ultracapacitors, are another key area of energy storage, offering rapid charge and discharge capabilities that complement batteries in many applications. Nanomaterials play a crucial role in the development of high-performance supercapacitors, particularly through the use of carbon-based materials. Graphene and carbon nanotubes, with their exceptionally high surface areas and good electrical conductivity, are being extensively studied as potential materials for supercapacitor electrodes. These materials could enable supercapacitors with higher energy densities while maintaining the characteristic high power densities for which they are known.[173,174]

Researchers are also focused on optimizing the pore size distributions of these carbon-based materials to maximize capacitance, as well as developing scalable production methods for high-quality graphene and carbon nanotubes. Another promising approach involves the creation of hybrid structures that combine different carbon nanomaterials to achieve superior performance.

In addition to carbon-based materials, transition metal oxides such as ruthenium oxide and manganese oxide are being explored for their pseudocapacitive properties. When structured at the nanoscale, these metal oxides can provide additional charge storage through fast, reversible redox reactions at the electrode surface. However, challenges persist, particularly in improving the electrical conductivity of metal oxide electrodes, enhancing cycle stability for high-capacity materials, and developing cost-effective alternatives to expensive noble metal oxides.[175]

8.6.3 Energy-efficient Buildings and Infrastructure

The built environment significantly contributes to global energy consumption and greenhouse gas emissions. Nanotechnology offers innovative solutions to enhance energy efficiency in buildings and infrastructure, which could result in substantial reductions in both energy use and environmental impact.

For instance, advanced thermal insulation materials are crucial in reducing the energy demands for heating and cooling buildings. Nanotechnology enables the development of highly efficient insulation materials that surpass the performance of traditional options. Aerogels, for example, are ultra-low-density materials composed of up to 99.8% air, making them excellent thermal insulators.[176]

Nanostructured aerogels provide numerous advantages, including extremely low thermal conductivity (0.01–0.02 $W\,m^{-1}\,K^{-1}$), lightweight properties (with densities ranging from 0.003 to 0.15 $kg\,m^{-3}$), and high porosity (85–99.9%). Additionally, they offer the potential for transparent insulation, allowing daylight to enter buildings while reducing thermal loss.[177]

Research in aerogel technology is currently focused on improving mechanical strength, developing hydrophobic variants to prevent moisture-related degradation, and reducing production costs to make them more accessible for widespread use.

Another promising nanotechnology application is Vacuum Insulated Panels (VIPs), which utilize nanostructured core materials like fumed silica to provide superior insulation. VIPs offer thermal conductivities as low as 0.0045 $W\,m^{-1}\,K^{-1}$ under pristine conditions, which makes them one of the most effective thermal insulation materials available. Their space-saving design enables thinner wall constructions without compromising on insulation performance, making them suitable for retrofitting existing buildings.

However, challenges such as maintaining vacuum integrity, addressing thermal bridging at panel edges, and the high costs associated with VIPs hinder their widespread adoption.[178]

Bio-based nanomaterials are also gaining attention as renewable resources for sustainable insulation.[179] One such example is cellulose nanofibers derived from plant materials that offer excellent insulative properties, with thermal conductivities as low as 0.023 $W\,m^{-1}\,K^{-1}$.[180] These fibers are biodegradable, possess significant mechanical strength (with tensile strengths comparable to Kevlar), and are a sustainable alternative to conventional insulation materials. Similarly, lignin, a byproduct of the paper industry, is being developed into foam-like insulation materials, presenting a renewable and efficient option. Chitin nanofibrils, extracted from crustacean shells, are also emerging as promising materials due to their biodegradable nature and inherent fire resistance.[181]

Research in bio-based nanomaterials is primarily aimed at improving moisture resistance and enhancing mechanical properties. For instance, surface treatments and chemical modifications are being explored to prevent water absorption and extend the lifespan of these materials. Additionally, scalable and cost-effective production methods are being developed to facilitate the large-scale adoption of bio-based nanomaterials in the construction industry.

Nanocoatings further enhance the energy efficiency of buildings by improving the insulative properties of construction materials. Reflective nanocoatings, often applied to roofs and facades, utilize materials like zinc oxide and titanium dioxide to reduce solar heat gain, thereby lowering cooling loads.[182,183]

Nanocoatings that incorporate phase-change materials can store thermal energy during the day and release it at night, stabilizing indoor temperatures and reducing the need for mechanical heating and cooling.[184] Self-cleaning nanocoatings, which use superhydrophobic and photocatalytic nanomaterials, help maintain the performance of insulative surfaces by keeping them free of dirt and contaminants.[185]

8.6.3.1 Smart and Photovoltaic Windows

Nanotechnology has enabled the development of advanced window systems that can dynamically control solar heat gain and light transmission while potentially generating electricity. These multifunctional windows are designed to optimize energy efficiency, enhance occupant comfort, and contribute to renewable energy generation in buildings.

One such innovation involves the integration of electrochromic coatings with transparent photovoltaic materials. Electrochromic windows can adjust their tint to manage daylight and optimize thermal conditions within a building. By selectively modulating solar transmittance, these windows reduce the need for artificial lighting and minimize cooling loads.[186]

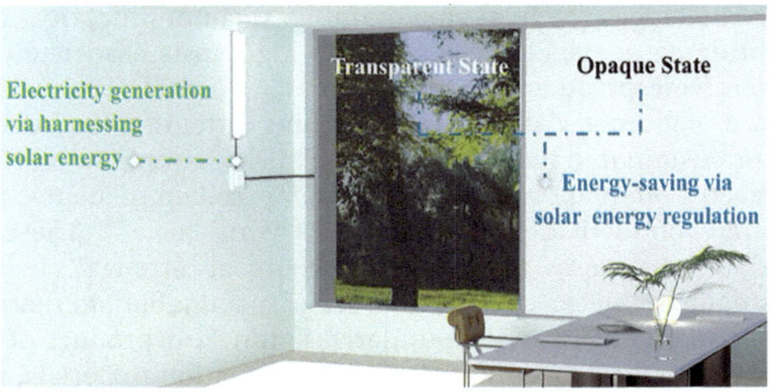

Figure 8.17 Key features of smart photovoltaic windows. Adapted from ref. 258, https://doi.org/10.1002/advs.202407177, under the terms of the CC BY 4.0 license, https://creativecommons.org/licenses/by/4.0/.

When combined with photovoltaic materials, these windows not only control light and heat but also generate electricity from sunlight (see Figure 8.17). This electricity can power the electrochromic functions, creating a self-sustaining system. However, challenges such as balancing transparency, energy generation, and tinting capabilities, optimizing the layer structure, and ensuring long-term stability must be addressed to make these systems viable on a large scale.[187]

Thermochromic windows change their optical properties in response to temperature. When combined with PV functionalities, these windows offer passive solar regulation, adjusting heat gain based on external conditions and reducing cooling loads in hot climates. In addition to regulating solar heat, thermochromic windows can generate electricity in both their clear and tinted states, potentially increasing efficiency at higher temperatures.[188] Ongoing research aims to fine-tune the transition temperatures for various climates, develop materials that maintain efficiency across a range of temperatures, and ensure the durability of these integrated systems.

Several nanostructured materials play a crucial role in enhancing the performance of smart and photovoltaic window technologies. Quantum dots enable spectrum-selective absorption, which improves photovoltaic efficiency without compromising the transparency of the windows.[189] Plasmonic nanoparticles enhance light trapping, potentially boosting both the tinting and energy generation capabilities of the windows.[190] Additionally, nanostructured transparent conductors improve electrical performance and optical clarity, which is critical for integrating multiple functions within a single window system.[191] Key research areas include the development of non-toxic alternatives to traditional quantum dot materials, optimization of nanostructures for balanced performance, and improved scalability of production techniques.

Smart and photovoltaic windows can also be integrated into broader building management systems to further optimize energy efficiency. These windows can be adjusted in real time based on occupancy, time of day, and weather conditions, coordinating with heating, ventilation, and air conditioning (HVAC) systems. Advanced algorithms can use weather forecasts and historical data to predict and regulate window settings, enhancing overall system efficiency. This integration involves developing reliable communication protocols, ensuring cybersecurity, and creating intuitive user interfaces for occupants.

While these advanced window technologies hold great potential to reduce building energy consumption and contribute to renewable energy generation, practical implementation still faces challenges. Overcoming issues related to cost, scalability, long-term performance, and seamless integration with existing building systems is critical for the widespread adoption of smart and photovoltaic windows in the construction industry.

8.6.3.2 Energy-efficient Lighting

Nanotechnology plays a crucial role in advancing energy-efficient lighting solutions, particularly in LED and OLED technologies. These innovations have led to significant improvements in both performance and energy savings.

Nanostructured materials have enhanced LED technology in several ways. Quantum dots, for instance, emit precise wavelengths of light, leading to more vibrant and accurate colors while maintaining high energy efficiency.[192] These nanoparticles allow for improved color rendering, reducing energy consumption by emitting specific wavelengths more effectively. Quantum dot-based LEDs (QLEDs) are particularly known for their high efficiency and stable emission, which contributes to superior color accuracy and energy savings in both lighting and display applications.

Another advancement comes from nanostructured surfaces, which improve light extraction from LEDs. These surfaces reduce internal reflection and allow more light to escape from the device, resulting in increased brightness without additional power consumption.[193,194]

Surface/interface engineering at the nanoscale plays a vital role in enhancing the performance of light-emitting devices by reducing energy losses. In addition, nanoscale phosphors have been shown to be highly efficient in wavelength conversion, further improving LED performance. These phosphors boost the luminous flux and color rendering index of LEDs, leading to higher brightness and better energy utilization without extra power consumption.[195]

Ongoing research in LED technology is focused on addressing environmental concerns by developing non-toxic alternatives to cadmium-based quantum dots, improving the efficiency of blue and green LEDs, and enhancing thermal management at the nanoscale. For example, researchers are exploring zinc oxide nanostructures as a safer alternative to cadmium, as well as other zinc-based compounds like ZnSe/ZnS quantum dots, which show promise in achieving pure-blue light emission while maintaining environmental safety.[196]

Overall, LEDs offer a highly efficient lighting solution, transforming up to 40% of electricity into light and boasting lifespans exceeding 100 000 hours. Their widespread adoption is projected to result in substantial energy savings by 2035, highlighting the critical role of nanotechnology in advancing LED efficiency.[197]

OLED technology, which offers flexible, large-area lighting solutions, also benefits significantly from nanotechnology. Nanostructured electrodes, such as those made from graphene and carbon nanotubes, improve charge injection in OLEDs by enhancing electrical conductivity and reducing operational voltages. For instance, graphene-based transparent electrodes have shown high conductivity and transparency, making them ideal for use in efficient, flexible OLEDs. Recent studies demonstrate that the incorporation of graphene with selective defect decoration further enhances light extraction and charge injection, positioning graphene-based OLEDs as a promising next-generation solution.[198]

Quantum dots are also employed in OLEDs to achieve high color purity and brightness. Their narrow emission spectra enable over 90% color purity, which is particularly valuable for high-definition displays. These QD-OLEDs provide superior color accuracy and efficiency compared to traditional emitters, making them highly suitable for applications that require vivid and accurate color reproduction.[199]

Nanocomposite materials have further improved the stability and operational lifetime of OLED devices. Hybrid organic–inorganic nanocomposites, when incorporated into the emissive layer, can enhance the thermal and mechanical properties of OLEDs, extending their operational lifetime by up to 50%.[200]

Despite these advances, OLED technology still faces challenges. One major issue is extending the operational lifespan of blue emitters, which degrade more quickly than red and green emitters. While red and green OLEDs have achieved operational lifetimes exceeding 20 000 hours, blue emitters often fall short, with typical lifetimes around 10 000 hours. Research is ongoing to develop more stable blue emitters and improve overall device longevity.[201]

Another challenge is scaling up the manufacturing processes for large-area OLED panels. Although small-scale OLED displays have found commercial success, producing large panels with uniform quality remains a significant hurdle. Researchers are exploring scalable manufacturing techniques, such as roll-to-roll processing, to facilitate the widespread adoption of large-area OLEDs.[202]

Additionally, improving light outcoupling efficiency remains a critical area of focus. A considerable amount of light generated within OLEDs is trapped inside the device due to internal reflection. To address this issue, techniques such as nanostructured surfaces and microlens arrays are being explored to increase light extraction. Studies show that these approaches can improve light outcoupling efficiency by up to 30%, leading to enhanced brightness and energy efficiency.

Nanotechnology has significantly advanced LED and OLED lighting by improving efficiency, color rendering, and durability. However, further research and development are needed to overcome remaining challenges in scalability, material stability, and light outcoupling efficiency, ensuring that these technologies continue to advance and play a pivotal role in energy-efficient lighting solutions.

8.6.3.3 Structural Materials

Incorporating nanomaterials into construction materials like concrete can improve strength, reduce environmental impacts, and enable the use of thinner products with shorter setting times. These advancements include using recycled materials and marginal aggregates, making concrete construction both economical and eco-friendly.[203]

Nanoengineered concrete benefits from the addition of nanomaterials such as carbon nanotubes and nanofibers, which improve compressive and flexural strength by up to 50% and 87%, respectively. These materials also enhance crack resistance, fracture energy, and toughness by creating a network within the concrete that bridges cracks and pores.[204]

Nanosilica further improves durability and reduces permeability, making the concrete more resistant to environmental factors, thereby extending its lifespan. Additionally, photocatalytic nanoparticles can create self-cleaning and air-purifying surfaces, contributing to healthier and more sustainable building environments.[205]

Challenges remain in ensuring the uniform dispersion of these nanomaterials within concrete mixtures, which is essential for maximizing their benefits. Techniques like carboxylation, COOH group functionalization, and the use of surfactants such as polycarboxylic acid-based superplasticizers are being explored to improve dispersion. Long-term studies are also necessary to assess the durability and environmental impact of nano-enhanced concrete, particularly in its interactions with environmental stressors over time. Research continues to focus on developing cost-effective, scalable production methods to make these advanced materials accessible for widespread construction use.

Nanotechnology has also paved the way for self-healing materials, which can autonomously repair minor damage. One approach involves microencapsulated healing agents that are released when cracks form, restoring the material's integrity.[206] Another method uses vascular networks embedded in the material, supplying healing agents continuously for ongoing repair. This system addresses various levels of damage and allows for the indefinite replenishment of the healing agent, a capability that intrinsic and capsule-based systems cannot provide.[207] Shape memory alloys offer yet another solution, with the ability to revert to their original shape and close cracks, providing a self-healing mechanism for structural materials.

Despite these innovations, there are challenges in ensuring the long-term functionality of self-healing mechanisms, particularly under varying environmental conditions. Scaling up production for large-scale applications presents another significant obstacle, as manufacturing methods need to preserve the materials' properties on a broader scale. Additionally, integrating self-healing capabilities without compromising other essential material properties, such as strength and durability, remains a key focus of ongoing research.

8.6.4 Energy-efficient Transportation

The transportation sector is a major consumer of energy and a significant source of greenhouse gas emissions. Advances in nanotechnology offer promising solutions to enhance the energy efficiency of vehicles and transportation systems. By incorporating lightweight materials, improving lubricants and coatings, and developing advanced propulsion systems, nanotechnology is paving the way for more sustainable and energy-efficient transportation options.

8.6.4.1 Lightweight Materials

Reducing the weight of vehicles is a critical step towards improving fuel efficiency and reducing emissions. Nanotechnology enables the creation of lightweight yet strong materials that can replace traditional heavy components in vehicles. These innovations contribute to significant energy savings and environmental benefits.

8.6.4.1.1 Carbon Nanotube Reinforced Polymers. These materials are renowned for their exceptional strength-to-weight ratio, offering mechanical strength that rivals or surpasses steel while maintaining a fraction of the weight. Carbon nanotube-reinforced polymer composites, in particular, exhibit outstanding tensile strength and stiffness, which make them ideal candidates for lightweight structural applications. Studies show that CNT-reinforced composites can reduce the weight of vehicle components significantly, with some reports indicating reductions of up to 25–30% when compared to conventional materials.[208] This substantial reduction in weight directly enhances fuel efficiency, as lighter vehicles require less energy to operate. In turn, this leads to lower fuel consumption and reduced greenhouse gas emissions, offering both economic and environmental benefits. The high tensile strength and elasticity of CNTs also enhance the durability and performance of these lightweight composites, making them ideal for automotive and aerospace applications.[209]

8.6.4.1.2 Nanostructured Metals. Grain refinement at the nanoscale leads to the creation of ultra-strong, lightweight alloys, offering superior mechanical properties and reduced weight compared to their bulk counterparts. Studies show that refining grains in magnesium and aluminum alloys results in enhanced strength and ductility, making them suitable for critical structural components in vehicles and aircraft.

For instance, aluminum alloys processed through methods like magnetron sputtering show significant improvements in strength due to refined grain structures. Similarly, magnesium alloys benefit from grain refinement techniques, leading to better strength-to-weight ratios and overall performance in critical applications.[210]

8.6.5 Improved Lubricants and Coatings

Enhancing the performance of lubricants and coatings is essential for reducing friction and wear in mechanical systems, leading to improved energy efficiency and longevity of transportation vehicles. Nanotechnology offers innovative solutions to achieve these improvements through the use of nanoparticle additives and nanostructured coatings.

8.6.5.1 Nanoparticle Additives

Incorporating graphene, MoS_2, or boron nitride into lubricants can significantly reduce friction coefficients, with studies showing reductions of up to 47%. These nanoparticle additives create a protective layer on the surfaces of moving parts, minimizing direct contact and reducing energy loss due to friction. Research demonstrates that these additives improve the dispersibility and stability of the lubricants, enhancing their tribological performance and ensuring more efficient lubrication under a variety of mechanical conditions.[211,212]

Graphene-based nanomaterials, in particular, exhibit excellent thermal conductivity and anti-wear properties, making them ideal for high-performance engines. The synthesis methods and tribological properties of various graphene-based nanomaterials have been extensively reviewed, revealing their ability to form a lubricating film that protects the contact surfaces during operation. This property significantly reduces friction and wear, thus enhancing the efficiency and longevity of mechanical systems.[213]

By incorporating these nanomaterials into lubricants, the overall efficiency of engines and machinery is improved, and the harmful effects of friction and wear are mitigated. The exceptional properties of these nanoparticles make them a valuable addition to modern lubricants, aligning with the growing demand for energy-efficient and durable mechanical components.

8.6.5.2 Nanostructured Coatings

Hard nanocomposite coatings, such as titanium nitride (TiN) and diamond-like carbon (DLC), are increasingly used to protect engine components from wear and corrosion. These coatings demonstrate exceptional hardness and low friction coefficients, which significantly reduce energy waste due to mechanical losses and extend the service life of critical engine parts. This leads to lower maintenance costs and improved fuel efficiency.

Research highlights the tribological properties of DLC and TiN coatings, showing their high wear resistance and low friction, making them suitable for heavy-duty aerospace transmission components.[214] In the automotive sector, the improvement of DLC coatings has been noted for their ultra-low friction and high wear resistance, contributing to enhanced durability and performance of engine parts.[215]

8.6.5.3 Advanced Propulsion Systems

Nanotechnology is playing a role in the transformation of propulsion systems in transportation, paving the way for more energy-efficient and sustainable solutions. By advancing batteries, supercapacitors, and fuel cells, it boosts the performance and efficiency of electric and hybrid vehicles, as well as conventional internal combustion engines.

8.6.5.3.1 Electric and Hybrid Vehicles. Nanomaterials are essential in advancing batteries and supercapacitors for electric and hybrid vehicles. Nanostructured electrodes, including graphene and carbon nanotubes, provide high surface areas and excellent conductivity, significantly enhancing the performance of energy storage devices.[216] Solid-state electrolytes with nanoscale architectures improve the safety and energy density of batteries, allowing for longer driving ranges and faster charging times. These advancements facilitate the widespread adoption of electric and hybrid vehicles, thereby reducing dependence on fossil fuels and lowering emissions.

Recent research highlights the impact of nanomaterials on battery technology. For instance, incorporating graphene into lithium-ion battery electrodes improves conductivity and durability, resulting in better overall battery performance. Graphene-enhanced electrodes show a significant increase in charge–discharge cycles and energy capacity compared to traditional materials.[217,218]

Carbon nanotubes increase the surface area and stability of supercapacitor electrodes, leading to a higher energy storage capacity. This improvement is crucial for applications requiring rapid energy delivery and high power density, such as electric vehicles.[219]

Nanostructured solid-state electrolytes in batteries enhance thermal stability and energy density, addressing safety concerns and extending battery life. The nanoscale architecture of the electrolytes allows for more efficient ion transport, which is key to improving the performance and longevity of batteries in electric and hybrid vehicles. The development of ultra-thin, flexible, and robust nanostructured solid electrolyte membranes showcases high electrochemical stability, excellent thermal endurance, and significant mechanical strength. This leads to better specific capacity, rate capability, and cycle life of lithium batteries, ensuring safer and more reliable operation.[220]

Additionally, advancements in high-energy all-solid-state lithium–metal batteries through the use of nanomaterials demonstrate enhanced energy density and improved safety by facilitating efficient ion transport and preventing lithium dendrite formation.[221] These improvements are crucial for the widespread adoption of electric and hybrid vehicles, making nanostructured solid-state electrolytes a promising solution for next-generation battery technologies.

These findings underscore the importance of nanomaterials in driving the innovation and efficiency of energy storage systems for electric and hybrid vehicles. The integration of advanced nanomaterials into battery and supercapacitor technologies is a critical step toward achieving sustainable transportation solutions and reducing the environmental impact of fossil fuel consumption.

8.6.5.3.2 Fuel Cells. Nanotechnology significantly enhances the performance and durability of fuel cells used in transportation. Nanostructured catalysts, such as platinum nanoparticles, are critical in increasing the efficiency of electrochemical reactions, thereby improving the power output of fuel cells. Research highlights that platinum-based catalysts, designed with controlled preparation and structural optimization, significantly boost the catalytic activity and stability of proton exchange membrane fuel cells (PEMFCs), addressing issues of high cost and poor durability.[222]

The development of alternative catalysts, including transition metal alloys and metal–organic frameworks, aims to reduce the dependency on expensive noble metals. These alternative catalysts are engineered to provide high catalytic activity and stability, which are essential for the long-term operation of fuel cells. Advances in nanostructured materials, such as carbon-supported catalysts, have shown promise in enhancing the performance and reducing the manufacturing costs of fuel cells.[223]

Nanostructured membranes with improved ionic conductivity and stability are being developed to enhance the efficiency and lifespan of fuel cells. These membranes offer high specific surface areas and unique nanoporous structures, which facilitate better ion transport and overall fuel cell performance.[224] Such innovations are paving the way for cleaner and more efficient transportation solutions, leveraging the benefits of nanotechnology to make fuel cells more viable for widespread use in the automotive industry.

By leveraging the unique properties of nanomaterials, significant strides are being made in improving the energy efficiency of transportation systems. The integration of lightweight materials, advanced lubricants and coatings, and innovative propulsion technologies highlights the transformative potential of nanotechnology in creating a sustainable and energy-efficient transportation future.

8.6.6 Industrial Processes and Manufacturing

This section explores how nanomaterials and nanoengineering techniques are transforming various aspects of industrial production.

8.6.6.1 Advanced Catalysis in Industrial Processes

While previous sections discussed catalysis in the context of energy generation and storage, this section focuses on how nanocatalysts are specifically revolutionizing large-scale industrial chemical processes. The emphasis here is on enhancing the efficiency and selectivity of complex, multi-step reactions that are common in the chemical manufacturing industry.

8.6.6.1.1 Process-specific Nanocatalysts. Process-specific nanocatalysts, tailored for distinct industrial applications, offer significant advancements over general-purpose catalysts. For example, TiO_2 nanoparticles are used in reducing nitrous oxide (N_2O) emissions in the production of nitric acid, a key industrial process. Cerium-doped TiO_2 nanoparticles have shown great promise in this area. These doped nanoparticles significantly improve photocatalytic performance, allowing for efficient decomposition of N_2O under optimized conditions, achieving reductions in emissions comparable to those in the adipic acid example. Research indicates that the application of cerium-doped TiO_2 can result in the decomposition of over 90% of N_2O, demonstrating a strong potential to mitigate the environmental impact of nitric acid production.[225]

The effectiveness of titanium dioxide as a nanocatalyst is attributed to its unique properties and the ability to enhance its performance through doping with metals such as platinum. Studies have demonstrated that precise control over the size and density of platinum nanoparticles deposited on titanium dioxide significantly boosts their photocatalytic capabilities. This process not only improves the catalytic activity but also ensures a higher efficiency in reducing harmful byproducts.[226]

Moreover, advancements in the synthesis and modification of titanium dioxide nanomaterials have further solidified their role in catalysis. For instance, research has highlighted the successful integration of titanium dioxide with other materials to enhance its photocatalytic and structural properties, enabling more sustainable and effective industrial applications.[227]

8.6.6.1.2 *In Situ* Generated Nanocatalysts. In the Fischer–Tropsch process for converting syngas to liquid fuels, recent research highlights the effectiveness of cobalt nanoparticles generated *in situ*. These *in situ* generated catalysts have shown improved selectivity and resistance to deactivation compared to pre-formed catalysts. Studies demonstrate that cobalt nanoparticles supported on graphene materials enhance catalytic efficiency and yield of liquid hydrocarbons. The high specific surface area and unique properties of the support materials contribute to this improved performance.[228]

Additionally, *in situ* characterization using advanced techniques like synchrotron-compatible X-ray diffraction has provided insights into the working state of cobalt catalysts, confirming their stability and activity under reaction conditions.[229]

8.6.6.1.3 Cascade Reaction Nanocatalysts. Cascade reaction nanocatalysts have emerged as a key innovation, facilitating multiple sequential reactions within a single process, which significantly reduces both energy consumption and waste typically associated with intermediate product isolation. Notably, gold-palladium nanoparticles supported on titanium dioxide are used for the direct synthesis of hydrogen peroxide from hydrogen and oxygen.[230] This method eliminates the traditional, energy-intensive anthraquinone process, showcasing a major advancement in catalytic efficiency and environmental impact.

Research in this field highlights the importance of developing nanocatalysts that can endure industrial-scale reaction conditions while maintaining their nanoscale properties. One example includes gold-palladium colloids which demonstrate high effectiveness in synthesizing hydrogen peroxide directly from hydrogen and oxygen, with notable stability and activity enhancements achieved through specific support materials and stabilizers. The integration of polyvinylpyrrolidone as a stabilizer has been shown to influence the catalytic properties significantly, optimizing the decomposition rates of hydrogen peroxide and improving overall reaction efficiency.[231]

Additionally, the focus has been on addressing challenges such as preventing catalyst sintering, which involves the loss of surface area and activity due to particle agglomeration under high temperatures. Managing heat transfer within large reactors and ensuring uniform distribution of nanocatalysts are also critical to maintaining the efficiency and scalability of these advanced catalytic systems. These improvements are pivotal for transitioning laboratory-scale innovations to practical, industrial applications.

8.6.6.2 Advanced Separation Technologies for Industrial Applications

This section expands on separation technologies beyond those mentioned in waste prevention, focusing on complex industrial separations:

8.6.6.2.1 Stimuli-responsive Nanostructured Membranes. Stimuli-responsive nanostructured membranes stand out from traditional static membranes by their ability to alter their separation properties in response to external stimuli. These advanced materials, particularly those incorporating thermo-responsive polymers, can dynamically switch between hydrophilic and hydrophobic states. This capability is particularly beneficial for pharmaceutical manufacturing and fine chemical production, where precise control over separation processes is crucial.

Recent research has shown that incorporating stimuli-responsive properties into membranes can dynamically tune their pore structure and surface properties, enhancing separation performance and efficiency. For instance, thermo-responsive hydrogels, when combined with membrane substrates, undergo temperature-induced volume-phase transitions, allowing for the fine-tuning of separation processes.[232]

Furthermore, the functionalization of porous membranes with stimuli-responsive polymers, such as by grafting or adsorption, has led to the development of membranes with smart gating capabilities. These membranes can self-regulate their permeability and selectivity, offering enhanced control in various separation applications.[233] The integration of nanoscale structures onto membrane surfaces further improves their performance by enabling switchable wettability and bioadhesion in response to environmental stimuli.

These advancements not only improve the efficiency and selectivity of separation processes but also open new avenues for their application in fields like drug delivery, where controlled release and protection of labile drugs are essential.[234]

8.6.6.2.2 Nanocomposite Membranes for Harsh Environments. Industrial separations often involve extreme pH, high temperatures, or organic solvents. Nanocomposite membranes incorporating robust materials like graphene oxide and MOFs into polymer matrices have shown exceptional stability under these conditions.

Advancements in this area include the integration of MOFs into polymeric membranes to enhance their durability and selectivity. MOFs provide high surface area and tunable pore sizes, advantageous for separating small molecules under harsh conditions. Additionally, graphene oxide has been employed to improve the mechanical strength and thermal stability of these membranes, making them ideal for use in the petrochemical industry. These innovations promise to make industrial separations more efficient and cost-effective by leveraging the unique properties of nanocomposite materials.[235]

8.6.6.2.3 Nanomaterial-enabled Molecular Sieving. Nanomaterials are significantly enhancing molecular sieving beyond basic size-based separation, enabling more precise molecular discrimination. MOFs, with their customizable pore sizes and chemical functionalities, are particularly noteworthy. These materials are being developed for the separation of structurally similar hydrocarbons in the petrochemical industry, traditionally a task requiring energy-intensive cryogenic distillation. Recent advances highlight the design and synthesis of MOFs for the selective adsorption and separation of light hydrocarbons such as alkanes, alkenes, and alkynes, improving both selectivity and efficiency in these processes.[236,237]

The separation of hydrocarbons using MOFs leverages the materials' highly tunable structures, allowing for specific interactions with target molecules. This approach addresses the limitations of traditional methods, such as high energy consumption and poor selectivity, by offering a more energy-efficient and effective solution. The innovative application of MOFs in this context underscores their potential to revolutionize separation technologies in the petrochemical industry, ensuring more sustainable and cost-effective operations.

8.6.6.2.4 Magnetic Nanoparticle-assisted Separations in Bioprocessing. Magnetic nanoparticles functionalized with specific ligands are increasingly employed in the pharmaceutical and biotechnology sectors for the targeted separation of biomolecules. This approach offers gentler and more efficient purification of sensitive biological products compared to traditional chromatography methods.

Current research focuses on scaling up these advanced separation technologies for industrial applications and addressing issues like long-term stability and fouling resistance. Recent studies highlight the effectiveness of these nanoparticles in enhancing bioanalysis performance and improving downstream processing efficiency. These advancements signify a promising future for magnetic nanoparticle-assisted separations in bioprocessing.[238]

8.6.6.3 Nanoengineered Surfaces for Industrial Equipment

This section explores how nanotechnology is improving the performance and longevity of industrial equipment.

8.6.6.3.1 Self-healing Nanocomposite Coatings. Self-healing nanocomposite coatings represent a significant advancement over traditional protective coatings due to their ability to autonomously repair minor damage. These coatings often incorporate microencapsulated healing agents and nanoparticle catalysts, enabling them to effectively self-repair scratches and microcracks in industrial pipelines and storage tanks. This capability extends equipment lifespan and reduces maintenance costs.

For instance, recent research has demonstrated the effectiveness of these coatings. Nanocomposites with microencapsulated agents and nanoparticle catalysts have been shown to autonomously repair damages, maintaining the structural integrity and functionality of industrial applications. These materials have been particularly beneficial in industries where continuous operation and minimal downtime are critical.[239]

For instance, these self-healing mechanisms not only extend the operational life of pipelines and storage tanks but also offer considerable cost savings by reducing the frequency and extent of maintenance required. This ability to autonomously address minor damages ensures continuous protection, making self-healing nanocomposite coatings a substantial improvement over conventional protective coatings.[240,241]

8.6.6.3.2 Superhydrophobic and Oleophobic Nanocoatings. Superhydrophobic and oleophobic nanocoatings, inspired by natural surfaces like lotus leaves, are revolutionizing industrial applications by preventing fouling and corrosion. In the oil and gas industry, nanostructured coatings that combine fluorinated compounds and silica nanoparticles have shown exceptional resistance to both water and oil adhesion. This innovation reduces pipeline clogging and enhances flow efficiency. Recent research highlights the effectiveness of these coatings in mitigating biofouling and corrosion, thus extending the lifespan of industrial equipment and reducing maintenance costs.[242]

For instance, the combination of fluorinated silane with alumina particles creates dual-function coatings that are both superhydrophobic and oleophobic, effectively minimizing the adhesion of contaminants.[243] Another study demonstrates the fabrication of superhydrophobic coatings using fluoroethylene-vinyl ether resin and modified SiO_2 nanoparticles, which exhibit remarkable oil and water repellency on stainless steel substrates.[244] These advancements underscore the critical role of nanotechnology in enhancing industrial coatings for better performance and sustainability.[245]

8.6.6.3.3 Nanostructured Surfaces for Enhanced Heat Transfer. Building on earlier discussions of heat transfer, this application focuses on creating nanoengineered surfaces directly on industrial equipment. For instance, laser-etched nanoscale patterns on heat exchanger surfaces have shown up to 200% improvement in boiling heat transfer coefficients, potentially revolutionizing the efficiency of industrial boilers and condensers. These nanoengineered surfaces significantly enhance nucleate boiling heat transfer and critical heat flux, leading to substantial improvements in heat exchanger performance.[246]

Research in this area is concentrated on developing durable nanoengineered surfaces that can withstand the harsh conditions of industrial environments while maintaining their functional properties over extended periods. Techniques such as electrochemical deposition and chemical etching methods have been employed to fabricate micro/nanostructured wicking surfaces, which exhibit remarkable heat transfer enhancements during quenching and boiling processes. These advancements are critical for improving thermal management in various industrial applications, including power generation and chemical processing.[247]

8.6.7 Case Study: Nanoengineered MXenes Boost Efficiency and Stability of Solar Cells

Researchers have developed a novel approach to enhancing colloidal quantum dot (CQD) photovoltaics, a promising technology that has faced persistent challenges in charge transport and stability. This work[248]

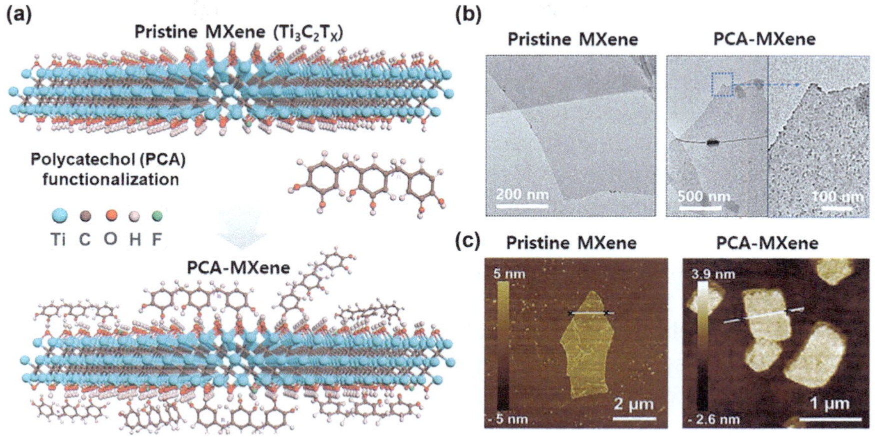

Figure 8.18 Synthesis and characterization of polycatechol functionalized MXene (PCA-MXene). (a) Schematic illustration of polycatechol functionalization on MXene. (b) TEM images of pristine MXene (left) and PCA-MXene (right). (c) AFM images of pristine MXene (left) and PCA-MXene (right). Adapted from ref. 248 with permission from John Wiley & Sons, Copyright © 2023 Wiley-VCH GmbH.

demonstrates how nanotechnology can contribute to reducing energy waste by improving the efficiency of solar energy generation.

The research team synthesized a polycatechol-functionalized MXene (PCA-MXene) (see Figure 8.18) that can be dispersed in organic solvents, allowing for seamless integration into CQD solar cell fabrication processes. This nanoengineered material serves dual functions in the device: as a dopant within the photoactive CQD layer and as an interlayer between the hole transport layer and the metal electrode.

When used as a dopant, the PCA-MXene modifies the work function of the CQD film, leading to improved energy band alignment for hole extraction. This enhancement in charge transport efficiency directly contributes to reducing energy waste within the solar cell. The researchers found that doping the CQD film with a small amount of PCA-MXene (0.02 mg mL^{-1}) resulted in a significant increase in the power conversion efficiency (PCE) of the solar cell, from 12.8% in the control device to 13.3% in the doped device.

The improvement in efficiency is primarily attributed to an increase in the short-circuit current density, indicating enhanced charge extraction. Transient photocurrent measurements revealed that the PCA-MXene doped devices exhibited faster charge decay times, further confirming the improved charge extraction capabilities and reduced energy losses.

In addition to its role as a dopant, the PCA-MXene also serves as an effective interlayer when deposited on top of the hole transport layer. This nanoengineered interlayer addresses another source of energy waste in conventional CQD solar cells: the presence of cracks in the hole transport layer, which can lead to direct contact between the photoactive layer and the metal electrode, resulting in increased charge recombination and reduced device performance.

The researchers demonstrated that the PCA-MXene interlayer effectively covers these crack sites, preventing metal electrode penetration and moisture intrusion. Time-of-flight secondary ion mass spectrometry (ToF-SIMS) depth profiling confirmed that the PCA-MXene layer acts as a physical barrier, reducing the penetration of gold from the electrode into the underlying CQD layers. This improvement in interface quality contributes to more efficient charge collection and reduced energy losses.

By combining the doping and interlayer functions of PCA-MXene, the research team achieved a remarkable PCE of 13.6% in their optimized devices, representing a significant improvement over the 12.8% efficiency of the control devices without PCA-MXene integration. This increase in efficiency translates directly to improved energy generation and reduced waste in solar energy conversion.

Perhaps even more importantly, the PCA-MXene integrated devices exhibited enhanced thermal stability. When subjected to thermal aging at 70 °C in air with 40% relative humidity, the PCA-MXene devices retained 76% of their initial PCE after 30 hours, compared to only 59% retention for the control devices. This improved stability is attributed to the hydrophobic nature of the PCA-MXene interlayer, which helps prevent moisture-induced degradation of the CQD surfaces. Enhanced stability ensures that the solar cells maintain their efficiency over time, reducing long-term energy waste due to device degradation.

This case study exemplifies how nanotechnology can contribute to reducing energy waste by improving the efficiency of energy generation in solar cells. The nanoengineered MXene material addresses multiple sources of inefficiency in CQD solar cells, including charge transport, interface quality, and long-term stability. By enhancing these aspects, the researchers have demonstrated a path toward more efficient and durable solar energy conversion devices, ultimately contributing to the goal of minimizing energy waste in renewable energy systems.

References

1. G. Busca, The industrial catalysis section: A place to publish applied catalysis research, *Catalysts*, 2024, **14**(3), 182.
2. European Research Institute of Catalysis (ERIC a.i.s.b.l.), *Science and Technology Roadmap for Catalysis in Europe*, 2016.
3. M. Velpandian, A. Sinhamahapatra and S. Basu, Environmental and economic benefits of single-atom catalysts in energy conversion and storage, *Oxford Open Energy*, 2023, **2**, oiad015.
4. C. Chen, J. Li, X. Tan, Y. Zhang, Y. Li and C. He, *et al.*, Harnessing single-atom catalysts for CO_2 electroreduction: a review of recent advances, *EES Catal.*, 2024, **2**, 71–93.

5. A. Hussain, I. Ghaffar, S. Sattar, M. Muneeb, A. Hasan and B. Deepanraj, Eco-friendly catalysts revolutionizing energy and environmental applications: An overview, *Top. Catal.*, 2025, **68**, 487–509.
6. M. Wu, F. Dong, Y. Yang, X. Cui, X. Liu and Y. Zhu, *et al.*, Emerging atomically precise metal nanoclusters and ultrasmall nanoparticles for efficient electrochemical energy catalysis: Synthesis strategies and surface/interface engineering, *Electrochem. Energy Rev.*, 2024, 7(1), 10.
7. P. Lakhani, D. Bhanderi and C. K. Modi, Nanocatalysis: recent progress, mechanistic insights, and diverse applications, *J. Nanopart. Res.*, 2024, **26**, 148.
8. C. Smith, A. K. Hill and L. Torrente-Murciano, Current and future role of Haber–Bosch ammonia in a carbon-free energy landscape, *Energy Environ. Sci.*, 2020, **13**(2), 331–344.
9. S. Li, E. Li, X. An, X. Hao, Z. Jiang and G. Guan, Transition metal-based catalysts for electrochemical water splitting at high current density: current status and perspectives, *Nanoscale*, 2021, **13**(30), 12788–12817.
10. M. Wang, L. Zhang, Y. He and H. Zhu, Recent advances in transition-metal-sulfide-based bifunctional electrocatalysts for overall water splitting, *J. Mater. Chem. A*, 2021, **9**(9), 5320–5363.
11. W. Kaminsky, The discovery and evolution of metallocene-based olefin polymerization catalysts, *Rend. Lincei Sci. Fis. Nat.*, 2017, **28**(S1), 87–95.
12. Y. Xiong, B. J. Wiley and Y. Xia, Nanocrystals with unconventional shapes–a class of promising catalysts, *Angew. Chem., Int. Ed.*, 2007, **46**(38), 7157–7159.
13. C. Li, N. Clament Sagaya Selvam and J. Fang, Shape-controlled synthesis of platinum-based nanocrystals and their electrocatalytic applications in fuel cells, *Nanomicro Lett.*, 2023, **15**(1), 83.
14. J. L. Obeso, J. G. Flores, C. V. Flores, M. T. Huxley, J. A. de Los Reyes and R. A. Peralta, *et al.*, MOF-based catalysts: insights into the chemical transformation of greenhouse and toxic gases, *Chem. Commun.*, 2023, **59**(68), 10226–10242.
15. X. Wang, Y. Ma, Q. Wu, Y. Wen and F.-S. Xiao, Zeolite nanosheets for catalysis, *Chem. Soc. Rev.*, 2022, **51**(7), 2431–2443.
16. M. Cheng, X. Zheng, F. Ma, Z. Zhu and Q. Xu, Structural engineering of metal-organic frameworks for efficient CO_2 reduction reaction, *ChemCatChem*, 2024, **16**, e202400438.
17. H. Eidsvåg, S. Bentouba, P. Vajeeston, S. Yohi and D. Velauthapillai, TiO_2 as a photocatalyst for water splitting-an experimental and theoretical review, *Molecules*, 2021, **26**(6), 1687.
18. C. Song, L. Xiao, Y. Chen, F. Yang, H. Meng and W. Zhang, *et al.*, TiO_2-based catalysts with various structures for photocatalytic application: A review, *Catalysts*, 2024, **14**(6), 366.
19. X. Zhong, E. Yuan, F. Yang, Y. Liu, H. Lu and J. Yang, *et al.*, Optimizing oxygen vacancies through grain boundary engineering to enhance electrocatalytic nitrogen reduction, *Proc. Natl. Acad. Sci. U. S. A.*, 2023, **120**(40), e2306673120.

20. A. For a discussion on the pros and cons of nanotechnology in food packing, read Ashfaq, N. Khursheed, S. Fatima, Z. Anjum and K. Younis, Application of nanotechnology in food packaging: Pros and cons, *J. Agric. Food Res.*, 2022, 7, 100270.
21. Q. Wang, W. Chen, W. Zhu, D. J. McClements, X. Liu and F. Liu, A review of multilayer and composite films and coatings for active biodegradable packaging, *npj Sci. Food*, 2022, **6**, 18.
22. K. Y. Perera, M. Hopkins, A. K. Jaiswal and S. Jaiswal, Nanoclays-containing bio-based packaging materials: Properties, applications, safety, and regulatory issues, *J. Nanostruct. Chem.*, 2024, **14**, 71–93.
23. N. Chausali, J. Saxena and R. Prasad, Recent trends in nanotechnology applications of bio-based packaging, *J. Agric. Food Res.*, 2022, 7, 100257.
24. A. Istiqola and A. Syafiuddin, A review of silver nanoparticles in food packaging technologies: Regulation, methods, properties, migration, and future challenges, *J. Chin. Chem. Soc.*, 2020, **67**(11), 1942–1956.
25. S. Sharma, L. Mulrey, M. Byrne, A. K. Jaiswal and S. Jaiswal, Encapsulation of essential oils in nanocarriers for active food packaging, *Foods*, 2022, **11**(15), 2337.
26. N. Tavker and M. Sharma, Nanosensors for Intelligent Food Packaging, *Nanosensors for Smart Agriculture*, 2022, pp. 737–756.
27. E. Assadpour and S. Mahdi Jafari, A systematic review on nanoencapsulation of food bioactive ingredients and nutraceuticals by various nanocarriers, *Crit. Rev. Food Sci. Nutr.*, 2018, **59**(19), 3129–3151.
28. W.-F. Lai, Design of polymeric films for antioxidant active food packaging, *Int. J. Mol. Sci.*, 2021, **23**(1), 12.
29. V. Suvarna, A. Nair, R. Mallya, T. Khan and A. Omri, Antimicrobial nanomaterials for food packaging, *Antibiotics*, 2022, **11**(6), 729.
30. L. Motelica, D. Ficai, A. Ficai, O. C. Oprea, D. A. Kaya and E. Andronescu, Biodegradable antimicrobial food packaging: Trends and Perspectives, *Foods*, 2020, **9**(10), 1438.
31. N. Mlalila, D. M. Kadam, H. Swai and A. Hilonga, Transformation of food packaging from passive to innovative *via* nanotechnology: Concepts and critiques, *J. Food Sci. Technol.*, 2016, **53**(9), 3395–3407.
32. T. Kasuga, T. Saito, H. Koga and M. Nogi, One-pot hierarchical structuring of nanocellulose by electrophoretic deposition, *ACS Nano*, 2022, **16**(11), 18390–18397.
33. S. Veličković, *et al.*, Application of nanocomposites in the automotive industry, *Mobility Vehicle Mech.*, 2019, **45**(3), 51–64.
34. S. Bhattacharya and A. T. Alpas, Tribology of aluminum and aluminum matrix composite materials for automotive components, *Lightweight and Sustainable Materials for Automotive Applications*, 2017, pp. 303–328.
35. E. Edwards, C. Brantley and P. B. Ruffin, Overview of nanotechnology in military and aerospace applications, *Nanotechnology Commercialization*, Wiley, 2017, pp. 133–176.

36. Hypergravity helping aircraft fly further, Esa.int, [cited 2024 Jul 3]. Available from: https://www.esa.int/Science_Exploration/Human_and_Robotic_Exploration/Research/Hypergravity_helping_aircraft_fly_further.
37. Commercial airlines: worldwide fuel consumption 2024, Statista, [cited 2024 Jul 3]. Available from: https://www.statista.com/statistics/655057/fuel-consumption-of-airlines-worldwide/.
38. J. U. Surjadi, *et al.*, Lightweight, ultra-tough, 3D-architected hybrid carbon microlattices, *Matter*, 2022, **5**(11), 4029–4046.
39. Y. Zeng, P. Gordiichuk, T. Ichihara, G. Zhang, E. Sandoz-Rosado and E. D. Wetzel, *et al.*, Irreversible synthesis of an ultrastrong two-dimensional polymeric material, *Nature*, 2022, **602**(7895), 91–95.
40. W. Chen, F. Kimura and Y. Kajihara, Effect of nanostructured zinc coating on high joining strength of polymer/galvanized high-strength steel composite *via* injection molding, *J. Manuf. Process.*, 2023, **85**, 295–305.
41. Z. Qin, G. S. Jung, M. J. Kang and M. J. Buehler, The mechanics and design of a lightweight three-dimensional graphene assembly, *Sci. Adv.*, 2017, **3**, e1601536.
42. Z. Jiang, M. O. Nayeem, K. Fukuda, S. Ding, H. Jin and T. Yokota, *et al.*, Highly stretchable metallic nanowire networks reinforced by the underlying randomly distributed elastic polymer nanofibers *via* interfacial adhesion improvement, *Adv. Mater.*, 2019, **31**(37), 1903446.
43. C. Choi, E. Schlenker, H. Ha, J. Y. Cheong and B. Hwang, Versatile Applications of Silver Nanowire-Based Electrodes and Their Impacts, *Micromachines*, 2023, **14**(3), 562.
44. J. Wang, X. Zhu, P. Xiong, J. Tu, Z. Yang and F. Yao, *et al.*, Flexible, robust and washable bacterial cellulose/silver nanowire conductive paper for high-performance electromagnetic interference shielding, *J. Mater. Chem. A*, 2022, **10**, 960–968.
45. Z. Li, Y. Peng, Q. Zhang and X. Xiao, The Interface Strengthening of Multi-Walled Carbon Nanotubes/Polylactic Acid Composites *via* the In-Loop Hybrid Manufacturing Method, *Polymers*, 2023, **15**(22), 4426.
46. A. Boczkowska, *et al.*, Nanocomposites Based on Thermoplastic Acrylic Resin with the Addition of Chemically Modified Multi-Walled Carbon Nanotubes, *Polymers*, 2024, **16**(3), 422.
47. M. R. H. Mazumder, L. D. Mathews, S. Mateti, N. V. Salim, J. Parameswaranpillai and P. Govindaraj, *et al.*, Boron nitride-based polymer nanocomposites for heat dissipation and thermal management applications, *Appl. Mater Today*, 2022, **29**, 101672.
48. J. Ma, X. Liu, R. Wang, C. Lu, X. Wen and G. Tu, Research progress and application of polyimide-based nanocomposites, *Nanomaterials*, 2023, **13**(4), 656.
49. B. Yu, R. Zhao, Z. Lu, H. Su, B. Liang and B. Liu, *et al.*, Thermal stability and crystallization processes of $Pd_{78}Au_4Si_{18}$ thin films visualized *via* *in situ* TEM, *Nanomaterials*, 2024, **14**(7), 635.

50. J. Wang, F. Ma and M. Sun, Graphene, hexagonal boron nitride, and their heterostructures: properties and applications, *RSC Adv.*, 2017, 7(27), 16801–16822.
51. Y. Wang, Z. Niu, Y. Dai, P. Mu and J. Li, Two-dimensional nanomaterial MXenes for efficient gas separation: a review, *Nanoscale*, 2023, **15**(9), 4170–4194.
52. Q. Qu, J. Xu, H. Wang, Y. Yu, Q. Dong and X. Zhang, *et al.*, Carbon nanotube-based intumescent flame retardants achieve high-efficiency flame retardancy and simultaneously avoid mechanical property loss, *Polymers*, 2023, **15**(6), 1406.
53. H. Ning, Z. Ma, Z. Zhang, D. Zhang and Y. Wang, A novel multi-functional flame retardant MXene/nanosilica hybrid for poly(vinyl alcohol) with simultaneously improved mechanical properties, *New J. Chem.*, 2021, **45**(9), 4292–4302.
54. A. K. Naskar, J. K. Keum and R. G. Boeman, Polymer matrix nano-composites for automotive structural components, *Nat. Nanotechnol.*, 2016, **11**(12), 1026–1030.
55. S. Wu, S. Peng and C. Wang, Multifunctional polymer nano-composites reinforced by aligned carbon nanomaterials, *Polymers*, 2018, **10**(5), 542.
56. G. M. Pinto, J. M. Cremonezzi, H. Ribeiro, R. J. Andrade, N. R. Demarquette and G. J. Fechine, From two-dimensional materials to polymer nanocomposites with emerging multifunctional applications: A critical review, *Polym. Compos.*, 2023, **44**(3), 1438–1470.
57. M. Pander, R. Gil-San-Millan, P. Delgado, C. Perona-Bermejo, U. Kostrzewa and K. Kaczkowski, *et al.*, MOF/polymer hybrids through *in situ* free radical polymerization in metal–organic frameworks, *Mater. Horiz.*, 2023, **10**(4), 1301–1308.
58. K. F. Chan, M. H. Zaid, M. S. Mamat, S. Liza, M. Tanemura and Y. Yaakob, Recent developments in carbon nanotubes-reinforced ceramic matrix composites: A review on dispersion and densification techniques, *Crystals*, 2021, **11**(5), 457.
59. S. A. Farooq, A. Raina, S. Mohan, R. Arvind Singh, S. Jayalakshmi and M. Irfan Ul Haq, Nanostructured coatings: Review on processing techniques, corrosion behaviour and tribological performance, *Nanomaterials*, 2022, **12**(8), 1323.
60. H. Lu, Y. Ren, Y. Chen, H. Zhu, Z. Xin and H. Wan, *et al.*, Wear resistance of 20CR2NI4A alloy steel treated by laser shock peening and implantation of diamond nanoparticles, *Surf. Coat. Technol.*, 2021, **412**, 127070.
61. Z. A. Ali, A. M. Takhakh and M. Al-Waily, A review of use of nanoparticle additives in lubricants to improve its tribological properties, *Mater. Today: Proc.*, 2022, **52**, 1442–1450.
62. A. Thakur, S. Kaya and A. Kumar, Recent trends in the characterization and application progress of nano-modified coatings in corrosion mitigation of metals and alloys, *Appl. Sci.*, 2023, **13**(2), 730.

63. S. E. Shin and D. H. Bae, Fatigue behavior of al2024 alloy-matrix nanocomposites reinforced with multi-walled carbon nanotubes, *Composites, Part B*, 2018, **134**, 61–68.
64. W. H. Danial and Z. Abdul Majid, Recent advances on the enhanced thermal conductivity of graphene nanoplatelets composites: a short review, *Carbon Lett.*, 2022, **32**(6), 1411–1424.
65. M. Wu, Z. Chen, C. Huang, K. Huang, K. Jiang and J. Liu, Graphene platelet reinforced copper composites for improved tribological and thermal properties, *RSC Adv.*, 2019, **9**(68), 39883–39892.
66. Y. Liu, S. Li, R. D. K. Misra, K. Geng and Y. Yang, Planting carbon nanotubes within Ti-6Al-4V to make high-quality composite powders for 3D printing high-performance Ti-6Al-4V matrix composites, *Scr. Mater.*, 2020, **183**, 6–11.
67. Z. Hu, F. Chen, J. Xu, Q. Nian, D. Lin and C. Chen, *et al.*, 3D printing graphene-aluminum nanocomposites, *J. Alloys Compd.*, 2018, **746**, 269–276.
68. M. Emerla, P. Bazarnik, Y. Huang, M. Lewandowska and T. G. Langdon, Using direct high-pressure torsion synthesis to produce aluminium matrix nanocomposites reinforced with carbon nanotubes, *J. Alloys Compd.*, 2023, **968**, 171928171928.
69. R. Saranu, R. Chanamala and S. Putti, Review of Magnesium Metal Matrix Composites, *IOP Conf. Ser.: Mater. Sci. Eng.*, 2020, **961**(1), 012001.
70. Y. Huang, Y. Xie, and X. Meng, Graphene nanoplatelet-reinforced aluminum matrix composites, in *Materials Forming, Machining and Tribology*, Springer Nature Singapore, Singapore, 2024, pp. 337–384.
71. V. K. Bommala, M. G. Krishna and C. T. Rao, Magnesium matrix composites for biomedical applications: A review, *J. Magnesium Alloys*, 2019, 7(1), 72–79.
72. P. Li, J. Dai, Y. Li, D. Alexander, J. Čapek and J. Geis-Gerstorfer, *et al.*, Zinc based biodegradable metals for bone repair and regeneration: Bioactivity and molecular mechanisms, *Mater. Today Bio*, 2024, **25**, 100932.
73. G. Gąsior, J. Szczepański and A. Radtke, Biodegradable iron-based materials—what was done and what more can be done?, *Materials*, 2021, **14**(12), 3381.
74. D. Svetlizky, B. Zheng, X. Wang, S. Jiang, L. Valdevit and J. M. Schoenung, *et al.*, Towards a self-healing aluminum metal matrix composite: Design, fabrication, and demonstration, *Appl. Mater. Today*, 2024, **37**, 102148.
75. Q. Ouyang, L. Liu and Z. Wu, Electrothermally Self-Healing Delamination Cracks in Carbon/Epoxy Composites Using Sandwich and Tough Carbon Nanotube/Copolymer Interleaves, *Polymers*, 2022, **14**(20), 4313.
76. X. Liu, F. Wang, W. Wang and H. Wu, Interfacial strengthening and self-healing effect in graphene-copper nanolayered composites under shear deformation, *Carbon*, 2016, **107**, 680–688.

77. Y. Wei, Q. Wu, H. Meng, Y. Zhang and C. Cao, Recent advances in photocatalytic self-cleaning performances of TiO_2-based building materials, *RSC Adv.*, 2023, **13**(30), 20584–20597.
78. M. Han, J.-K. Kim, Y.-I. Choi, J. P. Yun, G. S. Lee and S.-W. Kang, *et al.*, Carbon nanotube-based strain sensor for structural health monitoring, *Jpn. J. Appl. Phys.*, 2019, **58**, SDDJ07.
79. N. Roy, R. Sinha, H. B. Nemade and T. K. Mandal, Synthesis of MoS_2-CuO nanocomposite for room temperature acetone sensing application, *J. Alloys Compd.*, 2022, **910**, 164891.
80. O. Kanoun, T. Lazarević-Pašti, I. Pašti, S. Nasraoui, M. Talbi and A. Brahem, *et al.*, A review of nanocomposite-modified electrochemical sensors for water quality monitoring, *Sensors*, 2021, **21**(12), 4131.
81. L. Zhang, F. Jiang, L. Wang, Y. Feng, D. Yu and T. Yang, *et al.*, High performance flexible strain sensors based on silver nanowires/thermoplastic polyurethane composites for wearable devices, *Appl. Compos. Mater.*, 2022, **29**(4), 1621–1636.
82. R. Megha, F. A. Ali, Y. T. Ravikiran, C. H. V. V. Ramana, A. B. V. Kiran Kumar and D. K. Mishra, *et al.*, Conducting polymer nanocomposite-based temperature sensors: A review, *Inorg. Chem. Commun.*, 2018, **98**, 11–28.
83. S.-J. Lee, Y.-J. Jung, J. Park and S.-H. Jang, Temperature detectable surface coating with carbon nanotube/epoxy composites, *Nanomaterials*, 2022, **12**(14), 2369.
84. A. B. M. Supian, M. R. M. Asyraf, A. Syamsir, M. I. Najeeb, A. Alhayek and R. N. Al-Dala'ien, *et al.*, Thermochromic polymer nanocomposites for the heat detection system: Recent progress on properties, applications, and challenges, *Polymers*, 2024, **16**(11), 1545.
85. Y. Cui, H. Wang, K. Cao, Q. Zhou, W. Ding and J. Yin, Preparation and application of nanocomposite thin-film temperature sensor during the milling process, *Materials*, 2022, **15**(20), 7106.
86. P. G. Ingole, Application of sustainable nanocomposites in membrane technology, in *Sustainable Polymer Composites and Nanocomposites*, Springer International Publishing, Cham, 2019. pp. 935–960.
87. A. Sahu, R. Dosi, C. Kwiatkowski, S. Schmal and J. C. Poler, Advanced polymeric nanocomposite membranes for water and wastewater treatment: A comprehensive review, *Polymers*, 2023, **15**(3), 540.
88. C. I. Covaliu-Mierlă, O. Păunescu and H. Iovu, Recent advances in membranes used for nanofiltration to remove heavy metals from wastewater: A review, *Membranes*, 2023, **13**(7), 643.
89. J. Yang, S. Zhu and H. Zhang, Polycation-intercalated MXene membrane with enhanced permselective and anti-microbial properties, *Nanomaterials*, 2023, **13**(21), 2885.
90. G. Chen, G. Liu, Y. Pan, G. Liu, X. Gu and W. Jin, *et al.*, Zeolites and metal–organic frameworks for gas separation: the possibility of translating adsorbents into membranes, *Chem. Soc. Rev.*, 2023, **52**(14), 4586–4602.

91. N. Razmara, A. Kirch, J. R. Meneghini and C. R. Miranda, Efficient CH_4/CO_2 gas mixture separation through nanoporous graphene membrane designs, *Energies*, 2021, **14**(9), 2488.
92. A. Tayel, A. B. Abdelaal, A. M. K. Esawi and A. R. Ramadan, Thin-film nanocomposite (TFN) membranes for water treatment applications: Characterization and performance, *Membranes*, 2023, **13**(5), 477.
93. S. Prakash, M. Kumari and A. K. Chauhan, The intervention of nanotechnology in food packaging: a review, *J. Mater. Sci.*, 2024, **59**(7), 2585–2601.
94. X. Chen, R. Wreyford and N. Nasiri, Recent advances in ethylene gas detection, *Materials*, 2022, **15**(17), 5813.
95. K. Zhan, Y. Jiang, P. Qin, Y. Chen and L. Heinke, A colorimetric label-free sensor array of metal–organic-framework-based Fabry–Pérot films for detecting volatile organic compounds and food spoilage, *Adv. Mater. Interfaces*, 2023, **10**, 2300329.
96. H. M. Valenzuela-Amaro, A. Aguayo-Acosta, E. R. Meléndez-Sánchez, O. de la Rosa, P. G. Vázquez-Ortega and M. A. Oyervides-Muñoz, *et al.*, Emerging applications of nanobiosensors in pathogen detection in water and food, *Biosensors*, 2023, **13**(10), 922.
97. D. Barmpakos and G. Kaltsas, A review on humidity, temperature and strain printed sensors—current trends and future perspectives, *Sensors*, 2021, **21**(3), 739.
98. D. Barmpakos and G. Kaltsas, A review on humidity, temperature and strain printed sensors—current trends and future perspectives, *Sensors*, 2021, **21**(3), 739.
99. A. R. Bhatlawande, P. U. Ghatge, G. U. Shinde, R. K. Anushree and S. D. Patil, Unlocking the future of smart food packaging: biosensors, IoT, and nano materials, *Food Sci. Biotechnol.*, 2024, **33**(5), 1075–1091.
100. S. Khan, Z. Hosseinidoust, Y. Li, C. D. M. Filipe and T. F. Didar, Smart food packaging commercialization, *Nat. Rev. Bioeng.*, 2024, **2**(7), 535–537.
101. M. M. Calabretta, D. Gregucci, R. Desiderio and E. Michelini, Colorimetric paper sensor for food spoilage based on biogenic Amine monitoring, *Biosensors*, 2023, **13**(1), 126.
102. G. Lee, B. Kim, I. Jang, M. Il Kim, S. Shin and K. Kwon, Rapid detection of *Salmonella* using an aptamer-functionalized PDA liposome sensor with naked-eye colorimetric sensing, *Mater. Adv.*, 2024, **5**(6), 2400–2410.
103. S. Punia Bangar, Y. Phimolsiripol and M. Trif, Special issue "smart polymeric films and coatings for food packaging applications", *Polymers*, 2023, **15**(23), 4522.
104. G. A. Lanza, J. A. Perez-Taborda and A. Avila, Time temperature indicators (TTIs) based on silver nanoparticles for monitoring of perishables products, *J. Phys.:Conf. Ser.*, 2019, **1247**(1), 012055.
105. Y. Li, W. Zhang, Z. Cui, L. Shi, Y. Shang and Y. Ji, *et al.*, Machine learning-assisted nanosensor arrays: An efficiently high-throughput food detection analysis, *Trends Food Sci. Technol.*, 2024, **149**, 104564.

106. S. Roh, T. D. Nguyen and J. S. Lee, Applications of nanomaterials in RFID wireless sensor components, *Appl. Sci.*, 2024, **14**(3), 1216.
107. L. Atkinson, Smart shoppers? Using QR codes and 'green' smartphone apps to mobilize sustainable consumption in the retail environment, *Int. J. Consum. Stud.*, 2013, **37**(4), 387–393.
108. D. Schaefer and W. M. Cheung, Smart packaging: Opportunities and challenges, *Proc. CIRP*, 2018, **72**, 1022–1027.
109. S. Ahmadzadeh, T. Ajmal, R. Ramanathan and Y. Duan, A comprehensive review on food waste reduction based on IoT and big data technologies, *Sustainability*, 2023, **15**(4), 3482.
110. A.-A. A. Sharabati and E. R. Jreisat, Blockchain technology implementation in supply chain management: A literature review, *Sustainability*, 2024, **16**(7), 2823.
111. U.S. Government Accountability Office, Precision agriculture: Benefits and challenges for technology adoption and use. Gao.gov, [cited 2024 Aug 5]. Available from: https://www.gao.gov/products/gao-24-105962.
112. A. Yadav, K. Yadav, R. Ahmad and K. Abd-Elsalam, Emerging frontiers in nanotechnology for precision agriculture: Advancements, hurdles and prospects, *Agrochemicals*, 2023, **2**(2), 220–256.
113. A. Soussi, E. Zero, R. Sacile, D. Trinchero and M. Fossa, Smart sensors and smart data for precision agriculture: A review, *Sensors*, 2024, **24**(8), 2647.
114. V. S. Palaparthy, H. Kalita, S. G. Surya, M. S. Baghini and M. Aslam, Graphene oxide based soil moisture microsensor for *in situ* agriculture applications, *Sens. Actuators, B*, 2018, **273**, 1660–1669.
115. P. Parameswari, N. Belagalla, B. V. Singh, G. J. Abhishek, G. M. Rajesh and D. Katiyar, *et al.*, Nanotechnology-based sensors for real-time monitoring and assessment of soil health and quality: A review, *Asian J. Soil Sci. Plant Nutr.*, 2024, **10**(2), 157–173.
116. G. Archbold, C. Parra, H. Carrillo and A. M. Mouazen, Towards the implementation of ISFET sensors for *in-situ* and real-time chemical analyses in soils: A practical review, *Comput. Electron. Agric.*, 2023, **209**, 107828.
117. S. Patil, H. Ghadi, N. Ramgir, A. Adhikari and V. R. Rao, Monitoring soil pH variation using Polyaniline/SU-8 composite film based conductometric microsensor, *Sens. Actuators, B*, 2019, **286**, 583–590.
118. M. C.-Y. Ang, J. M. Saju, T. K. Porter, S. Mohaideen, S. Sarangapani and D. T. Khong, *et al.*, Decoding early stress signaling waves in living plants using nanosensor multiplexing, *Nat. Commun.*, 2024, **15**, 2943.
119. P. L. Kashyap, S. Kumar, P. Jasrotia, D. P. Singh and G. P. Singh, Nanosensors for plant disease diagnosis: Current understanding and future perspectives, in *Nanoscience for Sustainable Agriculture*, Springer International Publishing, Cham, 2019, pp. 189–205.
120. H. Y. Lau, H. Wu, E. J. H. Wee, M. Trau, Y. Wang and J. R. Botella, Specific and sensitive isothermal electrochemical biosensor for plant pathogen DNA detection with colloidal gold nanoparticles as probes, *Sci. Rep.*, 2017, **7**, 38896.

121. Z. Cebula, S. Żołędowska, K. Dziąbowska, M. Skwarecka, N. Malinowska and W. Białobrzeska, *et al.*, Detection of the plant pathogen Pseudomonas syringae pv. Lachrymans on antibody-modified gold electrodes by electrochemical impedance spectroscopy, *Sensors*, 2019, **19**(24), 5411.
122. D. S. Shaw and K. C. Honeychurch, Nanosensor applications in plant science, *Biosensors*, 2022, **12**(9), 675.
123. X. Li, T. Cui, X. Li, H. Liu, D. Li and J. Jian, *et al.*, Wearable temperature sensors based on reduced graphene oxide films, *Materials*, 2023, **16**(17), 5952.
124. Z. Hou, Z. Ma, X. Guan, H. Zhao, S. Liu and T. Fei, *et al.*, A cellulose nanofiber capacitive humidity sensor with high sensitivity and fast recovery characteristics, *Chemosensors*, 2022, **10**(11), 464.
125. H. Meng, Y. Gao, X. Wang, X. Li, L. Wang and X. Zhao, *et al.*, Quantum dot-enabled infrared hyperspectral imaging with single-pixel detection, *Light: Sci. Appl.*, 2024, **13**, 121.
126. S. W. Tong, W. P. Goh and C. Jiang, Review—recent advances in nanosensors for precision agriculture, *J. Electrochem. Soc.*, 2023, **170**(12), 127507.
127. B. Toksha, V. A. M. Sonawale, A. Vanarase, D. Bornare, S. Tonde and C. Hazra, *et al.*, Nanofertilizers: A review on synthesis and impact of their use on crop yield and environment, *Environ. Technol. Innovation*, 2021, **24**, 101986.
128. A. Dapkekar, P. Deshpande, M. D. Oak, K. M. Paknikar and J. M. Rajwade, Zinc use efficiency is enhanced in wheat through nanofertilization, *Sci. Rep.*, 2018, **8**, 6832.
129. M. S. Haydar, D. Ghosh and S. Roy, Slow and controlled release nanofertilizers as an efficient tool for sustainable agriculture: Recent understanding and concerns, *Plant Nano Biol.*, 2024, 7, 100058.
130. M. Pandorf, L. Pourzahedi, L. Gilbertson, G. V. Lowry, P. Herckes and P. Westerhoff, Graphite nanoparticle addition to fertilizers reduces nitrate leaching in growth of lettuce (*Lactuca sativa*), *Environ. Sci.: Nano*, 2020, 7(1), 127–138.
131. A. A. Sembada and I. W. Lenggoro, Transport of nanoparticles into plants and their detection methods, *Nanomaterials*, 2024, **14**(2), 131.
132. X. Lv, H. Sha, Z. Ye, Y. Wang and B. Mao, Nanomaterials in plant management: functions, mechanisms and prospects, *Environ. Sci.: Nano*, 2023, **10**(12), 3232–3252.
133. H. A. Yousef, H. M. Fahmy, F. N. Arafa, M. Y. Abd Allah, Y. M. Tawfik and K. K. El Halwany, *et al.*, Nanotechnology in pest management: advantages, applications, and challenges, *Int. J. Trop. Insect Sci.*, 2023, **43**(5), 1387–1399.
134. Industry, IEA, [cited 2024 Oct 5]. Available from: https://www.iea.org/energy-system/industry.
135. M. F. Alotibi, B. A. Alshammari, M. H. Alotaibi, F. M. Alotaibi, S. Alshihri and R. M. Navarro, *et al.*, ZSM-5 zeolite based additive in

FCC process: A review on modifications for improving propylene production, *Catal. Surv. Asia*, 2020, **24**(1), 1–10.
136. D. Strauch, P. Weiner, B. B. Sarma, A. Körner, E. Herzinger and P. Wolf, *et al.*, Bimetallic platinum rhenium catalyst for efficient low temperature dehydrogenation of perhydro benzyltoluene, *Catal. Sci. Technol.*, 2024, **14**(7), 1775–1790.
137. M. Martino, E. Meloni, G. Festa and V. Palma, Propylene synthesis: Recent advances in the use of Pt-based catalysts for propane dehydrogenation reaction, *Catalysts*, 2021, **11**(9), 1070.
138. S. Kim, S. Ga, H. Bae, R. Sluyter, K. Konstantinov and L. K. Shrestha, *et al.*, Multidisciplinary approaches for enzyme biocatalysis in pharmaceuticals: protein engineering, computational biology, and nanoarchitectonics, *EES Catal.*, 2024, **2**(1), 14–48.
139. D. S. R. Khafaga, M. G. Radwan, G. Muteeb, M. Aatif and M. Farhan, Green synthesis of biocatalysts based on nanocarriers promises an effective role in pharmaceutical and biomedical fields, *Catalysts*, 2023, **13**(11), 1448.
140. J.-L. Tu and B. Huang, Titanium in photocatalytic organic transformations: current applications and future developments, *Org. Biomol. Chem.*, 2024, **22**(33), 6650–6664.
141. H. He, H. H. Wang, J. Liu, X. Liu, W. Li and Y. Wang, Research progress and application of single-atom catalysts: A review, *Molecules*, 2021, **26**(21), 6501.
142. T. He, C. Zhang, L. Zhang and A. Du, Single Pt atom decorated graphitic carbon nitride as an efficient photocatalyst for the hydrogenation of nitrobenzene into aniline, *Nano Res.*, 2019, **12**(8), 1817–1823.
143. D. Alique, Reactors for process intensification: Recent advances and key applications, *Membranes*, 2021, **11**(10), 745.
144. N. Cerone, G. D. Zito, C. Florio, L. Fabbiano and F. Zimbardi, Recent advancements in Pd-based membranes for hydrogen separation, *Energies*, 2024, **17**(16), 4095.
145. A. Helmi and F. Gallucci, Latest developments in membrane (bio)-reactors, *Processes*, 2020, **8**(10), 1239.
146. V. Hakke, S. Sonawane, S. Anandan, S. Sonawane and M. Ashokkumar, Process intensification approach using microreactors for synthesizing nanomaterials—A critical review, *Nanomaterials*, 2021, **11**(1), 98.
147. M. Khan, S. Yan, M. Ali, F. Mahmood, Y. Zheng and G. Li, *et al.*, Innovative solutions for high-performance silicon anodes in lithium-ion batteries: Overcoming challenges and real-world applications, *Nanomicro Lett.*, 2024, **16**(1), 179.
148. C. Yang and K. S. R. Chandran, A critical review of silicon nanowire electrodes and their energy storage capacities in Li-ion cells, *RSC Adv.*, 2023, **13**(6), 3947–3957.
149. A. Murali, R. Ramesh, M. Sakar, S. Park and S. S. Han, Unveiling the potential of emergent nanoscale composite polymer electrolytes for

safe and efficient all solid-state lithium-ion batteries, *RSC Adv.*, 2024, **14**(42), 30618–30629.
150. A. Kausar, I. Ahmad, T. Zhao, O. Aldaghri, K. H. Ibnaouf and M. H. Eisa, Graphene nanocomposites as innovative materials for energy storage and conversion—design and headways, *Int. J. Mol. Sci.*, 2023, **24**(14), 11593.
151. S. Alwin and X. Sahaya Shajan, Aerogels: promising nanostructured materials for energy conversion and storage applications, *Mater. Renewable Sustainable Energy*, 2020, **9**, 7.
152. H. Zhang, C. Zhang, W. Ji, X. Wang, Y. Li and W. Tao, Experimental characterization of the thermal conductivity and microstructure of opacifier-fiber-aerogel composite, *Molecules*, 2018, **23**(9), 2198.
153. C. Thie, S. Quallen, A. Ibrahim, T. Xing and B. Johnson, Study of energy saving using silica aerogel insulation in a residential building, *Gels*, 2023, **9**(2), 86.
154. D. Božiček, J. Peterková, J. Zach and M. Košir, Vacuum insulation panels: An overview of research literature with an emphasis on environmental and economic studies for building applications, *Renewable Sustainable Energy Rev*, 2024, **189**, 113849.
155. Y. Xie, L. Wang, B. Liu, L. Zhu, S. Shi and X. Wang, Flexible, controllable, and high-strength near-infrared reflective Y_2O_3 nanofiber membrane by electrospinning a polyacetylacetone-yttrium precursor, *Mater. Des.*, 2018, **160**, 918–925.
156. C.-Y. Hsu, Z. H. Mahmoud, S. Abdullaev, F. K. Ali, Y. Ali Naeem and R. Mzahim Mizher, *et al.*, Nano titanium oxide (nano-TiO_2): A review of synthesis methods, properties, and applications, *Case Stud. Chem. Environ. Eng.*, 2024, **9**, 100626.
157. H. Kwon and S. J. Hong, Use of optical emission spectroscopy data for fault detection of mass flow controller in plasma etch equipment, *Electronics*, 2022, **11**(2), 253.
158. S. Rastogi, V. Kumari, V. Sharma and F. J. Ahmad, Gold nanoparticle-based sensors in food safety applications, *Food Anal. Methods*, 2022, **15**(2), 468–484.
159. V. Milo, G. Malavena, C. Monzio Compagnoni and D. Ielmini, Memristive and CMOS devices for neuromorphic computing, *Materials*, 2020, **13**(1), 166.
160. M. D. Bishop, G. Hills, T. Srimani, C. Lau, D. Murphy and S. Fuller, *et al.*, Fabrication of carbon nanotube field-effect transistors in commercial silicon manufacturing facilities, *Nat. Electron.*, 2020, **3**(8), 492–501.
161. A. Ucar, M. Karakose and N. Kırımça, Artificial intelligence for predictive maintenance applications: Key components, trustworthiness, and future trends, *Appl. Sci.*, 2024, **14**(2), 898.
162. M. C.-Y. Ang, J. M. Saju, T. K. Porter, S. Mohaideen, S. Sarangapani and D. T. Khong, *et al.*, Decoding early stress signaling waves in living plants using nanosensor multiplexing, *Nat. Commun.*, 2024, **15**, 2943.

163. G. Shilpa, P. M. Kumar, D. K. Kumar, P. R. Deepthi, V. Sadhu and A. Sukhdev, *et al.*, Recent advances in the development of high efficiency quantum dot sensitized solar cells (QDSSCs): A review, *Mater. Sci. Energy Technol.*, 2023, **6**, 533–546.
164. H. Goodwin, T. C. Jellicoe, N. J. L. K. Davis and M. L. Böhm, Multiple exciton generation in quantum dot-based solar cells, *Nanophotonics*, 2018, 7(1), 111–126.
165. R. Sharif, A. Khalid, S. W. Ahmad, A. Rehman, H. G. Qutab and H. H. Akhtar, *et al.*, A comprehensive review of the current progresses and material advances in perovskite solar cells, *Nanoscale Adv.*, 2023, **5**(15), 3803–3833.
166. S. Liu, Y. Sun, L. Chen, Q. Zhang, X. Li and J. Shuai, A review on plasmonic nanostructures for efficiency enhancement of organic solar cells, *Mater. Today Phys.*, 2022, **24**, 100680.
167. P. Baskaran and M. Rajasekar, Recent trends and future perspectives of thermoelectric materials and their applications, *RSC Adv.*, 2024, **14**(30), 21706–21744.
168. Z.-H. Hou, X. Qian, Q.-J. Cui, S.-F. Wang and L.-D. Zhao, Strategies to advance thermoelectric performance of PbSe and PbS materials, *Rare Met.*, 2024, **43**(9), 4099–4114.
169. M. Ibáñez, Z. Luo, A. Genç, L. Piveteau, S. Ortega and D. Cadavid, *et al.*, High-performance thermoelectric nanocomposites from nanocrystal building blocks, *Nat. Commun.*, 2016, 7(1), 10766.
170. C. E. Nwanno and W. Li, Aligned carbon nanotubes for lithium-ion batteries: A review, *Nano Res.*, 2023, **16**(11), 12384–12410.
171. A. Machín, C. Morant and F. Márquez, Advancements and challenges in solid-state battery technology: An in-depth review of solid electrolytes and anode innovations, *Batteries*, 2024, **10**(1), 29.
172. Z. Bitew, M. Tesemma, Y. Beyene and M. Amare, Nano-structured silicon and silicon based composites as anode materials for lithium ion batteries: recent progress and perspectives, *Sustainable Energy Fuels*, 2022, **6**(4), 1014–1050.
173. M. Fikry, M. Abbas, A. Sayed, A. Nouh, A. Ibrahim and A. S. Mansour, Using a novel graphene/carbon nanotubes composite for enhancement of the supercapacitor electrode capacitance, *J. Mater. Sci.: Mater. Electron.*, 2022, **33**(7), 3914–3924.
174. M. I. Ul Hoque, S. W. Donne and R. Holze, Graphene nanocomposite materials for supercapacitor electrodes, *Encyclopedia*, 2024, **4**(1), 101–116.
175. C. An, Y. Zhang, H. Guo and Y. Wang, Metal oxide-based supercapacitors: progress and prospectives, *Nanoscale Adv.*, 2019, **1**(12), 4644–4658.
176. D. Balaji, S. Sivalingam, V. Bhuvaneswari, V. Amarnath, J. Adithya and V. Balavignesh, *et al.*, Aerogels as alternatives for thermal insulation in buildings – A comparative teeny review, *Mater. Today*, 2022, **62**, 5371–5377.

177. L. Meliţă and C. Croitoru, Aerogel, a high performance material for thermal insulation - A brief overview of the building applications, *E3S Web Conf.*, 2019, **111**, 06069.
178. D. Božiček, J. Peterková, J. Zach and M. Košir, Vacuum insulation panels: An overview of research literature with an emphasis on environmental and economic studies for building applications, *Renewable Sustainable Energy Rev.*, 2024, **189**, 113849.
179. P. Raja, V. Murugan, S. Ravichandran, L. Behera, R. A. Mensah and S. Mani, *et al.*, A review of sustainable bio-based insulation materials for energy-efficient buildings, *Macromol. Mater. Eng.*, 2023, **308**, 2300086.
180. J. Fu, C. He, J. Huang, Z. Chen and S. Wang, Cellulose nanofibril reinforced silica aerogels: optimization of the preparation process evaluated by a response surface methodology, *RSC Adv.*, 2016, **6**(102), 100326–100333.
181. S. Nie, N. Hao, K. Zhang, C. Xing and S. Wang, Cellulose nanofibrils-based thermally conductive composites for flexible electronics: a mini review, *Cellulose*, 2020, **27**(8), 4173–4187.
182. R. C. Veloso, A. Souza, J. Maia, N. M. M. Ramos and J. Ventura, Nanomaterials with high solar reflectance as an emerging path towards energy-efficient envelope systems: a review, *J. Mater. Sci.*, 2021, **56**(36), 19791–19839.
183. K. W. Shah and T. Xiong, Nanomaterials and nanocomposites for energy-efficient building envelopes, in *Handbook of Nanomaterials and Nanocomposites for Energy and Environmental Applications*, Springer International Publishing, Cham, 2021, pp. 2621–2648.
184. R. Kumar, A. K. Thakur, L. R. Gupta, A. Gehlot and V. S. Sikarwar, Advances in phase change materials and nanomaterials for applications in thermal energy storage, *Environ. Sci. Pollut. Res.*, 2023, **31**(5), 6649–6677.
185. Y. Wei, Q. Wu, H. Meng, Y. Zhang and C. Cao, Recent advances in photocatalytic self-cleaning performances of TiO_2-based building materials, *RSC Adv.*, 2023, **13**(30), 20584–20597.
186. X. Y. Tan, H. Wang and T. G. Kim, Electrochromic smart windows: An energy-efficient technology, in *Composite Materials: Applications in Engineering, Biomedicine and Food Science*, Springer International Publishing, Cham, 2020, pp. 213–230.
187. M. S. Castillo, X. Liu, F. Abd-AlHamid, K. Connelly and Y. Wu, Intelligent windows for electricity generation: A technologies review, *Build. Simul.*, 2022, **15**(10), 1747–1773.
188. S. Wu, H. Sun, M. Duan, H. Mao, Y. Wu and H. Zhao, *et al.*, Applications of thermochromic and electrochromic smart windows: Materials to buildings, *Cell Rep. Phys. Sci.*, 2023, **4**(5), 101370.
189. J. Huang, J. Zhou, E. Jungstedt, A. Samanta, J. Linnros and L. A. Berglund, *et al.*, Large-area transparent "quantum dot glass" for building-integrated photovoltaics, *ACS Photonics*, 2022, **9**(7), 2499–2509.

190. L. Manai, B. D. Rezgui, A. Bou, D. Barakel, P. Torchio and R. B. Zaghouani, *et al.*, Enhanced light trapping using plasmonic nanoparticles, *J. Phys.: Conf. Ser.*, 2015, **596**, 012002.
191. S. Hazra, S. Dahiya, S. Veer Singh, U. Pandey, S. Suman and P. Swaminathan, *et al.*, Flexible transparent conductors with a percolated Ag nanostructure and its application as efficient self-bias plasmonic photodetector, *Chem. Eng. J.*, 2024, **498**, 155313.
192. J.-G. Lee, G. J. Lee, S. C. Hong, J.-H. Ko, T. Park and Y. W. Ko, Shape optimization of quantum-dot caps for high color-rendering white light-emitting diodes studied by optical simulation, *J. Korean Phys. Soc.*, 2021, **78**(9), 822–828.
193. L. Cao, X. Liu, Z. Guo and L. Zhou, Surface/interface engineering for constructing advanced nanostructured light-emitting diodes with improved performance: A brief review, *Micromachines*, 2019, **10**(12), 821.
194. F.-D. Lai, Effect of surface texture on light extraction efficiency for LEDs, *Crystals*, 2023, **13**(3), 491.
195. A. Fang, Z. Du, W. Guo, J. Liu, H. Xu and P. Tang, *et al.*, Advancements in micro-LED performance through nanomaterials and nanostructures: A review, *Nanomaterials*, 2024, **14**(11), 940.
196. Z. Bao, Z.-F. Jiang, Q. Su, H.-D. Chiu, H. Yang and S. Chen, *et al.*, ZnSe:Te/ZnSeS/ZnS nanocrystals: an access to cadmium-free pure-blue quantum-dot light-emitting diodes, *Nanoscale*, 2020, **12**(21), 11556–11561.
197. H. F. Chinchero, J. M. Alonso and H. Ortiz T, LED lighting systems for smart buildings: a review, *IET Smart Cities*, 2020, **2**(3), 126–134.
198. L.-P. Ma, Z. Wu, Y. Yan, D. Zhang, S. Dong and J. Du, *et al.*, Stably doped graphene transparent electrode with improved light-extraction for efficient flexible organic light-emitting diodes, *Nano Res.*, 2023, **16**(11), 12788–12793.
199. K. D. Patel, F.-S. Juang, H.-X. Wang, C.-Z. Jian and J.-Y. Chen, Quantum dot-based white organic light-emitting diodes excited by a blue OLED, *Appl. Sci.*, 2022, **12**(13), 6365.
200. S. Naqvi, M. Baig, T. Farid, Z. Nazir, S. Mohsan and Z. Liu, *et al.*, Unraveling degradation processes and strategies for enhancing reliability in organic light-emitting diodes, *Nanomaterials*, 2023, **13**(23), 3020.
201. I. Siddiqui, S. Kumar, Y.-F. Tsai, P. Gautam, Shahnawaz and K. Kesavan, *et al.*, Status and challenges of blue OLEDs: A review, *Nanomaterials*, 2023, **13**(18), 2521.
202. D. Wang, J. Hauptmann and C. May, OLED manufacturing on flexible substrates towards roll-to-roll, *MRS Adv.*, 2019, **4**(24), 1367–1375.
203. O. S. Olafusi, E. R. Sadiku, J. Snyman, J. M. Ndambuki and W. K. Kupolati, Application of nanotechnology in concrete and supplementary cementitious materials: a review for sustainable construction, *SN Appl. Sci.*, 2019, **1**, 580.
204. M. Krystek and M. Górski, Nanomaterials in structural engineering, in *New Uses of Micro and Nanomaterials*, InTech, 2018.

205. Y. Wei, H. Meng, Q. Wu, X. Bai and Y. Zhang, TiO_2-based photocatalytic building material for air purification in sustainable and low-carbon cities: A review, *Catalysts*, 2023, **13**(12), 1466.
206. S. Jadoun, Synthesis, mechanism, and applications of self-healing materials, *Biomed. Mater. Devices*, 2024, **2**(1), 225–240.
207. Y. Shields, N. De Belie, A. Jefferson and K. Van Tittelboom, A review of vascular networks for self-healing applications, *Smart Mater. Struct.*, 2021, **30**(6), 063001.
208. M. Ö. Seydibeyoğlu, A. Dogru, J. Wang, M. Rencheck, Y. Han and L. Wang, *et al.*, Review on hybrid reinforced polymer matrix composites with nanocellulose, nanomaterials, and other fibers, *Polymers*, 2023, **15**(4), 984.
209. A. Iqbal, A. Saeed and A. Ul-Hamid, A review featuring the fundamentals and advancements of polymer/CNT nanocomposite application in aerospace industry, *Polym. Bull.*, 2021, **78**(1), 539–557.
210. Q. Li, J. Wang, H. Wang and X. Zhang, Achieving strong and stable nanocrystalline Al alloys through compositional design, *J. Mater. Res.*, 2022, **37**(1), 183–207.
211. Z. Lu, Q. Lin, Z. Cao, W. Li, J. Gong and Y. Wang, *et al.*, MoS_2 nanomaterials as lubricant additives: A review, *Lubricants*, 2023, **11**(12), 527.
212. K. N. Sidh, D. Jangra and H. Hirani, An experimental investigation of the tribological performance and dispersibility of 2D nanoparticles as oil additives, *Lubricants*, 2023, **11**(4), 179.
213. Q. Gao, S. Liu, K. Hou, Z. Li and J. Wang, Graphene-based nanomaterials as lubricant additives: A review, *Lubricants*, 2022, **10**(10), 273.
214. L. Xiao, Y. Xu, Z. Chen and H. Xu, Tribological properties of Diamond-like carbon, TiAlN, and Diamond-like carbon/TiAlN coatings deposited on carburized 18CrNi4A steel for heavy-duty aerospace transmission components, *J. Mater. Eng. Perform.*, 2024, **33**(2), 601–617.
215. F. O. Kolawole, O. S. Kolade, S. A. Bello, S. K. Kolawole, A. T. Ayeni and T. F. Elijah, *et al.*, The improvement of diamond-like carbon coatings for tribological and tribo-corrosion applications in automobile engines: an updated review study, *Int. J. Adv. Manuf. Technol.*, 2023, **126**(5–6), 2295–2322.
216. S. Rajagopal, R. Pulapparambil Vallikkattil, M. Mohamed Ibrahim and D. G. Velev, Electrode materials for supercapacitors in hybrid electric vehicles: Challenges and current progress, *Condens. Matter*, 2022, 7(1), 6.
217. H.-H. Chang, T.-H. Ho and Y.-S. Su, Graphene-enhanced battery components in rechargeable lithium-ion and lithium metal batteries, *C*, 2021, 7(3), 65.
218. S. Lai, T. Huang, P. Liu, H.-B. Wang, S.-C. Yang and X.-H. Liu, *et al.*, Progress and prospects of graphene-based materials in lithium batteries, *Rare Met.*, 2024, **43**(5), 1886–1905.
219. S. Zhu, J. Ni and Y. Li, Carbon nanotube-based electrodes for flexible supercapacitors, *Nano Res.*, 2020, **13**(7), 1825–1841.

220. B. P. Dubey, A. Sahoo, V. Thangadurai and Y. Sharma, Fabrication of ultra-thin, flexible, dendrite-free, robust and nanostructured solid electrolyte membranes for solid-state Li-batteries, *J. Mater. Chem. A*, 2022, **10**(22), 12196–12212.
221. W. Liu, High-energy all-solid-state lithium–metal batteries by nanomaterial designs, in *Nanostructured Materials for Next-Generation Energy Storage and Conversion*, Springer Berlin Heidelberg, Berlin, Heidelberg, 2019. pp. 205–262.
222. H. Li, H. Zhao, B. Tao, G. Xu, S. Gu and G. Wang, *et al.*, Pt-based oxygen reduction reaction catalysts in proton exchange membrane fuel cells: Controllable preparation and structural design of catalytic layer, *Nanomaterials*, 2022, **12**(23), 4173.
223. A. A. Zasypkina, N. A. Ivanova, D. D. Spasov, R. M. Mensharapov, M. V. Sinyakov and S. A. Grigoriev, Recent advances in the development of nanocarbon-based electrocatalytic/electrode materials for proton exchange membrane fuel cells: A review, *Catalysts*, 2024, **14**(5), 303.
224. H. Helal, M. Ahrouch, A. Rabehi, D. Zappa and E. Comini, Nanostructured materials for enhanced performance of solid oxide fuel cells: A comprehensive review, *Crystals*, 2024, **14**(4), 306.
225. K. Kočí, L. Matějová, N. Ambrožová, M. Šihor, I. Troppová and L. Čapek, *et al.*, Optimization of cerium doping of TiO_2 for photocatalytic reduction of CO_2 and photocatalytic decomposition of N_2O, *J. Solgel Sci. Technol.*, 2016, **78**(3), 550–558.
226. O. Zakir, A. Ait-Karra, R. Idouhli, M. Khadiri, B. Dikici and A. Aityoub, *et al.*, A review on TiO_2 nanotubes: synthesis strategies, modifications, and applications, *J. Solid State Electrochem.*, 2023, **27**(9), 2289–2307.
227. J. Xia, L. Dong, H. Song, J. Yang and X. Zhu, Preparation of doped TiO_2 nanomaterials and their applications in photocatalysis, *Bull. Mater. Sci.*, 2023, **46**, 13.
228. R. P. Nippes, P. D. Macruz, M. H. N. O. Scaliante and L. Cardozo-Filho, Fischer–Tropsch synthesis using cobalt catalysts supported on graphene materials: a systematic review, *Res. Chem. Intermed.*, 2023, **49**(5), 2025–2052.
229. A. S. Hoffman, J. A. Singh, S. F. Bent and S. R. Bare, *In situ* observation of phase changes of a silica-supported cobalt catalyst for the Fischer–Tropsch process by the development of a synchrotron-compatible *in situ/operando* powder X-ray diffraction cell, *J. Synchrotron Radiat.*, 2018, **25**(6), 1673–1682.
230. J. Brehm, R. J. Lewis, D. J. Morgan, T. E. Davies and G. J. Hutchings, The direct synthesis of hydrogen peroxide over AuPd nanoparticles: An investigation into metal loading, *Catal Lett.*, 2022, **152**(1), 254–262.
231. S. J. Freakley, N. Agarwal, R. U. McVicker, S. Althahban, R. J. Lewis and D. J. Morgan, *et al.*, Gold–palladium colloids as catalysts for hydrogen peroxide synthesis, degradation and methane oxidation: effect of the PVP stabiliser, *Catal. Sci. Technol.*, 2020, **10**(17), 5935–5944.

232. T. Huang, Z. Su, K. Hou, J. Zeng, H. Zhou and L. Zhang, *et al.*, Advanced stimuli-responsive membranes for smart separation, *Chem. Soc. Rev.*, 2023, **52**(13), 4173–4207.
233. S. Uredat, A. Gujare, J. Runge, D. Truzzolillo, J. Oberdisse and T. Hellweg, A review of stimuli-responsive polymer-based gating membranes, *Phys. Chem. Chem. Phys.*, 2024, **26**(4), 2732–2744.
234. S. Chatterjee and P. C.-L. Hui, Review of applications and future prospects of stimuli-responsive hydrogel based on Thermo-responsive biopolymers in drug delivery systems, *Polymers*, 2021, **13**(13), 2086.
235. S. Tanaka, K. Fuku, N. Ikenaga, M. Sharaf and K. Nakagawa, Recent progress and challenges in the field of metal–organic framework-based membranes for gas separation, *Compounds*, 2024, **4**(1), 141–171.
236. X. Gao, W.-H. Yan, B.-Y. Hu, Y.-X. Huang and S.-M. Zheng, Porous metal–organic frameworks for light hydrocarbon separation, *Molecules*, 2023, **28**(17), 6337.
237. Y. Zhou, P. Li, Y. Wang, Q. Zhao and H. Sun, Progress in the separation and purification of carbon hydrocarbon compounds using MOFs and molecular sieves, *Separations*, 2023, **10**(10), 543.
238. Z. Zhong, J. He, G. Li and L. Xia, Recent advances in magnetic nanoparticles-assisted microfluidic bioanalysis, *Chemosensors*, 2023, **11**(3), 173.
239. S. K. Sethi, A. Verma, E. Akubueze and A. K. Maurya, Self-healing of polymer composites: Process and developments, in *Hybrid Composite Materials*, Springer Nature Singapore, 2024. pp. 253–271.
240. S. He, Y. Gao, X. Gong, C. Wu and H. Cen, Advance of design and application in self-healing anticorrosive coating: a review, *J. Coat. Technol. Res.*, 2023, **20**(3), 819–841.
241. A. Kausar, I. Ahmad, M. Maaza and P. Bocchetta, Self-healing nanocomposites—advancements and aerospace applications, *J. Compos. Sci.*, 2023, 7(4), 148.
242. N. Wang, Y. Zhao, M. Prodanović, M. T. Balhoff and C. Huh, 12012 fundamental mechanisms behind nanotechnology applications in oil and gas: Emerging nano-EOR processes, *Front. Nanotechnol.*, 2022, **4**, 887715.
243. A. Ali, M. I. Jamil, J. Jiang, M. Shoaib, B. U. Amin and S. Luo, *et al.*, An overview of controlled-biocide-release coating based on polymer resin for marine antifouling applications, *J. Polym. Res.*, 2020, **27**, 85.
244. D. Wang, H. Liu, J. Yang and S. Zhou, Seawater-induced healable underwater superoleophobic antifouling coatings, *ACS Appl. Mater. Interfaces*, 2019, **11**(1), 1353–1362.
245. Y. Si and Z. Guo, Superhydrophobic nanocoatings: from materials to fabrications and to applications, *Nanoscale*, 2015, 7(14), 5922–5946.
246. T. Hao, H. Yu, X. Ma and Z. Lan, Heat transfer enhancement of horizontal oscillating heat pipes with micro-/nanostructured surface, *J. Heat Transfer*, 2020, **142**(7), 072001.
247. J.-Y. Zhang, J.-Q. Li and L.-W. Fan, Enhanced pool boiling heat transfer during quenching on Micro/nanostructured wicking surface: Effects of

submicron-scale substructures, in *ASME 2020 Heat Transfer Summer Conference*, American Society of Mechanical Engineers; 2020.

248. H. R. You, S. Lee, D. H. Lee, G. Murali, A. S. Nissimagoudar and Y. Kim, *et al.*, Organic solvent dispersible MXene integrated colloidal quantum dot photovoltaics, *Adv. Energy Mater.*, 2023, **13**, 2301648.
249. Z. Li, S. Ji, Y. Liu, X. Cao, S. Tian and Y. Chen, *et al.*, Well-defined materials for heterogeneous catalysis: From nanoparticles to isolated single-atom sites, *Chem. Rev.*, 2020, **120**(2), 623–682.
250. X. Zhong, E. Yuan, F. Yang, Y. Liu, H. Lu and J. Yang, *et al.*, Optimizing oxygen vacancies through grain boundary engineering to enhance electrocatalytic nitrogen reduction, *Proc. Natl. Acad. Sci. U. S. A.*, 2023, **120**(40), e2306673120.
251. Q. Wang, W. Chen, W. Zhu, D. J. McClements, X. Liu and F. Liu, A review of multilayer and composite films and coatings for active biodegradable packaging, *Npj Sci. Food*, 2022, **6**(1), 18.
252. M. Thirupathi Vasuki, V. Kadirvel and G. Pejavara Narayana, Smart packaging—An overview of concepts and applications in various food industries, *Food Bioeng.*, 2023, **2**(1), 25–41.
253. T. Kasuga, T. Saito, H. Koga and M. Nogi, One-pot hierarchical structuring of nanocellulose by electrophoretic deposition, *ACS Nano*, 2022, **16**(11), 18390–18397.
254. J. U. Surjadi, Y. Zhou, S. Huang, L. Wang, M. Li and S. Fan, *et al.*, Lightweight, ultra-tough, 3D-architected hybrid carbon microlattices, *Matter*, 2022, **5**(11), 4029–4046.
255. Industrial energy consumption by fuel in the Net Zero Scenario, 2000-2030, IEA, [cited 2024 Oct 5]. Available from: https://www.iea.org/data-and-statistics/charts/industrial-energy-consumption-by-fuel-in-the-net-zero-scenario-2000-2030.
256. M. C.-Y. Ang, J. M. Saju, T. K. Porter, S. Mohaideen, S. Sarangapani and D. T. Khong, *et al.*, Decoding early stress signaling waves in living plants using nanosensor multiplexing, *Nat. Commun.*, 2024, **15**(1), 2943.
257. N. Farooq, Z. Rehman, M. I. Khan, S. Asghar, M. Saleem and R. Irshad, *et al.*, Nanomaterial-based energy conversion and energy storage devices: a comprehensive review, *New J. Chem.*, 2024, **48**(19), 8933–8962.
258. Q. Wang, Z. Na, L. Yu, S. Dai, M. K. Nazeeruddin and H. Yang, Smart photovoltaic windows for next-generation energy-saving buildings, *Adv. Sci.*, 2024, e2407177.

CHAPTER 9

Nanotechnology in Recycling and Waste Treatment

As we progress from examining nanotechnology's role in waste prevention and energy efficiency, we now turn our attention to its transformative potential in recycling and waste treatment. While Chapter 8 focused on proactive measures to minimize waste generation and optimize resource use, this chapter explores how nanotechnology can address the challenge of managing waste that has already been produced.

We begin with an overview of circular economy principles and their relationship to nanotechnology in waste management. This provides a crucial framework for understanding the broader implications of the advanced recycling and treatment technologies discussed throughout the chapter.

There are inevitably some areas of overlap between the topics covered in Chapter 8 and those we will discuss here. Both chapters deal with nanomaterials and their unique properties, and some of the fundamental principles and technologies—such as nanostructured membranes, nanocatalysts, and nanosensors—feature in both discussions. However, the context and applications of these technologies differ significantly when applied to recycling and waste treatment.

For instance, while Chapter 8 examined nanostructured membranes for process intensification and water purification in industrial settings, here we will explore how similar membrane technologies can be adapted for material recovery from complex waste streams or for advanced wastewater treatment. Similarly, nanocatalysts, which were previously discussed in the context of enhancing chemical reactions and energy efficiency, will now be considered for their potential in waste-to-energy conversion processes such as biomass gasification and anaerobic digestion.

Waste Not! How Nanotechnologies Can Increase Efficiencies Throughout Society
By Michael Berger

Published by the Royal Society of Chemistry, www.rsc.org

The overlap in base technologies underscores the versatility of nanotechnology and its potential to address multiple stages of the waste management hierarchy. However, this chapter will delve into the unique challenges posed by waste streams—their heterogeneity, variability, and often hazardous nature—and how nanotechnology can be specifically tailored to address these issues.

New topics that were not covered in Chapter 8 will also be introduced. These include nanomaterial-enhanced sorting and separation techniques for recycling, nanoparticle-based remediation strategies for contaminated sites, and the use of nanotechnology in specialized waste treatment processes like advanced oxidation.

Throughout this chapter, we will maintain a strong focus on the circular economy concept, examining how nanotechnology can enable more efficient recovery and reuse of materials from waste. We will also pay particular attention to the environmental implications of using nanomaterials in waste treatment, a critical consideration given the potential for these materials to enter the environment through treated waste streams.

By building upon the foundational knowledge established in Chapter 8 and extending it into recycling and waste treatment, this chapter aims to provide a comprehensive overview of how nanotechnology is revolutionizing our approach to waste management. From enhancing material recovery to enabling more efficient waste-to-energy conversion and advanced treatment technologies, nanotechnology offers promising solutions to some of our most pressing waste-related challenges.

9.1 Circular Economy and Nanotechnology in Waste Management

As we explore advanced recycling and waste treatment technologies enabled by nanotechnology, it's crucial to consider these innovations within the broader context of the circular economy. The circular economy represents a paradigm shift from the traditional linear "take-make-dispose" model to a more sustainable approach that aims to eliminate waste and maximize resource utilization (see Section 4.6 for a deeper discussion on this issue).

The Ellen MacArthur Foundation defines a circular economy as one that is restorative and regenerative by design, and which aims to keep products, components and materials at their highest utility and value at all times, distinguishing between technical and biological cycles.[1] In this model, waste is not just a problem to be managed, but a valuable resource to be reintegrated into production cycles.

Nanotechnology is critical in advancing circular economy principles in waste management and recycling. The techniques covered in this chapter—such as nanomaterial-enhanced sorting and separation, nanomaterial-based adsorbents, and nanosensors for detecting contaminants—improve the efficiency of resource recovery and reduce waste. These technologies support

Figure 9.1 Nanotechnology-enabled innovations driving circular economy in waste management.

more precise material separation for high-quality recycling, enable the recovery of valuable resources from waste streams that were previously uneconomical to process, allow for the detection and removal of contaminants with unprecedented precision, and foster the development of new materials designed for easier recycling and reuse (see Figure 9.1).

As we examine these nanotechnology-enabled processes, it's important to consider how they fit into the larger circular economy framework. Each innovation not only addresses immediate waste management challenges but also contributes to closing material loops and reducing the overall environmental impact of production and consumption cycles.

In the following sections, we will explore specific nanotechnology applications in recycling and waste treatment. As we do so, we'll consider how these technologies can be leveraged to support circular economy principles, creating more sustainable and resource-efficient systems. The advancements discussed here lay the groundwork for the more comprehensive resource monitoring and management strategies that will be explored in Chapter 10, where we'll delve deeper into how nanotechnology enables circular economy applications across various industries and processes.

9.2 Material Separation and Recovery

The efficient separation and recovery of valuable materials from waste streams is a cornerstone of effective recycling and waste management. This section explores how nanotechnology is revolutionizing these processes, enabling more precise, efficient, and cost-effective material recovery than ever before.

Nanotechnology offers unique advantages in material separation and recovery due to the high surface area-to-volume ratio of nanomaterials and their ability to be engineered for specific interactions with target substances.

These properties allow for enhanced selectivity, improved efficiency, and the ability to recover materials that were previously too difficult or costly to extract.

There are three key areas where nanotechnology is making significant impacts:

- Nanomaterial-enhanced sorting and separation techniques, including the use of magnetic nanoparticles, nanostructured membranes, and nanoparticle-enhanced flotation.
- Nanomaterial-based adsorbents, such as carbon nanotubes, graphene-based materials, metal–organic frameworks, and nanostructured inorganic adsorbents.
- Nanosensors for contaminant detection, including plasmonic, electrochemical, and aptamer-based nanosensors.

Each of these areas represents a significant advancement in our ability to identify, separate, and recover valuable materials from complex waste streams. By enabling more efficient resource recovery, these nanotechnology-based approaches not only improve the economics of recycling but also contribute to the broader goals of resource conservation and circular economy principles.

9.2.1 Nanomaterial-enhanced Sorting and Separation

Nanomaterials offer unique properties that can significantly enhance the efficiency and selectivity of sorting and separation processes in waste management and recycling.

9.2.1.1 Magnetic Nanoparticle-assisted Separation

Magnetic nanoparticles (MNPs) have emerged as powerful tools for selective separation in various waste treatment applications due to their high surface area, ease of functionalization, and ability to be manipulated by external magnetic fields. Key applications include:

9.2.1.1.1 Selective Recovery of Metals from Electronic Waste. Scientists have developed nanomaterials that can efficiently recover valuable metals, such as gold, from electronic waste.[2] These materials combine a specific covalent organic framework with MoS_2, creating powerful adsorbents for gold. The best-performing composite can extract large amounts of gold even from complex mixtures and electronic waste like old mobile phone circuit boards, and it works quickly, completing the process in just five minutes. Additionally, this material can be reused multiple times without losing its effectiveness. This innovation promises significant economic and environmental benefits by making it easier to recover precious metals from discarded electronics.

9.2.1.1.2 Oil–Water Separation in Wastewater Treatment. Oil contamination in wastewater is a pressing environmental issue, often arising from industrial activities such as petrochemical manufacturing and oil refining. One innovative solution involves the use of MNPs engineered with amphiphilic coatings. These coatings create surfaces that selectively adsorb oil while repelling water, making them highly effective for oil–water separation.

For instance, amphiphilic MNPs coated with a surfactant that has both hydrophilic and hydrophobic ends enable the nanoparticles to adsorb various types of oil. These MNPs can achieve a separation efficiency of over 95% in a continuous operation system. The process is rapid and allows for the nanoparticles to be easily regenerated and reused, maintaining their effectiveness over multiple cycles. This method offers significant potential for scalable industrial applications, addressing both environmental and economic concerns.[3]

9.2.1.1.3 Removal of Heavy Metals from Industrial Effluents. Heavy metal contamination in industrial effluents poses significant environmental and health risks. MNPs have proven to be highly effective in the removal of these contaminants due to their large surface area, ease of functionalization, and external magnetic field manipulation.

One advanced application involves MNPs functionalized with amino groups, enhancing their ability to adsorb heavy metals such as lead, cadmium, and chromium from wastewater. These amino-functionalized MNPs exhibit a strong affinity towards heavy metals, enabling rapid and efficient adsorption.

Researchers have demonstrated the efficacy of amino-functionalized MNPs in treating industrial wastewater.[4] The results showed that these nanoparticles could remove up to 99% of lead and cadmium and 95% of chromium within 30 minutes of treatment. The regeneration and reuse of the MNPs were successfully carried out over five cycles without a significant loss in adsorption capacity, highlighting their economic and sustainable application in industrial wastewater management.

This innovative approach not only offers a highly efficient solution for heavy metal removal but also aligns with sustainable practices by allowing repeated use of the MNPs, thus reducing the overall treatment costs and environmental impact.

9.2.1.2 Nanostructured Membranes for Molecular Separation

Nanostructured membranes have become pivotal in advancing molecular separation technologies due to their unique properties, including high surface area, tunable pore sizes, and enhanced selectivity.

Graphene oxide membranes are at the forefront, particularly for selective ion separation in wastewater treatment. These membranes, engineered with precise control over their pore structure, enable the efficient filtration of ions, offering significant improvements in desalination processes. They demonstrate excellent water permeance and ion selectivity, making them highly effective for removing contaminants from wastewater.[5]

Additionally, MOF-based membranes have shown great promise for gas separation, particularly in addressing emissions from landfills. The porous structure of MOFs allows for the selective adsorption and separation of various gases, thereby mitigating the environmental impact of landfill emissions. These membranes leverage the unique, tunable nature of MOFs, where the pore size and environment can be precisely controlled by connecting metal ions with organic linkers of various functionalities. This customization enables efficient separation of gases such as CO_2, CH_4, and others, essential for reducing greenhouse gas emissions from landfill sites.[6]

Research has highlighted the superior performance of MOF-based membranes in gas separation processes. For instance, these membranes have been shown to effectively separate CO_2 from other gases due to their high surface area and adjustable pore sizes, leading to significant advancements in reducing emissions and improving air quality.[7] Furthermore, the continuous development of MOF-based membranes emphasizes their potential for large-scale industrial applications, addressing both energy efficiency and environmental sustainability.[8]

MOF-based membranes are also effective in liquid separation applications, including water purification, organic solvent nanofiltration, and chiral resolution. Both pure MOF membranes and mixed-matrix membranes (MMMs), which combine MOF particles with polymer matrices, exhibit high selectivity and stability (see Figure 9.2). MMMs, in particular, enhance the

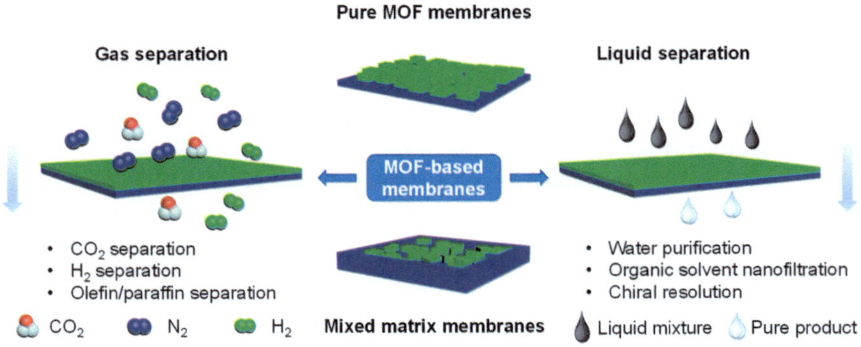

Figure 9.2 Schematic illustration showing the applications of pure MOF membranes and mixed-matrix membranes (MMMs) for both gas and liquid separations. Pure MOF membranes are used for gas separation, including CO_2, H_2, and olefin/paraffin separations, while MMMs combine MOF particles with polymers to enhance performance in both gas and liquid separations, such as water purification, organic solvent nanofiltration, and chiral resolution. Reproduced from ref. 6, https://doi.org/10.3390/membranes13050480, under the terms of the CC BY 4.0 license, https://creativecommons.org/licenses/by/4.0/.

processability and mechanical strength of MOFs, making them suitable for large-scale industrial applications in both gas and liquid separations.[6]

These innovations highlight the potential of nanostructured membranes in offering sustainable and efficient solutions for complex separation challenges in waste management and recycling. By continuously improving the selectivity, durability, and anti-fouling properties of these membranes, researchers are paving the way for more effective environmental protection and resource recovery strategies.

9.2.1.3 Nanoparticle-enhanced Flotation

Nanoparticle-enhanced flotation represents a significant advancement in separation technology, particularly in mineral processing and wastewater treatment. Flotation is a physical-chemical separation process that uses differences in surface properties to separate materials, typically by attaching air bubbles to specific particles to make them float to the surface of a slurry. This innovative approach leverages the unique properties of nanoparticles to improve the efficiency and selectivity of traditional flotation processes.

In mineral processing, nanoparticles are utilized as novel collectors or modifiers in froth flotation. Their high surface area-to-volume ratio and tunable surface properties allow for enhanced interaction with mineral particles. For instance, functionalized silica nanoparticles have shown promise in selectively adsorbing to specific mineral surfaces, thereby improving the separation of valuable minerals from gangue materials. This enhanced selectivity can lead to higher recovery rates and improved concentrate grades in mining operations. Recent studies have documented these benefits, highlighting the role of nanobubbles in improving the attachment of mineral particles during flotation processes.[9]

In oily wastewater treatment, nanoparticle-stabilized emulsions have emerged as an effective technique for oil removal. Amphiphilic nanoparticles, such as Janus nanoparticles with both hydrophobic and hydrophilic surfaces, can stabilize oil droplets in water, facilitating their separation through flotation. This approach has demonstrated superior oil removal efficiencies compared to conventional methods, particularly for emulsified oils that are traditionally challenging to separate.[10]

Research strongly supports the development of environmentally friendly and cost-effective nanoparticles for flotation processes. Studies have shown that bio-derived nanoparticles, synthesized from waste materials, are emerging as promising alternatives to synthetic nanoparticles in flotation applications. These bio-nanoparticles are not only effective in separation processes but also align with sustainability goals by utilizing waste resources, thus contributing to environmental conservation.[11]

Biogenically synthesized nanoparticles, derived from plant and microbial sources, have shown significant potential in removing contaminants from wastewater. These nanoparticles offer unique properties and environmentally friendly synthesis methods, making them effective for treating various pollutants.[12]

Researchers have explored the application of waste-derived nanomaterials, such as those obtained from biosolids and industrial effluents, for recovering valuable metals and enhancing wastewater treatment processes. These approaches not only improve the efficiency of pollutant removal but also reduce the environmental and health risks associated with waste disposal.[13]

Despite these advancements, the implementation of nanoparticle-enhanced flotation faces several challenges. One key concern is the potential environmental impact of nanoparticle release. To address this, researchers are developing strategies for nanoparticle recovery and recycling within flotation circuits. Additionally, the scalability of nanoparticle production and the long-term stability of nanoparticle-enhanced flotation systems in industrial settings remain areas of ongoing investigation.

The potential benefits of nanoparticle-enhanced flotation in improving separation efficiency and reducing chemical usage make it a promising area for further research and development in waste management and resource recovery applications.

While these nanomaterial-enhanced separation techniques show great promise, several common challenges need to be addressed for widespread implementation. These include scaling up production of nanomaterials, ensuring complete recovery of nanoparticles to prevent environmental release, improving long-term stability in complex waste environments, and establishing appropriate regulatory frameworks for their use in waste treatment applications.

As research in this field progresses, nanomaterial-enhanced sorting and separation technologies are expected to play an increasingly important role in advancing material recovery and waste treatment processes.

9.2.2 Nanomaterial-based Adsorbents

Nanomaterial-based adsorbents have revolutionized the field of material separation and recovery due to their exceptional surface area, tunable pore structures, and high adsorption capacities. These materials offer significant advantages over traditional adsorbents in terms of efficiency, selectivity, and reusability.

9.2.2.1 Carbon Nanotubes and Graphene-based Adsorbents

Carbon nanotubes and graphene-based materials have emerged as powerful adsorbents for various contaminants, particularly in water treatment applications. CNTs, with their hollow tubular structure and high aspect ratio, provide an extensive surface area for adsorption. Their surface can be functionalized to enhance selectivity towards specific contaminants. Oxidized CNTs have shown remarkable efficiency in removing heavy metals from water. The oxygen-containing functional groups on their surface act as binding sites for metal ions, enabling rapid and selective adsorption. Research indicates that functionalized CNTs can significantly improve the removal efficiency of heavy metals such as chromium and lead through various chemical interactions and modifications.[14,15]

Graphene and its derivatives offer unique adsorption properties due to their single-or few-layer structure and abundant functional groups. GO, in particular, has demonstrated exceptional capacity for removing organic dyes and pharmaceuticals from water. Its large surface area and oxygen-containing functional groups provide multiple interaction sites for contaminants. Studies have shown that graphene oxide can effectively adsorb a wide range of pollutants due to its versatile surface chemistry and large adsorption capacity.[16]

9.2.2.2 Metal–Organic Frameworks as Adsorbents

Metal–organic frameworks have attracted significant attention as advanced adsorbents due to their high surface areas, customizable pore sizes, and diverse chemical functionalities. In landfill gas purification, MOFs have demonstrated substantial potential for the selective adsorption of gases. By fine-tuning pore sizes and chemical environments, MOFs can effectively separate methane from carbon dioxide, a critical step in upgrading landfill gas to natural gas quality. Specific MOFs, such as those with fluorinated frameworks, have shown exceptional stability and selectivity for this application, making them promising candidates for efficient gas purification processes.[17,18]

MOFs also have shown promise in the recovery of rare earth elements from electronic waste. Researchers have developed MOFs with functional groups that have a high affinity for REEs, allowing for the selective adsorption of these valuable elements from complex waste streams. This approach provides a more environmentally friendly and potentially more cost-effective method for REE recovery compared to traditional methods.[19]

Despite these advancements, challenges remain in the large-scale application of MOFs. Water stability, scalability of synthesis, and regeneration efficiency are key issues that need to be addressed. Ongoing research is focused on developing water-stable MOFs and exploring more economical synthesis routes to make MOF-based adsorption technologies commercially viable.

9.2.2.3 Nanostructured Inorganic Adsorbents

Nanostructured inorganic materials, such as mesoporous silica nanoparticles (MSNs) and nanostructured metal oxides, offer distinct advantages for adsorption applications due to their unique properties. MSNs are notable for their high surface area and adjustable pore structures, making them highly effective in removing organic pollutants from water. Their surfaces can be functionalized with various groups to enhance selectivity for specific contaminants, such as amine-functionalized MSNs which have demonstrated significant capacity for CO_2 capture, useful in landfill gas treatment.[20]

Nanostructured titanium dioxide is recognized not only for its adsorption capabilities but also for its photocatalytic activity. Under UV light, TiO_2 nanoparticles can degrade adsorbed organic pollutants, thus providing dual functionality for adsorption and degradation, which is particularly valuable in water treatment applications.[21] Studies also discuss the production and enhanced photocatalytic activity of TiO_2-coated silicon carbide membranes for water treatment, showcasing the material's ability to degrade adsorbed pollutants under UV illumination.[22]

While integrating these nanostructured inorganic adsorbents into existing treatment systems offers superior performance, challenges such as nanoparticle agglomeration, recovery, and environmental impact need to be addressed. Research is ongoing to develop immobilization techniques and composite materials that maintain the high surface area and reactivity of these nanomaterials while facilitating their separation and reuse in practical applications.

As advancements continue, nanomaterial-based adsorbents are expected to play a growing role in addressing complex separation and purification challenges in waste management and resource recovery. Their ability to selectively remove contaminants at low concentrations, combined with potential for regeneration and reuse, positions them as promising candidates for developing more sustainable and efficient treatment technologies.

9.2.3 Nanosensors for Contaminant Detection

Nanosensors have emerged as powerful tools for the rapid and sensitive detection of contaminants in waste streams, offering significant advantages over traditional analytical methods. These sensors leverage the unique properties of nanomaterials to achieve high sensitivity, specificity, and real-time monitoring capabilities, crucial for improving waste treatment processes and environmental protection.

9.2.3.1 Plasmonic Nanosensors for Contaminant Detection

Plasmonic nanosensors, based on the localized surface plasmon resonance phenomenon, have made significant advancements in contaminant detection in waste management. Gold nanoparticle-based colorimetric sensors are notable for their simplicity and visual readout capabilities. These sensors utilize the optical properties of gold nanoparticles to detect various analytes, such as heavy metals in wastewater. When exposed to specific metal ions, these nanoparticles aggregate, causing a visible color change from red to blue. This change can be observed by the naked eye or quantified using simple spectrophotometric methods, facilitating rapid on-site detection of contaminants.[23]

Surface-enhanced Raman spectroscopy substrates have shown great promise in trace organic pollutant analysis. By enhancing the Raman signal of adsorbed molecules, SERS allows the detection of contaminants at extremely low concentrations (see Figure 9.3). Various nanostructures, such as perovskite quantum dot-graphene/nanogold composites, have been developed, enabling the ultrasensitive detection of pollutants like rhodamine 6G and *p*-nitrophenol.[24] Additionally, silver nanospheres loaded on cobalt-iron layered double hydroxides serve as robust SERS probes for detecting 4-nitrophenol, a significant environmental pollutant due to its toxicity and persistence.[25]

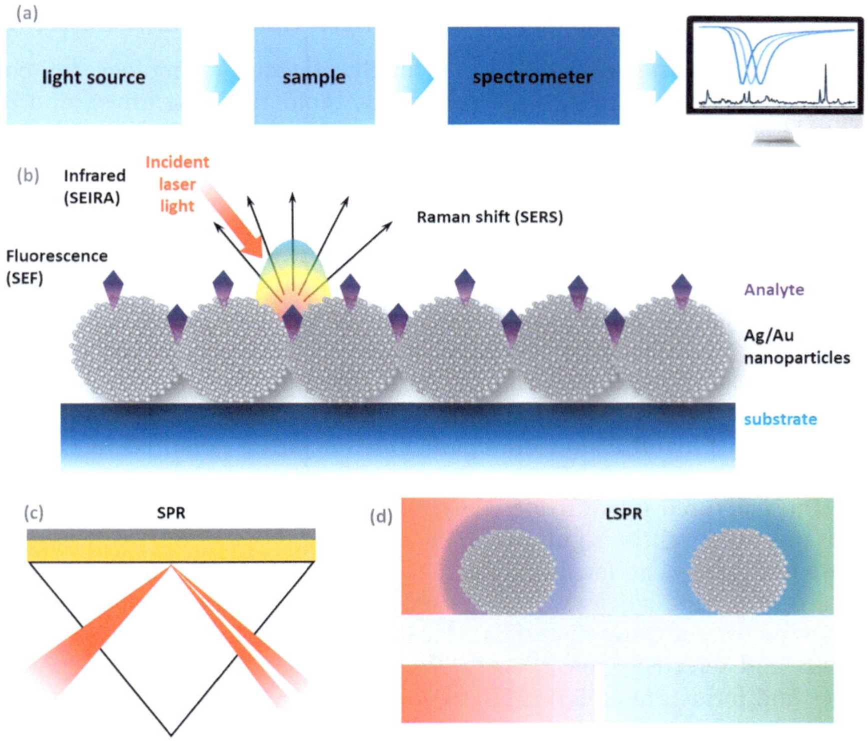

Figure 9.3 Basic principles of plasmonic-based detection methods. (a) General schematic of a plasmonic sensor workflow, showing the interaction between light, sample, and spectrometer; (b) illustration of SERS, SEF, and SEIRA mechanisms, highlighting how nanoparticles enhance signal detection through different optical techniques; (c) representation of surface plasmon resonance (SPR), demonstrating how light interacts with a metallic surface; (d) depiction of localized surface plasmon resonance (LSPR), illustrating the optical response of metal nanoparticles in proximity to analytes. Reproduced from ref. 23, https://doi.org/10.3390/chemosensors11010035, under the terms of the CC BY 4.0 license, https://creativecommons.org/licenses/by/4.0/.

Other optical techniques such as surface-enhanced fluorescence (SEF) and surface-enhanced infrared absorption (SEIRA) further expand the detection capabilities of plasmonic nanosensors. SEF amplifies the fluorescence signal of analytes, improving sensitivity in applications such as biological sensing. SEIRA, on the other hand, enhances infrared absorption for the analysis of molecular vibrations, offering a complementary approach to detecting molecular contaminants.

These advancements in SERS substrates facilitate the detection of pharmaceutical residues and pesticides in environmental samples at parts-per-billion levels, contributing significantly to environmental monitoring efforts.

9.2.3.2 Electrochemical Nanosensors for Contaminant Detection

Electrochemical nanosensors are a highly effective technology for contaminant detection due to their sensitivity, rapid response times, and potential for miniaturization, making them particularly suitable for continuous monitoring in waste treatment facilities.[26,27]

Carbon nanotube-modified electrodes excel in detecting a wide array of contaminants. Their high surface area and exceptional electrical conductivity facilitate enhanced electron transfer, thereby improving sensitivity. For instance, multi-walled carbon nanotube-modified electrodes have been developed to rapidly detect pharmaceutical residues in wastewater, capable of identifying multiple drug compounds simultaneously at nanomolar concentrations. This capability is crucial for monitoring emerging contaminants effectively. The enhanced sensitivity and selectivity of these electrodes in detecting bisphenol A and diazepam illustrate their potential in environmental monitoring and pharmaceutical applications.[28,29]

Nanostructured metal oxide sensors, such as those based on zinc oxide and tin oxide, show significant promise in monitoring volatile organic compounds in waste facilities. These sensors operate by detecting changes in electrical conductivity when exposed to VOCs. Through engineering the nanostructure of these materials, researchers have achieved improved sensitivity and selectivity towards specific VOCs commonly present in landfill gases and industrial emissions.[30]

Recent advancements have also enhanced sensor durability and long-term stability by developing protective coatings and self-cleaning mechanisms, ensuring consistent performance in harsh environments. For example, applying hydrophobic nanocoatings to electrochemical sensors helps prevent fouling, thus extending their operational lifetime in wastewater treatment plants.

Hydrophobic coatings with self-cleaning performances and UV durability have been evaluated under lab-simulated weathering conditions, demonstrating their effectiveness.[31] Conductive polymers have also been highlighted as an ideal choice for modifying electrochemical sensors, making them well-suited for biomedical and environmental applications.[32] The surface modification of working electrodes by carbon nanomaterials and metallic nanostructures has further improved the sensing performance of electrochemical sensors for water quality monitoring.

9.2.3.3 Aptamer-based Nanosensors for Contaminant Detection

Aptamer-based nanosensors represent a promising approach for the highly specific detection of contaminants, including pathogens and complex organic molecules. DNA aptamer-functionalized nanoparticles have been developed for specific pathogen detection in wastewater. These sensors utilize the high specificity of aptamers—short, single-stranded DNA or RNA molecules that can bind to specific targets—combined with the signal amplification properties of nanoparticles. For example, gold nanoparticles functionalized with aptamers specific to *E. coli* have demonstrated rapid and sensitive detection of this pathogen in water samples, with potential applications in monitoring the effectiveness of wastewater treatment processes.[33]

The incorporation of aptasensors into microfluidic devices has led to portable, on-site monitoring systems, integrating sample preparation, sensing, and signal readout into a single, compact platform. Recent efforts have been directed towards developing multiplexed aptasensors capable of simultaneously detecting multiple contaminants, thereby providing comprehensive water quality analysis in a single test.[34]

Challenges remain, particularly in the selection of aptamers and sensor regeneration. Developing aptamers with high specificity and affinity for target contaminants is often time-consuming and costly. Ensuring the reusability of aptamer-based sensors in complex waste matrices is another area requiring further research. Efforts are ongoing to enhance aptamer selection methods and incorporate regeneration mechanisms into sensor designs, which will enable long-term, cost-effective monitoring.[35]

As research in nanosensors for contaminant detection continues to advance, these technologies are poised to play a crucial role in improving the efficiency and effectiveness of waste treatment processes. Their ability to provide rapid, sensitive, and specific detection of contaminants in complex waste streams offers the potential for real-time process control and optimization in waste management facilities. However, further work is needed to address challenges related to sensor stability, interference from complex matrices, and cost-effective production before these nanosensors can be widely deployed in industrial settings.

9.2.3.4 Case Study: Protein-based Nanomaterials Enable Efficient Gold Recovery from E-waste

Gold recovery from electronic waste (e-waste) has become an increasingly critical issue as the global production of e-waste continues to surge at an alarming rate. According to the United Nations' fourth *Global E-waste*

Monitor (GEM), the world generated a record 62 million tons of e-waste in 2022, marking an 82% increase from 2010. This figure is projected to rise by another 32% to 82 million tons by 2030.[36] Even more concerning is that e-waste generation is outpacing documented recycling efforts by a factor of five. This rapid growth presents both environmental challenges and economic opportunities, as e-waste contains valuable metals, particularly gold, which is widely used in electronics due to its exceptional conductivity and stability.

The GEM report highlights that billions of dollars' worth of strategically valuable resources are being squandered, with only 1% of the demand for rare earth elements currently met by e-waste recycling. These statistics underscore the urgent need for more efficient and sustainable methods of recovering valuable materials from e-waste.

Historically, gold recovery from e-waste has relied on resource-intensive methods such as pyrometallurgy and hydrometallurgy. These processes often involve harsh chemicals and high energy consumption, raising concerns about their environmental impact. More recently, adsorption-based techniques using materials like activated carbon have gained attention as potentially more sustainable alternatives. However, these methods still face limitations in terms of efficiency and selectivity, particularly when dealing with the complex mixture of metals found in e-waste.

The search for more effective and environmentally friendly gold recovery methods has led researchers to explore novel nanomaterials and approaches. One promising avenue that has emerged in recent years is the use of protein-based nanostructures, specifically amyloid fibrils (AF). These self-assembling protein nanofibrils, with diameters ranging from 5 to 10 nm and lengths spanning several micrometers, have shown remarkable capabilities in heavy metal adsorption at the nanoscale, opening up new possibilities for selective metal recovery from complex waste streams.

Amyloid fibrils are supramolecular nanostructures characterized by their high content of β-sheet secondary structures. They possess unique nanoscale properties such as extremely high surface-to-volume ratios, modular surface functionality at the molecular level, and exceptional nanomechanical stability. What makes AF particularly intriguing for sustainable nanotechnology applications is that they can be produced from proteins extracted from food waste and industrial side streams, aligning with circular economy principles.

A research team from ETH Zurich explored the potential of AF nanofibrils derived from whey, a major dairy industry byproduct, for selective gold recovery from e-waste. Their study[37] describes a novel nanoadsorbent material in the form of AF aerogels and assesses its performance in gold recovery from solutions mimicking e-waste composition (see Figure 9.4).

The researchers began by preparing AF nanofibrils from whey protein through a process of denaturation and self-assembly under acidic conditions and high temperatures. They then used these nanofibrils to create lightweight, nanoporous aerogels by freeze-drying a 2 wt% aqueous dispersion of

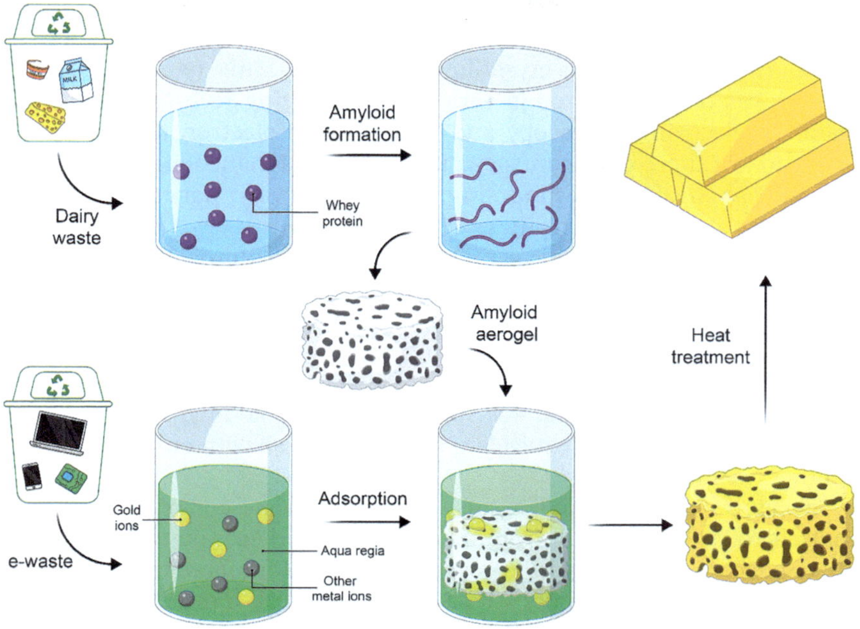

Figure 9.4 Schematic route followed to generate neat value *via* pure gold, extracted from food waste (whey) and e-waste as starting source materials. Reproduced from ref. 37, https://doi.org/10.1002/adma.202310642, under the terms of the CC BY 4.0 license, https://creativecommons.org/licenses/by/4.0/.

AF with added cross-linkers. The resulting nanostructured aerogels exhibited a remarkably low density of 33.18 mg cm^{-3} and a porosity of 97%, while maintaining excellent nanomechanical and water stability.

To evaluate the metal adsorption capacity of the AF nanoaerogels, the team conducted experiments using solutions containing a mixture of metals commonly found in e-waste, including gold, copper, nickel, lead, zinc, chromium, iron, and manganese. The results were striking: the AF nanoaerogels demonstrated a remarkable selectivity for gold at the nanoscale, with a removal efficiency of 93.3% and an adsorption capacity of 166.7 mg g^{-1}. In contrast, the removal efficiencies for other metals were significantly lower, ranging from 0% to 9%.

Further investigations into the gold adsorption properties of the AF aerogels revealed impressive performance across a wide range of concentrations. The maximum adsorption capacity reached as high as 1.887 grams of gold per gram of aerogel at an initial gold concentration of 100 000 ppm. This means that each gram of the aerogel could adsorb nearly twice its own weight in gold, demonstrating its exceptional efficiency. The nanoscale adsorption process was also found to be rapid, with 77% of gold removed within just 5 minutes of contact with the nanoaerogel.

One of the most intriguing aspects of the AF nanoaerogels' performance was their ability to not only adsorb gold ions but also convert them into gold nanoparticles and single crystalline nanoflakes. This nanoscale transformation was observed through various nanoimaging techniques, including scanning electron microscopy and transmission electron microscopy. The formation of these gold nanostructures on the AF nanoaerogels could have potential applications beyond simple metal recovery, particularly in nanotechnology and nanoelectronics.

To demonstrate the practical applicability of their approach, the researchers tested the AF aerogels on actual e-waste solutions obtained by dissolving computer motherboards in aqua regia, a powerful mixture of nitric and hydrochloric acids capable of dissolving gold and other noble metals. Even in this complex and highly acidic mixture, the AF aerogels maintained their high selectivity for gold. The team was able to recover a gold nugget with a purity of approximately 90.8 wt% (equivalent to 21–22 carats), with only trace amounts of other metals present.

The study also included a life cycle assessment and techno-economic analysis to evaluate the environmental impact and economic viability of the AF nanoaerogel-based gold recovery process. Compared to conventional activated carbon adsorbents, the AF nanoaerogels showed lower environmental impacts across several categories, including climate change, freshwater eutrophication, and fossil depletion. The economic analysis suggested that the cost of recovering 1 gram of gold using AF nanoaerogels would be around $1.1, which is a tiny fraction of the current market value of 22-carat gold (approximately $80 per gram in October 2024).

This research represents a significant step forward in sustainable gold recovery from e-waste using nanotechnology. By utilizing a waste product (whey) to create a highly efficient and selective nanoadsorbent for recovering a valuable resource (gold) from another waste stream, the study exemplifies the principles of circular economy and waste valorization at the nanoscale.

The development of AF nanoaerogels for gold recovery addresses several key challenges in the field of nanomaterials for resource recovery. First, it offers a more environmentally friendly alternative to traditional recovery methods, potentially reducing the use of harsh chemicals and energy-intensive processes. Second, the high selectivity for gold in complex metal mixtures at the nanoscale could greatly simplify the recovery process and improve overall efficiency. Finally, the ability of AF nanoaerogels to facilitate the formation of gold nanoparticles and crystalline nanostructures opens up possibilities for the direct synthesis of high-value gold nanomaterials from waste sources.

While the results are promising, it's important to note that this nanotechnology research is still at a relatively early stage. Scaling up the production of AF nanoaerogels and optimizing the nanoscale recovery process for industrial-scale applications will require further research and development. Additionally, the long-term stability and reusability of the AF nanoaerogels in real-world conditions will need to be thoroughly investigated.

9.3 Waste-to-energy Conversion

The transformation of waste into usable energy forms represents a critical frontier in sustainable resource management and renewable energy production. As global waste generation continues to rise, the development of efficient waste-to-energy (WtE) technologies becomes increasingly vital. This section explores the cutting-edge applications of nanotechnology in enhancing WtE conversion processes, focusing on three key areas: biomass gasification, anaerobic digestion, and fuel cells for waste-derived energy.

Nanotechnology offers unprecedented opportunities to optimize these processes at the molecular level, improving catalytic efficiency, increasing reaction rates, and enhancing overall system performance. By leveraging the unique properties of nanomaterials, researchers and engineers are pushing the boundaries of what's possible in waste valorization, paving the way for more sustainable and economically viable WtE solutions.

9.3.1 Nanocatalysts for Biomass Gasification

Biomass gasification is a thermochemical process that converts organic waste into a combustible gas mixture, primarily composed of carbon monoxide, hydrogen, and methane. This syngas can be used for heat and electricity generation or as a feedstock for the production of biofuels and chemicals. Nanocatalysts play a crucial role in enhancing the efficiency and selectivity of the gasification process, addressing key challenges such as tar formation and catalyst deactivation.

9.3.1.1 Metal Nanoparticle Catalysts

Biomass gasification is a promising technology for converting organic materials into useful gases, with catalysts playing a crucial role in enhancing the process efficiency. Nickel nanoparticles have emerged as highly effective catalysts for biomass gasification due to their excellent activity in C–C and C–H bond breaking. The catalytic performance of Ni nanoparticles is strongly influenced by their size, with an optimal range typically between 5–20 nm. Smaller Ni nanoparticles generally exhibit higher catalytic activity due to increased surface area and a higher proportion of edge and corner sites. However, particles below 5 nm may suffer from reduced stability and increased susceptibility to sintering.[38]

The formation of tar during biomass gasification is a significant challenge, as it can lead to equipment fouling and reduced gas quality. To address this issue, researchers have explored various catalyst formulations. Bimetallic nanoparticle catalysts have shown promise in enhancing tar cracking and reforming reactions. Ni–Co, Ni–Fe, and Ni–Cu nanoalloys often exhibit synergistic effects, combining the high activity of Ni with the improved stability or selectivity offered by the second metal. For example, Ni–Fe nanoalloys have demonstrated enhanced resistance to carbon deposition compared to monometallic Ni catalysts. Research has shown that Ni–Fe bimetallic catalysts, particularly when supported on SBA-15, significantly improve the conversion of biomass-derived tar into gaseous products and enhance hydrogen production while minimizing carbon deposition.[39]

Building on these findings, advanced catalyst designs have been developed to further improve performance. Additionally, hydrochar-supported Ni–Fe nanocatalysts, synthesized through a one-step hydrothermal strategy, exhibit strong metal–support interactions, resulting in superior performance in steam gasification of sewage sludge. These catalysts reduce tar yield and promote higher hydrogen selectivity compared to monometallic nickel catalysts.[40]

Designing core-shell bimetallic nanoparticles allows for fine-tuning of catalytic properties. A common approach is to use a core of a more abundant metal (*e.g.*, Fe or Cu) with a shell of Ni, optimizing catalyst cost while maintaining high activity. Techniques such as co-precipitation, wet impregnation, and sol–gel methods are employed to create bimetallic nanoparticles with precise composition and morphology. Advanced approaches like atomic layer deposition enable the creation of highly uniform bimetallic structures.[41]

While these catalysts show great promise, their long-term performance is a critical consideration. Understanding and mitigating catalyst deactivation is crucial for the economic viability of biomass gasification processes. Carbon deposition and metal sintering are the two primary modes of deactivation for metal nanoparticle catalysts in gasification. Carbon deposition can be mitigated by optimizing process conditions (*e.g.*, steam-to-carbon ratio) and through the use of promoters like CeO_2 that enhance carbon gasification. Sintering can be reduced by strong metal–support interactions and the use of advanced support materials.[42]

To maintain catalyst effectiveness over time, various regeneration strategies have been developed. Research supports the use of oxidation–reduction cycles to regenerate deactivated catalysts, which involves burning off deposited carbon in an oxidizing atmosphere followed by reduction to restore the active metal phase. Additionally, low-temperature plasma treatments have shown promise in efficiently removing carbon deposits while minimizing thermal stress on the catalyst.

Research on Mn–Cu catalysts supported by activated coke has demonstrated that water washing can effectively restore the catalyst's activity by removing sulfites and sulfates accumulated during the reaction process. This method has shown better results compared to thermal regeneration techniques.[43]

Cutting-edge research is focused on developing catalysts that can self-regenerate under gasification conditions. One approach involves the use of perovskite oxides that can exsolve metal nanoparticles under reducing conditions and reabsorb them under oxidizing conditions, providing a reversible mechanism for maintaining high catalytic activity. These advancements in metal nanoparticle catalysts are paving the way for more efficient and sustainable biomass gasification processes.

9.3.1.2 Nanostructured Oxide Catalysts

Nanostructured oxide catalysts significantly enhance the efficiency and selectivity of biomass gasification. These catalysts are favored due to their high surface area, controllable morphology, and tunable composition, which make them highly effective for gasification reactions.[44] Among the most promising types are cerium oxide (CeO_2) and perovskite oxides (ABO_3), which are key materials in biomass conversion.

9.3.1.2.1 Cerium Oxide in Gasification. CeO_2 nanoparticles serve a dual role in biomass gasification, acting both as a catalyst support and as an active participant in the reactions through their oxygen storage and release capabilities. The catalytic prowess of CeO_2 stems largely from its ability to form and repair oxygen vacancies, a crucial feature for gasification processes. Enhancing these capabilities, doping CeO_2 with lower-valence cations such as Gd^{3+} and Sm^{3+} increases the concentration of oxygen vacancies, thereby boosting the material's oxygen storage capacity and redox properties. Moreover, the introduction of dopants like Zr in CeO_2 ($Ce_{1-x}Zr_xO_2$) improves thermal stability and provides superior resistance to sintering at high temperatures.[45]

The activity of CeO_2 nanoparticles is also influenced by their shape, which determines the exposed crystal facets. For instance, nanorods and nanocubes, which expose the more reactive {110} and {100} facets, respectively, often show higher activity in gasification reactions than conventional spherical nanoparticles. This shape-controlled synthesis allows for tailored optimization of the catalytic performance.

9.3.1.2.2 Perovskite Oxides' Role. These catalysts are noted for their high thermal stability and tunable properties. Recent studies highlight the structural and chemical stability of these materials, making them suitable for various applications.[46] In reactions such as tar cracking and reforming, perovskite oxides exhibit excellent activity. Modifying these materials, such as partially substituting La with Sr or Ce, enhances their catalytic performance by altering the redox properties of the B-site cation.

Spray-flame-synthesized Sr- and Fe-substituted $LaCoO_3$ perovskite nanoparticles have shown enhanced oxygen evolution reaction activities due to these structural modifications, highlighting the adaptability and efficiency of these materials under various operational conditions.[47]

An intriguing phenomenon observed in certain perovskite materials is exsolution—a process where metal nanoparticles are expelled from the oxide lattice under reducing conditions, creating highly active and stable catalytic sites. This phenomenon has been substantiated by findings that demonstrate how cerium substitution in $LaCoO_3$ perovskite oxide not only enhances its bifunctional electrocatalytic activities for hydrogen and oxygen evolution reactions but also improves its structural stability and longevity under demanding environmental conditions.[48]

9.3.1.2.3 Synthesis and Applications. The performance of nanostructured oxide catalysts heavily depends on their size, morphology, and dispersion. Advanced synthesis methods such as sol–gel techniques and hydrothermal synthesis provide control over these parameters. Sol–gel techniques are particularly versatile, allowing for the synthesis of high surface area mixed oxide nanoparticles with controlled composition.[49]

Meanwhile, hydrothermal synthesis is ideal for producing uniformly shaped CeO_2 nanostructures and perovskite nanoparticles under relatively mild conditions. Additionally, spray pyrolysis, a scalable method, is well-suited for the continuous production of catalytic nanoparticles with precise stoichiometry and high dispersion.[50]

These advancements in nanocatalysts for biomass gasification represent a leap forward in enhancing the efficiency and economic viability of waste-to-energy conversion processes. Continuous research is expanding the boundaries of catalyst design, promising even more efficient and sustainable gasification technologies. This fusion of innovative materials, precise synthesis techniques, and a deep understanding of catalytic mechanisms is paving the way for more effective biomass utilization and cleaner energy production.

9.3.2 Nanomaterials for Anaerobic Digestion

Anaerobic digestion (AD) is a biological process that converts organic waste into biogas, primarily composed of methane and carbon dioxide. This process plays a crucial role in waste management and renewable energy production. The application of nanotechnology in AD has opened new avenues for enhancing process efficiency, stability, and biogas yield.

Nanoparticle additives have shown great potential in enhancing the efficiency and stability of AD processes. Various types of nanoparticles, such as iron oxide and zinc oxide, contribute uniquely to different aspects of AD performance.

9.3.2.1 Iron Oxide Nanoparticles

Iron oxide nanoparticles have shown remarkable potential in boosting methane production and overall AD performance. Both magnetite (Fe_3O_4) and hematite (Fe_2O_3) nanoparticles are effective but operate through different mechanisms. Magnetite nanoparticles are particularly effective due to their mixed valence state, facilitating electron transfer processes. They typically enhance methane production by 20–30%, as reported in some studies.

Hematite nanoparticles, while less effective in direct electron transfer, can still improve AD performance through iron reduction and subsequent formation of magnetite under anaerobic conditions.

One key benefit of iron oxide nanoparticles is their ability to facilitate direct interspecies electron transfer (DIET) between syntrophic bacteria and methanogenic archaea, bypassing the need for electron shuttles like hydrogen. This DIET mechanism can significantly accelerate the rate-limiting acetogenesis and methanogenesis steps in AD. Studies have shown that DIET facilitated by conductive iron oxide nanoparticles can lead to faster start-up times in AD reactors and improved resilience to organic loading shocks.[51]

9.3.2.2 *Zinc Oxide Nanoparticles*

Hydrogen sulfide (H_2S) is a corrosive and toxic component of biogas that needs to be removed for most applications. Zinc oxide nanoparticles offer an effective solution for H_2S mitigation. These nanoparticles can remove H_2S through physical adsorption and chemical reaction. ZnO reacts with H_2S to form zinc sulfide (ZnS), effectively capturing the sulfur. In the presence of oxygen, ZnO nanoparticles can also catalyze the partial oxidation of H_2S to elemental sulfur, allowing for potential sulfur recovery.[52]

The morphology of ZnO nanostructures significantly influences their H_2S removal efficiency. ZnO nanorods have shown high H_2S adsorption capacity due to their high aspect ratio and exposed reactive surfaces. Nanoflowers and hierarchical ZnO structures provide enhanced gas–solid contact, improving overall H_2S removal performance. Spherical ZnO nanoparticles, while effective, generally show lower capacity compared to more complex morphologies.[53]

9.3.2.3 *Economic Viability and Environmental Considerations*

The economic viability of ZnO nanoparticles for H_2S removal depends on their regeneration potential and overall lifecycle performance. Thermal regeneration at temperatures around 400–500 °C can convert ZnS back to ZnO, restoring H_2S removal capacity. Wet chemical methods using hydrogen peroxide have been developed for low-temperature regeneration, suitable for *in situ* applications. Lifecycle analyses show that despite the higher initial cost, the enhanced efficiency and regeneration potential of ZnO nanoparticles can make them economically competitive with traditional H_2S removal methods in the long term.[54]

Careful consideration must be given to dosage optimization and potential ecological consequences. Advanced modeling techniques, such as Monte Carlo simulations and machine learning approaches, are being developed to predict the behavior and fate of nanoparticles in complex anaerobic environments. These models can help optimize dosage strategies and predict long-term effects on AD performance.[55]

9.3.3 Nanotechnology in Fuel Cells for Waste-derived Energy

Fuel cells represent a promising technology for the efficient conversion of waste-derived energy carriers, such as biogas and hydrogen, into electricity. Nanotechnology plays a crucial role in enhancing the performance, durability, and cost-effectiveness of fuel cells, particularly in the context of waste-to-energy applications. This section explores the innovative use of nanomaterials in two key types of fuel cells: microbial fuel cells (MFCs) for direct electricity generation from wastewater, and conventional fuel cells for utilizing waste-derived gases. The working principle of a microbial fuel cell is illustrated in Figure 9.5.

9.3.3.1 Nanostructured Electrodes for Microbial Fuel Cells

Nanomaterials have emerged as exceptional components for MFC anodes due to their unique properties, including high conductivity, large surface area, and biocompatibility. These characteristics make nanomaterials particularly effective in enhancing electron transfer during wastewater treatment.[56,57]

Various types of nanomaterials have been explored for MFC anodes, including carbon nanotubes, graphene, metal nanoparticles, and nanostructured metal oxides. Each of these materials offers distinct advantages in MFC applications.

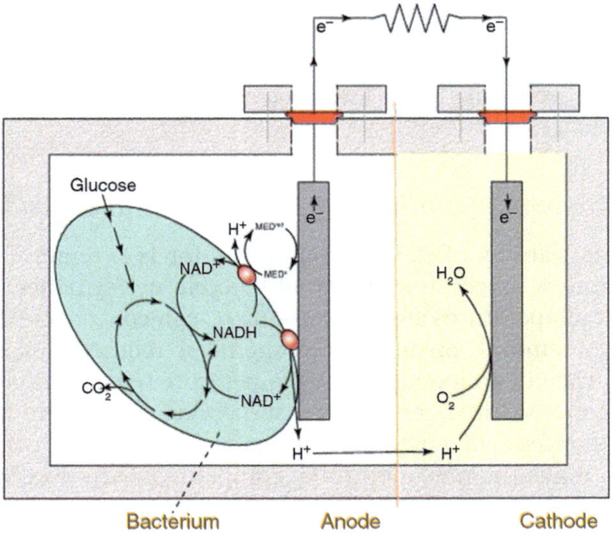

Figure 9.5 The working principle of a microbial fuel cell. Substrate is metabolized by bacteria, which transfer the gained electrons to the anode. This can occur either directly through the membrane or *via* mobile redox shuttles. MED, redox mediator; red oval, terminal electron shuttle in or on the bacterium. Reproduced from ref. 155 with permission from Elsevier, Copyright 2005.

Carbon nanotubes in particular have shown to be exceptional materials for microbial fuel cell anodes due to their high conductivity, large surface area, and biocompatibility. These characteristics make CNTs especially effective in enhancing electron transfer during wastewater treatment.[58]

The high electrical conductivity of many nanomaterials allows for efficient electron transfer from electroactive bacteria to the anode surface. This property is crucial in MFCs, as it directly impacts the power output of the cell. The enhanced conductivity reduces internal resistance, leading to improved overall MFC performance.[59]

The large surface area provided by nanomaterials creates an ideal environment for bacterial colonization and biofilm formation. This increased surface area allows for a higher density of microorganisms to attach to the anode, potentially increasing the rate of substrate oxidation and electron generation. The nanoscale structure of these materials also creates a three-dimensional network that can facilitate better mass transport of substrates and products within the biofilm.[60]

Biocompatibility is another key advantage of many nanomaterials in MFC applications. Carbon-based nanomaterials are generally well-tolerated by microorganisms, allowing for the development of robust and stable biofilms. Some metal nanoparticles and metal oxides can also exhibit good biocompatibility, though this can vary depending on the specific material and its concentration.

Nanomaterials can be incorporated into MFC anodes in various ways, including direct growth on electrode surfaces, mixing with other electrode materials, or as coatings. Each method of incorporation can affect the final properties of the anode and its interaction with the microbial community.

The use of nanomaterials in MFC anodes has shown promise in enhancing power output and coulombic efficiency in wastewater treatment applications. By improving electron transfer and bacterial attachment, nanomaterial-based anodes can potentially increase the efficiency of organic matter removal from wastewater while simultaneously generating electrical energy.[61]

However, challenges remain in the large-scale application of nanomaterials in MFCs, including cost considerations, potential long-term environmental impacts, and scalability of production. Ongoing research is focused on optimizing nanomaterial-based anodes, exploring hybrid materials, and developing scalable production methods to address these challenges.

9.3.3.2 Nanomaterial-enhanced Membrane Electrode Assemblies

Nanocomposite proton exchange membranes (PEMs) represent a significant advancement in the field of direct methanol fuel cells (DMFCs), particularly when utilizing methane-rich landfill gas as a fuel source. DMFCs are an attractive technology for converting this abundant and often wasted resource into electricity. However, traditional PEMs face several challenges, including methanol crossover, which significantly reduces the efficiency and lifespan of the fuel cells.

Nanocomposite membranes, which incorporate nanomaterials such as graphene oxide, silica nanoparticles, and carbon nanotubes, have shown great promise in addressing these issues. The incorporation of these nanomaterials enhances the mechanical strength, thermal stability, and proton conductivity of the membranes. This, in turn, leads to a significant reduction in methanol crossover, thereby improving the overall performance and durability of DMFCs.[62]

Graphene oxide-based nanocomposites, for instance, offer a unique combination of high surface area and functional groups that facilitate effective proton transport while acting as a barrier to methanol molecules. This dual function enhances the selectivity and efficiency of PEMs in DMFCs. Similarly, silica nanoparticles enhance the membrane's structural integrity and thermal stability, which are crucial for maintaining performance under the harsh conditions typical of DMFC operations.[63]

Recent studies have demonstrated that nanocomposite PEMs can significantly improve the power density and longevity of DMFCs. For example, membranes doped with sulfonated carbon nanotubes have shown a substantial increase in proton conductivity while effectively mitigating methanol crossover.[64] This enhancement in proton conductivity and reduction in methanol permeability results in higher power outputs and more efficient fuel utilization. Additionally, these nanocomposite membranes improve mechanical and thermal stability, contributing to the overall durability and extended performance of DMFCs.[65]

Moreover, the use of landfill gas in DMFCs aligns with sustainable energy goals by utilizing a readily available and renewable resource. Methane, a potent greenhouse gas, is effectively converted into useful electrical energy, reducing environmental impact and providing a cost-effective solution for waste management and energy production.

The integration of nanomaterials into PEMs for DMFCs for waste-derived energy represents a significant technological leap. By addressing critical challenges such as methanol crossover and enhancing membrane properties, nanocomposite PEMs pave the way for more efficient, durable, and sustainable fuel cell technologies.

9.3.3.3 Case Study: Harnessing Nanomaterials for Efficient Methane Conversion

The conversion of waste methane into valuable products stands as a critical challenge and opportunity. Methane, a potent greenhouse gas often produced as a byproduct in industrial and agricultural processes, has long been seen as both an environmental threat and a potential energy resource. The key lies in efficiently transforming this stable molecule into higher-value hydrocarbons, a process that has traditionally been hindered by methane's notorious chemical inertness under mild conditions. However, advancements in nanocatalysis are opening new avenues for this transformation, offering promising pathways for waste-to-energy applications.

One study[66] demonstrates the potential of carefully designed nanocatalysts in addressing this challenge. The researchers developed a system using platinum nanoparticles supported on gallium oxide (Ga_2O_3) for the photocatalytic nonoxidative coupling of methane. The work features precise control over the size of the Pt nanoparticles, ranging from 1.5 to 2.7 nanometers, achieved through a photodeposition method (see Figure 9.6).

This fine-tuning of nanoparticle size proved to be an important factor in optimizing catalyst performance. The study revealed a non-linear relationship between the Pt particle size and catalytic activity. As particle size increased, activity first improved, reached a peak, and then declined. The optimal size was found to be 1.9 nm Pt particles, which achieved the highest methane conversion rate of 134 μmol per gram of Pt per hour. This optimal size represents a balance in the nanoparticle's properties, maximizing benefits while minimizing drawbacks.

Investigating the mechanics of this enhanced activity, the researchers found that the corner atoms of the Pt nanoparticles serve as the primary active sites for methane activation. This insight explains why smaller particles, with their higher proportion of corner atoms, showed increased activity—but only to a point. The study revealed that particles below the optimal size suffered from stronger product adsorption, hampering overall efficiency.

The electronic properties of the nanoparticles also played a significant role. The presence of partially oxidized Pt species ($Pt^{\delta+}$), more prevalent in smaller particles, contributed to methane activation and facilitated charge transfer processes. However, the smallest particles (1.5 nm) with initially high $Pt^{\delta+}$ content underwent rapid reduction under reaction conditions, highlighting the balance between activity and stability in nanocatalyst design.

The optimized Pt/Ga_2O_3 nanocatalyst demonstrated methane conversion rates nearly 12 times higher than bare Ga_2O_3, showing the impact of the nanoengineered platinum particles. Moreover, the system achieved over

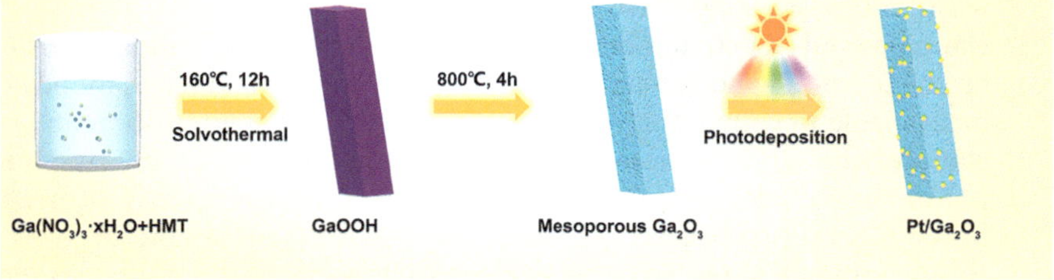

Figure 9.6 Schematic illustration for the synthesis of Pt/Ga_2O_3 nanocomposite. Reproduced from ref. 66 with permission from American Chemical Society, Copyright 2021.

93% selectivity towards ethane, indicating its potential for efficient and targeted methane conversion.

This case study illustrates the impact that nanomaterials can have in waste-to-energy processes. By controlling the size and properties of catalytic nanoparticles, researchers have improved efficiency in methane conversion. This approach offers potential for more effective utilization of waste methane and could contribute to mitigating greenhouse gas emissions while generating valuable energy products.

This study contributes to the development of more effective waste-to-energy technologies. It demonstrates how precise control at the nanoscale can lead to significant impacts at larger scales. While challenges remain in scaling up these technologies and ensuring long-term stability under real-world conditions, the potential benefits are considerable.

9.4 Advanced Waste Treatment Technologies

As we progress from examining nanotechnology's role in material recovery and waste-to-energy conversion, we now turn our attention to its applications in advanced waste treatment technologies. While the previous sections focused on extracting value from waste streams, this section explores how nanotechnology can enhance the degradation, removal, and stabilization of persistent pollutants and pathogens that resist conventional treatment methods in both liquid and solid waste streams.[67]

Advanced waste treatment technologies are crucial for addressing complex environmental challenges posed by recalcitrant contaminants, emerging pollutants, and the need for higher quality standards in waste management. Nanotechnology offers unique opportunities to enhance the efficiency, selectivity, and sustainability of these advanced treatment processes through the development of novel materials, catalysts, and engineered systems.[68]

In this section, we will examine four key areas where nanotechnology is making significant impacts on advanced waste treatment:

1. **Nanomaterial-enhanced advanced oxidation processes**, including photocatalytic nanoparticles, nanostructured electrodes for electrochemical oxidation, and nano-enabled Fenton-like processes for both wastewater and solid waste treatment.
2. **Nanomaterials for contaminant immobilization and stabilization**, focusing on the use of engineered nanoparticles to render pollutants less mobile or bioavailable in contaminated soil, groundwater, and solid waste matrices.
3. **Nanotechnology in membrane-based treatment systems**, exploring how nanomaterials can improve the performance and efficiency of membrane filtration processes for water and wastewater treatment.
4. **Nanocomposites for solid waste treatment**, examining the use of nanomaterial-enhanced composites for adsorption, degradation, and encapsulation of pollutants in solid waste streams, including municipal solid waste, industrial waste, and hazardous waste.

Each of these areas represents a significant advancement in our ability to treat complex waste streams, remove persistent contaminants, and manage environmental remediation challenges across various waste types. By enabling more efficient and effective treatment processes, these nanotechnology-based approaches not only improve environmental protection but also contribute to resource conservation and sustainable waste management practices.

As we explore these technologies, we will consider their current applications, ongoing research and development efforts, and the challenges that must be overcome for their widespread implementation. We will also examine the potential environmental implications of using nanomaterials in these processes, an important consideration for any technology deployed in waste treatment and environmental remediation.

9.4.1 Nanomaterial-enhanced Advanced Oxidation Processes

Advanced oxidation processes (AOPs) have gained significant attention due to their ability to degrade recalcitrant organic pollutants and inactivate pathogens in water and wastewater. Nanotechnology has opened new avenues for enhancing the efficiency and applicability of AOPs through the development of novel catalysts and reactor designs.[69,70]

9.4.1.1 Photocatalytic Nanoparticles

Titanium dioxide nanoparticles have long been recognized as effective photocatalysts for wastewater treatment due to their high stability, low toxicity, and strong oxidizing power under UV irradiation. Recent advances in nanotechnology have focused on enhancing the visible light activity of TiO_2 and improving its overall performance.[71]

Doping TiO_2 with non-metal elements such as nitrogen, sulfur, or carbon has been shown to narrow its bandgap, enabling visible light absorption and expanding its potential for solar-driven wastewater treatment.[72] N-doped TiO_2 nanoparticles have demonstrated enhanced degradation of pharmaceuticals and personal care products under simulated sunlight compared to undoped TiO_2. The improved visible light activity is attributed to the formation of mid-gap energy levels that facilitate electron-hole pair generation under lower energy irradiation.

Composite nanomaterials combining TiO_2 with other semiconductors or carbon-based materials have also shown promise in enhancing photocatalytic performance. TiO_2/graphene oxide composites exhibit improved charge separation and increased adsorption capacity for organic pollutants. These synergistic effects result in higher degradation rates and broader applicability across various contaminant classes.[73]

Reactor design plays a crucial role in optimizing the performance of photocatalytic nanoparticles.[74] Immobilization of nanoparticles on supports such as glass beads or polymer membranes can facilitate their recovery and reuse while maintaining a high surface area for catalytic activity.[75] Novel reactor configurations, such as optical fiber-based photoreactors, have been developed to improve light distribution and utilization efficiency in large-scale treatment systems.

9.4.1.2 *Nanostructured Electrodes for Electrochemical Oxidation*

Electrochemical advanced oxidation processes offer the advantage of *in situ* generation of oxidizing species without the need for chemical addition.[76] Nanotechnology has enabled the development of high-performance electrode materials with enhanced catalytic activity and stability.

Boron-doped diamond (BDD) nanoelectrodes have emerged as particularly effective for the degradation of persistent organic pollutants. The nanostructured surface of BDD electrodes provides a high density of active sites for hydroxyl radical generation, leading to rapid and complete mineralization of contaminants. Studies have demonstrated the efficacy of BDD nanoelectrodes in treating pharmaceutical residues, perfluorinated compounds, and other emerging contaminants resistant to conventional treatment methods.[77]

Nanostructured mixed metal oxide anodes, such as dimensionally stable anodes (DSAs) based on IrO_2 and RuO_2, have shown promise for chlorine-free disinfection of water and wastewater.[78] These electrodes generate reactive oxygen species directly from water molecules, avoiding the formation of harmful disinfection by-products associated with chlorine-based treatments. The high surface area and catalytic activity of nanostructured DSAs enable efficient inactivation of pathogens at lower energy inputs compared to conventional electrodes.

A significant challenge in this field is scaling up the fabrication of nanostructured electrodes for industrial-scale treatment systems. Researchers are exploring methods such as electrodeposition, sol–gel processes, and plasma spraying to develop cost-effective and reproducible manufacturing techniques for large-area nanostructured electrodes. The success of these efforts will be crucial for the widespread adoption of nanostructured electrodes in water treatment applications.

9.4.1.3 *Nano-enabled Fenton-like Processes*

The Fenton process, which utilizes iron-catalyzed decomposition of hydrogen peroxide to generate hydroxyl radicals, is a well-established AOP for wastewater treatment. Nanotechnology has opened new possibilities for enhancing the efficiency and applicability of Fenton-like processes through the development of novel catalysts.[79]

Iron-based nanoparticles, such as nanoscale zero-valent iron and iron oxide nanoparticles, have shown great potential as catalysts for activating hydrogen peroxide in wastewater treatment.[80] The high surface area and reactivity of these nanoparticles enable rapid generation of hydroxyl radicals at near-neutral pH, expanding the range of conditions under which Fenton-like processes can be applied. This broadens the applicability of Fenton processes to treat a wider range of wastewaters without the need for significant pH adjustment.

Strategies for minimizing nanoparticle loss and environmental release are crucial for the sustainable implementation of nano-enabled Fenton-like processes. Approaches such as immobilization on supports, encapsulation in porous matrices, and the development of self-regenerating catalysts are being explored to enhance the long-term stability and reusability of nanoparticle catalysts in wastewater treatment applications.

9.4.1.4 *Environmental Implications and Future Outlook*

While nanomaterial-enhanced AOPs show great promise for addressing challenging water treatment scenarios, several common challenges need to be addressed for widespread implementation. These include optimizing catalyst stability and longevity, minimizing energy inputs for process operation, and ensuring complete recovery of nanomaterials to prevent environmental release. A thorough understanding of the fate and potential ecological impacts of nanomaterials used in these processes is essential for their responsible development and application.

The environmental implications of using nanomaterials in waste treatment processes cannot be overlooked. Concerns exist about the potential release of nanoparticles into treated effluents and their subsequent impact on aquatic ecosystems.[81,82] Ongoing research is assessing the long-term effects of nanoparticle exposure on various organisms and developing strategies for mitigating any negative impacts. This includes improving nanoparticle recovery methods, designing more stable and less mobile nanostructures, and exploring biodegradable nanomaterials that can break down into harmless components after use.[83]

The field of nanomaterial-enhanced AOPs is likely to see continued innovation in catalyst design, with a focus on improving selectivity, stability, and recyclability. The development of multi-functional nanomaterials that can address multiple contaminants simultaneously is an exciting area of research. Additionally, the integration of nanotechnology with other advanced treatment technologies, such as membrane processes or biological treatment systems, may lead to more efficient and comprehensive water treatment solutions.

As research in this field progresses, nanomaterial-enhanced AOPs are expected to play an increasingly important role in addressing emerging contaminants and enabling water reuse in various industrial and municipal applications. The success of these technologies will depend not only on their technical performance but also on addressing economic considerations and ensuring their safe and sustainable implementation in real-world treatment scenarios.

9.4.2 Nanomaterials for Contaminant Immobilization and Stabilization

The immobilization and stabilization of contaminants in soil and solid waste represent critical strategies for managing environmental pollution and mitigating risks to human health and ecosystems.[84] Nanotechnology has emerged as a powerful tool in this field, offering innovative solutions for *in situ* remediation and waste stabilization. This section explores the application of nanomaterials in contaminant immobilization and stabilization, focusing on their unique properties and potential for enhancing remediation effectiveness in soil and solid waste matrices.

9.4.2.1 *Nanoparticles for* In Situ *Soil Remediation*

Metal oxide nanoparticles have shown considerable promise for heavy metal immobilization in contaminated soils. Nanoparticles such as nano-magnetite (Fe_3O_4), nano-maghemite (γ-Fe_2O_3), and nano-titanium dioxide (TiO_2) exhibit high surface areas and reactive sites that facilitate the adsorption and surface complexation of metal contaminants. The mechanisms of metal immobilization by these nanoparticles include electrostatic attraction, specific adsorption, and co-precipitation.[85]

Nano-magnetite has demonstrated high efficiency in immobilizing arsenic in contaminated soils.[86] The strong affinity of arsenic species for iron oxide surfaces, coupled with the high specific surface area of nano-magnetite, results in rapid and effective sequestration of both As(III) and As(V). Studies have shown that the application of nano-magnetite can significantly reduce the bioavailability and leachability of arsenic in contaminated soils, thereby mitigating environmental risks.[87]

Bimetallic nanoparticles offer enhanced reactivity and synergistic effects for the degradation and immobilization of organic pollutants in groundwater and soil.[88] Palladium–gold (Pd–Au) nanoparticles, for example, have shown remarkable efficiency in the dechlorination of persistent organic pollutants such as trichloroethylene (TCE).[89] The bimetallic structure allows for improved electron transfer and catalytic activity, enabling more complete degradation of target contaminants compared to monometallic nanoparticles.

Case studies have demonstrated the effectiveness of bimetallic nanoparticles in field-scale applications. In one instance, the injection of Pd–Au nanoparticles into PCB-contaminated soil resulted in a significant reduction of PCB concentrations within weeks, with the treated soil meeting regulatory cleanup standards.[90] The bimetallic nanoparticles not only facilitated the breakdown of PCBs but also helped immobilize the degradation products, reducing their mobility and bioavailability in the soil matrix. Studies have shown that Pd/Fe and Pd–Au bimetallic nanoparticles are highly effective in the catalytic dechlorination of PCBs, supporting these findings.[91]

Polymer-coated nanoparticles represent an innovative approach for the controlled release of remediation agents in soil environments.[92] By encapsulating reactive nanoparticles within biodegradable polymers, it is possible to modulate their release rate and extend their active lifespan in the subsurface. This approach addresses challenges associated with rapid aggregation and passivation of bare nanoparticles, which can limit their long-term effectiveness in soil remediation.

9.4.2.2 Nanocomposite Materials for Waste Stabilization

Clay–polymer nanocomposites have shown to be promising materials for heavy metal immobilization in contaminated soil and solid waste. These nanocomposites combine the high adsorption capacity of clay minerals with the structural integrity and processing advantages of polymers. The incorporation of nanoclays into polymer matrices creates tortuous pathways that significantly reduce the diffusion of contaminants, while the high surface area and reactive sites of nanoclays provide numerous binding locations for heavy metals.[93–95]

Studies have demonstrated the efficacy of clay–polymer nanocomposites in reducing the leachability of various heavy metals, including lead, cadmium, and zinc, from contaminated soils and industrial wastes. For instance, montmorillonite–polyacrylamide nanocomposites have shown superior performance in immobilizing lead and cadmium compared to conventional lime-based stabilization techniques.[96] The nanocomposite structure not only enhances metal binding capacity but also improves the mechanical properties of treated soils, potentially allowing for the beneficial reuse of remediated materials.

Carbon nanotube-reinforced geopolymers represent an innovative approach to hazardous waste encapsulation, applicable to both radioactive and chemical waste management.[97] Geopolymers, inorganic polymers formed through the alkali activation of aluminosilicate materials, have gained attention as alternatives to traditional cement-based waste forms due to their low permeability and high durability. The incorporation of carbon nanotubes into geopolymer matrices further enhances their mechanical properties and resistance to crack propagation.

The unique structure of carbon nanotubes allows them to bridge microcracks and dissipate stress within the geopolymer matrix, potentially improving the long-term integrity of waste forms exposed to environmental stressors. Additionally, the high aspect ratio and surface area of carbon nanotubes can contribute to increased adsorption of contaminants, providing an additional barrier against release. Research has shown that carbon nanotube-reinforced geopolymers exhibit superior compressive strength, flexural strength, and fracture toughness compared to conventional geopolymers, making them promising candidates for the encapsulation of both radioactive and hazardous chemical wastes.[98,99]

Nanoscale additives have also shown potential for improving the performance of cement-based waste forms. Nanosilica and nano-alumina, when incorporated into cementitious matrices, can enhance the microstructure and reduce porosity, leading to improved leach resistance and long-term durability. These nanomaterials participate in pozzolanic reactions, forming additional calcium silicate hydrate (C–S–H) gel that fills pore spaces and strengthens the cement matrix. Studies have demonstrated that the addition of just a small percentage of nanosilica (1–5% by weight of cement) can significantly reduce the leachability of heavy metals from cement-stabilized wastes and improve the overall mechanical properties of the waste form.[100,101]

9.4.2.3 Long-term Behavior and Performance Assessment

Understanding the long-term fate and transport of nanomaterials in soil and waste matrices is crucial for assessing the efficacy and environmental implications of nanomaterial-based remediation strategies. Factors such as aggregation, dissolution, and transformation processes can significantly influence the mobility and reactivity of nanoparticles over time. Research has shown that the behavior of nanoparticles in complex environmental matrices is governed by a combination of factors, including particle size, surface chemistry, soil properties, and geochemical conditions.

Studies on the long-term behavior of metal oxide nanoparticles in soils have revealed that processes such as aging and natural organic matter interactions can lead to changes in particle size distribution, surface charge, and reactivity. These transformations can affect the nanoparticles' ability to immobilize contaminants and their potential for migration in the subsurface. Advanced analytical techniques, such as synchrotron-based X-ray absorption spectroscopy and high-resolution transmission electron microscopy, are being employed to elucidate these complex interactions and transformations at the nanoscale.[102–104]

Durability testing of nanocomposite waste forms is essential for predicting their long-term performance in waste management applications. Accelerated aging techniques, involving exposure to elevated temperatures, radiation, and chemical stressors, are commonly employed to evaluate the stability of nanocomposite materials over extended periods. These studies aim to simulate the behavior of materials over decades or even centuries, a timescale relevant for the management of long-lived hazardous wastes.

Predictive modeling of nanocomposite behavior over extended timeframes is an active area of research. Multiscale modeling approaches, combining molecular dynamics simulations with continuum-level models, are being developed to predict the long-term stability and contaminant retention capacity of nanocomposite waste forms. These models incorporate factors such as chemical degradation kinetics, diffusion processes, and mechanical stress evolution to provide insights into the performance of nanocomposite materials under various environmental scenarios.[105,106]

9.4.2.4 Challenges and Future Directions

While nanomaterials offer significant potential for enhancing contaminant immobilization and waste stabilization, several challenges remain. Scaling up the production of nanomaterials for large-scale remediation applications presents technical and economic hurdles. Current research is focused on developing more cost-effective and environmentally friendly synthesis methods for nanomaterials, as well as exploring the use of waste-derived precursors to create value-added remediation products.

Addressing the potential environmental risks associated with the *in situ* application of nanoparticles is crucial for the sustainable implementation of nanomaterial-based remediation strategies. Ongoing studies are investigating the ecotoxicological impacts of various nanomaterials on soil microorganisms, plants, and higher organisms. This research aims to establish safe usage guidelines and develop mitigation strategies to minimize any adverse effects on ecosystem health.

The integration of nanomaterial-based approaches with other remediation technologies presents opportunities for synergistic treatments. For example, combining nanoparticle injection with electrokinetic remediation or phytoremediation techniques could enhance contaminant mobility and capture, leading to more efficient and comprehensive site cleanup.

9.4.3 Nanotechnology in Membrane-based Treatment Systems

Membrane-based treatment systems have become integral to water and wastewater purification processes, offering efficient separation of contaminants across various scales. While previous sections have touched on aspects of nanomaterials in water treatment, this section focuses specifically on the integration of nanotechnology with membrane systems, a field that has seen significant advancements in recent years.

9.4.3.1 Nanocomposite Membranes for Enhanced Separation

Nanocomposite membranes represent a significant leap forward in membrane technology, incorporating nanomaterials to enhance separation performance and membrane properties.

9.4.3.1.1 Carbon Nanotube-based Membranes. Carbon nanotubes have emerged as promising materials for creating high-flux membranes with precise selectivity. Vertically aligned CNT membranes exploit the smooth inner channels of nanotubes to facilitate rapid water transport while maintaining excellent rejection of contaminants (see Figure 9.7). Recent studies have demonstrated CNT membranes with water permeability orders of magnitude higher than conventional polymeric membranes, while maintaining high salt rejection rates.[107]

The incorporation of CNTs into polymer matrices has also shown promise in creating mixed-matrix membranes with enhanced antifouling properties

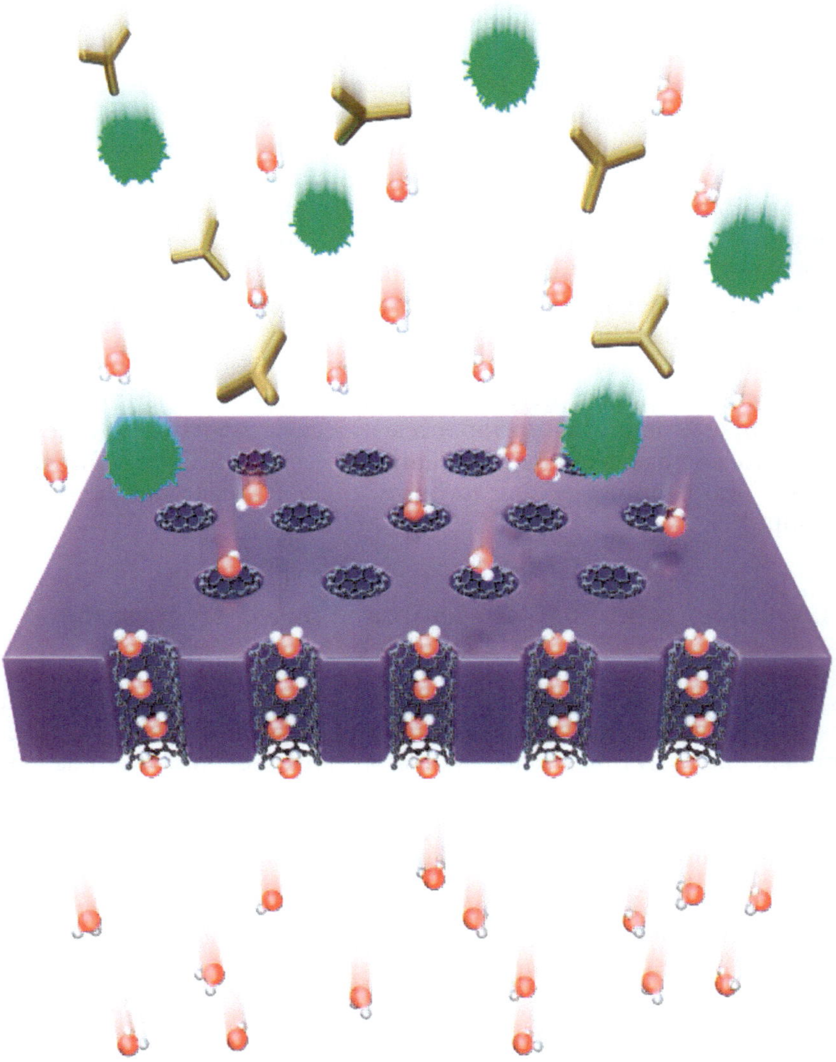

Figure 9.7 Schematic of a vertically aligned carbon nanotube (VACNT) membrane used for water treatment. The image shows water molecules, unwanted molecules, and an impermeable material filling the interstitial spaces between carbon nanotubes (CNTs) and VACNTs. Reproduced from ref. 107, https://doi.org/10.3390/membranes10100273, under the terms of the CC BY 4.0 license, https://creativecommons.org/licenses/by/4.0/.

and mechanical strength. For instance, polyamide thin-film composite membranes embedded with functionalized multi-walled CNTs have exhibited improved chlorine resistance and biofouling resistance compared to conventional reverse osmosis membranes.[108]

9.4.3.1.2 Graphene-based Membranes. Graphene oxide membranes have attracted significant research attention due to their unique layered structure, which allows for selective molecular sieving. The interlayer spacing in GO membranes can be tuned to achieve precise separation of ions and molecules. Recent advances have focused on creating stable GO membranes for practical applications, addressing challenges such as swelling in aqueous environments.[109,110]

Reduced graphene oxide has also been explored for creating electrically conductive membranes that can be electrochemically regenerated, potentially offering a solution to membrane fouling without the need for chemical cleaning.[111]

9.4.3.2 Surface-engineered Membranes

Nanotechnology enables precise control over membrane surface properties, leading to improved performance and new functionalities.

9.4.3.2.1 Superhydrophilic and Superhydrophobic Surfaces. By engineering membrane surfaces at the nanoscale, researchers have created membranes with extreme wetting properties. Superhydrophilic surfaces, often achieved through the incorporation of metal oxide nanoparticles or nanoscale surface texturing, enhance water permeability and reduce organic fouling. Conversely, superhydrophobic surfaces, inspired by natural structures like lotus leaves, have been developed for applications such as membrane distillation, where they prevent wetting of pores by feed water while allowing water vapor to pass through.[112]

9.4.3.2.2 Zwitterionic Polymer Brushes. The grafting of zwitterionic polymer brushes onto membrane surfaces represents a novel approach to creating ultra-low fouling membranes. These nanoscale polymer layers create a strong hydration layer that resists protein adsorption and bacterial adhesion. Recent studies have shown that membranes modified with zwitterionic brushes maintain high flux and low fouling propensity even in challenging environments like oily wastewater treatment.[113]

9.4.3.3 Stimuli-responsive Membranes

Nanotechnology has enabled the development of smart membranes that can respond to external stimuli, offering dynamic control over membrane properties.

9.4.3.3.1 Thermo-responsive Membranes. Membranes incorporating thermo-responsive polymers, such as poly(*N*-isopropylacrylamide) (PNIPAAm), can switch between hydrophilic and hydrophobic states in response to temperature changes. This property has been exploited to create self-cleaning membranes that can release foulants upon a temperature trigger, potentially reducing the need for chemical cleaning.[114]

9.4.3.3.2 Photo-responsive Membranes. The incorporation of photocatalytic nanomaterials, such as TiO_2 nanoparticles, into membrane matrices has led to the development of photo-responsive membranes. Under UV irradiation, these membranes can degrade organic contaminants and inactivate microorganisms, offering combined filtration and advanced oxidation capabilities.[115]

9.4.3.4 Nanostructured Ceramic Membranes

Advances in nanotechnology have revitalized interest in ceramic membranes, enabling the creation of membranes with precisely controlled pore sizes and distributions.[116]

9.4.3.4.1 Ordered Mesoporous Silica Membranes. Templated synthesis techniques have allowed for the creation of silica membranes with uniform, ordered mesopores. These membranes offer excellent thermal and chemical stability, making them suitable for harsh industrial separation processes. Recent research has focused on functionalizing the pore surfaces of these membranes to achieve selective separation of specific molecules or ions.[117]

9.4.3.4.2 Zeolite Membranes. Nanoscale zeolite crystals have been used to create molecular sieving membranes with exceptional selectivity. Advances in zeolite membrane synthesis, such as secondary growth methods and the use of 2D zeolite nanosheets, have addressed historical challenges related to membrane continuity and thickness control.[118]

9.4.3.5 Challenges and Future Directions

Challenges remain in harnessing nanotechnology for membrane-based treatment systems. Translating nanoscale phenomena observed in laboratory studies to practical, large-scale membrane modules is a significant hurdle. Ensuring the long-term stability of nanostructured and surface-modified membranes under real-world operating conditions is crucial for their practical implementation. Additionally, the potential release of nanomaterials from membranes during operation and disposal needs to be carefully assessed and mitigated. The economic viability of nano-enhanced membranes compared to conventional alternatives must also be demonstrated for widespread adoption.

Current research directions focus on developing multifunctional membranes that combine separation, catalytic degradation, and sensing capabilities. Bio-inspired and biomimetic approaches to membrane design, leveraging natural nanostructures for enhanced performance, are also being explored. The integration of artificial intelligence and machine learning for optimizing the design and operation of nano-enhanced membrane systems presents another promising avenue. Investigating sustainable and scalable production methods for nano-enhanced membranes, including the use of renewable and waste-derived materials, is equally important.

9.4.4 Nanomaterials for Solid Waste Treatment

Nanocomposites have emerged as a promising solution for addressing the complex challenges associated with solid waste treatment, complementing the nanotechnology applications discussed in previous sections such as advanced oxidation processes, contaminant immobilization, and membrane-based systems. These advanced materials, which incorporate nanoscale components into a larger matrix, offer unique properties that can significantly enhance the efficiency and effectiveness of solid waste management processes. While previous sections focused primarily on liquid waste and contaminated site remediation, this section expands our discussion to solid waste treatment, highlighting how nanotechnology principles are adapted for this distinct waste stream.

The use of nanomaterials in solid waste treatment offers several advantages. They significantly enhance the removal or degradation rates of contaminants compared to conventional materials, providing increased treatment efficiency.[119] Some nanocomposites can be designed to target specific pollutants, allowing for more precise treatment processes and improved selectivity. By improving treatment efficacy, nanocomposites can help minimize the environmental footprint of waste management facilities, resulting in a reduced environmental impact. Additionally, nanocomposite materials often exhibit improved durability and longevity, leading to more sustainable waste treatment solutions.

Certain nanocomposites facilitate the recovery of valuable materials from waste streams, promoting a circular economy approach and resource recovery. Here, waste materials present an opportunity as potential raw materials for the production of value-added products such as nanocomposites at low cost. Utilizing these waste materials in the composite industry offers a promising strategy for maintaining resource sustainability while providing economic and environmental benefits.[120]

Various types of nanocomposites have been developed for solid waste treatment, building upon the nanomaterial applications discussed in earlier sections. These nanocomposites adapt and enhance the properties of individual nanomaterials for the unique challenges of solid waste management.[121]

Nanocomposites for solid waste treatment come in various forms, each designed to address specific aspects of waste management. Nano-enhanced adsorbents incorporate materials like carbon nanotubes, graphene, or metal oxide nanoparticles into traditional adsorbents, improving their surface area and adsorption capacity.[122,123]

Photocatalytic nanocomposites combine nanoparticles such as TiO_2 and ZnO with supporting materials to enhance the degradation of organic pollutants.[124] Antimicrobial nanocomposites integrate silver nanoparticles or other antimicrobial nanomaterials into polymeric matrices to reduce pathogen levels in organic waste.[125] Nano-enabled stabilization matrices use nanoparticles to enhance the performance of stabilization and solidification techniques for hazardous waste containment.

These nanocomposites find applications across various aspects of solid waste treatment. In municipal solid waste management, they enhance biodegradation of organic waste in composting processes, improve the efficiency of landfill liners to prevent leachate contamination, and augment plastic waste degradation through synergistic action of microbes and nanoparticles.[126]

For industrial waste treatment, nanocomposites are employed to remove heavy metals and organic pollutants from industrial sludges, stabilize and immobilize hazardous components in solid waste, and enhance the treatment of electronic waste to recover valuable materials.[127]

In hazardous waste remediation, they enable more effective encapsulation techniques for radioactive waste, create advanced filtration systems for contaminated soils, and improve the degradation of persistent organic pollutants in contaminated sites.[128]

Despite their potential, several challenges need to be addressed for the widespread adoption of nanocomposites in solid waste treatment. Developing cost-effective methods for large-scale production remains significant. The potential long-term effects of nanoparticles on the environment and human health require further investigation. Establishing appropriate regulations and standards for the use of nanomaterials in waste treatment is crucial. Continued research is needed to optimize the performance of nanocomposites under varied waste compositions and environmental conditions.

9.4.4.1 Comparison with Conventional Solid Waste Treatment Techniques

Nanocomposite-based approaches offer several distinct advantages over conventional solid waste treatment techniques.[129] Conventional methods, such as landfilling, incineration, and traditional chemical treatments, often suffer from inefficiencies, high costs, and significant environmental impacts.

- **Efficiency**: Conventional methods like landfilling and incineration have limited efficiency in terms of resource recovery and pollutant degradation. Nanocomposites, however, can significantly enhance the removal or degradation rates of contaminants, leading to more efficient treatment processes.
- **Selectivity**: Traditional treatment methods often lack the ability to target specific pollutants. Nanocomposites can be engineered to selectively interact with particular contaminants, allowing for more precise and effective treatment.

- **Environmental impact:** Conventional techniques can contribute to significant environmental pollution, such as greenhouse gas emissions from landfills and incinerators. Nanocomposites can reduce the environmental footprint of waste management facilities by improving treatment efficacy and enabling the recovery of valuable resources.
- **Resource recovery:** Traditional methods usually do not prioritize resource recovery, leading to the loss of potentially valuable materials. Nanocomposites facilitate the recovery of valuable materials from waste streams, promoting a circular economy approach and reducing the need for virgin materials.
- **Durability and longevity:** Conventional materials used in waste treatment can degrade over time, reducing their effectiveness. Nanocomposite materials often exhibit improved durability and longevity, providing more sustainable and long-lasting waste treatment solutions.

9.4.4.2 Case Study: Prussian Blue Nanoparticles for Recycling Rare Metals from Nuclear Waste

The recovery and recycling of precious metals from nuclear and electronic waste has long been a critical challenge at the intersection of environmental sustainability, resource management, and technological advancement. As global demand for rare metals continues to rise, driven by their essential roles in electronics, catalysts, and other high-tech applications, the need for efficient recycling methods has become increasingly urgent. However, traditional recovery techniques have often been limited by issues of cost, efficiency, and environmental impact.

Among the promising solutions, Prussian blue nanoparticles (PBNPs) have emerged as particularly effective candidates. Prussian blue, a pigment discovered in the early 18th century, has found applications ranging from artists' paints to medical treatments. Its unique crystal structure, resembling a three-dimensional jungle gym, allows it to trap certain ions within its nanoscale pores.

Researchers also demonstrated the effectiveness of PBNPs in removing radioactive cesium from contaminated soils following nuclear accidents.[130] This success sparked interest in applying similar principles to the recovery of valuable metals from waste streams. However, the precise mechanisms by which PBNPs interact with different metal ions, particularly those with multiple valence states, remained poorly understood. This knowledge gap has hindered the development of optimized PBNP-based recovery systems for a broader range of metals.

A team of researchers from Nagoya University and the Tokyo Institute of Technology has taken a significant step forward in unraveling these mechanisms. Their study[127] provides detailed insights into how PBNPs capture platinum group metals (PGMs) and molybdenum from nitric acid solutions—conditions that mimic those found in nuclear waste processing.

The researchers focused on four metals of particular interest: ruthenium (Ru), rhodium (Rh), palladium (Pd), and molybdenum (Mo). These elements are not only valuable in their own right but also pose challenges in nuclear waste management. In spent nuclear fuel reprocessing, PGMs can cause problems by settling on the walls of glass melters used to vitrify high-level radioactive waste. Molybdenum can form compounds that reduce the quality of the vitrified waste form. By developing efficient methods to remove these metals before vitrification, the researchers aim to improve both the economics and safety of nuclear waste disposal.

The team employed a multi-faceted approach to investigate the uptake of these metals by PBNPs. They conducted sorption experiments, exposing the nanoparticles to metal-containing nitric acid solutions for 24 hours. The researchers then used a combination of analytical techniques to probe the resulting materials. Inductively coupled plasma atomic emission spectroscopy (ICP-AES) allowed them to quantify the amount of metal absorbed by the PBNPs and the amount of iron released from the particles. X-ray diffraction (XRD) provided information about changes in the crystal structure of the PBNPs, while ultraviolet-visible-near infrared (UV-Vis-NIR) spectroscopy offered insights into electronic changes within the particles.

To complement their experimental work, the team also performed extensive computational modeling. They used density functional theory calculations to simulate the energetics of different potential uptake mechanisms, including surface adsorption, diffusion into the PBNP structure, and substitution of iron atoms within the Prussian blue framework.

The results revealed a complex picture of metal uptake, with each element behaving somewhat differently. Ruthenium (Ru^{4+}) and palladium (Pd^{2+}) ions were found to substitute directly for iron atoms in the Prussian blue structure, with Ru^{4+} replacing Fe^{3+} and Pd^{2+} replacing Fe^{2+}. Rhodium (Rh^{3+}) showed a more nuanced behavior, primarily substituting for Fe^{3+} but also replacing some Fe^{2+} to a lesser extent. Molybdenum (Mo^{6+}) demonstrated the most complex behavior, substituting for both Fe^{2+} and Fe^{3+} ions.

These substitution processes were found to maintain the overall crystal structure of the PBNPs, an important factor for their stability and potential reusability. The researchers were able to estimate the substitution efficiency for each metal, finding that 39.0% of Ru, 47.8% of Rh, 87% of Pd, and 17.1% of Mo^{6+} ions replaced iron atoms in the PBNP structure.

The computational studies provided further insights into the energetics of these processes. The researchers found that the surface adsorption energy of the metal ions onto PBNPs increased with the ion's valence, explaining why Pd^{2+} was more readily taken up than Mo^{6+}. The calculations also revealed that the metal ions tended to become trapped at certain sites within the Prussian blue structure, rather than freely diffusing through the material.

These findings have significant implications for the development of PBNP-based metal recovery systems. The high uptake efficiency for palladium is particularly noteworthy, suggesting that PBNPs could be an excellent material for recovering this valuable catalyst metal from waste streams. The researchers estimate that 1 gram of PBNPs could potentially recover 0.30 grams of palladium, 0.16 grams of rhodium, 0.13 grams of ruthenium, and 0.11 grams of molybdenum.

The study also highlights the importance of tailoring PBNP properties to target specific metals. By understanding the precise mechanisms of metal uptake, researchers may be able to modify the Prussian blue structure or synthesis conditions to enhance selectivity for particular elements.

Beyond their immediate application to nuclear waste processing, these findings open up new possibilities for recovering precious metals from electronic waste. As the researchers point out, the concentration of gold in discarded mobile phones can be 10–80 times higher than in natural ore deposits. Developing efficient, nanoparticle-based recovery methods for these "urban mines" could significantly reduce the environmental impact of electronics manufacturing while providing a new source of valuable materials.

The study also demonstrates the power of combining experimental and computational approaches in materials science. The researchers' use of advanced spectroscopic techniques alongside DFT calculations allowed them to build a comprehensive picture of metal uptake processes that would have been difficult to achieve through either method alone.

While the results are promising, several challenges remain before PBNP-based metal recovery can be implemented on an industrial scale. The researchers note that their experiments were conducted in simplified systems, and real-world waste streams are likely to contain a complex mixture of elements that could interfere with metal uptake. Additionally, the long-term stability of metal-loaded PBNPs and methods for extracting the captured metals will need to be thoroughly investigated.

9.5 CO_2 Capture and Transformation into Functional Materials

As global challenges such as climate change and resource sustainability intensify, there is a growing need for strategies that address both carbon emissions and resource management. One promising approach is the capture and transformation of carbon dioxide into valuable materials. Traditionally seen as a major greenhouse gas, CO_2 can also be considered a resource that can be repurposed into high-performance materials through advanced chemical processes. Transforming CO_2 into carbon nanotubes, graphene, and other nanomaterials represents an innovative form of recycling that reduces atmospheric CO_2 levels while generating valuable products.

CO_2 capture and transformation technologies, particularly when combined with advancements in nanotechnology, offer the potential to revolutionize sustainable material development, support climate mitigation efforts, and promote the principles of a circular economy.[131]

9.5.1 Overview and Strategic Relevance of CO_2 Capture and Utilization

CO_2 capture and utilization technologies provide innovative solutions to reduce greenhouse gas emissions while promoting sustainable resource management. By repurposing CO_2 from industrial emissions or directly from the atmosphere, we can convert it into high-performance nanomaterials such as carbon nanotubes and graphene. This process transforms CO_2 from a pollutant into a valuable resource, reducing reliance on carbon-intensive raw materials and supporting the development of sustainable materials.

The conversion of CO_2 into stable, solid forms like CNTs and graphene involves advanced chemical processes that break down CO_2 molecules and reassemble them into complex structures. These nanomaterials are valued for their unique properties, including exceptional strength, electrical conductivity, and thermal stability, making them suitable for a variety of applications, from electronics to structural composites.

9.5.1.1 Strategic Importance of CO_2 Utilization

The strategic significance of converting CO_2 into functional materials is twofold. Firstly, it directly contributes to climate change mitigation by reducing the amount of CO_2 in the atmosphere or preventing it from being released in the first place. Some approaches capture CO_2 at the source, such as from industrial plant chimneys, before it can even enter the atmosphere. Every ton of CO_2 that is captured and converted into a stable material represents a reduction in greenhouse gas emissions. Utilizing CO_2 as a feedstock not only mitigates its impact as a greenhouse gas but also effectively prevents it from contributing to atmospheric CO_2 levels.

Secondly, this approach aligns with the principles of a circular economy. In a circular economy, the goal is to minimize waste and optimize resource use by recycling and repurposing materials. By turning CO_2, a waste product, into valuable nanomaterials, industries can reduce the need for virgin raw materials, which often require energy-intensive extraction and processing. This practice not only lowers the environmental impact associated with material production but also reduces costs, enhances sustainability, and supports more resilient supply chains.

Moreover, transforming CO_2 into advanced materials can reduce reliance on fossil fuels and conventional raw materials. The production of materials like CNTs and graphene typically relies on energy-intensive processes, such as chemical vapor deposition, which have a high carbon footprint. Using CO_2 as a raw material offers a more sustainable and potentially less expensive pathway, promoting greener industrial practices and driving innovation in materials science.

9.5.1.2 The Role of Nanotechnology in CO_2 Utilization

Nanotechnology is pivotal in enhancing both the capture and conversion of CO_2 into functional materials. Nanomaterials and nanostructures provide unique properties that can significantly improve the efficiency and effectiveness of CO_2 capture and its subsequent transformation into valuable products.[132]

For CO_2 capture, nanomaterials such as metal–organic frameworks and carbon-based adsorbents offer high surface areas and tunable pore structures that can selectively adsorb CO_2 from various gas mixtures. These nanomaterials can be engineered to exhibit high affinities for CO_2, making them highly effective for capturing emissions directly from industrial sources or the atmosphere. Nanotechnology thus enables more efficient and cost-effective CO_2 capture solutions, which are critical for scaling these technologies.[133]

In terms of conversion, the principles outlined in Section 8.1 on catalysis are highly relevant. Nanotechnology facilitates the transformation of captured CO_2 into high-performance carbon nanomaterials by enhancing catalytic processes. Electrochemical reduction, a process that uses electricity to convert CO_2 into solid carbon forms, is particularly effective when nanocatalysts are employed. Nanostructured materials offer high surface areas, tunable pore sizes, and optimized active sites, which significantly boost the catalytic activity. These properties reduce the energy requirements of chemical reactions, increasing both the yield and selectivity of the desired products.[134] Consequently, processes that convert CO_2 into carbon nanotubes, graphene, and other nanostructures become more viable and efficient, paving the way for broader adoption.

9.5.1.3 Opportunities and Challenges in CO_2 Transformation

While the potential benefits of transforming CO_2 into functional materials are significant, several challenges must be addressed to make these technologies commercially viable and widely adopted.

One of the main technical challenges is the efficiency of the conversion processes. Electrochemical and other chemical methods for converting CO_2 into nanomaterials often require substantial energy inputs, which can negate some environmental benefits unless powered by renewable energy sources. Improving the efficiency of these processes and developing catalysts that can operate under less energy-intensive conditions are key areas of ongoing research.[135]

Economic considerations are also critical. The initial capital investment needed for CO_2 capture and conversion facilities, including specialized equipment and reactors, can be substantial.[136] Moreover, the cost-effectiveness of producing nanomaterials from CO_2 must be competitive with traditional methods to encourage widespread adoption. This challenge is especially relevant in industries with thin profit margins and established supply chains that may resist change.

Environmental and regulatory factors present additional challenges. Large-scale CO_2 conversion facilities must effectively manage by-products and ensure their operations do not create new environmental issues. Furthermore, the regulatory landscape for handling and using nanomaterials is still evolving, with concerns about potential health and environmental impacts requiring careful management and oversight.

9.5.2 Technological Approaches and Challenges in CO_2 Transformation

Transforming CO_2 into valuable nanomaterials involves several advanced technological approaches that capitalize on chemical processes to convert this greenhouse gas into high-performance materials. Developing these technologies is essential for improving the efficiency and scalability of CO_2 utilization strategies, thereby making them more economically viable and environmentally sustainable. This section explores the primary technological methods used in CO_2 transformation, their potential applications, and the challenges that need to be addressed for broader adoption.

9.5.2.1 Electrochemical Reduction

One of the most promising methods for converting CO_2 into carbon nanomaterials is electrochemical reduction.[137] This process involves the use of an electrochemical cell where CO_2 is reduced to solid carbon forms, such as carbon nanotubes or graphene, through the application of an electrical current. Typically, the reaction takes place in an electrolyte solution, which can be aqueous or non-aqueous, with catalysts—often nanoparticles—used to enhance reaction rates and selectivity.

Electrochemical reduction offers several advantages. It is highly scalable, capable of being adapted from small laboratory setups to larger industrial operations. The method also allows precise control over the product characteristics, such as the morphology and properties of the resulting nanomaterials, by adjusting the electrochemical conditions, including voltage, current density, and electrolyte composition. Moreover, electrochemical reduction can potentially be integrated with renewable energy sources like solar or wind power, enhancing its sustainability.

However, there are significant challenges associated with electrochemical reduction. The process requires substantial energy input, which can limit its environmental benefits if the energy is sourced from non-renewable resources. Efforts are underway to develop more efficient electrochemical cells and catalysts that can reduce the energy requirements. Another challenge lies in the durability and cost of the catalysts. Over time, these catalysts can degrade, reducing the efficiency of the conversion process and increasing operational costs. Research is focused on finding cost-effective and durable catalysts to make the process more viable for large-scale applications.

9.5.2.2 Other Transformation Processes

In addition to electrochemical methods, thermochemical and photochemical processes are also being explored for CO_2 transformation. Thermochemical processes typically involve the reaction of CO_2 with a reductant, such as hydrogen or methane, at high temperatures to produce carbon monoxide, which can be further processed into carbon nanomaterials. These reactions often take place in advanced reactors, such as those using plasma or solar-thermal energy, which are designed to improve the efficiency and selectivity of the processes.[138]

Photochemical processes,[139] on the other hand, use light in combination with a photocatalyst to drive the reduction of CO_2 into useful carbon forms. This method mimics natural photosynthesis, offering a potentially sustainable pathway for CO_2 utilization. Nanostructured photocatalysts, such as titanium dioxide doped with various metals, are being developed to enhance the efficiency of CO_2 photoreduction.

Both thermochemical and photochemical methods present unique advantages and challenges. They utilize abundant energy sources—heat and light—which can be derived from renewable resources, thereby reducing dependence on fossil fuels. These processes also have the potential to produce a variety of high-value products, including different types of carbon nanomaterials, which can be tailored to specific applications. However, thermochemical processes often require high temperatures, which demand significant energy input and can compromise the overall sustainability unless managed carefully. Photochemical processes are often limited by the efficiency and stability of the photocatalysts, which can degrade under prolonged exposure to light and CO_2. Overcoming these limitations is a key focus of current research, with efforts concentrated on developing more robust and efficient catalysts and optimizing reaction conditions.

Catalytic hydrogenation offers another promising approach for CO_2 transformation.[140] This process involves reacting CO_2 with hydrogen in the presence of a catalyst to produce hydrocarbons, such as methane, which can then be used as precursors for more complex carbon nanomaterials. Catalytic hydrogenation integrates well with the hydrogen economy, where hydrogen produced *via* water electrolysis or other methods serves as both a fuel and a reducing agent. This synergy can create opportunities for more sustainable industrial processes. Additionally, catalytic hydrogenation can yield a variety of carbon nanostructures, offering flexibility in terms of the types of products that can be generated. However, the supply and cost of hydrogen pose significant challenges. Producing hydrogen is often energy-intensive and costly, particularly if not derived from renewable sources. Furthermore, the development of efficient and cost-effective catalysts remains a critical hurdle to making catalytic hydrogenation a feasible option for large-scale CO_2 transformation.

Despite the promising potential of these technologies, several challenges must be addressed to realize their full potential:

- **Technical challenges** include improving the efficiency of conversion processes, developing catalysts that can operate under less energy-intensive conditions, and enhancing the durability and cost-effectiveness of these catalysts.
- **Economic challenges** revolve around the initial capital investment required for setting up CO_2 capture and conversion facilities and ensuring that the cost of producing nanomaterials from CO_2 is competitive with traditional methods.
- **Environmental and regulatory challenges** involve managing by-products effectively and ensuring that operations do not introduce new environmental risks. Furthermore, the regulatory framework for handling and using nanomaterials is still evolving, requiring careful oversight to mitigate potential health and environmental impacts.

The future of CO_2 transformation technologies, particularly those involving the production of nanomaterials, looks promising as research continues to advance. Innovations in catalyst development, reactor design, and the integration of renewable energy sources are expected to drive down costs and improve overall sustainability. As global efforts to mitigate climate change intensify, the demand for technologies that provide both environmental and economic benefits will likely grow. CO_2 utilization strategies, especially those that result in the production of high-value nanomaterials, are well-positioned to meet this demand and contribute to a more sustainable future.

By transforming CO_2 from a waste product into a valuable resource, these technologies offer a unique opportunity to address some of the most pressing environmental challenges of our time, supporting both climate action and sustainable development goals.

9.5.2.3 Case Study: Polymer Composites with Carbon Nanotubes Made from CO_2

Plastics are ubiquitous in modern life, from food packaging to components in electronic devices. However, their widespread use comes with significant environmental costs, including substantial greenhouse gas emissions during production and pervasive plastic waste polluting oceans and landscapes. As global plastic consumption is projected to double by 2040, the urgency to address these environmental challenges has never been greater.[141]

Carbon nanotubes have emerged as a potential solution to enhance the properties of plastics. Known for their extraordinary strength, electrical conductivity, and thermal properties, CNTs can reinforce plastics, potentially creating stronger, lighter materials that require less raw material to achieve comparable performance. However, the broad adoption of CNTs in plastic composites has been limited by high production costs and the energy-intensive nature of conventional manufacturing processes. Ironically, producing CNTs through traditional methods often results in a significant carbon footprint, seemingly counteracting the environmental benefits of creating more sustainable plastics.

Recent advancements in carbon capture and utilization technologies, discussed earlier in this chapter, have opened new possibilities for overcoming these obstacles. Innovative methods have been developed to convert carbon dioxide into valuable carbon nanomaterials, including CNTs. This approach not only offers a pathway for reducing atmospheric CO_2 levels but also presents a more economical and environmentally friendly method for producing CNTs.

Combining these technologies—carbon capture, CNT synthesis, and polymer science—has set the stage for breakthroughs in developing stronger, more sustainable plastics with reduced environmental impact. One study demonstrated the use of CNTs derived from CO_2 to enhance the properties of various plastics, including both thermoset epoxies and thermoplastics.[142] This research explores the potential of these CO_2-derived CNTs to strengthen plastics while simultaneously addressing environmental concerns, representing a significant step forward in developing next-generation, eco-friendly materials.

The study utilized a process known as molten carbonate electrolysis to convert CO_2 into CNTs. In this method, CO_2 is dissolved in molten carbonate salt and subjected to electrolysis, resulting in the formation of carbon nanomaterials at the cathode. The product, termed a carbanogel, consists of intertwined CNTs retaining some of the electrolytes within their structure. This carbanogel can be further processed to isolate the CNTs for use as additives in plastic composites.

The effects of adding these CO_2-derived CNTs to various epoxy resins and thermoplastics were then investigated, focusing on tensile strength—a measure of how much force a material can withstand before breaking when stretched. The findings revealed significant improvements in mechanical properties across different materials. For instance, adding 1.5% by weight of CNTs to a deep-pour epoxy resin resulted in a 30% increase in tensile strength compared to the pure epoxy. Similar enhancements were observed in other epoxies and thermoplastics, demonstrating the versatility and potential of CO_2-derived CNTs as reinforcing agents.

These results suggest that incorporating CNTs derived from CO_2 can reduce the amount of plastic required to achieve desired strength levels, thereby lowering the carbon footprint associated with plastic production. The environmental benefits are twofold: utilizing CO_2 as a raw material and reducing the overall use of virgin polymers.

Economically, the CO_2-to-CNT process could significantly reduce the cost of producing carbon nanomaterials. While conventional methods, such as chemical vapor deposition, are expensive, the electrolysis-based approach offers a potential reduction in production costs—estimated at around \$1000 per ton for bulk production, compared to about \$1 million per ton for traditional methods. However, these estimates are preliminary and would require further validation at larger scales.

The implications of this technology extend beyond reducing material use. Enhanced electrical and thermal conductivity in plastics could open up new applications in electronics and heat management, although these applications require additional research to confirm their feasibility in CNT-enhanced plastics. Improved mechanical properties could lead to lighter, stronger materials for transportation and construction, potentially reducing fuel consumption and enhancing durability.

While the initial results are promising, further studies are needed to fully understand the long-term performance and environmental impacts of CNT-enhanced plastics. Questions remain about the scalability of the CO_2-to-CNT process and the behavior of these materials during recycling or disposal. Nonetheless, this case study illustrates the potential for CO_2-derived nanomaterials to play a crucial role in developing sustainable, high-performance materials that align with the goals of climate mitigation and a circular economy.

9.6 Future Frontiers in Sustainable Materials – Engineered Living Materials

In the context of this book, we highlight the importance of synergies and interdisciplinary approaches in applying nanotechnology across diverse industries and applications. To fully appreciate the potential of these advancements, it is essential to recognize the broader landscape of innovative materials science. Engineered living materials (ELMs) represent a rapidly emerging field that, while distinct from traditional nanotechnology, shares common goals and can complement nanotechnology solutions. By leveraging living cells as building blocks, ELMs introduce dynamic properties and sustainability potential that align well with the goals of nanotechnology (the concept of ELMs is illustrated in Figure 9.8). Exploring ELMs allows us to see how these innovative materials could intersect with nanotechnological advancements, offering new pathways for creating high-performance, environmentally responsive materials.[143]

One of nature's most remarkable achievements is morphogenesis—the process by which a single cell develops into a complex organism through a series of genetic and physical patterns that guide growth into specific, functional structures. Whether forming the intricate spikes of a snail's shell, the smooth hydrodynamic skin of a shark, or the branched airways of human lungs, biological systems inherently possess the ability to self-assemble into a vast diversity of forms optimized for their environment. This natural process contrasts sharply with traditional manufacturing methods, which are often rigid and lack the capacity for self-assembly and adaptation.

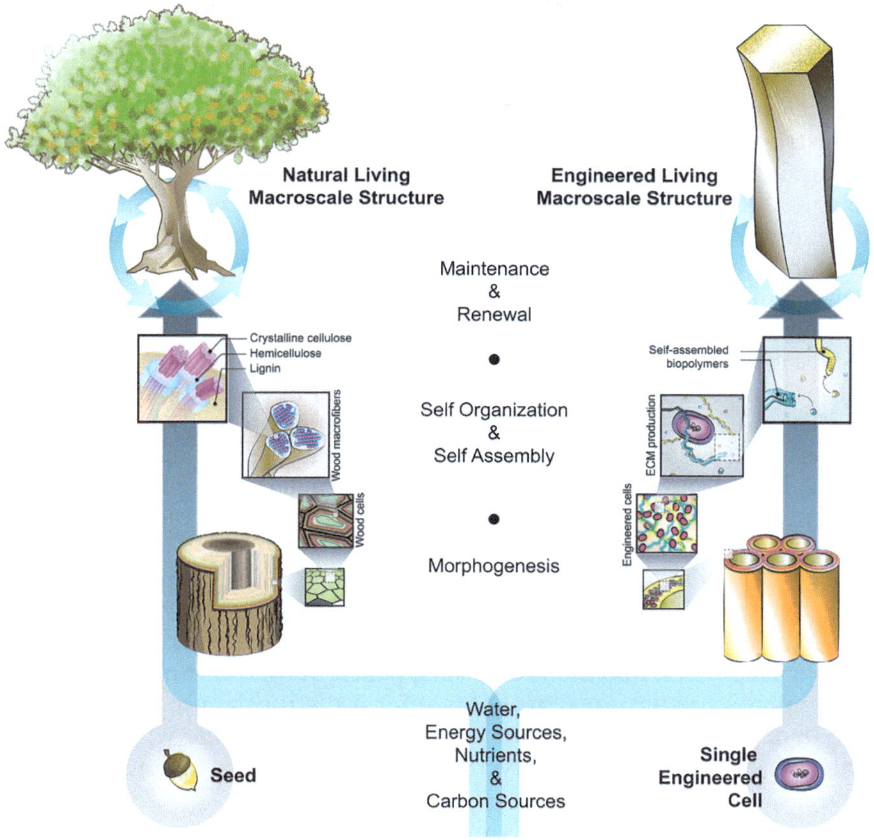

Figure 9.8 ELMs as bioinspired materials engineering. A single seed contains all the essential information needed to create a tree, and it does so by genetically encoding the metabolic functions required for acquiring energy (photosynthesis) and mechanisms for utilizing local resources (carbon dioxide and water). It also contains the information necessary for successful multiscale hierarchical morphogenesis based on principles of cellular self-organization and molecular self-assembly. These living structures are capable of dynamically responding to the environment and can maintain and renew themselves over time. Synthetic ELMs attempt to replicate these characteristics, by using synthetic biology to engineer cells to fabricate, assemble, and maintain autonomously produced materials composed of one or more biomaterials from a library of potential biomaterials. The cells could be metabolically engineered to utilize local energy sources and materials. By also incorporating programmed synthetic morphogenesis, large-scale structures can be designed with complex organization over multiple length scales. Reproduced from ref. 143 with permission from John Wiley & Sons, Copyright © 2018 WILEY-VCH Verlag GmbH & Co. KGaA, Weinheim.

Unlike conventional manufacturing, which relies on static components assembled into fixed structures, living systems can dynamically self-organize and adapt to environmental cues, enabling the creation of materials with highly specialized functions. This is a frontier that artificial systems have struggled to replicate; the ability to autonomously remodel and adapt in response to changing conditions remains a challenge for current technologies. Engineered living materials, however, offer a promising new direction by mimicking these natural processes, potentially allowing us to develop materials that can self-create optimized, application-specific architectures.[144]

Researchers at the intersection of biology, engineering, and materials science are exploring ELMs as a new paradigm. These materials utilize living cells as microscopic factories to fabricate, assemble, and maintain structures with desirable properties. By leveraging the inherent abilities of biological systems—such as self-assembly, self-healing, and responsiveness—ELMs aim to reduce our reliance on unsustainable materials derived from fossil fuels while offering new capabilities that are difficult to achieve with purely synthetic materials.

In the broader context of materials science, ELMs represent one of several exciting frontiers. Alongside advancements in nanomaterials, which manipulate matter at the atomic or molecular scale to achieve unique properties, and metamaterials, which derive their properties from their microstructural design, ELMs offer a distinct approach. They provide dynamic properties that are challenging to replicate with non-living materials, such as the ability to self-heal or adapt to environmental changes.

The development of ELMs is inherently interdisciplinary, requiring the convergence of biology, materials science, and engineering. This integration opens up opportunities to create hybrid materials that combine the adaptability and functionality of living systems with the robustness and durability of traditional materials. For instance, integrating nanomaterials into ELMs could enhance their mechanical properties or introduce new functionalities, like electrical conductivity or magnetic responsiveness. These hybrid systems represent a promising avenue for future research, potentially leading to innovations that neither biological nor synthetic materials could achieve alone.

Several startups are now commercializing early living material technologies, but they remain beset by challenges such as high production costs and inferior mechanical strength compared to conventional materials. However, if these hurdles can be overcome, living materials may one day replace unsustainable conventional materials in applications ranging from packaging to infrastructure construction. By harnessing the synergies between nanotechnology and the unique capabilities of ELMs, researchers are paving the way for a future where biology and nanotechnology converge to create sustainable, high-performance materials that are adaptable and environmentally responsive.

9.6.1 Natural Living Materials: Taking Cues from Billions of Years of Evolution

All living organisms employ a diverse repertoire of functional materials composed of proteins, polysaccharides, and minerals. For instance, woody plants biosynthesize lignin, cellulose, and hemicellulose to build sturdy tree trunks, while marine mussels secrete underwater adhesive proteins for attaching to surfaces. Most intriguingly, these living materials exhibit dynamic properties that synthetic counterparts lack, such as the ability to self-assemble, self-heal after injury, adapt to environmental stimuli, and undergo continuous self-renewal.

The field of bioinspired materials aims to mimic such natural structures, but replicating their living attributes remains challenging. Now, an emerging approach is to engineer the organisms themselves to serve as microbial "factories" for producing functional materials. As experts in biochemical synthesis after billions of years of evolution, living cells potentially offer a sustainable way to manufacture a huge diversity of tailored biopolymers.

Researchers categorize living material systems based on their designs:

- **Self-organizing living materials:** Built solely from living components such as engineered bacteria, fungi, or mammalian cells. They aim to recapitulate natural self-assembly and environmentally responsive behaviors.
- **Hybrid living materials:** Merge living components with abiotic scaffolds such as hydrogels and electronic devices. The non-living parts enhance manufacturability and augment the functionality of embedded organisms.

Both self-organizing and hybrid living materials provide unique opportunities to integrate with nanotechnology. Self-organizing systems could be enhanced with nanoscale sensors or catalysts, while hybrid systems could leverage nanomaterials to bolster structural integrity or introduce novel functionalities.

9.6.2 Programming Living Materials Using Synthetic Biology

Building on the foundational role of synthetic biology mentioned earlier, researchers are now leveraging this toolkit to explore advanced strategies for developing self-organizing living materials with tailored functionalities. Using libraries of well-characterized DNA parts that encode basic genetic functions, synthetic biologists can introduce artificial gene circuits to give cells computer-like capabilities. For instance, engineered gene networks allow microbes to sense chemical signals, perform logic computations, or synchronize their behaviors across populations.[145]

9.6.3 Leveraging Living–Nonliving Hybrids

Although composed purely of life, the materials fabricated by engineered organisms currently suffer from limitations like weak mechanical strength.

To address this, researchers are exploring hybrid systems that combine living cells with robust abiotic components while taking advantage of manufacturing techniques from materials science. These hybrid systems are particularly promising when combined with nanotechnology. For instance, 3D printing techniques can incorporate nanomaterials alongside living cells to enhance mechanical properties or introduce novel capabilities like electrical conductivity, bridging the gap between biological adaptability and synthetic strength.[146]

Techniques such as 3D printing and microfluidics enable the controlled encapsulation of living cells within customizable polymer hydrogels.[147] These gels provide a soft, aquatic environment to maintain cell viability while enhancing the physical characteristics of the overall hybrid material. In other cases, researchers have incorporated functional non-living components that synergize with microbial metabolism to enable new material capabilities.

9.6.4 Realizing the Sustainability Potential of Living Materials

Proponents argue that, compared to conventional manufacturing, fabricating materials from living systems may offer sustainability benefits such as:

- Using genetically modified microbes as cell factories to produce renewable bioplastics, leather substitutes, and pigments, reducing reliance on petrochemical feedstocks.[148,149]
- Employing organisms for active bioremediation of pollutants and waste. Engineered microbes show promise for capturing carbon from the air or degrading plastic waste.[150]
- Designing probiotic living coatings that prolong food shelf life, reducing spoilage and waste.[151]
- Using nitrogen-fixing or mineral-depositing bacteria as microbe-based fertilizers for more sustainable agriculture, lowering requirements for synthetic fertilizers.[152]

However, researchers still face major obstacles in optimizing living materials and fully realizing their sustainability potential. Current products remain expensive compared to conventional options, require costly feedstocks (often sugars), and exhibit inferior material properties like low mechanical strength. There are also biosafety concerns surrounding the environmental release of genetically engineered organisms.

Nonetheless, living materials represent an exciting intersection between synthetic biology and materials science. With continued progress in engineering cells and managing microbial communities, living technologies may one day provide sustainable solutions for manufacturing chemicals, treating wastewater, sequestering carbon from the air, and much more.[153] As research advances, the integration of nanotechnology and ELMs could drive innovations that reshape our approach to sustainable material design, blending the best of both biological and synthetic worlds.

9.6.5 Case Study: Self-organizing Nanomaterials on Cell Surfaces for Precise Engineered Living Materials

Traditionally, scientists have relied on engineered proteins to create one-dimensional materials, such as filaments, that can be displayed on cell surfaces. While these methods have shown promise, they lack the ability to assemble materials in two dimensions with precise control over spatial positioning and density. This gap in technology has limited the structural versatility of engineered living materials, preventing them from reaching their full potential. Yet, the development of surface-layer (S-layer) proteins presents a compelling new approach. These proteins form a dense, periodic lattice on the outermost surface of many prokaryotic cells, providing a promising platform for 2D material assembly.

For instance, researchers are engineering the S-layer of the bacterium *Caulobacter crescentus* to create stable, high-density, 2D living materials.[154] The S-layer in *Caulobacter* consists of a protein called RsaA, which naturally forms a hexagonal lattice structure on the cell surface. By exploiting this unique property, scientists can display engineered proteins, nanocrystals, and biopolymers in a controlled manner on the cell surface, paving the way for developing new materials with customizable properties.

To achieve this, the researchers developed a method to engineer the S-layer of *Caulobacter* by inserting specific peptides called SpyTags into various locations of the RsaA protein. SpyTags can form strong, covalent bonds with a complementary protein called SpyCatcher. By modifying the RsaA protein to display SpyTags (see Figure 9.9), the researchers created a modular platform that can bind a wide range of materials—such as proteins, quantum dots, and polymers—directly to the cell surface.

Flow cytometry and confocal microscopy showed that the attachment of materials to the cell surface was uniform, specific, and covalent. This specificity is crucial for the development of engineered living materials because it allows for the precise arrangement of different components, which can influence the material's overall properties. Furthermore, the density of attached materials could be controlled by the insertion location within the S-layer protein, allowing for a high degree of customization.

One of the key achievements of this work was demonstrating the high density of materials that can be displayed on the cell surface without affecting cell viability. The engineered S-layer could display up to one attachment site per 288 square nanometers, which is among the highest densities reported for cell-surface displayed proteins. This capability is important for creating materials with specific properties, such as enhanced strength or

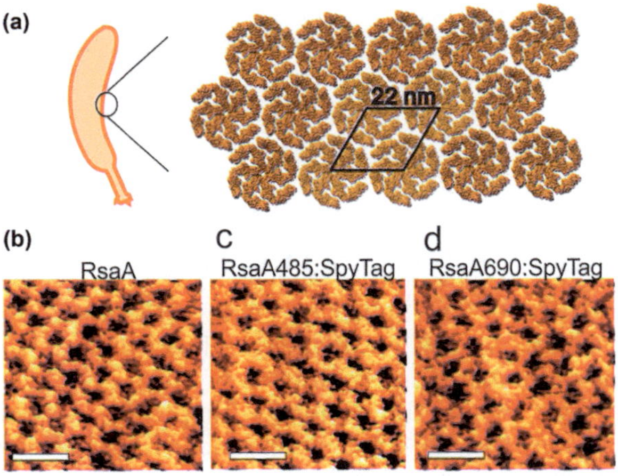

Figure 9.9 RsaA forms a 2D hexameric lattice on the surface of *C. crescentus*. (a) Structure of the RsaA lattice. (b) High resolution AFM images of the wild-type RsaA lattice (strain MFm111), (c) RsaA485:SpyTag (strain MFm 118), and (d) RsaA690:SpyTag (strain MFm 120) on the surface of *C. crescentus* cells. In all three cases, a well-ordered, hexagonal protein lattice is observed. The unit cell length (center-to-center distance between adjacent hexagons) is 22 ± 1 nm, which is the same as reported in the literature. Scale bar is 40 nm. Reproduced from ref. 154 with permission from American Chemical Society, Copyright 2023.

conductivity, which require a high concentration of functional groups on the material's surface.

The study also explored the potential applications of this technology by attaching quantum dots to the surface of the engineered cells. Quantum dots have unique optical properties that make them useful for a range of applications, including biosensing and bioimaging. The researchers showed that these quantum dots could be attached to the cell surface without impairing cell viability, and the resulting composite material remained stable for up to two weeks. This stability is essential for the development of durable materials that can function over extended periods.

Another interesting aspect of this research is its potential for creating hybrid materials that combine the best of both biological and synthetic worlds. By using living cells as a scaffold for the assembly of synthetic materials, it is possible to create materials that are not only strong and durable but also capable of self-repair and adaptation—traits that are challenging to achieve with purely synthetic materials. This approach could open up new possibilities in various fields, such as bioelectronics, where materials that can interact with biological systems are in high demand.

The ability to control the spatial arrangement of different components on the cell surface also has implications for creating multifunctional materials. By arranging enzymes in a specific pattern, it is possible to create materials that can catalyze complex chemical reactions, which could be useful in areas such as biocatalysis and environmental remediation. Moreover, the ability to tune the density of attached materials allows for the optimization of these reactions by ensuring that the enzymes are positioned at the correct distances from each other.

The significance of this study extends beyond the immediate applications of the engineered materials. It also provides a foundational platform for future research in the field of synthetic biology and materials science. The modular nature of the engineered S-layer system means that it can be adapted to display a wide range of different materials, depending on the desired application. This versatility makes it a valuable tool for researchers looking to explore new ways of creating materials with tailored properties.

References

1. The circular economy in detail, Ellenmacarthurfoundation.org, [cited 2024 Oct 8]. Available from: https://www.ellenmacarthurfoundation.org/the-circular-economy-in-detail-deep-dive.
2. Q. Zhou, X. Jin, Y. Wei, Y. Wu, F. Jiang and S. Wang, *et al.*, Highly selective and effective gold recovery from electronic wastewater by synergistic effect between covalent organic framework (COF) and MoS_2: Performance and mechanisms, *Sep. Purif. Technol.*, 2024, **338**, 126599.
3. S. Liu, G. Peng, S. Chen, H. Huang, H. Fan and S. Zhou, *et al.*, Efficient oil–water separation with amphipathic magnetic nanoparticles of Fe_3O_4@TiO_2, *J. Dispersion Sci. Technol.*, 2023, **44**(10), 1965–1971.
4. A. F. P. Allwin Mabes Raj, M. Bauman, M. Lakić, N. Dimitrušev, A. Lobnik and A. Košak, Removal of Pb^{2+}, CrT, and Hg^{2+} ions from aqueous solutions using amino-functionalized magnetic nanoparticles, *Int. J. Mol. Sci.*, 2022, **23**(24), 16186.
5. Y. Wu, C.-F. Fu, Q. Huang, P. Zhang, P. Cui and J. Ran, *et al.*, 2D heterostructured nanofluidic channels for enhanced desalination performance of graphene oxide membranes, *ACS Nano*, 2021, **15**(4), 7586–7595.
6. Y. Duan, L. Li, Z. Shen, J. Cheng and K. He, Engineering metal–organic-framework (MOF)-based membranes for gas and liquid separation, *Membranes*, 2023, **13**(5), 480.
7. Y. Cheng, S. J. Datta, S. Zhou, J. Jia, O. Shekhah and M. Eddaoudi, Advances in metal–organic framework-based membranes, *Chem. Soc. Rev.*, 2022, **51**(19), 8300–8350.
8. X. Xu, Y. Hartanto, J. Zheng and P. Luis, Recent advances in continuous MOF membranes for gas separation and pervaporation, *Membranes*, 2022, **12**(12), 1205.
9. G. Z. Kyzas, A. C. Mitropoulos and K. A. Matis, From microbubbles to nanobubbles: Effect on flotation, *Processes*, 2021, **9**(8), 1287.
10. J. Huang, X. Ran, L. Sun, H. Bi and X. Wu, Recent advances in membrane technologies applied in oil–water separation, *Discover Nano*, 2024, **19**, 66.

11. M. Sati, V. Sharma, A. J. Goswami, K. Giri and G. Mishra, Prospects and challenges of bio-nanomaterials for wastewater treatment, in *Nanobiotechnology for Waste Water Treatment*, Springer International Publishing, Cham, 2022, pp. 381–397.
12. S. M. Banihashem, A. Moradi, B. Evazzadeh, F. Namvar and Z. N. Fang, Biogenically synthesized nanoparticles in wastewater treatment; a greener approach: a review, *Clean Technol. Environ. Policy*, 2024, **26**(6), 1731–1754.
13. S. M. Abdelbasir, K. M. McCourt, C. M. Lee and D. C. Vanegas, Waste-derived nanoparticles: Synthesis approaches, environmental applications, and sustainability considerations, *Front. Chem.*, 2020, **8**, 782.
14. M. M.-A. Aslam, H.-W. Kuo, W. Den, M. Usman, M. Sultan and H. Ashraf, Functionalized carbon nanotubes (CNTs) for water and wastewater treatment: Preparation to application, *Sustainability*, 2021, **13**(10), 5717.
15. C. Rodríguez, S. Briano and E. Leiva, Increased adsorption of heavy metal ions in multi-walled carbon nanotubes with improved dispersion stability, *Molecules*, 2020, **25**(14), 3106.
16. M. Bassyouni, A. E. Mansi, A. Elgabry, B. A. Ibrahim, O. A. Kassem and R. Alhebeshy, Utilization of carbon nanotubes in removal of heavy metals from wastewater: a review of the CNTs' potential and current challenges, *Appl. Phys. A: Mater. Sci. Process.*, 2020, **126**, 38.
17. W. Li, D. Wang, Y. Wang, Z. Shi, J. Liu and L. Zhang, *et al.*, Biological metal–organic frameworks for natural gas purification and MTO product separation, *Chem. Synth.*, 2024, **4**, 22.
18. S. Zhou, O. Shekhah, A. Ramírez, P. Lyu, E. Abou-Hamad and J. Jia, *et al.*, Asymmetric pore windows in MOF membranes for natural gas valorization, *Nature*, 2022, **606**(7915), 706–712.
19. W. Qin, A. Yu, X. Han, J. Wang, J. Sun and J. Zhang, *et al.*, Postsynthetic of MIL-101-NH_2 MOFs supported on PVDF membrane for REEs recovery from waste phosphor, *RSC Adv.*, 2022, **12**(38), 24670–24680.
20. H. Li, X. Chen, D. Shen, F. Wu, R. Pleixats and J. Pan, Functionalized silica nanoparticles: classification, synthetic approaches and recent advances in adsorption applications, *Nanoscale*, 2021, **13**(38), 15998–16016.
21. M.-A. Gatou, A. Syrrakou, N. Lagopati and E. A. Pavlatou, Photocatalytic TiO2-based nanostructures as a promising material for diverse environmental applications: A review, *Reactions*, 2024, **5**(1), 135–194.
22. S. Trepte, C. Kutzer-Schulze, U. Langklotz and M. Krug, Electrochemically produced nano-TiO2-coated SiC membranes for photocatalytic water treatment: Preparation, characterization, and hydroxyl radical formation, *Water Sci. Technol.*, 2024, **89**(10), 2783–2795.
23. T. Lazarević-Pašti, T. Tasić, V. Milanković and N. Potkonjak, Molecularly imprinted plasmonic-based sensors for environmental contaminants—current state and future perspectives, *Chemosensors*, 2023, **11**(1), 35.

24. J. Wang, C. Qiu, H. Pang, J. Wu, M. Sun and D. Liu, High-performance SERS substrate based on perovskite quantum dot–graphene/nano-Au composites for ultrasensitive detection of rhodamine 6G and *p*-nitrophenol, *J. Mater. Chem. C*, 2021, **9**(28), 9011–9020.
25. T. Kokulnathan, K. M. Arun Kumar, T.-J. Wang, E. Ashok Kumar, A. Joseph Anthuvan and K.-J. Chen, *et al.*, Plasmonic coupling of silver nanospheres loaded on cobalt–iron layered double hydroxides: a robust SERS probe for 4-nitrophenol detection, *J. Mater. Chem. A*, 2024, **12**(12), 7287–7299.
26. S. O. Fakayode, C. Walgama, V. E. Fernand Narcisse and C. Grant, Electrochemical and colorimetric nanosensors for detection of heavy metal ions: A review, *Sensors*, 2023, **23**(22), 9080.
27. O. Kanoun, T. Lazarević-Pašti, I. Pašti, S. Nasraoui, M. Talbi and A. Brahem, *et al.*, A review of nanocomposite-modified electrochemical sensors for water quality monitoring, *Sensors*, 2021, **21**(12), 4131.
28. W. Liu, M. Li, P. Zhang, H. Jiang, W. Liu and J. Guan, *et al.*, One-step growth of Cu-doped carbon dots in amino-modified carbon nanotube–modified electrodes for sensitive electrochemical detection of BPA, *Mikrochim. Acta*, 2024, **191**, 309.
29. M. R. Milani Hosseini and A. Motaharian, Electroanalytical determination of diazepam in tablet and human serum samples using a multiwalled carbon nanotube embedded molecularly imprinted polymer-modified carbon paste electrode, *RSC Adv.*, 2015, **5**(99), 81650–81659.
30. M. Benamara, A. Ly, S. Soltani, M. Essid, H. Dahman and R. Dhahri, *et al.*, Enhanced detection of low concentration volatile organic compounds using advanced doped zinc oxide sensors, *RSC Adv.*, 2023, **13**(43), 30230–30242.
31. S. Lu, Y. Zhao and E. A. P. Hellerman, UV-durable self-cleaning coatings for autonomous driving, *Sci. Rep.*, 2024, **14**, 8066.
32. Y. Pan, J. Zhang, X. Guo, Y. Li, L. Li and L. Pan, Recent advances in conductive polymers-based electrochemical sensors for biomedical and environmental applications, *Polymers*, 2024, **16**(11), 1597.
33. E. A. Flores-Contreras, R. B. González-González, E. González-González, E. M. Melchor-Martínez, R. Parra-Saldívar and H. M. N. Iqbal, Detection of emerging pollutants using aptamer-based biosensors: Recent advances, challenges, and outlook, *Biosensors*, 2022, **12**(12), 1078.
34. H. Yoo, H. Jo and S. S. Oh, Detection and beyond: challenges and advances in aptamer-based biosensors, *Mater. Adv.*, 2020, **1**(8), 2663–2687.
35. M. Saad and S. P. Faucher, Aptamers and aptamer-coupled biosensors to detect water-borne pathogens, *Front. Microbiol.*, 2021, **12**, 643797.
36. The Global E-waste Monitor 2024, E-Waste Monitor. 2024 [cited 2024 Jul 28]. Available from: https://ewastemonitor.info/the-global-e-waste-monitor-2024/.
37. M. Peydayesh, E. Boschi, F. Donat and R. Mezzenga, Gold recovery from E-waste by food-waste amyloid aerogels, *Adv. Mater.*, 2024, **36**, 2310642.

38. Q. Pan, Y. Chen, H. Li, G. Ma, S. Jiang and X. Cui, *et al.*, Size effect of nickel from nanoparticles to clusters to single atoms for electrochemical CO_2 reduction, *J. Mater. Chem. A*, 2024, **12**, 20035–20044.
39. Y. Kathiraser, J. Ashok and S. Kawi, Synthesis and evaluation of highly dispersed SBA-15 supported Ni–Fe bimetallic catalysts for steam reforming of biomass derived tar reaction, *Catal. Sci. Technol.*, 2016, **6**(12), 4327–4336.
40. C. Gai, F. Zhang, T. Yang, Z. Liu, W. Jiao and N. Peng, *et al.*, Hydrochar supported bimetallic Ni–Fe nanocatalysts with tailored composition, size and shape for improved biomass steam reforming performance, *Green Chem.*, 2018, **20**(12), 2788–2800.
41. C. Wang, Y. Shi, D. Qin and Y. Xia, Bimetallic core–shell nanocrystals: opportunities and challenges, *Nanoscale Horiz.*, 2023, **8**(9), 1194–1204.
42. S. L. Narnaware and N. L. Panwar, Catalysts and their role in biomass gasification and tar abetment: a review, *Biomass Convers. Biorefin.*, 2021, DOI: 10.1007/s13399-021-01981-1.
43. X. Ren, Z. Ou and B. Wu, Low-temperature selective catalytic reduction DeNOX and regeneration of Mn–Cu catalyst supported by activated coke, *Materials*, 2021, **14**(20), 5958.
44. S. Ferreira, E. Monteiro, P. Brito and C. Vilarinho, A holistic review on biomass gasification modified equilibrium models, *Energies*, 2019, **12**(1), 160.
45. S. Pansambal, R. Oza, S. Borgave, A. Chauhan, P. Bardapurkar and S. Vyas, *et al.*, Bioengineered cerium oxide (CeO_2) nanoparticles and their diverse applications: a review, *Appl. Nanosci.*, 2023, **13**(9), 6067–6092.
46. S. Niu, J. Milam-Guerrero, Y. Zhou, K. Ye, B. Zhao and B. C. Melot, *et al.*, Thermal stability study of transition metal perovskite sulfides, *J. Mater. Res.*, 2018, **33**(24), 4135–4143.
47. M. Shafiullah, S. Ullah, M. A. Ali, M. D. Albaqami, S. Mohammad and M. Faizan, A study of the structural, electronic, optical, and thermoelectric properties of the stable double perovskite Ba_2KIO_6 using density functional theory, *Opt. Quantum Electron.*, 2024, **56**, 1149.
48. D. Ji, C. Liu, Y. Yao, L. Luo, W. Wang and Z. Chen, Cerium substitution in $LaCoO_3$ perovskite oxide as bifunctional electrocatalysts for hydrogen and oxygen evolution reactions, *Nanoscale*, 2021, **13**(22), 9952–9959.
49. D. Navas, S. Fuentes, A. Castro-Alvarez and E. Chavez-Angel, Review on Sol–Gel synthesis of perovskite and oxide nanomaterials, *Gels*, 2021, **7**(4), 275.
50. Y. Tao and P. P. Pescarmona, Nanostructured oxides synthesised *via* $scCO_2$-assisted sol–gel methods and their application in catalysis, *Catalysts*, 2018, **8**(5), 212.
51. N. Hoffmann, P. Fincheira, G. Tortella and O. Rubilar, The role of iron nanoparticles on anaerobic digestion: mechanisms, limitations, and perspectives, *Environ. Sci. Pollut. Res.*, 2022, **29**(55), 82619–82631.

52. M. R. Zaeri and F. Esmaeilzadeh, Hydrogen sulfide removal from normal heptane using zinc oxide, silicon dioxide and zeolite 13X: adsorption capacity, kinetics, selectivity, breakthrough and regeneration, *Environ. Sci. Pollut. Res.*, 2023, **30**(35), 84314–84333.
53. B. R. Dearden, A. C. Edwards, Z. T. Evans, B. Woolsey, C. R. Blair and N. G. Harrison, *et al.*, Synthesis of zinc oxide nanoplates and their use for hydrogen sulfide adsorption, *J. Solgel Sci. Technol.*, 2022, **101**(1), 279–286.
54. J. H. Yang, Hydrogen sulfide removal technology: A focused review on adsorption and catalytic oxidation, *Korean J. Chem. Eng.*, 2021, **38**(4), 674–691.
55. T. Yu, Z. Chen, Z. Liu, J. Xu and Y. Wang, Review of hydrogen sulfide removal from various industrial gases by zeolites, *Separations*, 2022, **9**(9), 229.
56. N. Dey, G. V. Samuel, D. S. Raj and B. Gajalakshmi, Nanomaterials as potential high performing electrode materials for microbial fuel cells, *Appl. Nanosci.*, 2023, **13**(4), 2625–2640.
57. N. K. Abd-Elrahman, N. Al-Harbi, N. M. Basfer, Y. Al-Hadeethi, A. Umar and S. Akbar, Applications of nanomaterials in microbial fuel cells: A review, *Molecules*, 2022, **27**(21), 7483.
58. N. A. M. Barakat, S. Gamal, H. Y. Kim, N. M. Abd El-Salam, H. Fouad and O. A. Fadali, *et al.*, Synergistic advancements in sewage-driven microbial fuel cells: novel carbon nanotube cathodes and biomass-derived anodes for efficient renewable energy generation and wastewater treatment, *Front. Chem.*, 2023, **11**, 1286572.
59. R. H. Mahmoud, F. A. Samhan, M. K. Ibrahim, G. H. Ali and R. Y. A. Hassan, Formation of electroactive biofilms derived by nanostructured anodes surfaces, *Bioprocess Biosyst. Eng.*, 2021, **44**(4), 759–768.
60. A. Banerjee, R. Calay and M. Mustafa, Review on material and design of anode for microbial Fuel Cell, *Energies*, 2022, **15**(6), 2283.
61. C. Bi, Q. Wen, Y. Chen and H. Xu, Boosting power output in microbial fuel cell with sulfur-doped MXene/polypyrrole hydrogel anode for enhanced extracellular electron transfer process, *J. Appl. Electrochem.*, 2024, **54**, 2729–2743.
62. S. Zhang, W. Li, H. Wu and J. Jiao, Multi-optimized flexible graphene oxide/multi-walled carbon nanotubes/ferroferric oxide nanopaper with enhanced electromagnetic wave absorption performance, *Adv. Compos. Hybrid Mater.*, 2023, **6**, 154.
63. H. Rasuli and R. Rasuli, Nanoparticle-decorated graphene/graphene oxide: synthesis, properties and applications, *J. Mater. Sci.*, 2023, **58**(7), 2971–2992.
64. H. Lin, W. Sun, C. Zhao and H. Na, Self-assembly of multiwall carbon nanotubes on sulfonated poly (arylene ether ketone) as a proton exchange membrane, *J. Polym. Res.*, 2013, **20**, 306.
65. P. Molla-Abbasi, K. Janghorban and M. S. Asgari, A novel heteropolyacid-doped carbon nanotubes/Nafion nanocomposite membrane

for high performance proton-exchange methanol fuel cell applications, *Iran. Polym. J.*, 2018, **27**(2), 77–86.

66. J. Ma, X. Tan, Q. Zhang, Y. Wang, J. Zhang and L. Wang, Exploring the size effect of pt nanoparticles on the photocatalytic nonoxidative coupling of methane, *ACS Catal.*, 2021, **11**(6), 3352–3360.
67. N. Asghar, A. Hussain, D. A. Nguyen, S. Ali, I. Hussain and A. Junejo, *et al.*, Advancement in nanomaterials for environmental pollutants remediation: a systematic review on bibliometrics analysis, material types, synthesis pathways, and related mechanisms, *J Nanobiotechnol.*, 2024, **22**, 26.
68. R. Julius, A. Suman and A. V. Samrot, Advancements in nanocomposites for wastewater treatment addressing emerging pollutants and contaminants, *Nanotechnol. Environ. Eng.*, 2024, **9**(1), 99–110.
69. Y. Shu, M. Hu, M. Zhou, H. Yin, P. Liu and H. Zhang, *et al.*, Emerging electrocatalysts for electrochemical advanced oxidation processes (EAOPs): recent progress and perspectives, *Mater. Chem. Front.*, 2023, **7**(13), 2528–2553.
70. N. N. Roslan, H. L. H. Lau, N. A. A. Suhaimi, N. N. M. Shahri, S. B. Verinda and M. Nur, *et al.*, Recent advances in advanced oxidation processes for degrading pharmaceuticals in wastewater—A review, *Catalysts*, 2024, **14**(3), 189.
71. R. Tyagi, A. Maurya, Samriti, J. Prakash and S. Kohli, Role of TiO_2-based photocatalysts in waste water treatment, in *Multifunctional Hybrid Semiconductor Photocatalyst Nanomaterials*, Springer International Publishing, Cham, 2023, pp. 201–220.
72. A. Piątkowska, M. Janus, K. Szymański and S. Mozia, C-,N- and S-doped TiO_2 photocatalysts: A review, *Catalysts*, 2021, **11**(1), 144.
73. L. Yu, W. Xu, H. Liu and Y. Bao, Titanium dioxide–reduced graphene oxide composites for photocatalytic degradation of dyes in water, *Catalysts*, 2022, **12**(11), 1340.
74. A. Enesca, The influence of photocatalytic reactors design and operating parameters on the wastewater organic pollutants removal—A mini-review, *Catalysts*, 2021, **11**(5), 556.
75. A. Ashley, B. Thrope, M. R. Choudhury and A. H. Pinto, Emerging investigator series: photocatalytic membrane reactors: fundamentals and advances in preparation and application in wastewater treatment, *Environ. Sci.*, 2022, **8**(1), 22–46.
76. Y. Shu, M. Hu, M. Zhou, H. Yin, P. Liu and H. Zhang, *et al.*, Emerging electrocatalysts for electrochemical advanced oxidation processes (EAOPs): recent progress and perspectives, *Mater. Chem. Front.*, 2023, 7, 2528–2553.
77. P. Brosler, A. V. Girão, R. F. Silva, J. Tedim and F. J. Oliveira, In-house *vs.* commercial boron-doped diamond electrodes for electrochemical degradation of water pollutants: A critical review, *Front. Mater.*, 2023, **10**, 1020649.
78. D.-Y. Kuo, H. Paik, J. N. Nelson, K. M. Shen, D. G. Schlom and J. Suntivich, Chlorine evolution reaction electrocatalysis on RuO_2(110)

and IrO_2(110) grown using molecular-beam epitaxy, *J. Chem. Phys.*, 2019, **150**, 041726.
79. M. Xu, Q. Tian, Y. Quan, L. Xu and M. Namadchian, Preparation of a new and effective heterogeneous catalyst for treatment of organic pollutant using Fenton process, *Top. Catal.*, 2025, 670–676.
80. T. de Oliveira Guidolin, M. A. P. Cechinel and S. Arcaro, Iron-Based Nanomaterials for Fenton Reaction, in *Environmental Applications of Nanomaterials*, Springer International Publishing, Cham, 2022, pp. 133–152.
81. D. B. Olawade, O. Z. Wada, O. Fapohunda, B. I. Egbewole, O. Ajisafe and A. O. Ige, Nanoparticles for microbial control in water: mechanisms, applications, and ecological implications, *Front. Nanotechnol.*, 2024, **6**, 1427843.
82. R. Thirumalaisamy, R. Suriyaprabha, M. Prabhu and A. S. Thesai, Role of nanomaterials in environmental remediation: Recent advances—A review, in *Strategies and Tools for Pollutant Mitigation*, Springer International Publishing, Cham, 2022, pp. 51–68.
83. R. T. Prasad and J. Gupta, Recent advances in the application of nanomaterials for environmental sustainability, in *Hybrid Composite Materials*, Springer Nature Singapore, Singapore, 2024, pp. 333–348.
84. W. Cui, X. Li, W. Duan, M. Xie and X. Dong, Heavy metal stabilization remediation in polluted soils with stabilizing materials: a review, *Environ. Geochem. Health*, 2023, **45**(7), 4127–4163.
85. M. Gol-Soltani, R. Ghasemi-Fasaei, A. Ronaghi, M. Zarei, S. Zeinali and S. B. Haderlein, Efficient Immobilization of heavy metals using newly synthesized magnetic nanoparticles and some bacteria in a multi-metal contaminated soil, *Environ. Sci. Pollut. Res.*, 2024, **31**(27), 39602–39624.
86. R. Jain, Recent advances of magnetite nanomaterials to remove arsenic from water, *RSC Adv.*, 2022, **12**(50), 32197–32209.
87. Q. Jin, Y. Dong, H. Pan and H. Lin, Remediation performance of As-contaminated water and soil using a novel Fe–Mn bimetallic (oxyhydr) oxide coated on natural magnetite, *Environ. Sci. Pollut. Res.*, 2023, **30**(23), 64233–64245.
88. Z. Cai, X. Zhao, J. Duan, D. Zhao, Z. Dang and Z. Lin, Remediation of soil and groundwater contaminated with organic chemicals using stabilized nanoparticles: Lessons from the past two decades, *Front. Environ. Sci. Eng.*, 2020, **14**, 84.
89. K. Meduri, C. Stauffer, W. Qian, O. Zietz, A. Barnum and G. O'Brien Johnson, *et al.*, Palladium and gold nanoparticles on carbon supports as highly efficient catalysts for effective removal of trichloroethylene, *J. Mater. Res.*, 2018, **33**(16), 2404–2413.
90. X. Chen, X. Yao, C. Yu, X. Su, C. Shen and C. Chen, *et al.*, Hydrodechlorination of polychlorinated biphenyls in contaminated soil from an e-waste recycling area, using nanoscale zerovalent iron and Pd/Fe bimetallic nanoparticles, *Environ. Sci. Pollut. Res.*, 2014, **21**(7), 5201–5210.

91. R. Jing, S. Fusi and B. V. Kjellerup, Remediation of polychlorinated biphenyls (PCBs) in contaminated soils and sediment: State of knowledge and perspectives, *Front. Environ. Sci.*, 2018, **6**, 79.
92. P. Mandal, D. Palai and P. Goel, Polymer composites for environmental solutions and remediation, in *Hybrid Composite Materials*. Singapore: Springer Nature Singapore; 2024. pp. 67–90.
93. A. Anjum, D. Gupta, B. Singh, R. Garg, B. Pani and M. Kashif, *et al.*, Clay–polymer nanocomposites for effective water treatment: opportunities, challenges, and future prospects, *Environ. Monit. Assess.*, 2024, **196**, 666.
94. C. Arun, A. Sethupathy and R. Selvasembian, Clay-based nanocomposites and hybrid membrane for the treatment of wastewater, in *Clay Composites*. Singapore: Springer Nature Singapore; 2023. pp. 437–450.
95. G. Mohammadnezhad and A. Keikavousi Behbahan, Polymer matrix nanocomposites for heavy metal adsorption: a review, *J. Iran. Chem. Soc.*, 2020, **17**(6), 1259–1281.
96. J. Hızal and M. Yılmazoğlu, Montmorillonite clay composite for heavy metal removal from water, in *Environmental Chemistry for a Sustainable World*, Springer International Publishing, Cham, 2021, pp. 93–112.
97. R. Singh and S. Budarayavalasa, Solidification and stabilization of hazardous wastes using geopolymers as sustainable binders, *J. Mater. Cycles Waste Manage.*, 2021, **23**(5), 1699–1725.
98. D. Jia, P. He, M. Wang and S. Yan, Geopolymers and their matrix composites: A state-of-the-art review, in *Geopolymer and Geopolymer Matrix Composites*, Springer Singapore, Singapore, 2020, pp. 7–34.
99. Z. Su, W. Hou and Z. Sun, Recent advances in carbon nanotube-geopolymer composite, *Constr. Build. Mater.*, 2020, **252**, 118940.
100. H. AlTawaiha, F. Alhomaidat and T. Eljufout, A review of the effect of nano-silica on the mechanical and durability properties of cementitious composites, *Infrastructures*, 2023, **8**(9), 132.
101. M. Wang, Y. Jiang, S. Liu, Z. Chen and H. Su, Experimental study on mechanical properties of nano-alumina-modified cement composites exposed to high temperatures, *Processes*, 2023, **12**(1), 94.
102. B. Anegbe and I. H. Ifijen, Recent advances in the application of manganese oxide nanoparticles for remediation of soil contaminated with organic pollutants, in *The Minerals, Metals & Materials Series*, Springer Nature Switzerland, Cham, 2024, pp. 1358–1374.
103. X. Wang, D. Liang, Y. Wang, W. J. G. M. Peijnenburg, F. A. Monikh and X. Zhao, *et al.*, A critical review on the biological impact of natural organic matter on nanomaterials in the aquatic environment, *Carbon Res.*, 2022, **1**, 13.
104. A. M. Vindedahl, J. H. Strehlau, W. A. Arnold and R. L. Penn, Organic matter and iron oxide nanoparticles: aggregation, interactions, and reactivity, *Environ. Sci.: Nano*, 2016, **3**(3), 494–505.
105. B. Arash, D. Abliz and R. Rolfes, Modeling and simulation of nanocomposites and their manufacturing processes, in *Acting Principles of*

Nano-Scaled Matrix Additives for Composite Structures, Springer International Publishing, Cham, 2021. pp. 27–54.
106. G. Chen, W. Xian, Q. Wang and Y. Li, Molecular simulation-guided and physics-informed mechanistic modeling of multifunctional polymers, *Acta Mech. Sin.*, 2021, **37**(5), 725–745.
107. J. H. Lee, H.-S. Kim, E.-T. Yun, S.-Y. Ham, J.-H. Park and C. H. Ahn, *et al.*, Vertically aligned carbon nanotube membranes: Water purification and beyond, *Membranes*, 2020, **10**(10), 273.
108. E. Valamohammadi, F. Behdarvand, T. Mohammadi, M. A. Tofighy and Z. Moghiseh, Effects of carbon nanotubes on structure, performance and properties of polymer nanocomposite membranes for water/wastewater treatment applications: a comprehensive review, *Polym. Bull.*, 2023, **80**(11), 11589–11632.
109. F. Jia, X. Xiao, A. Nashalian, S. Shen, L. Yang and Z. Han, *et al.*, Advances in graphene oxide membranes for water treatment, *Nano Res.*, 2022, **15**(7), 6636–6654.
110. W. Liu, Y. Li, R. Li, H. Xu, X. Lu and W. Dong, *et al.*, Constructing 2D ultrathin graphene oxide membranes with supramolecular-assembled nano-adhesives for enhancing water stability, *Nano Res.*, 2023, **16**(7), 8505–8511.
111. X. Wang, H. Zhang, X. Wang, S. Chen, H. Yu and X. Quan, Electroconductive RGO-MXene membranes with wettability-regulated channels: improved water permeability and electro-enhanced rejection performance, *Front. Environ. Sci. Eng.*, 2023, **17**, 1.
112. X. Mao, Y. Wang, X. Yan, Z. Huang, Z. Gao and Y. Wang, *et al.*, A review of superwetting membranes and nanofibers for efficient oil/water separation, *J. Mater. Sci.*, 2023, **58**(1), 3–33.
113. P. Kaner, A. V. Dudchenko, M. S. Mauter and A. Asatekin, Zwitterionic copolymer additive architecture affects membrane performance: fouling resistance and surface rearrangement in saline solutions, *J. Mater. Chem. A*, 2019, 7(9), 4829–4846.
114. H. Ghasemi, N. Abu-Zahra, U. Baig, A. Waheed and I. H. Aljundi, Surface-engineered polyethersulfone membranes using Thermo-responsive poly(*N*-isopropyl acrylamide) with antibiofouling properties for water treatment applications, *J. Polym. Environ.*, 2024, **32**(6), 2589–2605.
115. M. N. Subramaniam, P. S. Goh, D. Kanakaraju, J. W. Lim, W. J. Lau and A. F. Ismail, Photocatalytic membranes: a new perspective for persistent organic pollutants removal, *Environ. Sci. Pollut. Res.*, 2022, **29**(9), 12506–12530.
116. N. Chauhan, R. Sharma and S. P. Singh, Synthesis of ceramic membranes and their application in wastewater treatment and emerging contaminants removal, in *Energy, Environment, and Sustainability*, Springer Nature Singapore, Singapore, 2023, pp. 413–446.
117. S. Borcănescu, A. Popa, O. Verdeș and M. Suba, Functionalized ordered mesoporous MCM-48 silica: Synthesis, characterization and adsorbent for CO_2 capture, *Int. J. Mol. Sci.*, 2023, **24**(12), 10345.

118. C. Y. Chuah, K. Goh and T.-H. Bae, Enhanced performance of carbon molecular sieve membranes incorporating zeolite nanocrystals for air separation, *Membranes*, 2021, **11**(7), 489.
119. *Nanomaterials for soil remediation*, ed. A. Amrane, D. Mohan, T. A. Nguyen, A. A. Assadi and G. Yasin, Elsevier Science Publishing, Philadelphia, PA, 2020.
120. L. O. Ejeta and E. A. Elimian, Solid waste recycling and organic particulate hybrid nanocomposite technologies for sustainable infrastructure—A comprehensive review, *J. Polym. Sci. Eng.*, 2024, **7**(1), 3440.
121. V. Dave, P. Sharma, A. Kuila and S. S. Sahay, Revolutionizing environmental sustainability: exploring the transformative potential of nanocomposites/nanomaterials from recycled waste, *Environ. Sci. Pollut. Res.*, 2024, **31**(29), 41621–41623.
122. G. Yu, Y. Lu, J. Guo, M. Patel, A. Bafana and X. Wang, *et al.*, Carbon nanotubes, graphene, and their derivatives for heavy metal removal, *Adv. Compos. Hybrid Mater.*, 2018, **1**(1), 56–78.
123. Y. Wang, C. Pan, W. Chu, A. K. Vipin and L. Sun, Environmental remediation applications of carbon nanotubes and graphene oxide: Adsorption and catalysis, *Nanomaterials*, 2019, **9**(3), 439.
124. H. AlMohamadi, S. A. Awad, A. K. Sharma, N. Fayzullaev, A. Távara-Aponte and L. Chiguala-Contreras, *et al.*, Photocatalytic activity of metal- and non-metal-anchored ZnO and TiO_2 nanocatalysts for advanced photocatalysis: Comparative study, *Catalysts*, 2024, **14**(7), 420.
125. A. Uthaman, H. M. Lal and S. Thomas, Silver nanoparticle on various synthetic polymer matrices: Preparative techniques, characterizations, and applications, in *Polymer Nanocomposites Based on Silver Nanoparticles*, Springer International Publishing, Cham, 2021, pp. 109–138.
126. A. Kandpal, A. Singh, D. Jain, S. K. Upadhyay, A. J. Siddiqui and K. Sharma, *et al.*, Managing plastic waste with nanotechnology: current sustainability prospects, *Nanotechnol. Environ. Eng.*, 2023, **8**(4), 1015–1031.
127. S. Watanabe, Y. Inaba, M. Harigai, K. Takeshita and J. Onoe, The uptake characteristics of Prussian-blue nanoparticles for rare metal ions for recycling of precious metals from nuclear and electronic wastes, *Sci. Rep.*, 2022, **12**, 5135.
128. D. Nath, Nanomaterial for the management of radioactive waste, in *Handbook of Ecomaterials*, Springer International Publishing, Cham, 2018, pp. 1–18.
129. R. Remya, A. Julius, T. Y. Suman and A. V. Samrot, Advancements in nanocomposites for wastewater treatment addressing emerging pollutants and contaminants, *Nanotechnol. Environ. Eng.*, 2024, **9**(1), 99–110.
130. P. C. Pandey, H. P. Yadav, A. K. Tiwari, S. N. Sawant, P. Sinharoy and D. Banerjee, *et al.*, Prussian blue nanoparticles–mediated sensing and removal of ^{137}Cs, *Front. Environ. Sci.*, 2023, **11**, 1230983.
131. Z. Shen, H. Zhang and Y. Wang, Editorial: CO_2 capture and conversion: advanced materials and processes, *Front. Chem.*, 2023, **11**, 1310024.

132. A. Kumar and L. Kumar, Nanostructured catalysts for CO_2 reduction: systematic insights and emerging strategies, *Res. Chem. Intermed.*, 2024, **50**(1), 195–217.
133. Z. Sun, Y. Liao, S. Zhao, X. Zhang, Q. Liu and X. Shi, Research progress in metal–organic frameworks (MOFs) in CO_2 capture from post-combustion coal-fired flue gas: characteristics, preparation, modification and applications, *J. Mater. Chem. A*, 2022, **10**(10), 5174–5211.
134. A. K. Buckley, S. Ma, Z. Huo, T. Z. Gao and K. P. Kuhl, Nanomaterials for carbon dioxide conversion at industrial scale, *Nat. Nanotechnol.*, 2022, **17**(8), 811–813.
135. Y. Kwon, B. Wu, N. Zhang, D. Hand, T. Mou and X. Han, *et al.*, Recent advances in integrated capture and electrochemical conversion of CO_2, *MRS Commun.*, 2024, **14**, 728–740.
136. T. Zhang, M. Zhang, L. Jin, M. Xu and J. Li, Advancing carbon capture in hard-to-abate industries: technology, cost, and policy insights, *Clean Technol. Environ. Policy*, 2024, **26**(7), 2077–2094.
137. M. M. Barcelos, M. L. S. de Vasconcellos and J. Ribeiro, Recent progress in electrochemical CO_2 reduction at different electrocatalyst materials, *Processes*, 2024, **12**(2), 303.
138. K. Gao, X. Liu, Q. Wang, Z. Jiang, C. Tian and N. Sun, *et al.*, Remarkable solar thermochemical CO_2 splitting performances based on Ce- and Al-doped $SrMnO_3$ perovskites, *Sustainable Energy Fuels*, 2023, 7(4), 1027–1040.
139. F. Fresno, A. Iglesias-Juez and J. M. Coronado, Photothermal catalytic CO_2 conversion: Beyond catalysis and photocatalysis, *Top. Curr. Chem.*, 2023, **381**, 21.
140. Q. Qian and B. Han, Transformation of CO_2 and H_2 to C_{2+} chemicals and fuels, *Natl. Sci. Rev.*, 2023, **10**(9), nwad160.
141. Global Plastics Outlook, OECD, [cited 2024 Aug 29]. Available from: https://www.oecd-ilibrary.org/environment/data/global-plastic-outlook_c0821f81-en.
142. G. Licht, K. Hofstetter and S. Licht, Polymer composites with carbon nanotubes made from CO_2, *RSC Sustainable*, 2024, **2**(9), 2496–2504.
143. P. Q. Nguyen, N.-M. D. Courchesne, A. Duraj-Thatte, P. Praveschotinunt and N. S. Joshi, Engineered living materials: Prospects and challenges for using biological systems to direct the assembly of smart materials, *Adv. Mater.*, 2018, **30**, 1704847.
144. S. Molinari, R. F. Tesoriero Jr and C. M. Ajo-Franklin, Bottom-up approaches to engineered living materials: Challenges and future directions, *Matter*, 2021, **4**(10), 3095–3120.
145. J. Luo, J. Chen, Y. Huang, L. You and Z. Dai, Engineering living materials by synthetic biology, *Biophys. Rev.*, 2023, **4**, 011305.
146. G. Kannayiram, S. Sendilvelan and M. Priya, Importance of nanocomposites in 3D bioprinting: An overview, *Bioprinting*, 2023, **32**, e00280.

147. Z. U. Arif, M. Y. Khalid, A. Tariq, M. Hossain and R. Umer, 3D printing of stimuli-responsive hydrogel materials: Literature review and emerging applications, *Giant*, 2023, 100209.
148. J. S. Cho, G. B. Kim, H. Eun, C. W. Moon and S. Y. Lee, Designing microbial cell factories for the production of chemicals, *JACS Au*, 2022, **2**(8), 1781–1799.
149. X. Han, J. Liu, S. Tian, F. Tao and P. Xu, Microbial cell factories for bio-based biodegradable plastics production, *iScience*, 2022, **25**(11), 105462.
150. G. D. Barone, A. Rodríguez-Seijo, M. Parati, B. Johnston, E. Erdem and T. Cernava, *et al.*, Harnessing photosynthetic microorganisms for enhanced bioremediation of microplastics: A comprehensive review, *Environ. Sci. Ecotechnol.*, 2024, **20**, 100407.
151. S. Sáez-Orviz, M. Rendueles and M. Díaz, Impact of adding prebiotics and probiotics on the characteristics of edible films and coatings- a review, *Food Res. Int.*, 2023, **164**, 112381.
152. T. Wang, J. Xu, J. Chen, P. Liu, X. Hou and L. Yang, *et al.*, Progress in microbial fertilizer regulation of crop growth and soil remediation research, *Plants*, 2024, **13**(3), 346.
153. B. An, Y. Wang, Y. Huang, X. Wang, Y. Liu and D. Xun, *et al.*, Engineered living materials for sustainability, *Chem. Rev.*, 2023, **123**(5), 2349–2419.
154. M. Charrier, D. Li, V. R. Mann, L. Yun, S. Jani and B. Rad, *et al.*, Engineering the S-layer of *Caulobacter crescentus* as a foundation for stable, high-density, 2D living materials, *ACS Synth. Biol.*, 2019, **8**(1), 181–190.
155. K. Rabaey and W. Verstraete, Microbial fuel cells: novel biotechnology for energy generation, *Trends Biotechnol.*, 2005, **23**(6), 291–298.

CHAPTER 10

Nanotechnology for Monitoring and Managing Waste

This chapter builds upon the foundational concepts and applications discussed in Chapters 8 and 9, expanding our focus to explore cutting-edge developments in monitoring and managing resource efficiency enabled by nanotechnology. While previous chapters focused primarily on waste reduction, prevention, and treatment technologies, Chapter 10 takes a broader perspective, examining how nanotechnology can support the transition to a circular economy through advanced monitoring and management of resources throughout their lifecycle.

The integration of nanotechnology into resource monitoring and management systems represents a significant leap forward in our ability to track, analyze, and optimize resource use in real-time. By leveraging the unique properties of nanomaterials and nanodevices, we can now detect and quantify pollutants, valuable materials, energy flows, and process parameters at unprecedented levels of sensitivity and specificity. This capability, when combined with advanced data processing and communication technologies, opens up new possibilities for intelligent and adaptive strategies to enhance resource efficiency and minimize waste generation.

In this chapter, we will explore several key areas where nanotechnology is revolutionizing resource monitoring and management in the context of circular economy principles:[1]

- Nanosensors for resource monitoring and efficiency, examining how these miniature devices are pushing the boundaries of detection and enabling more precise control of material flows and energy use.

Waste Not! How Nanotechnologies Can Increase Efficiencies Throughout Society
By Michael Berger

Published by the Royal Society of Chemistry, www.rsc.org

- Advanced nanosensors for comprehensive resource monitoring, exploring multi-parameter detection systems that provide holistic insights into complex industrial processes and environmental conditions.
- The integration of nanotechnology with IoT and AI for optimized resource management, investigating how these technologies combine to create smart, connected systems capable of dynamic response and optimization.
- Smart systems for circular economy applications, focusing on how nanotechnology enables tracking of materials throughout product lifecycles, enhances recycling and material recovery processes, and facilitates the implementation of circular economy strategies.
- The role of nanotechnology in environmental impact assessment and mitigation, exploring how nanomaterials and nanosensors contribute to more effective monitoring and remediation of environmental issues.

Throughout this chapter, we will explain how these nanotechnology-enabled monitoring and management systems complement and enhance the waste reduction, prevention, and treatment strategies explored in earlier chapters. By providing a more comprehensive view of resource flows and enabling more precise control over material and energy use, these technologies support the broader goals of a circular economy, including minimizing waste, optimizing resource utilization, and reducing environmental impacts.

As we examine these advanced applications of nanotechnology, we will also consider the challenges and potential risks associated with their implementation, including issues of data privacy, environmental impacts of nanomaterials, and the need for standardization and regulatory frameworks. By addressing these concerns alongside the potential benefits, we aim to provide a balanced perspective on the role of nanotechnology in shaping a more sustainable and resource-efficient future.

10.1 Nanosensors for Resource Monitoring and Efficiency

Nanosensors are devices operating at the nanoscale, designed to detect and measure physical, chemical, or biological properties of substances or systems. These sophisticated sensors harness unique nanoscale phenomena to achieve remarkable sensitivity, selectivity, and rapid response times. In the context of resource monitoring and efficiency, nanosensors offer unprecedented capabilities for identifying and quantifying pollutants, contaminants, valuable materials, energy flows, and process parameters at extremely low concentrations or with high precision.[2,3]

The field of nanosensors encompasses various types, as illustrated in Figure 10.1, each operating through distinct mechanisms.

Optical nanosensors detect changes in properties like absorption, fluorescence, or surface plasmon resonance. For example, quantum dot-based

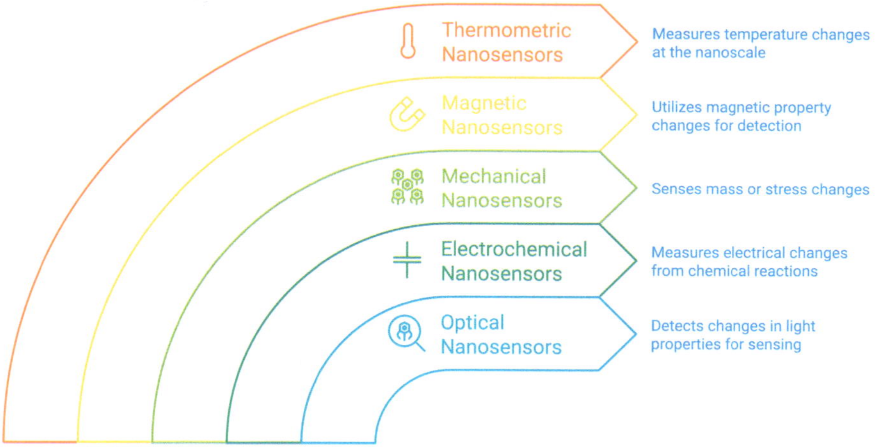

Figure 10.1 Overview of nanosensor technologies.

fluorescent sensors exhibit high sensitivity in detecting heavy metals in water and monitoring material quality during manufacturing processes.[4] **Electrochemical nanosensors** measure electrical changes resulting from chemical reactions at the nanoscale, with graphene and carbon nanotube-based sensors excelling in detecting organic pollutants in wastewater and monitoring energy consumption in industrial processes.[5] **Mechanical nanosensors** detect changes in mass or stress, and cantilever-based sensors measure specific molecules by detecting subtle shifts in vibration frequency, useful for monitoring material properties and process conditions.[6] **Magnetic nanosensors** detect changes in magnetic properties, such as in magneto-resistive sensors, which can sense magnetic nanoparticles that are functionalized to bind specific contaminants or track material flows in complex systems.[7] Lastly, **thermometric nanosensors** measure temperature changes at the nanoscale, which can indicate certain chemical reactions, energy inefficiencies, or the presence of specific substances in industrial processes.[8]

The role of nanosensors in advancing resource monitoring and efficiency cannot be overstated. Their enhanced sensitivity allows for the detection of pollutants, contaminants, and valuable materials at remarkably low concentrations, often at parts per billion or even parts per trillion levels. This exquisite sensitivity enables early detection of potential environmental hazards, process inefficiencies, and opportunities for resource recovery, providing a crucial window for preventative action and optimization.[9]

The improved selectivity of nanosensors is equally valuable, particularly in complex industrial streams where multiple components may be present. Many nanosensors can be designed to target specific molecules or compounds, reducing false positives and providing more accurate data on material composition and energy flows.[10] This precision is essential for developing targeted resource management strategies, optimizing processes, and ensuring regulatory compliance.

Real-time monitoring capabilities represent a significant advantage of micro- and nanotechnology-enabled sensors in addressing environmental and waste management challenges. These sensors' rapid response times and high sensitivity enable continuous monitoring of various critical environmental parameters.[11]

In environmental applications, these sensors facilitate real-time detection of pollutants in air, water, and soil, allowing for prompt identification and mitigation of environmental hazards. For instance, they can provide immediate data on air quality parameters such as particulate matter, greenhouse gases, and other harmful emissions. In water systems, nanosensors can detect contaminants, monitor eutrophication processes, and track water quality in real-time, enabling swift responses to potential threats.

For waste management, these sensors can continuously monitor landfill conditions, detect the presence of toxic substances, and assess the efficiency of waste treatment processes. The integration of these sensors with wireless communication technologies allows for immediate data transmission, supporting remote environmental monitoring and enabling rapid response to emerging environmental issues. This real-time data is invaluable for early warning systems in environmental protection, allowing for quick identification of pollution sources, unexpected events, or opportunities for improvement in waste reduction strategies. The ability to collect and analyze environmental data in real-time not only enhances our understanding of complex ecological systems but also enables more efficient and targeted interventions, ultimately leading to more effective environmental management and conservation strategies.

The miniature size of nanosensors offers remarkable flexibility in their deployment. They can be integrated into compact, portable devices or distributed in large numbers across wide areas, enabling comprehensive monitoring of resources and environmental conditions in various settings. This miniaturization, coupled with low power consumption, makes nanosensors ideal for point-of-care applications and long-term deployment in remote or hard-to-reach locations, expanding our ability to monitor and optimize resource use across diverse ecosystems and industrial environments.

Advanced nanosensor arrays take these capabilities further by enabling multi-parameter detection.[12] These sophisticated systems can simultaneously monitor multiple process parameters, contaminants, or environmental conditions, providing a holistic view of resource flows, energy consumption, and potential impacts. This comprehensive data is crucial for developing nuanced, effective strategies for resource management and circular economy applications.

The ability to perform *in situ* monitoring is another key advantage of nanosensors. By allowing direct deployment in industrial processes, waste streams, or environmental settings, nanosensors eliminate the need for time-consuming and costly sample collection and laboratory analysis. This capability not only speeds up the monitoring process but also provides more accurate, real-time data on resource use and environmental conditions.

The integration of nanosensors with IoT platforms and AI systems is opening new frontiers in comprehensive resource management. These integrated systems can collect and analyze vast amounts of data from distributed nanosensor networks, providing insights for predictive maintenance, process optimization, and holistic resource flow management across entire supply chains or industrial ecosystems.

As nanosensor technology continues to evolve, its transformative impact on resource monitoring and efficiency is becoming increasingly evident across various sectors, from manufacturing and energy production to waste management and environmental remediation. By offering advanced capabilities for precise, real-time, and comprehensive monitoring, nanosensors are paving the way for more efficient, effective, and environmentally friendly resource management strategies.

10.1.1 Case Study: Portable Electrochemical Sensor Enhances Lead Detection

Traditional methods for detecting lead (Pb^{2+}) contamination in water often rely on complex laboratory-based procedures such as inductively coupled plasma mass spectrometry and atomic absorption spectroscopy. These techniques, while highly sensitive, are not practical for *in situ* or on-site analysis due to their requirement for sophisticated instrumentation, skilled operators, and time-consuming sample preparation. This limitation has posed significant challenges for rapid environmental monitoring, particularly in regions where industrial discharge frequently contaminates water supplies with heavy metals like lead.

In recent years, efforts to improve lead detection have focused on developing portable and efficient sensing devices. However, many of these devices face limitations, including insufficient sensitivity, slow response times, and the need for intricate procedures. An ideal solution would be highly sensitive, selective for lead ions, quick to respond, and easy to operate in the field.

To address these challenges, a research team has developed an innovative approach using a portable microfluidic electrochemical sensing platform that employs a novel material, 3D Ag-rGO-f-$Ni(OH)_2$/NF, to enhance signal amplification. This platform represents a breakthrough in the rapid detection of lead ions in water samples, offering a new solution to traditional detection methods.[13]

The researchers designed a microfluidic device that utilizes thermocapillary convection to improve fluid dynamics within the sensing channels. This approach allows for efficient interaction between the target ions and the sensor surface, increasing the sensitivity and speed of detection. The key to this platform is the use of a 3D nanocomposite material consisting of silver (Ag), reduced graphene oxide (rGO), nickel hydroxide ($Ni(OH)_2$), and nickel foam (NF). This material is engineered to have a high surface area and excellent conductivity, crucial for amplifying electrochemical signals.

Experiments revealed that the sensor could detect lead ions at very low concentrations, even in complex water samples, with minimal interference from other metals. At ambient temperature, the sensor displayed remarkable sensitivity, detecting Pb^{2+} concentrations down to the nanomolar range. The unique properties of the 3D Ag-rGO-f-$Ni(OH)_2$/NF material allow for the rapid and selective detection of lead ions, significantly surpassing the capabilities of existing portable sensors.

Compared to other portable lead detection methods, this electrochemical sensing platform demonstrates superior performance. The integration of thermocapillary convection ensures rapid sample processing, and the high sensitivity of the nanocomposite material enables the detection of trace amounts of lead. This combination results in a device that is both efficient and user-friendly, making it suitable for on-site environmental monitoring.

One of the standout features of this platform is its selectivity. The sensor can accurately detect lead ions in the presence of various other metal ions, thanks to the tailored properties of the 3D nanocomposite material. This high selectivity is crucial for real-world applications, where water samples often contain a mix of different contaminants.

The detection mechanism is based on the high surface area and conductivity of the 3D Ag-rGO-f-$Ni(OH)_2$/NF material, which facilitates the adsorption and electrochemical reduction of lead ions. This process results in a significant change in the electrochemical signal, proportional to the concentration of lead, enabling precise quantification.

The practical applicability of this method was demonstrated through successful field tests, where the sensor rapidly detected lead contamination in various water sources. This capability to perform real-time monitoring is a significant advantage, allowing for immediate responses to contamination events.

Economically, the production of this sensing platform is promising. The materials used are cost-effective, and the fabrication process is scalable, making it feasible for widespread adoption. The portability and ease of use of the device further enhance its economic appeal, as it reduces the need for extensive laboratory infrastructure and personnel training.

Beyond lead detection, the researchers envision potential applications of this technology in detecting other hazardous metal ions, such as cadmium and mercury, in environmental samples. The versatility and robustness of the 3D nanocomposite material make it adaptable for various sensing applications.

10.2 Advanced Nanosensors for Comprehensive Resource Monitoring

Building upon the foundational concepts of nanosensors, advanced nanosensors represent the cutting edge of resource monitoring technology. These sophisticated devices push the boundaries of sensitivity, selectivity, and multi-parameter detection. By leveraging the unique properties of nanomaterials and advanced fabrication techniques, these sensors offer unprecedented capabilities for comprehensive resource monitoring across various industrial and environmental applications.

10.2.1 Multi-parameter Nanosensors for Industrial Processes

One of the most significant advancements in nanosensor technology is the development of multi-parameter sensors capable of simultaneously monitoring multiple variables within industrial processes. These sensors integrate various sensing modalities into a single, compact device, enabling the real-time tracking of parameters such as refractive index, magnetic field, temperature, and voltage.

In the chemical industry, multi-parameter nanosensors based on selectively filled D-shaped photonic crystal fibers can be deployed to monitor refractive index, magnetic field strength, temperature, and voltage levels simultaneously within a single reaction vessel. This comprehensive monitoring allows for precise control and optimization of reaction conditions, leading to improved yield, reduced waste, and enhanced energy efficiency. The ability to correlate multiple parameters in real-time provides invaluable insights into process dynamics, enabling rapid adjustments and fine-tuning of industrial operations.

Similarly, in various industrial applications, these multi-parameter nanosensors can monitor factors such as refractive index changes (indicating chemical composition), magnetic field variations (useful in electromagnetic processes), temperature fluctuations, and voltage levels throughout production processes. This holistic approach to monitoring ensures product quality, optimizes process conditions, and enhances safety by enabling early detection of potential issues.

The effectiveness of these sensors lies in their high sensitivity across multiple parameters. For instance, recent developments have achieved remarkable sensitivities: 4600 nm per RIU for refractive index, allowing precise detection of chemical changes; 1.375 nm per Oe for magnetic field, enabling accurate monitoring of electromagnetic processes; 15.143 nm $^{\circ}C^{-1}$ for temperature, ensuring tight control of thermal conditions; and 0.971 nm V^{-1} for voltage, facilitating precise electrical measurements. These high sensitivities enable the detection of subtle changes in industrial processes, allowing for fine-tuned control and optimization across diverse manufacturing environments.[14]

10.2.2 Real-time Monitoring of Energy Efficiency and Emissions

Advanced sensors play a crucial role in improving energy efficiency and reducing emissions across various industries. By providing real-time data on energy consumption, heat loss, and emissions, these sensors enable industries to identify inefficiencies, optimize processes, and reduce their carbon footprint.

In agriculture, nanosensors can potentially monitor soil moisture levels, nutrient content, and crop health. Real-time data would then allow for precise irrigation and fertilization, reducing water and chemical use while optimizing crop yields. Research has shown that integrating sensor data with irrigation systems can significantly improve water use efficiency and reduce environmental impact.[17]

In the healthcare sector, wearable sensors can monitor patients' vital signs, physical activity, and environmental exposures.[15] Real-time health data can lead to early detection of health issues, personalized treatment plans, and improved patient outcomes. Studies have confirmed the benefits of wearable health technology in managing chronic diseases and reducing healthcare costs.

In the oil and gas industry, sensors are deployed in pipelines to detect leaks, monitor pressure, and assess the integrity of infrastructure. Real-time monitoring can prevent catastrophic failures, minimize environmental damage, and ensure efficient operation. Nanosensors help to illuminate detailed information about the characterization of formation, fluid flow, and fluid type of materials. This enhanced level of detail allows for more precise monitoring and analysis.[16]

We can envision how nanosensor technology might transform real-time monitoring of energy efficiency and emissions across various industries. While some applications are already in use, many remain on the horizon. In the following, we explore potential developments and applications of nanosensors specifically related to energy efficiency and emissions monitoring, along with possible reasons for delayed implementation in certain sectors. It's important to note that this information is largely speculative and based on current trends and technological trajectories.

In the **manufacturing sector**, nanosensors have the potential to revolutionize energy efficiency monitoring and emissions control. Nanoscale sensors could be integrated directly into production lines to monitor energy consumption at a granular level, detect heat loss, and measure emissions in real-time. This could lead to more precise process optimization and significant reductions in energy waste. However, widespread adoption may be hindered by factors such as high implementation costs, concerns about sensor reliability in harsh industrial environments, and the need for significant infrastructure changes to integrate nanosensor data into existing manufacturing processes.

The **electric vehicle industry** could benefit significantly from nanosensor technology for energy efficiency monitoring. Advanced battery management systems could use nanosensors to monitor energy flow and heat generation at the molecular level, optimizing charging cycles and overall energy use. This could lead to extended battery life and more accurate range predictions, ultimately improving the energy efficiency of EVs. The delay in practical implementation is due to challenges in scaling up production of reliable nanosensors for automotive use, concerns about long-term durability in varied driving conditions, and the need for extensive testing to meet stringent automotive safety standards.

In **building management**, networks of nanosensors could offer unprecedented insights into energy efficiency. Nanoscale thermal sensors distributed throughout a building could create highly detailed thermal maps, enabling ultra-precise climate control and significant energy savings. These sensors could detect heat loss through walls, windows, and roofs with extreme accuracy, allowing for targeted improvements in insulation and overall building energy performance. The slow adoption in this sector can be attributed to the conservative nature of the construction industry, concerns about the long-term stability of embedded sensors in building materials, and the complexity of retrofitting existing structures with nanosensor networks.

These advancements in nanosensor technology for **energy efficiency and emissions monitoring** promise to provide more detailed, accurate, and timely data than traditional sensor networks. By detecting changes at the molecular or atomic level, nanosensors can offer earlier warnings of inefficiencies, finer control over energy-consuming processes, and more precise emissions measurements. This could potentially lead to significant improvements in energy efficiency and emissions reduction across multiple sectors.

However, common barriers to practical implementation may include high initial costs, technical challenges in sensor miniaturization scaling up production volumes, data management and privacy concerns, and the need for industry-wide standards and regulations for nanosensor use in energy monitoring.

10.2.3 Nanosensors for Water Quality and Resource Management

Water scarcity and quality issues are growing concerns worldwide, and advanced nanosensors offer powerful tools for addressing these challenges. Multi-parameter nanosensors can simultaneously monitor various water quality indicators, including pH, dissolved oxygen, turbidity, specific ions, and microbial contamination.[17]

In wastewater treatment plants, these sensors enable real-time monitoring of influent and effluent quality, allowing for dynamic adjustment of treatment processes to optimize efficiency and ensure regulatory compliance. The ability to detect emerging contaminants, such as pharmaceuticals or microplastics, at trace levels provides valuable data for developing more effective treatment strategies and assessing potential environmental impacts.[18]

In agriculture, soil moisture nanosensors combined with nutrient sensors can provide farmers with detailed, real-time information about soil conditions. This data enables precision agriculture practices, optimizing irrigation and fertilizer use to maximize crop yields while minimizing water consumption and reducing nutrient runoff.[19]

10.2.4 Challenges and Future Directions

While advanced nanosensors offer significant potential for comprehensive resource monitoring, several challenges must be addressed to fully realize their benefits:

- **Sensor stability and longevity**: Ensuring consistent performance over extended periods, especially in harsh industrial environments, remains a key challenge.
- **Data integration and management**: The vast amounts of data generated by multi-parameter nanosensor networks require sophisticated data management and analysis systems to extract meaningful insights.
- **Calibration and standardization**: Developing standardized calibration methods for multi-parameter nanosensors is crucial for ensuring data reliability and comparability across different systems and industries.
- **Cost-effectiveness**: While nanosensor technologies have become more affordable, further cost reductions are necessary to enable widespread adoption, particularly in small and medium-sized enterprises.
- **Ethical and privacy considerations**: The comprehensive monitoring capabilities of advanced nanosensors raise important questions about data privacy and potential misuse, necessitating the development of robust ethical frameworks and data protection measures.

As research in this field continues to advance, we can expect to see further improvements in sensor sensitivity, selectivity, and multi-parameter detection capabilities. The integration of advanced nanosensors with artificial intelligence and machine learning algorithms promises to unlock new possibilities for predictive maintenance, process optimization, and resource management across various industries.

10.2.5 Case Study: Multi-parameter Integrated Sensor Transforms Resource Monitoring

Traditional resource monitoring in industrial processes typically involves separate sensors for different parameters, leading to complex systems requiring multiple devices and data integration efforts. This conventional approach presents significant limitations, including increased costs, spatial constraints, and difficulties in achieving real-time monitoring due to the need for various sensing instruments and synchronization of their outputs.

Recent advancements in sensor technology have sought to overcome these challenges by developing multi-parameter sensors capable of monitoring several variables simultaneously. However, many existing solutions still struggle with issues such as cross-sensitivity between parameters, limited sensitivity, and the need for intricate calibration procedures. An ideal solution would be a highly sensitive, selective sensor capable of monitoring multiple parameters independently and efficiently.

To address these challenges, a research team has developed an innovative approach using a multi-parameter integrated sensor based on selectively filled D-shaped photonic crystal fiber (PCF).[20] This sensor utilizes a novel design that leverages surface plasmon resonance and directional coupling effects to achieve independent and simultaneous detection of refractive index, magnetic field, temperature, and voltage. This development represents a significant advancement in comprehensive resource monitoring, offering a new solution to the limitations of traditional sensors.

The researchers designed the sensor by filling different air holes in the D-shaped PCF with distinct materials, such as magnetic fluid, toluene, and liquid crystal, to enable multi-parameter sensing. This approach allows each parameter to be measured independently, minimizing cross-sensitivity and enhancing the sensor's overall performance. The key to this technology is the selective filling technique, which optimizes the interaction between the sensor's core and the external environment, thereby improving sensitivity and specificity.

The design incorporates nanotechnology through the use of a gold nano-film and nanoparticles to enhance the sensor's responsiveness. The gold nano-film acts as a crucial component in the sensor's surface plasmon resonance, increasing the sensitivity to changes in the surrounding environment. Nanoparticles within the magnetic fluid also play a critical role, responding dynamically to changes in magnetic fields and contributing to the sensor's ability to provide accurate readings across multiple parameters.

Experiments revealed that the sensor achieved remarkable sensitivities for multiple parameters: 4600 nm per RIU for refractive index, 1.375 nm per Oe for magnetic field, 15.143 nm $^{\circ}C^{-1}$ for temperature, and 0.971 nm V^{-1} for voltage. These high sensitivities enable the detection of subtle changes in industrial processes, providing valuable insights for optimizing operations and ensuring product quality. The sensor's design also ensures that changes in one parameter do not affect the others, maintaining measurement accuracy and reliability.

Compared to traditional multi-parameter sensors, this integrated sensor demonstrates superior performance and versatility. The use of selectively filled PCF allows for the simultaneous monitoring of multiple variables with high precision, making it suitable for diverse industrial applications. This capability significantly reduces the need for separate sensing devices, simplifying system design and reducing costs.

One of the standout features of this sensor is its ability to provide real-time data for various industrial processes. The sensor's high sensitivity and multi-parameter capabilities enable rapid adjustments and fine-tuning of operations, leading to improved efficiency and reduced waste. This real-time monitoring capability is crucial for industries aiming to enhance sustainability and reduce their environmental impact.

The detection mechanism is based on the unique properties of the D-shaped PCF and the selective filling technique, which facilitate precise measurements of multiple parameters. The nanoscale integration of materials ensures the sensor can operate effectively in challenging industrial environments, providing reliable data for process optimization and resource management.

The practical applicability of this sensor was demonstrated through successful laboratory tests, where it accurately monitored multiple parameters in complex industrial settings. This capability to perform comprehensive monitoring makes it a valuable tool for industries seeking to enhance process control and efficiency.

Economically, the production of this sensor is promising. The materials used are cost-effective, and the fabrication process is scalable, making it feasible for widespread adoption across various industries. The sensor's integrated design further enhances its economic appeal by reducing the need for multiple devices and associated maintenance costs.

Beyond industrial applications, the researchers envision potential uses of this technology in environmental monitoring and healthcare, where multiparameter sensing can provide valuable insights. The versatility and robustness of the D-shaped PCF sensor make it adaptable for a wide range of sensing applications, potentially transforming how resources are monitored across different sectors. By utilizing nanoscale innovations, such as gold nanofilms and nanoparticles, this sensor highlights the transformative potential of nanotechnology in advancing sensor capabilities.

10.3 IoT and AI Integration for Optimized Resource Management

The convergence of nanotechnology, IoT, and AI is creating smart, connected systems that enhance resource management. While many of these applications are still in development, they represent a logical evolution from today's technology, spanning from optical and terahertz frequencies to radio frequencies using innovative technologies like magnetoelectric antennas.

Studies already have shown the viability of using nanosensor networks for smart healthcare systems, where they continuously monitor patient vitals and environmental conditions in hospital settings.[21,22]

10.3.1 Industrial Applications and Smart Manufacturing

In industrial settings, nanosensors have demonstrated the capability to monitor manufacturing processes and equipment health, enhancing predictive maintenance efforts. They offer real-time insights and optimize operations across various sectors.[23] In smart production, the integration of products and machines with sensors allows the system to immediately collect current operating parameters, predict product quality, and feed optimal parameters back to machines in the production line. Smart sensors and their wireless networks are thus important components of smart manufacturing.[24]

Currently, industries rely on conventional sensors combined with AI and machine learning to forecast equipment failures and optimize processes.[25] In research settings, nanosensors provide more detailed data that enhance these predictive maintenance models. As this technology matures, it is poised to integrate seamlessly with AI, offering improved precision and earlier detection of potential issues, significantly reducing downtime and maintenance costs.

AI and machine learning are used with data from traditional sensors to improve decision-making processes, but nanosensors are being tested to provide richer datasets for more complex analyses. For instance, researchers have developed early pathogen detection systems for crops using carbon nanotube-based nanosensors combined with machine learning.[26] These nanosensors can detect minute changes in plant volatile organic compounds associated with pathogen infection. Neural networks are then trained on the nanosensor data to classify healthy *versus* infected plants with high accuracy (see Figure 10.2). This nanosensor-AI integration allows for much earlier and more sensitive detection of crop diseases compared to traditional visual inspection methods.

The rich, multidimensional data from nanosensors enables more complex analysis of subtle chemical signatures that would not be possible with conventional sensors alone. Such systems demonstrate how nanosensors can provide more detailed data for AI analysis, enabling more sophisticated and sensitive detection capabilities in agriculture.

The future integration of nanosensors across industries will enable AI systems to detect subtle changes that might indicate early signs of wear or

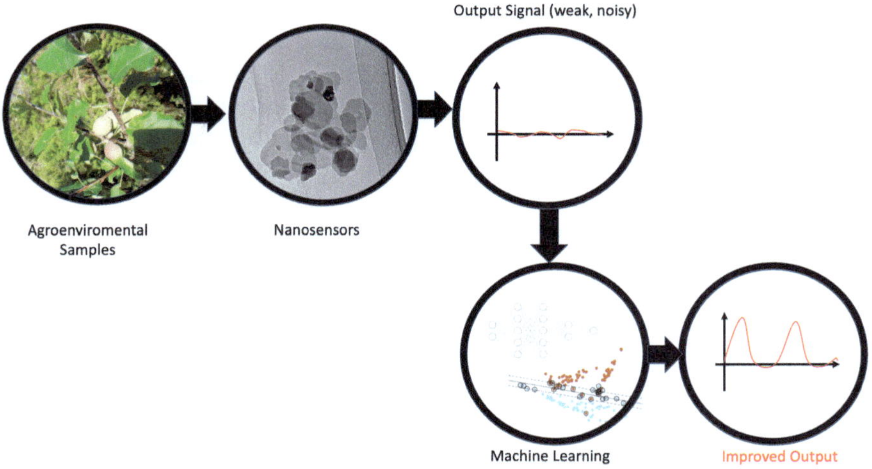

Figure 10.2 Nanosensor-based pathogen detection in agroenvironmental samples: From raw data to enhanced signal processing through machine learning. Reproduced from ref. 26, https://doi.org/10.3390/agronomy14020341, under the terms of the CC BY 4.0 license, https://creativecommons.org/licenses/by/4.0/.

inefficiencies. This will facilitate automated, real-time adjustments in operational parameters, optimizing resource use and minimizing waste.

10.3.2 Digital Twins and the Internet of NanoThings

Digital twins, which simulate and optimize various industrial processes, primarily rely on data from conventional sensors.[27] While current implementations do not yet incorporate nanosensor data, the potential evolution towards integrating such advanced sensing technologies into these models is a logical and promising development. Though speculative at this stage, the incorporation of nanosensor data could potentially enhance the accuracy and effectiveness of digital twins. This advancement may allow operators to simulate more complex scenarios and develop more refined resource allocation strategies, potentially leading to significant improvements in efficiency and productivity.

Beyond traditional resource management, nanonetworks are opening up exciting possibilities in various domains, particularly in computing and the Internet of Nano-Things (IoNT). The IoNT concept envisions individual nanomachines, comprising nano-processors, nano-memory, nano-radios, nanosensors, and energy-harvesting nano-systems, communicating with each other to create nanonetworks. These nanonetworks can then connect to macro-scale networks and ultimately the Internet through nano-routers serving as nano-to-macro interfaces.[28]

In computing applications, wireless nanonetworks within computing packages are being explored to enhance chip-to-chip communication, potentially revolutionizing computer architecture. These nanonetworks can significantly improve resource utilization by reducing latency up to an order of magnitude, providing dynamically shared bandwidth, and enabling more scalable architectures by overcoming physical wiring constraints.

In quantum computing, nanonetworks could play a crucial role in scaling up quantum systems and managing qubit communications. They address critical efficiency challenges such as the ultra-low latency requirements needed to prevent qubit decoherence, the high bandwidth demands of large-scale quantum systems (potentially reaching Tb/s for millions of qubits), and more efficient control and readout of quantum cores.

By tackling these issues across various scales—from individual nanomachines to global networks—nanonetworks aim to maximize computational resources while minimizing physical constraints, energy consumption, and communication overhead in both conventional and quantum computing paradigms, as well as in broader Internet of Nano-Things applications.

10.3.3 Challenges in Widespread Nanosensor Adoption

Despite their potential, several challenges need to be addressed for the widespread adoption of nanosensors. One major hurdle is manufacturing scalability. Producing nanosensors at scale while maintaining their precision and reliability presents significant technical and economic challenges. The fabrication processes for nanosensors often involve complex and costly techniques that need to be optimized for large-scale production. Overcoming these manufacturing challenges is crucial for enabling the widespread deployment of nanosensors in resource management applications, potentially leading to more efficient use of resources across various industries.

Integration with existing systems is another challenge. Nanosensors must be compatible with current IoT and data infrastructure, which requires seamless communication and data processing capabilities. Ensuring interoperability between nanosensors and traditional systems is crucial for their successful implementation. The development of standards for nanosensor technology could facilitate this integration and interoperability, accelerating their adoption in resource management systems.

Long-term reliability in various environments also poses a challenge. Nanosensors need to operate consistently and accurately under diverse conditions, from industrial settings with harsh chemicals to outdoor environments exposed to weather extremes. Ensuring their durability and stability over time is essential to maintain their effectiveness and sustain long-term improvements in resource efficiency and waste reduction.

Economic factors also play a significant role in the widespread implementation of nanosensor technology for resource management. While the potential for long-term cost savings through improved efficiency is substantial, the initial investment required for research, development, and implementation of nanosensor systems can be a barrier for many organizations. Balancing these short-term costs with long-term benefits will be crucial for driving adoption across various sectors.

10.3.4 Case Study: IoT-enabled Wearable Sensors for Optimized Plant Health Monitoring

Traditional methods of plant health monitoring rely on visual inspections and laboratory analyses, which can be time-consuming and may not detect diseases until they are well-established. This approach often results in delayed interventions, leading to decreased crop yields and increased resource usage. The complexity of plant physiology and the subtlety of early disease symptoms pose significant challenges to timely and effective monitoring.

Recent advancements in IoT and AI technology have sought to overcome these limitations by developing wearable nanosensors capable of real-time monitoring of plant health parameters. However, many existing solutions struggle with issues such as limited sensitivity, data processing challenges, and the inability to provide continuous monitoring in diverse environmental conditions. An ideal solution would involve a highly sensitive, integrated sensor system capable of detecting early signs of stress and disease in plants, enabling prompt interventions.

To address these challenges, a research team developed an innovative IoT-enabled wearable nanosensor system using a combination of Au/Ag nanowires, multiwalled carbon nanotubes, and sol–gels to monitor volatile organic compounds, temperature, and humidity directly on plant leaves. This system leverages the nanoscale properties of the materials to enhance sensitivity and specificity in detecting physiological changes in plants.[29]

The wearable nanosensor is designed to attach to the underside of tomato leaves, where it continuously measures environmental and physiological parameters. The nanoscale integration of materials enables the sensor to detect subtle changes associated with plant stress, such as those induced by infections with pathogens like the tomato spotted wilt virus. The key to this technology is its ability to provide real-time data on plant health, which is processed using principal component analysis and Euclidean distance algorithms to detect changes in plant physiology.

Experiments revealed that the sensor could accurately identify stress signals related to viral infection before visual symptoms appeared, demonstrating its potential for early disease detection. The integration of PCA and machine learning techniques allowed for precise classification of stress-related signals, facilitating timely and effective interventions.

Compared to traditional monitoring methods, this IoT-enabled wearable nanosensor system offers superior performance by providing continuous, real-time monitoring of plant health. Its ability to detect early signs of disease allows farmers to optimize resource use and reduce the need for chemical interventions, thereby enhancing the sustainability of agricultural practices.

One of the standout features of this sensor is its potential to transform plant disease management. By providing early warnings of disease outbreaks, the sensor enables farmers to implement targeted interventions, reducing crop losses and improving yield quality. This capability is crucial for enhancing food security and promoting sustainable farming practices in the face of global challenges such as climate change and resource scarcity.

The detection mechanism is based on the unique properties of the integrated nanosensors, which facilitate the precise measurement of multiple parameters. The system's high sensitivity and specificity make it adaptable to various agricultural settings, providing reliable data for optimizing crop management.

The practical applicability of this sensor was demonstrated through successful field trials, where it accurately monitored plant health under different environmental conditions. This capability to perform comprehensive, real-time monitoring makes it a valuable tool for precision agriculture, enhancing resource efficiency and crop productivity.

Economically, the production of this wearable nanosensor system is promising. The materials used are cost-effective, and the fabrication process is scalable, making it feasible for widespread adoption across the agricultural sector. The sensor's integrated design further enhances its economic appeal by reducing the need for multiple devices and associated maintenance costs.

Beyond agriculture, the researchers envision potential applications of this technology in environmental monitoring and horticulture, where continuous, real-time data can provide valuable insights. The versatility and robustness of the wearable nanosensor make it adaptable for a wide range of applications, potentially transforming how plant health is monitored and managed.

10.4 Smart Systems for Circular Economy Applications

The intersection of nanotechnology and circular economy principles represents a frontier of innovation with immense potential to revolutionize sustainable resource management.[30] While many applications are still in development, nanotechnology offers unprecedented opportunities to enhance resource efficiency, reduce waste, and support the transition to a more circular economic model. By leveraging nanotechnology, we can make significant strides toward addressing global environmental challenges while creating new business opportunities and breakthroughs in sustainability.

Key areas where nanotechnology can advance circular economy goals include:

- **Enhancing resource efficiency**: Nanotechnology offers unprecedented precision in material tracking, sorting, and recycling, potentially dramatically improving resource utilization across industries.
- **Waste reduction**: Advanced nanosensors and nanomaterials could revolutionize waste management, enabling more effective recycling and resource recovery processes.
- **Enabling circular design**: Integrating nanotechnology into product design could facilitate easier disassembly, reuse, and recycling, supporting core circular economy principles.
- **Data-driven decision making**: Nanosensor networks can provide real-time, granular data on material flows and product lifecycles, informing more sustainable business models and policy decisions.
- **Catalyzing innovation**: The convergence of nanotechnology and circular economy thinking can spur new technological breakthroughs and business opportunities in sustainability.
- **Addressing global challenges**: This field has the potential to contribute significantly to addressing pressing environmental issues such as resource depletion, pollution, and climate change.

The integration of nanotechnology also raises critical ethical and environmental considerations. The widespread deployment of nanosensors, for example, could lead to privacy concerns, as constant tracking of product use and material flows may compromise individual and corporate confidentiality.

The presence of nanomaterials in products introduces uncertainties regarding their recyclability and potential environmental impacts. If nanomaterials are not adequately managed, they could interfere with current recycling processes or introduce harmful effects during degradation. Additionally, regulatory frameworks will need to adapt to ensure safe deployment and standardized oversight of nano-enabled products in recycling systems.

Given these challenges, the development of nanotechnology must adhere to principles of responsible innovation. This involves ensuring the long-term safety and sustainability of nanomaterials throughout their lifecycle, from production to disposal. Additionally, economic feasibility remains a key factor, as the adoption of these technologies requires significant investment in research, infrastructure, and cross-sector collaboration.

As we explore the potential of nanotechnology in circular economy applications, it's crucial to consider a set of key questions throughout the development and implementation process. Figure 10.3 illustrates these critical considerations, spanning from nanomaterial development to nanoproduct life-cycle management.

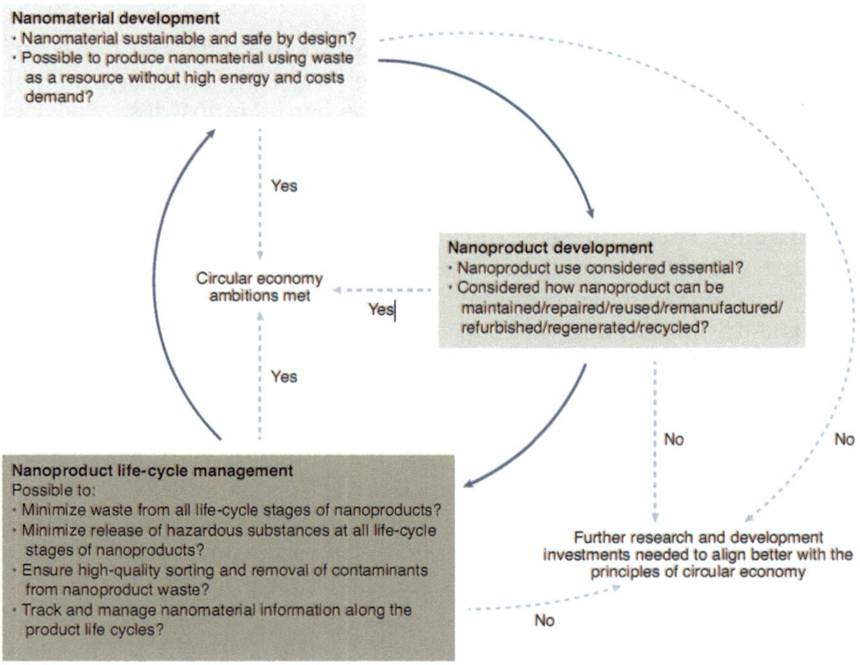

Figure 10.3 Critical questions to be addressed for nanomaterials in a circular economy from development, over nanoproduct development, to nanoproduct life-cycle management. Reproduced from ref. 30 with permission from Springer Nature, Copyright 2022.

This framework emphasizes the importance of considering sustainability, safety, and circularity at every stage of nanomaterial and nanoproduct development. It highlights the need for a holistic approach that addresses not only the technical aspects of nanotechnology but also its broader implications for resource efficiency, waste reduction, and environmental impact.

By examining the potential applications of nanotechnology in circular economy systems through this lens, we can anticipate future developments, identify research priorities, and prepare for the integration of these technologies into sustainable industrial practices. This forward-looking approach is essential for staying at the forefront of technological advancements in waste management and resource efficiency.

This chapter will examine both the promising advancements and the complex hurdles in applying nanotechnology to circular economy practices. By addressing these challenges head-on, we can harness the full potential of nanotechnology to build a more sustainable and resilient future.

10.4.1 Potential Applications of Nanotechnology

Despite the current lack of widespread application, nanotechnology holds significant potential to transform circular economy practices in the following ways:

10.4.1.1 Enhanced Material Tracking

10.4.1.1.1 Possibility. Embedding nanosensors in products could enable precise tracking of materials throughout their lifecycle. These sensors could provide real-time data on product usage, degradation, and recycling needs, optimizing lifecycle management and resource efficiency. For instance, nanosensors integrated into electronic devices could monitor battery health, component wear, and overall device performance, alerting users and manufacturers to optimal repair or recycling timelines.

10.4.1.1.2 Current Research. Recent advances in nanosensor technology have demonstrated significant potential in enhancing the performance and efficiency of photovoltaic cells.[31] By embedding nanosensors made from nanoparticles or nanowires within solar cells, researchers can monitor critical parameters such as temperature, light intensity, and voltage in real-time. This capability allows for precise control and optimization of the solar cells' efficiency, stability, and lifespan. The integration of nanosensors facilitates the early detection of potential issues, such as material defects or performance degradation, enabling proactive maintenance and troubleshooting. As nanosensor technology continues to evolve, its application could expand beyond energy systems to enable predictive maintenance across various industrial assets and operations, allowing for real-time monitoring and optimization of equipment performance.[32] This advancement would not only reduce downtime and maintenance costs but also enhance overall operational efficiency, supporting the goals of a circular economy by minimizing resource use and waste.

10.4.1.1.3 Future Potential. As technology advances, nanosensors could be integrated into a broader range of products, facilitating more efficient recycling and reducing waste. This could lead to the development of "smart products" that communicate their composition, condition, and recycling requirements to automated sorting systems. Additionally, these sensors could enable predictive maintenance in industrial equipment, extending product lifespans and reducing waste.

10.4.1.1.4 Challenges. Key hurdles include ensuring the durability of nanosensors throughout a product's lifecycle, addressing privacy concerns related to product tracking, and developing standardized communication protocols for nanosensor data.

10.4.1.2 Intelligent Sorting and Recycling

10.4.1.2.1 Possibility. Nanotechnology can significantly enhance recycling processes. Nanosensors have the potential to accurately identify and sort different materials in waste streams, ensuring effective recycling and reducing contamination. This could dramatically improve the efficiency and quality of recycled materials.

10.4.1.2.2 Current Applications. Some recycling facilities are already experimenting with near-infrared spectroscopy combined with AI for material identification.[33–35] Nanotechnology could take this a step further by enabling more precise material characterization at the molecular level.

10.4.1.2.3 Future Potential. The development of nanosensors that can distinguish between various polymers, metals, and composites could lead to more sophisticated recycling systems, improving material recovery rates. This could enable the separation of complex multi-material products that are currently challenging to recycle, such as multi-layer packaging or electronic components.

10.4.1.2.4 Current Research. The development of nanomaterials like reduced graphene oxide could lead to improving material recovery rates. Researchers have demonstrated that rGO can extract gold with high capacity at concentrations as low as 1 ppm.[36] The rGO-based method demonstrates good selectivity, extracting gold without significant uptake of other metals commonly found in e-waste. The extracted gold is reduced to its metallic form on the rGO surface, simplifying recovery. Nanomaterial-based approaches like these could address efficiency and selectivity challenges in current recycling processes.

10.4.1.2.5 Challenges. Scaling up nanosensor technology for high-volume waste processing and ensuring the cost-effectiveness of these advanced sorting systems are key areas that need addressing.

10.4.2 Challenges in Nanomaterial Management and Nanosensor Integration

While nanotechnology, particularly in the form of nanosensors, offers significant benefits for circular economy applications, it also presents challenges that need to be addressed. These challenges arise both from the use of nanosensors for monitoring and from the presence of nanomaterials in products and waste streams.[37]

10.4.2.1 Nanosensor Applications

The integration of nanosensors into products and recycling processes promises enhanced tracking and sorting capabilities. However, challenges include ensuring the durability of these sensors throughout a product's lifecycle, addressing potential interference with recycling processes, and managing the large volumes of data generated by these sensors.

10.4.2.2 Nanomaterials in Products and Waste Streams

Beyond nanosensors, the broader use of nanomaterials in products presents its own set of challenges:

- The presence of nanomaterials in certain waste streams may pose obstacles to high recycling rates and multiple material cycles.[38]
- Current recycling systems are mostly open loops, and the presence of nanomaterials in existing products could potentially impair the functionality of new products made from recycled materials.
- There is a lack of traceability for nanomaterials in recycled products, as little information is typically available about material properties in open-loop recycling operations.[39]

This knowledge gap hinders the effective management of nanomaterials throughout product lifecycles and in recycling processes. It also complicates the implementation of nanosensor-based tracking systems, as these systems need to account for the presence and behavior of nanomaterials in the products they're monitoring.

Addressing these challenges requires a holistic approach that considers both the potential of nanosensor technologies and the implications of nanomaterials in the circular economy. This includes developing methods to detect and manage nanomaterials in waste streams, creating protocols for the safe recycling of nano-enabled products, and ensuring that nanosensor technologies are designed with the full lifecycle of products in mind.

10.4.3 Recycling Rare Elements in Nanomaterial Production

While the challenges in nanomaterial management and nanosensor integration are significant, they also highlight opportunities for innovation in circular economy practices. One such opportunity, which addresses both resource efficiency and waste management concerns, is the recycling of rare elements used in nanomaterial production. This approach not only alleviates some of the challenges associated with nanomaterial waste but also opens up new avenues for sustainable nanomaterial synthesis.

Researchers have highlighted[40] this often-overlooked area of research, which has significant implications for resource efficiency and sustainable nanomaterial production: Commercially available nanomaterials include at least 51 rare elements, such as silver, gold, platinum group metals, and rare earth elements. These elements are generally more valuable than abundant elements, making their recycling economically attractive. Non-product outputs that may contain substantial amounts of rare elements include off-specification nanomaterials, sacrificed nano-templates, soot from electric-arc-based synthesis, and media remaining after synthesis and separation of nanomaterials.

Despite the potential benefits, research on recycling these rare elements is surprisingly scarce. Published studies on this topic cover only 6 out of the 51 rare elements used in commercial nanomaterials. For instance, in the case of gold nanoparticle synthesis, only about 0.01% of published papers address the recycling of non-product outputs.

However, the few studies that do exist show promising results. Examples include:

- Recycling scandium oxide from waste soot in the synthesis of metallic nitride fullerenes
- Recovering gold from waste nanoparticles for the synthesis of new gold nanoparticles
- Recycling silver from sacrificed templates used in the synthesis of other noble metal nanostructures
- Reusing tellurium from sacrificed nanowire templates

These recycling processes generally led to the synthesis of nanomaterials with acceptable functionality, demonstrating the feasibility of incorporating such practices into nanomaterial production.

The lack of research in this area presents both a challenge and an opportunity. There is a clear need for increased research efforts aimed at developing low-cost recycling methods for non-product outputs containing rare elements. Such research could significantly improve the economics and sustainability of nanomaterial production, aligning it more closely with circular economy principles.

10.4.4 Safe and Sustainable Design

To fully realize the potential of nanotechnology in circular economy applications, it's crucial to adopt a "safe and sustainable by design" approach. This involves considering the entire lifecycle of nanomaterials and nano-enabled products from the design phase onwards. Key considerations include:

- Minimizing the environmental impact of nanomaterial production and use
- Ensuring the recyclability or safe degradation of nano-enabled products
- Developing standardized methods for detecting and quantifying nanomaterials in complex waste streams
- Establishing regulatory frameworks for the use and disposal of nanomaterials

Research is ongoing to develop nanomaterials that are both highly functional and environmentally benign. This includes exploring biodegradable nanomaterials and designing nano-enabled products for easy disassembly and material recovery.

10.4.5 Economic and Practical Considerations

The implementation of nanotechnology in circular economy applications faces economic challenges. While the long-term benefits in terms of resource efficiency and waste reduction are significant, the initial investments in research, development, and infrastructure can be substantial. Balancing these short-term costs with long-term benefits is crucial for driving adoption across various sectors.

Moreover, the integration of nanotechnology into existing circular economy strategies requires collaboration across different industries and stakeholders. This includes developing standardized communication protocols for nanosensor data and establishing best practices for the use of nanomaterials in recyclable products.

10.4.6 Case Study: Graphene-based Method Improves Electronic Waste Recycling

Traditional methods for gold recovery from e-waste have relied heavily on energy-intensive and environmentally hazardous techniques, such as pyrometallurgy and hydrometallurgy. These processes often involve the use of toxic chemicals and generate significant pollution. Moreover, they frequently fail to recover gold efficiently, especially when dealing with low concentrations of the metal. This inefficiency has led to a situation where less than 20% of e-waste is currently recycled, resulting in a significant loss of valuable resources and potential environmental damage.

Researchers have explored various nanomaterials for gold extraction, including metal-organic frameworks, covalent organic polymers, and nanoporous materials. While these approaches have shown promise, they have been limited by factors such as low extraction capacity at concentrations relevant to e-waste, poor selectivity, and the need for additional processing steps to isolate the extracted gold. The ideal solution would combine high extraction capacity, especially for trace amounts of gold, with excellent selectivity and simplicity of use.

Against this backdrop, a team of researchers from Tsinghua University, the University of Manchester, and the Chinese Academy of Sciences has developed a new approach to gold extraction using reduced graphene oxide (rGO). Their work[36] demonstrates a method that addresses many of the limitations of previous gold extraction techniques and offers a potential breakthrough in e-waste recycling.

By carefully controlling the reduction process of graphene oxide, the researchers created a material with unique properties that make it exceptionally effective at extracting gold from complex solutions.

The team's experiments revealed that rGO exhibits an extraordinarily high capacity for gold extraction, even at extremely low concentrations. At room temperature (25 °C), one gram of rGO could extract 1850 milligrams of gold from a solution containing just 10 ppm of gold. This extraction capacity increased dramatically at higher temperatures, reaching 9059 milligrams per gram at 60 °C. Most remarkably, rGO showed significant extraction capability at concentrations as low as parts per trillion. For instance, the researchers observed notable gold extraction even at 10 parts per trillion, a level of sensitivity that far surpasses most existing methods and opens up possibilities for extracting gold from sources previously considered too dilute for economic recovery.

The extraction efficiency of rGO varies with the concentration of gold in the solution. While it shows exceptional performance at higher concentrations, its efficiency decreases in more dilute solutions, which is a common challenge in extraction processes. However, even at very low concentrations, rGO still demonstrates significant extraction capabilities, making it potentially useful for processing large volumes of dilute solutions.

Compared to other novel gold adsorbents, rGO demonstrates superior performance, especially at low concentrations. For example, at 1 ppm gold concentration, rGO showed an extraction capacity of 1180 $mg\,g^{-1}$, significantly outperforming other materials such as metal-organic frameworks or covalent organic polymers, which typically achieve less than 250 $mg\,g^{-1}$ at similar concentrations. Moreover, rGO's selectivity for gold in the presence of other metals is exceptional. Using pH control, the researchers achieved over 99% gold extraction with less than 3% extraction of other metals from complex e-waste leachates.

One of the key advantages of the rGO-based extraction method is its selectivity. The researchers demonstrated that by controlling the pH of the solution, they could extract gold with high specificity, even in the presence of 14 other elements typically found in e-waste (see Figure 10.4). This selectivity is crucial for practical applications, as it eliminates the need for complicated separation processes after extraction.

The mechanism behind this highly efficient and selective extraction process involves the unique structure of rGO. The material contains both graphene areas and oxidized regions. The graphene areas spontaneously reduce gold ions to metallic gold, while the oxidized regions allow for good dispersibility of the rGO in solution. This combination enables efficient adsorption and reduction of gold ions in the graphene areas.

Importantly, the researchers found that over 95% of the extracted gold was reduced to its metallic form directly on the rGO surface. This feature eliminates the need for additional steps to convert the extracted gold into a usable form, further simplifying the recycling process.

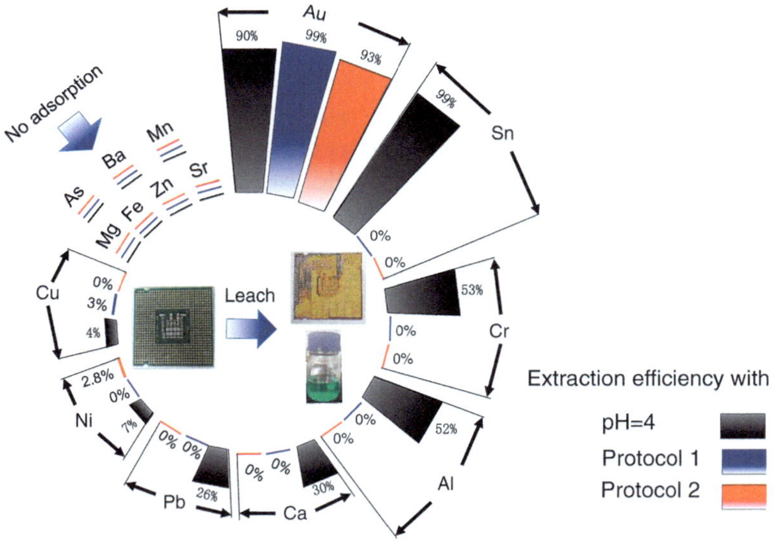

Figure 10.4 Comparison of ion selectivity using different extraction protocols. Protocol 1, performed at pH ∼ 1, achieves 99.3% gold extraction with less than 3% aluminum and copper co-extracted. Protocol 2, starting at pH ∼ 4, then reducing to ∼1, yields 93% gold extraction with minimal co-extraction of other metals (less than 2% for nickel). Both protocols demonstrate high gold selectivity from e-waste leachate. The photos inside the circle show the discarded, leached CPU and the resulting leachate. All the experiments were performed at 25 °C. Reproduced from ref. 36, https://doi.org/10.1038/s41467-022-32204-4, under the terms of the CC BY 4.0 license, https://creativecommons.org/licenses/by/4.0/.

To demonstrate the practical applicability of their method, the team developed a continuous flow-through process using rGO membranes. This approach allows for the efficient extraction of gold from large volumes of dilute solutions, making it potentially suitable for industrial-scale applications. The researchers successfully extracted gold from real-world e-waste leachates, achieving high efficiency and selectivity.

The economic viability of the process is promising. The researchers estimate that a 1 square meter rGO membrane, weighing less than 3 grams, could extract about 1.6 grams of gold from approximately 20 tons of a 100 ppb leachate.

The production of rGO itself is scalable and cost-effective. It involves the chemical reduction of commercially available, mass-produced graphene oxide using ascorbic acid as a reductant. This process is relatively simple and can be performed at mild temperatures, making it potentially suitable for large-scale industrial applications. The researchers also demonstrated the possibility of regenerating used rGO, further enhancing its economic attractiveness.

Beyond e-waste recycling, the researchers speculate that their method could have applications in other areas, such as extracting trace amounts of gold from seawater. While the concentration of gold in seawater is extremely low (estimated at less than 20 parts per trillion), the total amount of gold in the world's oceans is vast. In their experiments with simulated seawater containing 10 parts per trillion of gold, the researchers estimated complete gold extraction, suggesting potential for this seemingly far-fetched application.

The development of this rGO-based gold extraction method represents a significant step forward in addressing the global e-waste challenge. By combining high extraction capacity, excellent selectivity, and operational simplicity, it offers a promising approach to recovering valuable resources from electronic waste more efficiently and sustainably.

References

1. J. Kirchherr, D. Reike and M. Hekkert, Conceptualizing the circular economy: An analysis of 114 definitions, *Resour., Conserv. Recycl.*, 2017, **127**, 221–232.
2. S. O. Fakayode, C. Walgama, V. E. Fernand Narcisse and C. Grant, Electrochemical and colorimetric nanosensors for detection of heavy metal ions: A review, *Sensors*, 2023, **23**(22), 9080.
3. E. T. Athira, S. Dutta, M. M. Singh, N. Madaboosi, V. V. R. Sai and J. Satija, Recent progress in optical nanosensors for antibiotics detection, *Appl. Nanosci.*, 2023, **13**(9), 6519–6538.
4. Z. Wang, B. Yao, Y. Xiao, X. Tian and Y. Wang, Fluorescent quantum dots and its composites for highly sensitive detection of heavy metal ions and pesticide residues: A review, *Chemosensors*, 2023, **11**(7), 405.
5. C.-Y. Hsu, A. M. Rheima, M. S. Mohammed, M. M. Kadhim, S. H. Mohammed and F. H. Abbas, *et al.*, Application of carbon nanotubes

and graphene-based nanoadsorbents in water treatment, *Bionanoscience*, 2023, **13**(4), 1418–1436.
6. B. O. Alunda and Y. J. Lee, Review: Cantilever-based sensors for high-speed atomic force microscopy, *Sensors*, 2020, **20**(17), 4784.
7. D. Su, K. Wu, R. Saha, C. Peng and J.-P. Wang, Advances in magneto-resistive biosensors, *Micromachines*, 2019, **11**(1), 34.
8. L. Đačanin Far and M. D. Dramićanin, Luminescence thermometry with nanoparticles: A review, *Nanomaterials*, 2023, **13**(21), 2904.
9. *Nanosensor technologies for environmental monitoring*, ed. Inamuddin and A. M. Asiri, Springer Nature, Cham, Switzerland, 1st edn, 2021.
10. A. Munawar, Y. Ong, R. Schirhagl, M. A. Tahir, W. S. Khan and S. Z. Bajwa, Nanosensors for diagnosis with optical, electric and mechanical transducers, *RSC Adv.*, 2019, **9**(12), 6793–6803.
11. F. J. Tovar-Lopez, Recent progress in micro- and nanotechnology-enabled sensors for biomedical and environmental challenges, *Sensors*, 2023, **23**(12), 5406.
12. S. X. Leong, Y. X. Leong, C. S. L. Koh, E. X. Tan, L. B. T. Nguyen and J. R. T. Chen, *et al.*, Emerging nanosensor platforms and machine learning strategies toward rapid, point-of-need small-molecule metabolite detection and monitoring, *Chem. Sci.*, 2022, **13**(37), 11009–11029.
13. S. Ma, W. Zhao, Q. Zhang, K. Zhang, C. Liang and D. Wang, *et al.*, A portable microfluidic electrochemical sensing platform for rapid detection of hazardous metal Pb^{2+} based on thermocapillary convection using 3D Ag-rGO-f-$Ni(OH)_2$/NF as a signal amplifying element, *J. Hazard. Mater.*, 2023, **448**, 130923.
14. D. Yang, T. Wu, Y. Wang, W. Cao, H. Zhang and Z. Liu, *et al.*, A multi-parameter integrated sensor based on selectively filled D-shaped photonic crystal fiber, *Materials*, 2022, **15**(8), 2811.
15. J. Gu, Y. Shen, S. Tian, Z. Xue and X. Meng, Recent advances in nanowire-based wearable physical sensors, *Biosensors*, 2023, **13**(12), 1025.
16. Z. Zhe and A. Yuxiu, Nanotechnology for the oil and gas industry – an overview of recent progress, *Nanotechnol. Rev.*, 2018, **7**(4), 341–353.
17. M. N. Abd El-Ghany, R. A. Yahia and H. A. Fahmy, Nanosensors for agriculture, water, environment, and health, in *Handbook of Nanosensors*, Springer Nature Switzerland, Cham, 2024, pp. 1–29.
18. K. Jain, A. S. Patel, V. P. Pardhi and S. J. S. Flora, Nanotechnology in wastewater management: A new paradigm towards wastewater treatment, *Molecules*, 2021, **26**(6), 1797.
19. H. M. M. Abdel-Aziz and Y. M. Heikal, Nanosensors for the detection of fertilizers and other agricultural applications, in *Nanosensors for Environment, Food and Agriculture*, Springer International Publishing, Cham, 2021, vol. 1, pp. 157–168.
20. D. Yang, T. Wu, Y. Wang, W. Cao, H. Zhang and Z. Liu, *et al.*, A multi-parameter integrated sensor based on selectively filled D-shaped photonic crystal fiber, *Materials*, 2022, **15**(8), 2811.

21. K. Yang, D. Bi, Y. Deng, R. Zhang, M. M. U. Rahman and N. A. Ali, *et al.*, A comprehensive survey on hybrid communication in context of molecular communication and terahertz communication for body-centric nanonetworks, *IEEE Trans. Mol. Biol. Multi-Scale Commun.*, 2020, **6**(2), 107–133.
22. A. Rizwan, A. Zoha, R. Zhang, W. Ahmad, K. Arshad and N. Abu Ali, *et al.*, A review on the role of nano-communication in future healthcare systems: A big data analytics perspective, *IEEE Access*, 2018, **6**, 41903–41920.
23. M. Javaid, A. Haleem, R. P. Singh, S. Rab and R. Suman, Significance of sensors for industry 4.0: Roles, capabilities, and applications, *Sens. Int.*, 2021, **2**, 100110.
24. *Nanosensors for smart manufacturing*, ed. S. Thomas, T. A. Nguyen, M. Ahmadi, A. Farmani and G. Yasin, Elsevier Science Publishing, Philadelphia, PA, 2021.
25. A. Ucar, M. Karakose and N. Kırımça, Artificial intelligence for predictive maintenance applications: Key components, trustworthiness, and future trends, *Appl. Sci.*, 2024, **14**(2), 898.
26. C. L. Arellano Vidal and J. E. Govan, Machine learning techniques for improving nanosensors in agroenvironmental applications, *Agronomy*, 2024, **14**(2), 341.
27. D. Renard, R. Saddem, D. Annebicque and B. Riera, From sensors to digital twins toward an iterative approach for existing manufacturing systems, *Sensors*, 2024, **24**(5), 1434.
28. S. Abadal, C. Han, V. Petrov, L. Galluccio, I. F. Akyildiz and J. M. Jornet, Electromagnetic nanonetworks beyond 6G: From wearable and implantable networks to on-chip and quantum communication, *IEEE J. Sel. Areas Commun.*, 2024, **42**(8), 2122–2142.
29. G. Lee, O. Hossain, S. Jamalzadegan, Y. Liu, H. Wang and A. C. Saville, *et al.*, Abaxial leaf surface-mounted multimodal wearable sensor for continuous plant physiology monitoring, *Sci. Adv.*, 2023, **9**, eade2232.
30. S. F. Hansen, R. Arvidsson, M. B. Nielsen, O. F. H. Hansen, L. P. W. Clausen and A. Baun, *et al.*, Nanotechnology meets circular economy, *Nat. Nanotechnol.*, 2022, **17**(7), 682–685.
31. T. S. Dhahi, A. K. Yousif Dafhalla, O. E. Tayfour, A. Mubarakali, A. S. Alqahtani and A. E. Tayfour Ahmed, *et al.*, Advances in nano sensors for monitoring and optimal performance enhancement in photovoltaic cells, *iScience*, 2024, **27**(4), 109347.
32. S. X. Leong, Y. X. Leong, C. S. L. Koh, E. X. Tan, L. B. T. Nguyen and J. R. T. Chen, *et al.*, Emerging nanosensor platforms and machine learning strategies toward rapid, point-of-need small-molecule metabolite detection and monitoring, *Chem. Sci.*, 2022, **13**(37), 11009–11029.
33. J. Xia, Y. Huang, Q. Li, Y. Xiong and S. Min, Convolutional neural network with near-infrared spectroscopy for plastic discrimination, *Environ. Chem. Lett.*, 2021, **19**(5), 3547–3555.
34. Plasticmakers.org. [cited 2024 Aug 4]. Available from: https://plasticmakers.org/stories/polymer-scientist-recycling-innovation/.

35. Near-infrared spectroscopy in recycling, *NIRLAB – Handheld NIR analysis in the cloud*, NIRLAB, 2023, [cited 2024 Aug 4]. Available from: https://www.nirlab.com/blog/near-infrared-spectroscopy-in-recycling/.
36. F. Li, J. Zhu, P. Sun, M. Zhang, Z. Li and D. Xu, *et al.*, Highly efficient and selective extraction of gold by reduced graphene oxide, *Nat. Commun.*, 2022, **13**(1), 4472.
37. European Chemicals Agency, Z. Manžuch, R. Akelytė, R. Perez García, G. Kriščiūnaitė, *et al.*, *Study on the product lifecycles, waste recycling and the circular economy for nanomaterials – November 2021*, European Chemicals Agency, 2021.
38. Z. Zahra, Z. Habib, S. Hyun and M. Sajid, Nanowaste: Another future waste, its sources, release mechanism, and removal strategies in the environment, *Sustainability*, 2022, **14**(4), 2041.
39. L. Heggelund, S. F. Hansen, T. F. Astrup and A. Boldrin, Semi-quantitative analysis of solid waste flows from nano-enabled consumer products in Europe, Denmark and the United Kingdom – Abundance, distribution and management, *Waste Manag.*, 2016, **56**, 584–592.
40. L. Reijnders, Recycling of non-product outputs containing rare elements originating in nanomaterial syntheses, *Environ. Sci.: Nano*, 2024, **11**(3), 684–687.

CHAPTER 11

Nanotechnology for Environmental Protection and Resource Recovery

Building on the concepts introduced in Chapters 8 through 10, this chapter explores nanotechnology's pivotal role in advancing environmental protection and resource recovery. Previous chapters examined waste reduction through innovative materials, recycling and waste treatment processes, and the integration of advanced sensors and IoT for waste monitoring.

Emphasizing proactive environmental protection, Chapter 11 explores nanotechnology's potentially transformative impact through applications in water purification, air quality improvement, and energy systems. These solutions integrate insights from materials science, environmental engineering, and renewable energy to address complex environmental challenges.

The impact of these technologies varies by region—from providing clean water in water-scarce areas to mitigating pollution in industrialized regions. This chapter examines how these solutions can be adapted to different socioeconomic and environmental contexts, offering scalable approaches to sustainability.

By building on principles established earlier, this chapter presents a comprehensive view of nanotechnology's contribution to environmental stewardship and resource recovery. It extends beyond waste management to demonstrate the synergy between technological innovation and environmental protection in creating a more resilient world.

11.1 Water Purification and Desalination

As water scarcity and pollution become increasingly urgent global issues, traditional waste management practices are insufficient to meet these challenges. Nanotechnology offers promising solutions through advanced

Waste Not! How Nanotechnologies Can Increase Efficiencies Throughout Society
By Michael Berger

Published by the Royal Society of Chemistry, www.rsc.org

water purification and desalination technologies. This section introduces innovations like nanostructured membranes and photocatalytic nanomaterials, which enhance water treatment efficiency by leveraging the unique properties of nanomaterials.

Nanostructured membranes, with their high permeability and selectivity, allow for more effective removal of contaminants such as heavy metals, pathogens, and organic pollutants. Photocatalytic nanomaterials enable pollutant degradation through light-activated reactions, further improving water quality. These technologies not only contribute to cleaner water but also address global sustainability goals by reducing the environmental footprint of water treatment processes.

With millions of people affected by water scarcity and pollution worldwide, nanotechnology's role in water purification and desalination is crucial. These technologies offer scalable, innovative solutions that align with broader environmental sustainability efforts while improving access to clean water.

11.1.1 Advanced Membrane Technologies

Membrane-based treatment systems have become integral to water and wastewater purification processes, offering efficient separation of contaminants across various scales. Nanotechnology plays a pivotal role in enhancing membrane performance, leveraging the unique properties of nanomaterials to improve permeability, selectivity, and durability (see Figure 11.1). This section explores how these technologies extend beyond traditional waste treatment

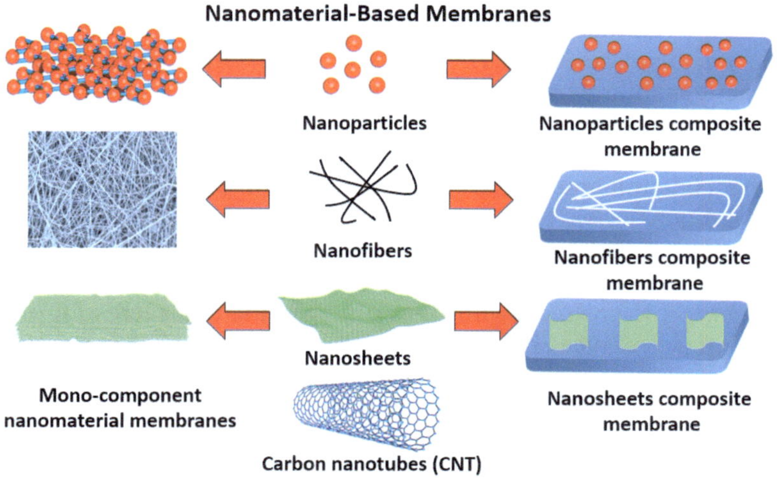

Figure 11.1 The most commonly used nanomaterial-based membrane structures. Reproduced from ref. 111, https://doi.org/10.3390/membranes11120995, under the terms of the CC BY 4.0 license, https://creativecommons.org/licenses/by/4.0/.

applications, offering groundbreaking solutions for environmental protection and resource recovery.

Previously, in Section 9.4.3, we discussed the application of nanotechnology in enhancing membrane-based systems for recycling and waste treatment. There, we focused on improving the efficiency of existing waste management processes using nanostructured membranes to facilitate the treatment of various waste streams.

Now we broaden our focus to emerging and advanced technologies that emphasize proactive environmental protection and resource recovery, highlighting the potential of cutting-edge membrane technologies to transform water purification and desalination beyond the scope of waste treatment.

11.1.1.1 Innovations in Nanostructured Membranes

Nanostructured membranes offer superior selectivity, permeability, and mechanical stability, paving the way for more efficient and sustainable solutions. Aquaporin-based biomimetic membranes leverage natural proteins to facilitate water transport with minimal energy input, while hybrid organic–inorganic membranes integrate the strengths of both material types to withstand harsh conditions. Advanced metal–organic framework membranes further expand the possibilities with their customizable porosity and surface properties, enabling targeted separation processes. These innovations exemplify the potential of nanostructured membranes to transform industries and contribute to global sustainability efforts.

11.1.1.1.1 Aquaporin-based Biomimetic Membranes. Inspired by nature, aquaporin-based membranes utilize proteins that facilitate water transport across cell membranes, offering high selectivity and permeability. These biomimetic membranes are designed to mimic natural processes, enabling efficient desalination and water purification with reduced energy consumption. Their unique structure allows for rapid water passage while blocking contaminants, making them ideal for sustainable water treatment solutions.[1]

11.1.1.1.2 Hybrid Organic–Inorganic Membranes. These membranes combine organic polymers with inorganic nanomaterials, creating a hybrid structure that enhances mechanical stability and chemical resistance.[2] For instance, zeolite nanoparticles can be incorporated into polymer matrices to improve ion selectivity and water flux. This approach is particularly beneficial for applications in harsh environments, where traditional membranes may degrade or lose efficiency.[3]

11.1.1.1.3 Advanced Metal–Organic Framework Membranes. MOF-based membranes offer exceptional selectivity and permeability, making them suitable for applications in gas separation, water purification, and resource recovery.[4] By adjusting the pore size and surface chemistry of MOFs, these membranes can be tailored to target specific contaminants, enhancing their versatility and effectiveness.[5]

11.1.1.2 Applications and Impact Beyond Waste Treatment

The applications and impact of advanced membrane technologies extend beyond traditional waste treatment, offering transformative solutions across various sectors. In zero-liquid discharge systems, these membranes are pivotal in achieving complete water recovery and resource reclamation, turning wastewater into a valuable asset while minimizing environmental harm. Decentralized water treatment systems utilize portable, high-performance membranes to bring clean water to remote or infrastructure-lacking areas, effectively addressing water scarcity and fostering community resilience. Additionally, in industrial settings, these membranes enable the recovery and reuse of valuable byproducts from wastewater, promoting resource efficiency and supporting a circular economy. These diverse applications highlight the profound potential of advanced membranes to drive sustainable practices across multiple domains.

11.1.1.2.1 Zero-liquid Discharge Systems. Advanced membrane technologies play a crucial role in zero-liquid discharge (ZLD) systems, which aim to eliminate wastewater discharge by recovering water and valuable resources. These systems utilize high-performance membranes to concentrate and recover salts, minerals, and other materials from wastewater, turning waste streams into valuable resources and minimizing environmental impact.[6]

11.1.1.2.2 Decentralized Water Treatment. In areas lacking centralized water infrastructure, advanced membranes enable the development of decentralized water treatment systems. These systems provide communities with access to clean water by leveraging portable and modular membrane technologies that can be deployed in remote locations. This approach not only addresses water scarcity but also empowers communities by providing sustainable water solutions.[7,8]

11.1.1.2.3 Industrial Wastewater Reuse. Industrial processes often generate large volumes of wastewater containing valuable byproducts. Advanced membranes facilitate the selective recovery of these byproducts, such as metals and chemicals, enabling their reuse within industrial processes. This not only reduces waste but also enhances resource efficiency, contributing to a circular economy.[9]

11.1.1.3 Future Directions

11.1.1.3.1 Energy-efficient Membrane Systems. Future research focuses on reducing the energy requirements of membrane systems through innovations such as low-pressure and forward osmosis technologies. By improving energy efficiency, these systems can further minimize the environmental footprint of water treatment processes.[10,11]

11.1.1.3.2 Smart Membranes with Adaptive Properties. Smart membranes that respond to environmental stimuli, such as temperature and pH changes, are being developed to optimize performance in real-time. These adaptive membranes can adjust their filtration properties based on water quality, enhancing efficiency and effectiveness.[12,13]

11.1.1.3.3 Integration with Renewable Energy Sources. Integrating advanced membrane systems with renewable energy sources, such as solar and wind power, offers a sustainable approach to water treatment. This synergy can reduce reliance on non-renewable energy, making water purification processes more environmentally friendly.[14,15]

The exploration of advanced membrane technologies highlights their transformative potential in water treatment, environmental protection, and resource recovery. While innovations like aquaporin-based membranes and metal–organic frameworks show promise in laboratory settings, many remain in the research phase and need further development before widespread use. Continued interdisciplinary collaboration between academia, industry, and government is crucial for overcoming the technical and regulatory challenges of bringing these technologies to market. As research progresses, these membrane technologies have the potential to address critical issues like water scarcity and environmental sustainability, driving more resource-efficient practices in the future.

11.1.2 Nanocatalysts for Pollutant Degradation

Building upon the foundational concepts of nanocatalysis discussed in Section 8.1, this section explores the latest innovations in nanocatalyst design specifically tailored for pollutant degradation. These advancements represent a significant leap forward in addressing environmental challenges through the active elimination of contaminants. By harnessing the unique properties of nanomaterials, these catalysts provide improved efficiency, selectivity, and sustainability in pollutant degradation processes.

In addition to the innovations previously discussed, recent advancements have also highlighted the transformative potential of metal–organic frameworks in pollutant degradation. These nanomaterials, particularly those that have undergone post-synthetic modifications, represent another significant leap forward in nanocatalysis. For instance, researchers have developed chemically tailored MOFs, such as TFA-MOF-808, that demonstrate exceptional efficacy in removing persistent pollutants like the PFAS "forever chemicals" from water sources. By enhancing the interactions between the MOF structure and the target pollutants through hydrophobic interactions and coordinative bonding, these modified MOFs achieve remarkable adsorption capacities, further expanding the toolkit available for effective environmental remediation.[16]

11.1.2.1 Advanced Photocatalytic Nanoparticles

Advanced photocatalytic nanoparticles are a promising area of development in nanocatalysis. While Section 8.1 introduced TiO_2 nanoparticles for photocatalysis, recent innovations have expanded the potential of this technology. Nitrogen-doped TiO_2 has been developed to enhance visible light absorption, extending the usable spectrum for pollutant degradation and increasing the efficiency of removing contaminants like dyes and pharmaceuticals from water sources. Similarly, composites such as bismuth tungstate (Bi_2WO_6) and graphitic carbon nitride (g-C_3N_4) have shown improved charge separation and visible light activity, providing a sustainable approach to water purification under sunlight.[17,18]

Plasmonic photocatalysts also represent a groundbreaking advancement. By coupling gold or silver nanoparticles with semiconductors, researchers can utilize localized surface plasmon resonance to enhance light absorption and charge separation, leading to increased degradation rates of persistent pollutants.[19] Au–TiO_2 core–shell nanostructures, for instance, significantly improve the degradation of pharmaceutical pollutants under visible light by leveraging the plasmonic effect.[20,21]

11.1.2.2 Bimetallic Nanocatalysts

Bimetallic nanocatalysts expand on nanostructured catalysts by offering synergistic effects that enhance catalytic performance. Palladium–gold nanoparticles provide enhanced activity for the hydrodehalogenation of chlorinated pollutants, allowing for tunable surface compositions that optimize catalytic activity and selectivity.[22]

Silver–nickel nanoparticles encapsulated in smart polymer microgels offer a compelling approach for the catalytic degradation of azo dyes, expanding on the concept of bimetallic nanocatalysts. The combination of silver and nickel provides enhanced activity for breaking down these prevalent industrial pollutants. This system allows for precise control over the surface interactions within the polymer matrix, optimizing both catalytic activity and selectivity. The innovative use of smart polymer microgels not only supports the nanoparticles but also enhances their stability and effectiveness, showcasing the potential of bimetallic nanocatalysts in diverse environmental remediation applications.[23]

11.1.2.3 Magnetic Nanocatalysts

Magnetic nanocatalysts provide unique advantages in pollutant degradation processes. Building on the stability improvements mentioned in Section 8.1, these catalysts combine a magnetic core with a catalytic shell, such as in

Fe_3O_4-based core–shell structures. This design enables easy recovery of the catalyst after use, making them particularly suitable for the catalytic reduction of nitroaromatic compounds[24] and degradation of azo dyes.[25]

Magnetically separable metal–organic framework composites, such as Fe_3O_4@UiO-66, integrate magnetic nanoparticles with MOFs to enhance catalytic activity and facilitate the recovery and reuse of catalysts. These composites are highly effective in degrading organic dyes[26] and pharmaceutical pollutants.[27]

11.1.2.4 2D Nanomaterial-based Catalysts

Beyond traditional nanoparticles, 2D nanomaterials offer unique catalytic properties that make them highly effective for pollutant degradation. Graphene-based nanocatalysts, including graphene oxide and reduced graphene oxide, serve as excellent supports for metal nanoparticles, enhancing electron transfer and increasing active site exposure. This improves the efficiency of pollutant degradation processes.[28,29]

Similarly, MXenes, a family of 2D transition metal carbides and nitrides, show promise in catalytic oxidation. For example, $Ti_3C_2T_x$ MXene is effective in generating reactive oxygen species for the catalytic degradation of organic pollutants, and its tunable surface chemistry allows for selective pollutant adsorption and degradation.[30,31]

11.1.2.5 Enzyme-mimicking Nanocatalysts

Enzyme-mimicking nanocatalysts, or nanozymes, represent a novel approach inspired by biological systems. Iron oxide nanoparticles exhibit peroxidase-like activity that catalyzes the decomposition of hydrogen peroxide to generate hydroxyl radicals, which are highly effective for pollutant degradation.[32] This approach is particularly useful for removing phenolic compounds and pharmaceutical residues from water.[33]

Cerium oxide nanozymes offer multienzyme-mimicking properties, including superoxide dismutase, catalase, and peroxidase-like activities, making them effective for degrading organic dyes and phenolic compounds.[34,35]

11.1.2.6 Comparative Analysis of Nanocatalyst Types

Each type of nanocatalyst offers distinct advantages and potential drawbacks, making them suitable for different applications based on specific needs:

- **Photocatalytic nanoparticles:** These are highly effective for applications requiring light-driven processes. Their ability to utilize sunlight makes them energy-efficient, but their performance can be limited by light availability and the need for specific light wavelengths.
- **Bimetallic nanocatalysts:** Offering enhanced catalytic activity through synergistic effects, these catalysts are ideal for complex pollutant

degradation processes. However, their synthesis can be more complex and costly compared to monometallic catalysts.

- **Magnetic nanocatalysts:** These catalysts provide easy recovery and reuse, reducing waste and operational costs. However, the magnetic properties may sometimes interfere with catalytic performance, requiring careful design.
- **2D nanomaterials:** With their large surface area and high electron mobility, these materials are excellent for fast and efficient catalysis. Their potential limitations include scalability and production costs.
- **Nanozymes:** Mimicking natural enzymes, nanozymes offer a biologically inspired approach that can operate under mild conditions. Their activity may be less specific than natural enzymes, but they offer robustness and longevity in harsh conditions.

Nanocatalyst design advances bridge materials science and environmental engineering, enabling the development of tailored nanostructures for efficient pollution control and resource recovery systems. These innovations leverage unique nanomaterial properties and biomimetic approaches to enhance pollutant degradation efficiency and selectivity. As the technology matures, these catalysts could address diverse global pollution challenges, advancing both water treatment and environmental protection.

Future research and development efforts will focus on overcoming challenges related to scaling these technologies for industrial applications and reducing production costs. Additionally, smart, adaptive catalysts that can respond dynamically to varying pollution levels will enhance the applicability of these technologies across different environmental scenarios.

11.1.2.7 Case Study: Advanced MOFs for the Removal of Forever Chemicals

Building upon the foundational concepts of nanocatalysis discussed earlier, this section explores the latest innovations in nanocatalyst design specifically tailored for pollutant degradation. These advancements represent a significant leap forward in addressing environmental challenges through the active elimination of contaminants. By harnessing the unique properties of nanomaterials, these catalysts provide improved efficiency, selectivity, and sustainability in pollutant degradation processes.

Recent research has highlighted the potential of metal–organic frameworks as highly effective nanocatalysts for pollutant degradation, particularly in addressing persistent contaminants like per- and polyfluoroalkyl substances (PFAS). MOFs, such as the chemically tailored TFA-MOF-808, have demonstrated exceptional adsorption capacities due to their high porosity and the ability to undergo post-synthetic modifications. These modifications enhance the interactions between the MOFs and the pollutants, promoting efficient capture through mechanisms like hydrophobic interactions and coordinative bonding. The development of these advanced MOFs represents a promising approach in the field of environmental remediation, offering a new avenue for the effective removal of both short- and long-chain pollutants from water sources.

In the context of nanotechnology for environmental protection and resource recovery, the application of advanced MOFs stands out as a compelling example. The TFA-MOF-808, a chemically tailored MOF, has been developed to address the persistent issue of PFAS contamination in water. The research behind this innovation focused on synthesizing and testing various MOFs, specifically DUT-67 and MOF-808, which were selected for their high water stability, suitable porosity, and ease of synthesis through green methodologies.[16]

The study explored the functionalization of these MOFs by incorporating fluorinated ligands, specifically 2-fluorobenzoic acid (FBA), 2,6-difluorobenzoic acid (DFBA), and trifluoroacetic acid (TFA), into the MOF structures through a post-synthetic modification process. This functionalization was critical in enhancing the hydrophobicity and pollutant binding capabilities of the MOFs. Among the tested materials, TFA-MOF-808 stood out due to its remarkable ability to adsorb both short- and long-chain PFAS, such as perfluorooctanoic acid (PFOA) and perfluorobutanoic acid (PFBA). TFA-MOF-808 achieved an adsorption capacity of up to 2496 mg of PFOA per gram of MOF, significantly outperforming other materials, including its unmodified counterparts and traditional adsorbents like activated carbon and ion-exchange resins.

The practicality of using TFA-MOF-808 for large-scale applications was also addressed through regeneration and recyclability studies. The researchers demonstrated that the material could be regenerated using a simple ligand exchange process, where TFA-MOF-808, after PFAS capture, was reintroduced into a solution containing excess TFA. This process efficiently reversed the ligand exchange that had occurred during pollutant adsorption, restoring the material's functionality without loss of crystallinity or stability. TFA-MOF-808 maintained its full adsorption capacity over multiple cycles, highlighting its potential for repeated use in real-world applications.

Mechanistic investigations provided insights into the PFAS adsorption process. The study revealed that the adsorption mechanism was driven by a combination of hydrophobic interactions and coordinative bonding. Specifically, the fluorinated ligands within TFA-MOF-808 enhanced the hydrophobic environment inside the MOF pores, facilitating the strong interaction with PFAS molecules (see Figure 11.2). Additionally, the PFAS were found to replace the non-structural ligands in the MOF, forming coordinative bonds with the Zr_6 nodes of the framework. This ligand exchange process was a key factor in the material's high adsorption capacity and selectivity for PFAS.

When compared to conventional PFAS removal methods, such as activated carbon filtration and ion-exchange resins, TFA-MOF-808 offers several advantages. Conventional methods often suffer from limited adsorption capacities, slow kinetics, and reduced effectiveness for short-chain PFAS. In

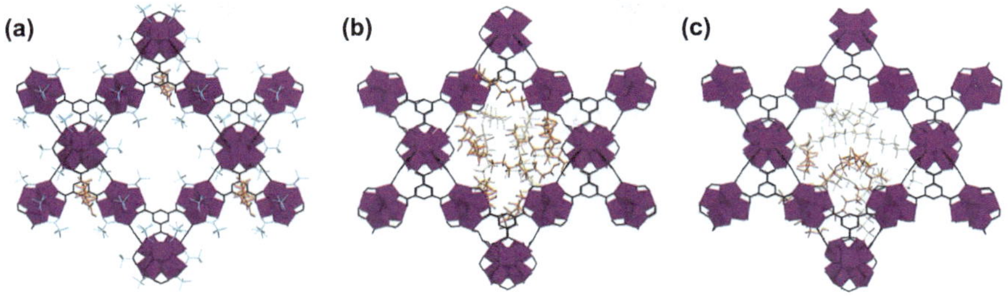

Figure 11.2 Grand canonical Monte Carlo-simulated snapshots of (a) 4TFA-MOF-808, (b) 2PFOA-MOF-808, and (c) 3PFOA-MOF-808 at 298 K and 0.1 Pa. Color scheme: Zr – purple polyhedra, main organic linkers – black, TFA – turquoise. PFOA molecules physically adsorbed in the pores are shown in orange color, and PFOA molecules chemically adsorbed in the cluster nodes are shown in light yellow color. Reproduced from ref. 16 with permission from John Wiley & Sons, Copyright © 2024 The Authors. Advanced Functional Materials published by Wiley-VCH GmbH.

contrast, TFA-MOF-808 not only provides superior adsorption capacities but also captures a broader range of PFAS molecules more efficiently. Additionally, the ability to regenerate the MOF through a straightforward process further enhances its practicality and cost-effectiveness for long-term use in water treatment applications.

This work underscores the significant potential of chemically tailored MOFs, such as TFA-MOF-808, in advancing nanotechnology for environmental protection and resource recovery. By integrating sophisticated design strategies and understanding the underlying mechanisms of pollutant adsorption, researchers are developing next-generation materials that offer substantial improvements over conventional methods. As these technologies evolve, they are poised to play a critical role in addressing global challenges related to water purification and environmental sustainability.

11.2 Air Purification and Greenhouse Gas Reduction

This section explores the potential applications of nanotechnology in addressing air pollution and atmospheric remediation, building upon the foundational concepts of environmental protection introduced earlier in this chapter. It's important to note that while nanotechnology shows promise in this field, many of the technologies discussed are still in developmental stages and their large-scale effectiveness and safety require further research.

11.2.1 Nanostructured Materials for Air Filtration

Nanostructured materials are being investigated for their potential to enhance air filtration systems. These materials generally offer high surface area-to-volume ratios, which could theoretically improve filtration efficiency.

11.2.1.1 Nanofiber Membranes

Researchers are exploring the use of nanofiber membranes in air filtration. These membranes are typically made from polymers and are characterized by their small fiber diameters. The potential benefits of these membranes include high porosity and potentially improved particle capture rates, particularly for fine particulate matter. They can also be designed with complex structures such as core-sheath and Janus structures to enhance their filtration performance.[36]

Furthermore, these membranes can incorporate additional functionalities like antibacterial and flame-retardant properties, making them versatile for various applications. However, the long-term durability and cost-effectiveness of these membranes in real-world applications require further study.

To ensure these membranes meet practical requirements, ongoing research focuses on optimizing fabrication methods like electrospinning and developing composite and hybrid nanofiber structures. These efforts aim to balance the high performance of nanofiber membranes with the economic feasibility and scalability necessary for widespread adoption.[37]

11.2.1.2 Carbon Nanotube Filters

Carbon nanotubes are being investigated for their potential in air filtration due to their unique structural properties, including high surface area, excellent mechanical strength, and unique electronic properties, which allow them to efficiently capture various airborne pollutants. Recent studies have shown that carbon nanotube-based filters can achieve filtration efficiencies exceeding 90% for particulate matter, particularly for particle sizes ranging from 0.3 to 2.5 micrometers. These filters are lightweight and flexible, making them suitable for a variety of applications, including wearable air filters.[38]

However, challenges remain in scaling up the production of carbon nanotubes and integrating them into existing filtration systems. The high cost and complexity of producing high-quality carbon nanotubes present significant barriers. Additionally, there are concerns about the environmental and health impacts of carbon nanotubes throughout their lifecycle. Ensuring the safety and sustainability of these materials is crucial.[39]

Research is ongoing to address these challenges, with efforts focused on improving the synthesis methods for carbon nanotubes to reduce costs and increase production efficiency. As these advancements continue, carbon nanotubes hold promise for transforming air filtration technology, offering enhanced performance and safety.[40]

11.2.1.3 Metal–Organic Framework Based Filters

MOFs have gained attention not only for water purification[41] but also for their potential in air filtration applications. Their highly porous nature allows for a large surface area, making them ideal for capturing gaseous pollutants such as CO_2, volatile organic compounds, and other harmful gases. The customizable chemistry of MOFs enables the selective adsorption of specific gases, which is crucial for effective air filtration.[42]

However, the practical application of MOFs in air filtration systems faces challenges. One major concern is their stability under various environmental conditions, such as humidity and temperature fluctuations, which can affect their performance and lifespan. Additionally, while MOFs offer excellent selectivity and adsorption capacity, their scalability for industrial applications requires further research to optimize synthesis methods and reduce production costs.

Current research emphasizes the need for developing MOFs with enhanced stability and exploring scalable production techniques. This includes creating composites with polymers or other materials to improve mechanical strength and environmental resilience. As these challenges are addressed, MOFs hold promise as effective materials for air purification, contributing to cleaner and healthier air quality.[43]

11.2.2 Photocatalytic Air Purification

Photocatalytic materials are gaining attention for their ability to degrade pollutants through light-activated chemical reactions, offering promising applications in air purification. These materials, particularly nanoscale titanium dioxide, are well-studied for their photocatalytic properties, which enable them to break down pollutants such as volatile organic compounds and nitrogen oxides.[44] However, while this technology shows potential, its effectiveness under real-world conditions, especially in environments with low light intensities, requires further investigation.

Studies have highlighted several advancements and challenges associated with TiO_2-based photocatalytic materials. One major area of research focuses on enhancing the efficiency of TiO_2, particularly under visible light, which is more prevalent in real-world scenarios compared to ultraviolet light.[45] Researchers are exploring techniques such as doping TiO_2 with other elements and developing composite materials to improve its ability to absorb visible light and increase its photocatalytic activity.[46]

Despite these advancements, the practical application of TiO_2 for large-scale air purification faces significant challenges. The efficiency of these materials can vary greatly depending on environmental factors like humidity, pollutant concentration, and light intensity. Ensuring long-term stability and maintaining high photocatalytic efficiency in varied conditions are critical obstacles that researchers are striving to overcome.

The development of TiO_2-based building materials is a promising direction for integrating photocatalytic properties into urban environments.[47] These materials aim to provide sustainable solutions for air quality improvement, aligning with goals for energy-saving and emission-reduction in modern cities. By incorporating photocatalytic technology into construction materials, buildings can actively contribute to reducing air pollution, offering a dual function of structural integrity and environmental remediation.

The potential environmental and health benefits of photocatalytic technology are substantial. It offers a way to mitigate air pollution without significant energy consumption or secondary pollution, making it a sustainable option for improving air quality.

11.2.3 Nanomaterial-based Sensors for Air Quality Monitoring

Nanotechnology offers unique opportunities for developing highly sensitive and compact sensors for detecting and measuring air pollutants. These sensors leverage the properties of nanomaterials to achieve high sensitivity and potentially fast response times. Various nanomaterials, including metal oxides, carbon-based materials, and composite structures, are being investigated for their gas-sensing capabilities.

11.2.3.1 Advantages of Nanomaterial-based Air Quality Sensors

The primary advantages of nanomaterial-based air quality sensors include their high surface area-to-volume ratio, allowing for increased interaction with gas molecules; potential for miniaturization, enabling the development of portable and wearable air quality monitors; lower power consumption compared to traditional gas sensors; and the possibility of room-temperature operation, reducing energy requirements.[48,49]

Compared to traditional air quality sensors, which often rely on bulky and less sensitive components, nanomaterial-based sensors offer a significant improvement in terms of sensitivity and selectivity, enabling the detection of lower concentrations of pollutants. Traditional sensors, like electrochemical and metal oxide sensors, can suffer from issues like cross-sensitivity and slower response times, which nanomaterial-based sensors can potentially overcome through the use of advanced materials and surface modifications.[50]

11.2.3.2 Types and Applications of Nanomaterial Gas Sensors

Graphene-based sensors possess exceptional sensitivity to gases such as nitrogen dioxide and ammonia, primarily due to graphene's high electron mobility and extensive surface area.[51] These properties allow graphene to detect even trace amounts of gases, making it highly effective for environmental monitoring and industrial applications. Advances in graphene sensor technology, such as functionalization with metallic nanoparticles and the integration of graphene with other materials, have further enhanced their sensitivity and selectivity.[52] These improvements have expanded the applications of graphene sensors, enabling their use in various fields, including healthcare, where they can detect ammonia as a biomarker for certain medical conditions.

Similarly, studies have demonstrated that sensors utilizing zinc oxide nanowires are highly effective for detecting hydrogen gas due to their remarkable sensitivity and selectivity. The unique structural and electronic properties of zinc oxide, such as high surface area and electron mobility, make it an excellent material for gas sensing applications.[53] Various strategies, including surface modifications and additive doping with materials like noble metals and carbon-based compounds, have been employed to enhance the performance of ZnO nanowire sensors. These modifications improve sensitivity and enable operation at lower temperatures, which is crucial for real-time air quality monitoring.[54] Additionally, the development of composite materials and advanced fabrication techniques has further increased the sensors' efficiency, making zinc oxide nanowires a promising candidate for integration into portable and cost-effective air quality monitoring systems.

The development of hybrid sensors that combine carbon nanotubes with metal oxides, such as tin oxide (SnO_2), has significantly improved sensitivity and selectivity in gas detection. These hybrid sensors leverage the high surface area and excellent conductivity of CNTs along with the high reactivity of metal oxides to achieve enhanced performance compared to single-material sensors.[55] By integrating CNTs with metal oxides, these sensors benefit from synergistic effects that increase gas molecule interactions and improve electron transport, leading to more accurate and sensitive gas detection.[56] This makes them highly effective for applications in environmental monitoring and industrial safety.

11.2.3.3 Challenges and Limitations

Despite the promising results in laboratory settings, the transition of nanomaterial-based air quality sensors to practical, real-world applications faces several challenges.

- **Selectivity** remains a significant issue, as many nanomaterial sensors respond to multiple gas species, making it difficult to accurately identify specific pollutants in complex urban or industrial environments.[57,58]
- Ensuring **consistent sensor performance** over extended periods, especially when exposed to varying environmental conditions, is another crucial challenge.
- Developing **reliable calibration methods** for nanomaterial-based sensors and establishing industry standards for their use in air quality monitoring is essential for widespread adoption.
- **Interference** from environmental factors such as humidity, temperature fluctuations, and the presence of other airborne substances can affect sensor performance, necessitating the development of compensation techniques.

- Additionally, translating laboratory-scale fabrication processes to **large-scale, cost-effective production methods** while maintaining sensor quality and consistency is an ongoing challenge.

11.2.3.4 Future Directions and Potential Impact

Advances continue pushing the boundaries of what nanomaterial-based sensors can achieve. One such development is the integration of metal–organic frameworks with nanostructured sensors. MOFs, known for their high porosity and tunable chemistry, can enhance the selectivity and sensitivity of sensors, enabling the detection of specific gases even in the presence of complex mixtures.[59,60]

Another promising area is the use of surface plasmon resonance techniques, which can further enhance the sensitivity of nanosensors by exploiting the interaction of light with metal nanoparticles, allowing for real-time monitoring of pollutants at very low concentrations.[61]

Nanomaterial-based sensors for air quality monitoring have significant potential to contribute to waste avoidance, management, and resource efficiency across various sectors. These advanced sensors can play a crucial role in optimizing industrial processes, reducing emissions, and improving overall environmental management strategies. In the context of waste avoidance and management, these sensors can provide real-time data on air pollutants generated during manufacturing processes or waste treatment operations. This immediate feedback allows industries to quickly identify and address sources of pollution, potentially preventing the release of harmful substances into the environment. By detecting even trace amounts of pollutants, these sensors can help industries fine-tune their processes to minimize waste generation at the source.

For resource efficiency, nanomaterial-based air quality sensors can contribute in several ways. In energy-intensive industries, these sensors can help monitor combustion processes in great detail, allowing for optimized fuel consumption and reduced emissions. This not only conserves resources but also improves overall energy efficiency. In waste management facilities, these sensors can monitor the effectiveness of air filtration systems, ensuring that resources invested in pollution control are being used efficiently. Moreover, the ability of these sensors to provide detailed, localized air quality data can inform urban planning and resource allocation decisions. Cities can use this information to optimize traffic flow, reduce vehicle emissions, and improve public transportation systems, all of which contribute to more efficient use of energy resources.

In the broader perspective of circular economy principles, these sensors can aid in creating closed-loop systems by monitoring air quality throughout the product lifecycle. From raw material extraction to manufacturing, use, and eventual recycling or disposal, these sensors can help identify opportunities for reducing waste and improving resource utilization at each stage. As research progresses and these technologies continue to develop, nanomaterial-based air quality sensors hold the promise of transforming how we monitor and manage air quality, contributing to a more sustainable and resource-efficient future.

11.2.4 Nanotechnology Solutions for Greenhouse Gas Capture

Addressing the rising levels of greenhouse gases in the atmosphere is a critical component of combating climate change. Traditional methods of GHG capture, such as chemical absorption and physical adsorption, face challenges related to efficiency, cost, and scalability. Nanotechnology offers promising alternatives by enabling the development of advanced materials that can effectively capture and store greenhouse gases, such as carbon dioxide and methane.

This section explores the role of nanotechnology in enhancing the capture of greenhouse gases through the use of novel nanomaterials. Metal–organic frameworks, carbon nanotubes, and nanoporous polymers are among the key materials investigated for their high surface areas, tunable pore structures, and selective adsorption capabilities. These properties make them highly effective for trapping GHGs under various conditions, potentially improving the efficiency and feasibility of large-scale deployment.

Advancements in nanocatalysts facilitate the conversion of captured greenhouse gases into useful chemicals and fuels, thus integrating capture with resource recovery. This approach not only mitigates GHG emissions but also contributes to a circular carbon economy by transforming waste gases into valuable products.

Despite their potential, the widespread application of nanotechnology solutions for GHG capture faces challenges, including material stability, regeneration efficiency, and environmental impact. Research efforts are focused on developing more robust and sustainable nanomaterials, optimizing their performance, and evaluating their long-term effects on the environment.

11.2.4.1 Recent Innovations in Nanomaterial Sensors for Air Quality

Recent innovations in nanomaterial sensors are paving the way for future advancements in air quality monitoring by improving multi-gas detection systems in laboratory settings, with the potential for simultaneous identification of various air pollutants with enhanced accuracy and efficiency.

11.2.4.1.1 Multi-gas Detection Systems. Recent advancements in multi-gas detection systems have leveraged nanomaterial-based sensor arrays to simultaneously detect various air pollutants, offering a more comprehensive approach to air quality monitoring. Key developments include:

- **Nanocomposite arrays**: These systems combine metal oxide nanoparticles, such as SnO_2, ZnO, and WO_3, with different dopants to create unique response patterns for complex gas mixtures. This enhances the selectivity and sensitivity of sensors, allowing them to accurately identify pollutants in urban environments.[62]
- **Graphene-based sensor arrays**: By utilizing functionalized graphene integrated with metal nanoparticles, these sensors improve detection capabilities for a range of gases, benefiting from graphene's high electron mobility and surface area.[63,64]
- **Machine learning integration**: Advanced artificial intelligence algorithms are employed to analyze data from nanosensor arrays, improving accuracy in pollutant identification within complex atmospheric conditions. This integration helps in distinguishing between similar gases, reducing false positives.[65]

These systems significantly enhance the ability to monitor air quality in real-time, providing detailed insights into pollutant interactions and concentrations. For example, cities implementing these systems can track pollution sources and patterns, informing more effective environmental regulations and urban planning.

Recent research efforts continue to explore innovative strategies to further improve nanosensor platforms for air quality monitoring, focusing on enhancing sensitivity, multiplex detection, and handling complex datasets. Figure 11.3 outlines some of the cutting-edge approaches currently being developed to achieve these goals.

11.2.4.1.2 Self-powered Nanosensors. Self-powered nanosensors have the potential to revolutionize autonomous air quality monitoring systems in the future by minimizing the need for external power sources and reducing maintenance requirements. Key areas of development include:

- **Triboelectric nanogenerators**: These devices harness static electricity generated from the friction between nanomaterials, potentially using the movement of air itself as an energy source to power sensors.[66,67]
- **Photovoltaic integration**: Ultra-thin solar cells are combined with nanomaterial sensors to create self-sustaining devices that can operate continuously under sunlight, ensuring uninterrupted monitoring.[68]
- **Thermoelectric nanosystems**: By utilizing temperature differentials and nanomaterials with high thermoelectric properties, these systems generate power for sensors, particularly useful in industrial settings with significant heat fluctuations.[69]
- **Energy harvesting and storage**: Nanomaterial-based supercapacitors or thin-film batteries are developed to store energy collected from the environment, ensuring continuous operation of sensors even when power generation is variable.[70,71]

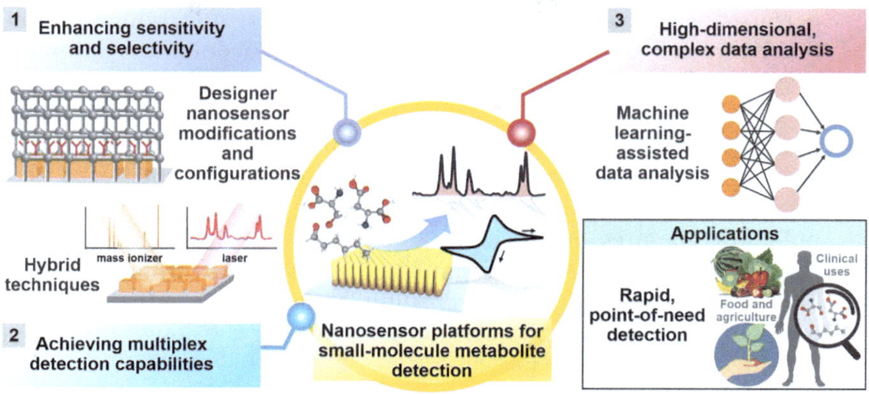

Figure 11.3 Current and emerging research strategies to enhance sensitivity and selectivity, achieve multiplex detection capabilities, and facilitate analysis of complex, high-dimensional datasets for pollutant and small-molecule detection using various nanosensor platforms across diverse applications. These strategies are broadly categorized into (1) customization of platform modifications and designer platform configurations, (2) development of hybrid techniques involving two or more analytical methods, and (3) complementary use of machine learning algorithms for data analysis. Reproduced from ref. 65 with permission from the Royal Society of Chemistry.

These self-powered systems could revolutionize air quality monitoring networks, enabling long-term deployment in diverse environments without frequent maintenance or battery replacement. For example, remote areas with limited infrastructure can benefit from deploying these sensors to provide critical air quality data for research and public health initiatives.

Each of these innovations addresses key challenges in air quality monitoring, such as the need for comprehensive pollutant detection, personal exposure assessment, and long-term, widespread sensor deployment. As these technologies continue to develop, they promise to provide more accurate, accessible, and sustainable methods for monitoring and managing air quality, with significant implications for public health and environmental policy.

11.2.4.2 Challenges Specific to Air Quality Monitoring

The application of nanotechnology in air quality monitoring brings with it specific challenges that must be addressed to ensure effective deployment and operation:

- **Calibration in complex atmospheres**: The development of reliable calibration methods for nanomaterial-based sensors in real-world atmospheric conditions remains challenging. Varying humidity levels and the presence of interfering gases can significantly affect sensor performance. Advances in materials like metal–organic frameworks and functionalized graphene are being explored to enhance calibration accuracy by providing more consistent response patterns across diverse conditions.

- **Long-term stability in outdoor environments:** Ensuring consistent sensor performance over extended periods, especially when exposed to varying weather conditions, temperature fluctuations, and complex air pollutant mixtures, is crucial for reliable air quality monitoring. The robustness and durability of nanomaterials such as carbon nanotubes and nanoporous polymers are being enhanced to withstand these environmental stressors, potentially increasing the longevity and reliability of monitoring systems.
- **Data interpretation and integration:** As nanomaterial sensors become more widespread, the need for advanced data processing algorithms to interpret the large volumes of data generated is growing. The integration of machine learning techniques with nanosensor arrays is a promising approach to improve accuracy in pollutant identification, allowing for more effective management of air quality data. This technological advancement also aids in distinguishing between similar gases and reducing false positives, thereby enhancing the precision of air quality assessments.

11.2.4.3 Future Directions and Challenges

As research in nanotechnology for air pollution control and atmospheric remediation continues, several key areas require attention:

- **Scalability and cost-effectiveness:** Transitioning from laboratory-scale results to practical, large-scale applications remains a significant challenge. Developing cost-effective production methods for nanomaterials is essential to make these technologies economically viable for widespread use.
- **Long-term stability and performance:** Ensuring that nanomaterials maintain their performance over extended periods in real-world conditions is crucial. Research efforts are focusing on enhancing the durability and resilience of materials like metal–organic frameworks and carbon nanotubes to withstand environmental stressors.
- **Safety and environmental impact:** Comprehensive studies on the potential effects of nanomaterials throughout their lifecycle are necessary. Assessing the environmental and health impacts of nanomaterials will ensure their safe and sustainable deployment.
- **Integration with existing systems:** Developing methods to incorporate nanotechnology-based solutions into current air quality management infrastructure is important for practical implementation. This includes adapting technologies to work alongside traditional air quality monitoring and control systems.

- **Standardization and regulation:** Establishing standards for testing and evaluating nanomaterials for environmental applications is crucial for their responsible development and deployment. Regulatory frameworks need to be developed to ensure safety and efficacy.

Much of the research is still in the early stages, and the transition from laboratory success to real-world application involves overcoming significant challenges. Continued research, development, and rigorous testing are necessary to fully understand and responsibly harness the potential of nanotechnology in this field.

11.2.4.4 Case Study: Multi-scale Nanofiber Membranes for Enhanced Air Filtration

Multi-scale membranes that combine nanofibers and microfibers are being developed to improve air filtration performance, addressing the ongoing challenge of balancing high filtration efficiency with low air resistance in air purification systems. This innovative approach demonstrates how nanotechnology can significantly enhance the performance of air filtration materials when integrated with traditional filtration methods.[72]

The system employs a multi-scale membrane where a nanofiber layer, composed of polyurethane (PU) and polyvinylidene fluoride (PVDF) created *via* electrospinning, is combined with a microfiber layer of polypropylene (PP) produced through melt-blowing. The nanofiber layer provides high filtration efficiency due to its small fiber diameters and high surface area, while the microfiber layer offers structural support and helps maintain low air resistance.

The nanotechnology aspect of this system is centered on the use of electrospun nanofibers. These nanofibers, with an average diameter of about 590 nm, significantly increase filtration efficiency by capturing fine particles through mechanisms such as interception and diffusion. Moreover, the integration of these nanofibers with microfibers creates a hierarchical structure that balances high filtration performance with manageable pressure drop, addressing a key challenge in air filtration technology.

The performance of the multi-scale membrane system was rigorously evaluated, with results highlighting significant improvements. The system achieved a filtration efficiency of 96.72% for $PM_{0.3}$ particles, with a quality factor of 0.04167 Pa^{-1}. These results represent a substantial improvement over conventional melt-blown PP filters, which typically achieve 65–80% efficiency for $PM_{0.3}$ with quality factors below 0.03 Pa^{-1}. Notably, this performance was achieved with just 60 minutes of electrospinning time, suggesting potential for efficient production. The multi-scale membrane demonstrated high filtration efficiency while maintaining a relatively low pressure drop, a combination that is difficult to achieve with traditional filtration materials.

The membrane also demonstrated good breathability, with an air permeability of 18.67 mm s^{-1} and a water vapor transmission rate of 263.09 g $(m^2 h)^{-1}$, comparable to commercial surgical masks but offering superior filtration. This balance of high filtration efficiency, relatively low pressure drop, and good breathability addresses key challenges in air filtration technology.

However, implementing this technology presents certain challenges. The long-term stability and durability of the nanofiber layer under various environmental conditions require further investigation. Additionally, while the fabrication method shows promise for scalability, the cost implications of integrating electrospinning with traditional melt-blowing processes need to be assessed. Current estimates suggest a 20–30% increase in production costs compared to standard melt-blown filters, which could potentially be offset by the membrane's superior performance and longer lifespan.

From an environmental perspective, the improved efficiency of these membranes could reduce energy consumption in air purification systems by an estimated 15–20%, particularly in applications like building HVAC systems. This could translate to significant energy savings and reduced carbon emissions over time.

Future research may focus on optimizing nanofiber composition for enhanced durability, exploring eco-friendly polymer alternatives, and investigating the membrane's performance against specific pollutants such as volatile organic compounds. There's also potential to incorporate additional functionalities, such as antimicrobial properties, by doping the nanofibers with metal nanoparticles.

This work illustrates the potential of nanotechnology to enhance the performance of air filtration systems. The multi-scale membrane approach not only improves air quality but also promises more energy-efficient and potentially cost-effective air purification processes. As air quality concerns continue to grow globally, such innovations could play a crucial role in developing next-generation filtration technologies for applications ranging from personal protective equipment to industrial air cleaning systems.

11.3 Energy Savings, Storage, and Harvesting

As global energy demands continue to rise and the imperative to transition from fossil fuels to renewable energy sources intensifies, the development of sustainable, efficient, and scalable energy solutions is more critical than ever. Nanotechnology offers promising avenues for addressing these challenges, complementing the water and air purification technologies discussed earlier in this chapter.

11.3.1 Nanomaterials for Energy Efficiency

Enhancing energy efficiency is crucial for reducing resource consumption and mitigating environmental impact. Nanotechnology plays a pivotal role in improving the performance of various energy systems, particularly in renewable energy applications.[73]

11.3.1.1 Solar Cell Efficiency

Nanostructured materials are revolutionizing solar cell technology.[74] For instance, perovskite solar cells incorporating nanoscale crystalline structures have achieved remarkable efficiency improvements in recent years (see Figure 11.4). These cells can be fabricated using low-temperature processes, potentially reducing manufacturing costs and energy input.[75]

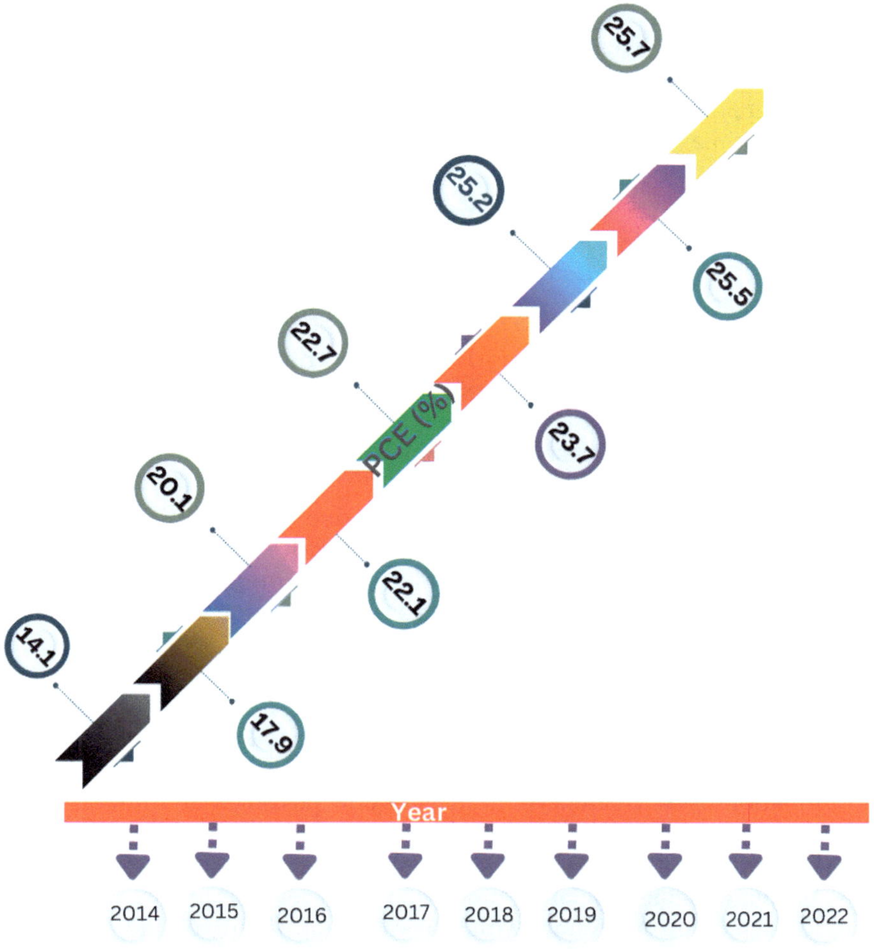

Figure 11.4 Progress in the power conversion efficiency (PCE) of perovskite solar cells in the last decade. Reproduced from ref. 75, https://doi.org/10.1039/D3RA07518D, under the terms of the CC BY 3.0 license, https://creativecommons.org/licenses/by/3.0/.

Additionally, quantum dot solar cells leverage nanoscale semiconductor particles to capture a broader spectrum of light, potentially surpassing the theoretical efficiency limits of traditional silicon cells.[76]

Research into other nanoscale quantum materials is pushing solar cell efficiency even further. For instance, by inserting copper atoms between atomic-thin layers of germanium selenide and tin sulfide, researchers have created a material—a two-dimensional Cu_xGeSe/SnS heterostructure—that can absorb both infrared and visible light. Traditional single-junction silicon solar cells have a theoretical maximum efficiency of about 32%, meaning they can convert up to 32% of incoming sunlight into electricity. In contrast, simulations show that solar cells using this new nanoscale material could potentially convert up to 190% of incoming light into electricity across a broad range of solar wavelengths—nearly six times the theoretical limit of silicon cells.[77] This seemingly impossible efficiency is achieved through a process called *multiple exciton generation*, where a single high-energy photon can create multiple electron–hole pairs, effectively multiplying the energy conversion.

This breakthrough highlights the transformative potential of engineered nanomaterials in solar energy technology.

11.3.1.2 Nanocoatings for Improved Insulation

Advanced nanocoatings are being developed to enhance building insulation. These coatings, often composed of ceramic nanoparticles[78] or silica aerogels, can significantly reduce heat transfer through walls and windows. By improving insulation, these materials contribute to reduced energy consumption for heating and cooling.

Silica aerogels are particularly impressive due to their unique nanoscale structure.[79] These materials consist of a network of interconnected silica nanoparticles, typically 2–5 nm in diameter, forming a highly porous structure. This nanostructure gives silica aerogels remarkable properties:

- **Transparent thermal insulation**: Despite being excellent insulators, silica aerogels can be used to create transparent, insulating windowpanes. These materials have a thermal conductivity as low as 0.02 $W\,m^{-1}\,K^{-1}$ at ambient pressure, while still allowing light transmission. This combination of transparency and insulation is possible due to the aerogel's nanoporous structure, where pores are smaller than the wavelength of visible light.
- **Ultralight, superinsulating materials**: Silica aerogels are often called the lightest solid material, with 80–99.8% of their volume being air. A typical silica aerogel has a density between 3 and 350 $kg\,m^{-3}$, compared to traditional insulation materials like fiberglass (20–100 $kg\,m^{-3}$). Despite this ultra-low density, silica aerogels provide exceptional insulation performance, with thermal conductivity reaching as low as 0.01 $W\,m^{-1}\,K^{-1}$ when evacuated—lower than still air (0.025 $W\,m^{-1}\,K^{-1}$).

These nanoscale-engineered materials demonstrate how nanotechnology can create insulation with properties that seem contradictory—transparent yet insulating, ultralight yet highly effective—offering exciting possibilities for energy-efficient building design.

11.3.1.3 Thermoelectric Materials

Thermoelectrocatalysis (TECatal) is emerging as a strategy for converting waste heat into chemical energy, with various material structures, including nanostructures, being explored in this field.[80] This approach combines the thermoelectric effect with surface catalysis processes.

The performance of these materials is measured by a factor called *ZT* (figure of merit), where higher values indicate better efficiency. Through nanostructuring, scientists have pushed *ZT* values from around 1 (which was considered good for bulk materials) to exceeding 2 in some nanostructured materials. This represents a substantial improvement in thermoelectric performance. For instance, p-type doped magnesium–silicon monolayers have demonstrated high *ZT* values of 2.51 at room temperature and up to 3.92 at 500 K, which are exceptional cases and represent significant advancements.[81]

These materials can operate in several modes, including hybrid structures and single-phase nanostructures. One example is a thermoelectric-gel-based system developed for solar-driven water splitting to produce hydrogen, achieving a solar-to-hydrogen efficiency of 0.4%.[82] While this efficiency might seem low, it is notable because the system utilizes waste heat—energy that would otherwise be lost—while simultaneously generating a thermopower of 8.2 mV per degree Kelvin. This dual functionality demonstrates the potential of nanostructured thermoelectric materials to recover and utilize waste energy in multiple ways.

The development of nanostructured thermoelectrocatalytic materials focuses on optimizing factors such as high *ZT*, appropriate band alignment, and surface modulation at the nanoscale. While promising, these technologies are still under research and development, particularly for industrial applications where waste heat is abundant. If successfully scaled, this technology could revolutionize waste heat recovery, contributing to overall energy efficiency and resource utilization.

11.3.1.4 Nanofluids for Enhanced Thermal Performance

Additionally, nanofluids, which are engineered by dispersing nanoparticles into base fluids, have shown great potential in enhancing the thermal performance of energy systems. When used in solar collectors, nanofluids can directly absorb solar radiation and improve heat transfer efficiency by up to 28.3%, as demonstrated by the integration of alumina-water nanofluids.[83] This improvement not only enhances the overall efficiency of solar collectors but also contributes to more stable and efficient energy systems, aligning with the broader goals of energy conservation and resource optimization.

11.3.2 Energy Storage Innovations

Efficient energy storage is critical for the widespread adoption of renewable energy sources and the development of sustainable energy systems. Nanotechnology is driving significant advancements in this field,[84] complementing the resource recovery themes explored earlier in this chapter (see Figure 11.5).

11.3.2.1 Nanostructured Batteries

Lithium-ion batteries, the current standard for many applications, are being enhanced through nanotechnology. Nanostructured electrode materials, such as nanostructured silicon[85] and advanced nanofiber materials[86] offer increased surface area and improved ion transport, leading to higher energy density and faster charging capabilities.[87]

Moreover, emerging battery technologies like lithium-sulfur and solid-state batteries benefit from nanomaterial innovations.[88,89] For instance, nanostructured sulfur cathodes and solid electrolyte interfaces at the nanoscale are addressing challenges related to cycle life and safety, potentially revolutionizing energy storage capabilities.[90]

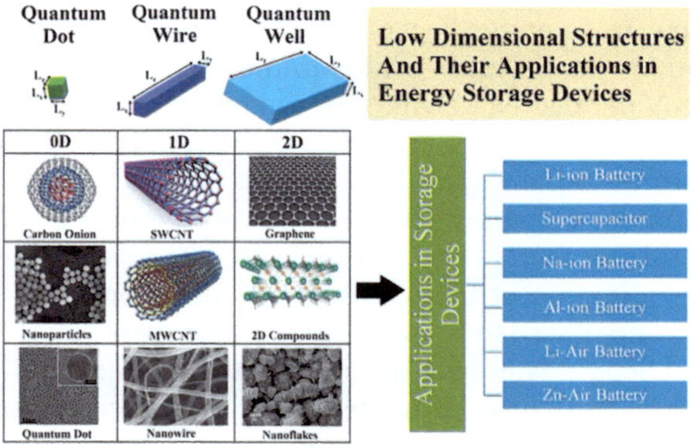

Figure 11.5 This figure illustrates the classification of low-dimensional nanostructures—such as 0D quantum dots, 1D quantum wires, and 2D quantum wells—and highlights their applications in various energy storage devices. These nanomaterials, including carbon onions, graphene, and nanowires, are being utilized in lithium-ion batteries, supercapacitors, sodium-ion batteries, and other next-generation storage technologies. Their unique properties at the nanoscale enable improved energy efficiency, capacity, and longevity, making them crucial components in the evolution of modern energy storage systems. Reproduced from ref. 84 with permission from Springer Nature, Copyright 2022.

Graphene-based composites in particular are revolutionizing lithium-ion battery technology, offering significant improvements in performance and longevity.[91] A striking example of this is the development of a crumpled nitrogen-doped graphene-wrapped phosphorus composite as an anode material. This innovative nanocomposite addresses several key challenges in battery design simultaneously.[92]

The most notable advancement is in capacity. This graphene-based composite anode achieves a remarkably high capacity of 2522.6 mA h g^{-1} at a current density of 130 mA g^{-1}, which is very close to the theoretical capacity of red phosphorus (2596 mA h g^{-1}). This far surpasses conventional graphite anodes that typically max out around 372 mA h g^{-1}. Such a dramatic increase in capacity could lead to batteries that store much more energy in the same volume or weight.

Equally impressive is the composite's rate capability and stability. It maintains a high capacity of 1340.5 mA h g^{-1} even at a high current density of 3900 mA g^{-1}, showing excellent rate performance. Furthermore, it exhibits great cyclability, maintaining 1470.1 mA h g^{-1} at 1300 mA g^{-1} for 300 cycles. This improved cycling stability is crucial for developing batteries that can maintain their performance over extended use, a key requirement for practical applications.

The nitrogen-doping of the graphene and its three-dimensional crumpled structure contribute to improved electron conduction within the anode material. This enhanced conductivity can lead to faster charge and discharge rates, addressing one of the main limitations of current battery technology.

Another significant advantage of this composite is its protective qualities. The graphene wrapping encapsulates the phosphorus, which is known for its high theoretical capacity but tends to degrade due to volume changes during cycling. The flexible graphene wrapper accommodates these changes, preventing degradation and extending battery life.

Furthermore, the crumpled structure of the graphene provides a high surface area, maximizing contact between the electrolyte and the active material. This increased surface area facilitates faster ion transfer, potentially leading to even quicker charge and discharge rates.

By addressing multiple challenges simultaneously—increasing capacity, improving conductivity, enhancing stability, and protecting active materials—these graphene-based composites represent a significant step forward in lithium-ion battery technology. Such advancements are essential for developing next-generation batteries with higher energy density and longer lifespans, crucial for applications ranging from electric vehicles to grid-scale energy storage.

While these advancements in nanostructured materials offer significant potential, several challenges remain before they can be widely implemented in commercial batteries. Manufacturing complexity, scalability, and the high cost of producing nanomaterials are significant hurdles. Additionally, long-term reliability and durability under real-world conditions are areas that require further research to ensure these innovative materials can perform consistently over the lifespan of a battery. Addressing these challenges is crucial for transitioning from laboratory successes to commercially viable products.

11.3.2.2 Supercapacitors

Nanotechnology is enabling the development of high-performance supercapacitors, which offer rapid charge/discharge cycles and long operational lifetimes. Carbon-based nanomaterials are being used to create electrodes with exceptionally high surface areas, dramatically increasing energy storage capacity.[93] These devices are particularly useful for applications requiring quick bursts of power, complementing batteries in hybrid energy storage systems.

Carbon nanotubes offer several key benefits for supercapacitor electrodes.[94] Their high surface area, excellent electrical conductivity, and good mechanical strength make them ideal materials for enhancing energy storage capacity and performance. CNT-based electrodes, such as carbon nanotube fiber fabrics, have demonstrated impressive areal capacitances of up to 1988 mF cm^{-2} at 2 mA cm^{-2} and energy densities reaching 143 μW h cm^{-2} at 1000 μW cm^{-2}.[95] The versatility of CNTs allows them to be incorporated into various electrode architectures, including 1D carbon fibers, 2D carbon films, and 3D carbon composites. When combined with other materials like graphene, metal oxides, or MXenes, CNTs can create hybrid electrodes with synergistic effects, further improving electrochemical performance.[96,97]

11.3.3 Energy Harvesting Technologies

Innovative energy harvesting technologies at the nanoscale present opportunities to capture and utilize energy from various environmental sources, furthering the resource recovery goals outlined in this chapter.

11.3.3.1 Piezoelectric Nanogenerators

Piezoelectric nanomaterials, such as zinc oxide nanowires, lead zirconate titanate (PZT) nanofibers, and barium titanate ($BaTiO_3$) nanostructures, are increasingly recognized for their ability to convert mechanical energy from vibrations, human motion, or fluid flow into electrical energy. Already, ZnO nanowires have been successfully integrated into wearable devices to harness energy from the movement of the human body, providing a sustainable power source for small electronics. Similarly, $BaTiO_3$ nanogenerators have demonstrated effectiveness in harvesting energy from environmental sources such as ocean waves, offering a renewable energy solution for remote monitoring systems.

Studies have highlighted the high piezoelectric efficiency of ZnO nanowires and $BaTiO_3$ nanogenerators, making them ideal for applications in self-powered sensors, wearable electronics, and other small-scale energy harvesting systems. In one instance, $BaTiO_3$-based nanogenerators were used to power autonomous sensors in environmental monitoring, showcasing their practical utility. Advances in the design of these nanogenerators, particularly in materials like PZT and $BaTiO_3$, have significantly improved power output. This enhancement broadens their potential for practical deployment in various energy harvesting applications, including those from environmental sources like ocean waves or wind, where even small amounts of captured energy can be utilized effectively.[98,99]

Moreover, PZT nanofibers continue to be a focus of research due to their superior piezoelectric properties, which are critical for the development of efficient energy conversion systems. An example of their application is in microelectromechanical systems (MEMS), where PZT nanofibers have been employed to generate power for miniature devices, enabling advancements in medical implants and wireless sensors.[100] These developments underscore the versatility and importance of piezoelectric nanogenerators in both current and future technologies.

11.3.3.2 Triboelectric Nanogenerators

Building on the principles of triboelectricity, triboelectric nanogenerators (TENGs) utilize nanomaterials to enhance charge generation and separation. For instance, polymer nanofibers, such as those made from polydimethylsiloxane (PDMS) or polyvinylidene fluoride (PVDF), can be engineered with micro- and nano-patterned surfaces to significantly increase the contact area, thereby enhancing the surface charge density and boosting energy conversion efficiency. Similarly, nanostructured metal oxides like zinc oxide or titanium dioxide are often incorporated into these devices to optimize charge transfer mechanisms due to their high dielectric properties and ability to form well-defined nanostructures.[101–103]

Triboelectric nanogenerators have shown great promise for harvesting energy from a wide range of human activities and environmental sources. TENGs integrated into wearable textiles can capture biomechanical energy from walking or other movements, providing a sustainable power source for portable electronics.[104,105] Additionally, TENGs have been successfully deployed in environmental settings, such as on rooftops to harvest energy from raindrops or in the ocean to capture energy from waves.[106]

Advancements in the design and materials of TENGs have significantly bolstered their efficiency and broadened their scope of application, particularly in the context of future eco-smart cities.[107] TENGs are now recognized as pivotal components in the development of decentralized and sustainable urban energy systems. The integration of TENGs into smart infrastructure, such as renewable energy systems (harvesting energy from water, wind, solar, and raindrop sources), smart homes, and intelligent transportation networks, underscores their versatility and potential. Additionally, TENGs serve as key enablers of smart healthcare and environmental protection, powering self-sustaining sensors and devices essential for real-time monitoring and data collection. These innovations, particularly the development of hybrid TENGs combining triboelectricity with other energy harvesting mechanisms, highlight the potential for TENGs to play a crucial role in addressing the energy and environmental challenges of modern cities. TENGs are expected to play a significant role in the evolution of eco-smart cities, potentially enhancing both the sustainability and intelligence of urban environments.

11.3.3.3 Plasmonic Hot Electron Devices

Leveraging the unique properties of metallic nanostructures, plasmonic hot electron devices can harvest energy from light, including the infrared spectrum. This technology has the potential to enhance solar energy capture beyond traditional photovoltaics, utilizing a broader range of the solar spectrum and potentially even harvesting waste heat.

Studies highlight how plasmonic nanostructures can be engineered to maximize the generation and transfer of hot electrons, which are key to converting light into electrical energy more efficiently.[108]

Additionally, the research outlines how tuning the properties of these metallic nanostructures, such as particle size and distribution, can optimize their effectiveness in solar energy devices. This points to the potential of plasmonic hot electron devices to surpass the capabilities of conventional solar cells by utilizing energy from both sunlight and waste heat.[109]

These advancements underline the significant role that plasmonic nanostructures could play in future solar energy technologies, offering a promising path to more efficient and comprehensive solar energy systems.

11.3.3.4 Future Directions and Challenges

While these nanotechnology-based energy solutions show great promise, several significant challenges remain. One of the primary hurdles is scaling up the production of nanomaterials while preserving their unique properties, such as high efficiency and stability. The complexity of manufacturing at scale often leads to variations in material quality, which can affect performance and reliability.

Another critical challenge is ensuring the long-term stability and safety of these materials in various applications. Issues such as material degradation, potential toxicity, and environmental impact during the lifecycle of these nanomaterials need to be thoroughly addressed to ensure they can be deployed safely and sustainably.

To overcome these challenges, research is increasingly focusing on the development of more sustainable and abundant materials for energy applications. Reducing reliance on rare or toxic elements is particularly important, as it aligns with global efforts to minimize environmental harm and ensure the accessibility of these technologies. In this context, the integration of artificial intelligence and machine learning in materials discovery and optimization is accelerating progress, enabling researchers to predict and design materials with enhanced properties and lower environmental footprints.

The environmental impact of nanomaterial production and disposal must be carefully considered. Ensuring that these technologies contribute to sustainability goals requires comprehensive life cycle assessments and the adoption of responsible development practices.

11.3.3.5 Case Study: Boosting Desalination Efficiency with Nanofluid Solar Collectors

Integrating advanced nanotechnology-enabled systems with renewable energy sources offers a promising approach to sustainable water treatment. A compelling example of this synergy is found in the development and application of a nanofluid-based direct absorption solar collector (DASC) coupled with a brine recirculation multistage flash (BR-MSF) desalination system. This system demonstrates how nanotechnology can significantly enhance the efficiency of desalination processes when combined with solar energy.[110]

The system employs a DASC where the working fluid, a nanofluid composed of a base fluid with suspended aluminum nanoparticles, directly absorbs solar energy (see Figure 11.6). The nanofluid flows through a solar collector, where it is heated by incident solar radiation. The heated

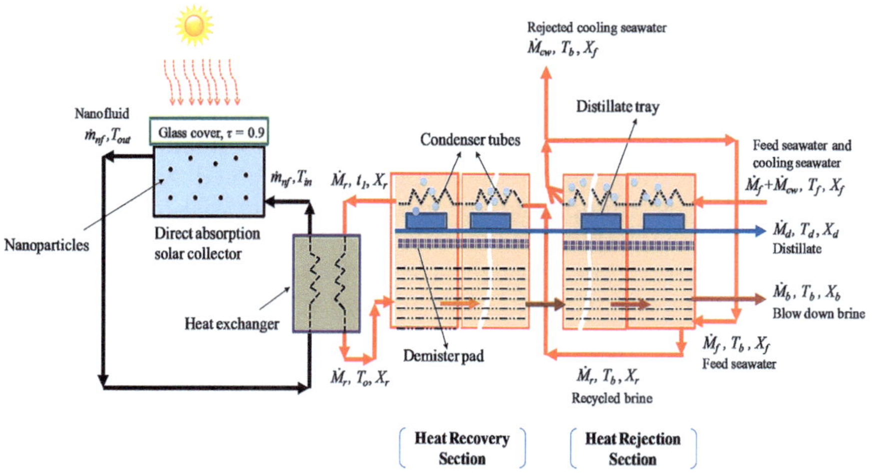

Figure 11.6 Schematic of a nanofluid-based direct absorption solar collector (DASC) integrated with a brine recirculation multistage flash (BR-MSF) desalination system. The DASC heats the nanofluid using solar energy, which then transfers this heat to the recycled brine *via* a counter-flow heat exchanger. The brine, now at an elevated temperature (top brine temperature, TBT), undergoes successive flashing stages where water vapor is produced, condensed, and collected as distillate, with the brine recirculated or rejected. The system is divided into heat recovery and heat rejection sections to maximize energy efficiency by reusing latent heat and rejecting excess heat. Reproduced from ref. 110 with permission from Elsevier, Copyright 2018.

nanofluid then transfers its thermal energy to the recycled brine within the BR-MSF desalination unit through a counter-flow heat exchanger. This process raises the temperature of the brine, enabling more efficient water vapor production in the desalination stages.

The nanotechnology aspect of this system is centered on the use of aluminum nanoparticles dispersed in a base fluid to create the nanofluid. These nanoparticles enhance the thermal conductivity and solar absorption capabilities of the fluid. The aluminum nanoparticles significantly increase the absorption of solar radiation due to their high absorptivity for solar wavelengths, allowing the fluid to capture more energy compared to conventional fluids. Moreover, the nanoparticles improve the thermal conductivity of the fluid, enabling more efficient heat transfer from the nanofluid to the brine. This increased heat transfer rate is crucial for raising the top brine temperature (TBT), which directly influences the desalination process's efficiency.

The performance of the DASC-based BR-MSF system was rigorously evaluated, and the results highlighted significant improvements. The system achieved a Gained Output Ratio (GOR) ranging from 11 to 14, depending on factors such as nanoparticle volume fraction and incident solar flux. This represents a substantial improvement over conventional systems. Specifically, the DASC-based system demonstrated a GOR of 12.64, compared to 11.41 for a parabolic trough collector (PTC)-based BR-MSF system under identical conditions. This enhanced performance is attributed to the higher TBT achieved, which could reach up to 160 °C, significantly above the typical range of 110–120 °C for conventional systems. The higher TBT leads to a greater flashing range, allowing for more efficient desalination.

However, implementing this nanotechnology-enabled system presents certain challenges. One primary concern is the potential for scaling and fouling within the system, particularly at the elevated temperatures enabled by the DASC. While the higher TBT improves system performance, it also increases the risk of scale formation, which can reduce efficiency over time. The study suggests that pretreatment methods, such as nanofiltration to remove scale-forming compounds, can mitigate this issue, but these add complexity and cost to the system. Additionally, as the incident solar flux increases, the efficiency of the solar collector can be affected by higher heat losses to the environment, which, while offset by the overall system's higher GOR, still represents a challenge in optimizing the system's performance.

The environmental benefits of this system are notable. By harnessing solar energy, the DASC-based BR-MSF system significantly reduces reliance on non-renewable energy sources, thereby lowering the carbon footprint of the desalination process. This not only makes the process more sustainable but also more cost-effective in the long term, particularly in regions with abundant solar resources. The improved efficiency of the system further enhances its environmental credentials, as more water can be desalinated with less energy input, reducing the overall environmental impact.

This case study illustrates the potential of nanotechnology to enhance the performance of renewable energy-powered water treatment systems. The use of a nanofluid in a DASC significantly improves the thermal performance of a BR-MSF desalination unit, making it a viable option for sustainable and efficient water desalination, especially in regions with abundant solar energy resources. Integrating nanotechnology with renewable energy not only reduces reliance on non-renewable energy sources but also makes water purification processes more environmentally friendly and cost-effective.

References

1. A. Fuwad, H. Ryu, E. D. Han, J.-H. Lee, N. Malmstadt and Y.-R. Kim, *et al.*, Highly permeable and shelf-stable aquaporin biomimetic membrane based on an anodic aluminum oxide substrate, *npj Clean Water*, 2024, 7, 11.
2. C. Mendes-Felipe, A. Veloso-Fernández, J. L. Vilas-Vilela and L. Ruiz-Rubio, Hybrid organic–inorganic membranes for photocatalytic water remediation, *Catalysts*, 2022, **12**(2), 180.
3. A. S. M. Ali, M. M. Soliman, S. H. Kandil and M. M. A. Khalil, Emerging mixed matrix membranes based on zeolite nanoparticles and cellulose acetate for water desalination, *Cellulose*, 2021, **28**(10), 6417–6426.
4. A. Felix Sahayaraj, H. Joy Prabu, J. Maniraj, M. Kannan, M. Bharathi and P. Diwahar, *et al.*, Metal–organic frameworks (MOFs): The next generation of materials for catalysis, gas storage, and separation, *J. Inorg. Organomet. Polym. Mater.*, 2023, **33**(7), 1757–1781.
5. T. Fu, H. Liu, X. Wang and Y. Liu, Fine tuning of pore size in metal–organic frameworks for superior removal of U(vi) from aqueous solution, *New J. Chem.*, 2024, **48**(14), 6211–6220.
6. A. Kumar, A. S. Reddy and S. P. Singh, Near-zero liquid discharge for wastewater through membrane technology, in *Energy, Environment, and Sustainability*, Springer Nature Singapore, Singapore, 2023, pp. 447–477.
7. W.-S. Yu, H.-C. Liu, H.-C. Lin, M. Lee and C.-H. Hou, Synergizing MBR and MCDI systems as a sustainable solution for decentralized wastewater reclamation and reuse, *Sustainable Environ. Res.*, 2024, **34**, 12.
8. G. E. Clayton, R. M. S. Thorn, B. G. Fox and D. M. Reynolds, Long-term trial of a community-scale decentralized point-of-use drinking water treatment system, *PLOS Water*, 2024, **3**(4), e0000187.
9. K. Staszak and K. Wieszczycka, Recovery of metals from wastewater—state-of-the-art solutions with the support of membrane technology, *Membranes*, 2023, **13**(1), 114.
10. M. Mohammadifakhr, J. de Grooth, H. D. W. Roesink and A. J. B. Kemperman, Forward osmosis: A critical review, *Processes*, 2020, **8**(4), 404.
11. Y. Xu, Y. Zhu, Z. Chen, J. Zhu and G. Chen, A comprehensive review on forward osmosis water treatment: Recent advances and prospects of

membranes and draw solutes, *Int. J. Environ. Res. Public Health*, 2022, **19**(13), 8215.
12. T. Huang, Z. Su, K. Hou, J. Zeng, H. Zhou and L. Zhang, *et al.*, Advanced stimuli-responsive membranes for smart separation, *Chem. Soc. Rev.*, 2023, **52**(13), 4173–4207.
13. R. Mills, K. C. Baldridge, M. Bernard and D. Bhattacharyya, Recent advances in responsive membrane functionalization approaches and applications, *Sep. Sci. Technol.*, 2023, **58**(6), 1202–1236.
14. M. A. Al-Obaidi, S. Alsadaie, A. Alsarayreh, M. T. Sowgath and I. M. Mujtaba, Integration of renewable energy systems in desalination, *Processes*, 2024, **12**(4), 770.
15. M. K. Shahid, B. Mainali, P. R. Rout, J. W. Lim, M. Aslam and A. E. Al-Rawajfeh, *et al.*, A review of membrane-based desalination systems powered by renewable energy sources, *Water*, 2023, **15**(3), 534.
16. E. Loukopoulos, S. Marugán-Benito, D. Raptis, E. Tylianakis, G. E. Froudakis and A. Mavrandonakis, *et al.*, Chemically tailored metal-organic frameworks for enhanced capture of short- and long-chain per- and polyfluoroalkyl substances from water, *Adv. Funct. Mater.*, 2024, **34**, 2409932.
17. V. S. Manikandan, K. George, M. Shimna, A. Thirumurugan, S. Harish and M. Navaneethan, Electrochemical performance of B_2WO_6/g-C_3N_4 (2D–2D) hybrid heterostructure as negative electrode materials for supercapacitor application, *J. Mater. Sci.: Mater. Electron.*, 2024, **35**(1), 51.
18. Y. Zhang, H. Yu, R. Zhai, J. Zhang, C. Gao and K. Qi, *et al.*, Recent progress in photocatalytic degradation of water pollution by Bismuth tungstate, *Molecules*, 2023, **28**(24), 8011.
19. F. Dingenen, N. Blommaerts, M. Van Hal, R. Borah, D. Arenas-Esteban and S. Lenaerts, *et al.*, Layer-by-layer-stabilized plasmonic gold-silver nanoparticles on TiO_2: Towards stable solar active photocatalysts, *Nanomaterials*, 2021, **11**(10), 2624.
20. P. Martins, S. Kappert, H. Nga Le, V. Sebastian, K. Kühn and M. Alves, *et al.*, Enhanced photocatalytic activity of Au/TiO_2 nanoparticles against ciprofloxacin, *Catalysts*, 2020, **10**(2), 234.
21. J. Yang, S. He, H. Liu, E. Jaatinen, E. Waclawik and J. Quan, *et al.*, Enhancing visible-light photocatalytic performance of Au/TiO_2catalysts through light reflection-promoted optical absorption with oriented anatase mesocrystals, *J. Mater. Chem. A*, 2023, **11**(9), 4751–4757.
22. J. Pinto, A. Weilhard, L. T. Norman, R. W. Lodge, D. M. Rogers and A. Gual, *et al.*, Unravelling synergistic effects in bi-metallic catalysts: deceleration of palladium–gold nanoparticle coarsening in the hydrogenation of cinnamaldehyde, *Catal. Sci. Technol.*, 2023, **13**(14), 4082–4091.
23. M. Arif, Catalytic degradation of azo dyes by bimetallic nanoparticles loaded in smart polymer microgels, *RSC Adv.*, 2023, **13**(5), 3008–3019.
24. S. Sun, Z. Zhang, S. Li, J. Le, H. Qian and X. Yin, *et al.*, Fabrication of recyclable magnetic Fe_3O_4–SiO_2–Cu_2O/Cu core–shell nano-catalyst for

high-efficient reduction of aromatic nitro compound, *J. Mater. Sci.*, 2023, **58**(13), 5587–5598.
25. E. Alzahrani, Photodegradation of binary azo dyes using core–shell $Fe_3O_4/SiO_2/TiO_2$ Nanospheres, *Am. J. Anal. Chem.*, 2017, **08**(01), 95–115.
26. Y. Fu, Y. Xu, B. Lou, X. Qin, L. Zhang and H. Yuan, *et al.*, Magnetically recyclable core–shell MOF nanoparticles of Fe_3O_4@PDA@UIO-66-NH_2 grafted by organic acids for intensified cationic dye adsorption, *New J. Chem.*, 2022, **46**(23), 11071–11081.
27. S. Ramu, I. Kainthla, L. Chandrappa, J. M. Shivanna, B. Kumaran and R. G. Balakrishna, Recent advances in metal organic frameworks–based magnetic nanomaterials for waste water treatment, *Environ. Sci. Pollut. Res. Int.*, 2023, **31**(1), 167–190.
28. N. C. Joshi and P. Gururani, Advances of graphene oxide based nanocomposite materials in the treatment of wastewater containing heavy metal ions and dyes, *Curr. Res. Green Sustainable Chem.*, 2022, **5**, 100306.
29. I. J. Budiarso, V. A. Dabur, R. Rachmantyo, H. Judawisastra, C. Hu and A. Wibowo, Carbon nitride- and graphene-based materials for the photocatalytic degradation of emerging water pollutants, *Mater. Adv.*, 2024, **5**(7), 2668–2688.
30. X. Li, Y. Bai, X. Shi, N. Su, G. Nie and R. Zhang, *et al.*, Applications of MXene ($Ti_3C_2T_x$) in photocatalysis: a review, *Mater. Adv.*, 2021, **2**(5), 1570–1594.
31. H. Yu, H. Jiang, S. Zhang, X. Feng, S. Yin and W. Zhao, Review of two-dimensional MXenes ($Ti_3C_2T_x$) materials in photocatalytic applications, *Processes*, 2023, **11**(5), 1413.
32. L. Gao, Enzyme-like property (nanozyme) of iron oxide nanoparticles, in *Iron Oxide Nanoparticles*, IntechOpen, 2022.
33. H. A. Elkomy, S. A. El-Naggar, M. A. Elantary, S. M. Gamea, M. A. Ragab and O. M. Basyouni, *et al.*, Nanozyme as detector and remediator to environmental pollutants: between current situation and future prospective, *Environ. Sci. Pollut. Res. Int.*, 2023, **31**(3), 3435–3465.
34. S. Singh, Cerium oxide based nanozymes: Redox phenomenon at biointerfaces, *Biointerphases*, 2016, **11**, 04B202.
35. M. M. Sozarukova, E. V. Proskurnina, A. L. Popov, A. L. Kalinkin and V. K. Ivanov, New facets of nanozyme activity of ceria: lipo- and phospholipoperoxidase-like behaviour of CeO_2 nanoparticles, *RSC Adv.*, 2021, **11**(56), 35351–35360.
36. Y. Zhou, Y. Liu, M. Zhang, Z. Feng, D.-G. Yu and K. Wang, Electrospun nanofiber membranes for air filtration: A review, *Nanomaterials*, 2022, **12**(7), 1077.
37. F. Sun, Y. Zheng, E. Liu, T. Chen and Q. Xiao, Three-dimensional multi-scale membrane for efficient air filtration performances: preparation, optimization and characterization, *J. Polym. Res.*, 2024, **31**, 75.
38. L. Wang, F. Zhu, E. Liu, Y. Yang, Q. Yu and Y. He, *et al.*, Progress in advanced carbon nanotubes composites for air purification, *Adv. Compos. Hybrid Mater.*, 2024, 7, 100.

39. M. Chitranshi, D. R. Chen, P. Kosel, M. Cahay and M. Schulz, Flexible and lightweight carbon nanotube composite filter for particulate matter air filtration, *Nanomaterials*, 2022, **12**(22), 4094.
40. Z. Tao, Y. Zhao, Y. Wang and G. Zhang, Recent advances in carbon nanotube technology: Bridging the gap from fundamental science to wide applications, *C*, 2024, **10**(3), 69.
41. M. Kiteto, B. Vidija, C. A. Mecha, R. Mrosso and M. N. Chollom, Advances in metal–organic frameworks as adsorbents, photocatalysts and membranes: a new frontier in water purification, *Discover Water*, 2024, **4**, 54.
42. A. Felix Sahayaraj, H. Joy Prabu, J. Maniraj, M. Kannan, M. Bharathi and P. Diwahar, *et al.*, Metal–organic frameworks (MOFs): The next generation of materials for catalysis, gas storage, and separation, *J. Inorg. Organomet. Polym. Mater.*, 2023, **33**(7), 1757–1781.
43. J. Liu, E. Tian, S. Zhang, D. Kong, K. Liu and X. Bai, *et al.*, The rise of two-dimensional-material-based filters for airborne particulate matter removal, *Adv. Fiber Mater.*, 2023, **5**(2), 461–483.
44. Q. Geng, H. Wang, R. Chen, L. Chen, K. Li and F. Dong, Advances and challenges of photocatalytic technology for air purification, *Natl. Sci. Open*, 2022, **1**(2), 20220025.
45. B. Muzikova, I. Martiniakova, E. Mikyskova, M. Mergl, M. Kalbac and R. Zouzelka, *et al.*, Composite TiO_2-based photocatalyst with enhanced performance, *Photochem. Photobiol. Sci.*, 2022, **22**(1), 73–86.
46. I. Papailias, N. Todorova, T. Giannakopoulou, N. Plakantonaki, M. Vagenas and P. Dallas, *et al.*, Enhancing the visible light photocatalytic activity of TiO_2-based coatings by the addition of exfoliated g-C_3N_4, *Catalysts*, 2024, **14**(5), 333.
47. Y. Wei, H. Meng, Q. Wu, X. Bai and Y. Zhang, TiO_2-based photocatalytic building material for air purification in sustainable and low-carbon cities: A review, *Catalysts*, 2023, **13**(12), 1466.
48. K. Pradip, Nanomaterials based sensors for air pollution control, in *Environmental Chemistry for a Sustainable World*, Springer International Publishing, Cham, 2020, pp. 349–403.
49. N. A. Isaac, I. Pikaar and G. Biskos, Metal oxide semiconducting nanomaterials for air quality gas sensors: operating principles, performance, and synthesis techniques, *Mikrochim. Acta*, 2022, **189**, 196.
50. K. Singh, K. K. Maurya and M. Malviya, Review of electrochemical sensors and biosensors based on first-row transition metals, their oxides, and noble metals nanoparticles, *J. Anal. Test.*, 2024, **8**(2), 143–159.
51. X. Tang, M. Debliquy, D. Lahem, Y. Yan and J.-P. Raskin, A review on functionalized graphene sensors for detection of ammonia, *Sensors*, 2021, **21**(4), 1443.
52. A. Chakraborthy, S. Nuthalapati, A. Nag, N. Afsarimanesh, M. E. E. Alahi and M. E. Altinsoy, A critical review of the use of graphene-based gas sensors, *Chemosensors*, 2022, **10**(9), 355.

53. K. G. Krishna, G. Umadevi, S. Parne and N. Pothukanuri, Zinc oxide-based gas sensors and their derivatives: a critical review, *J. Mater. Chem. C*, 2023, **11**(12), 3906–3925.
54. P. Chen, Q. Shao, X. Qu and L. Ma, Sensitivity enhancement of lateral zinc oxide nanowire gas sensors, *Appl. Phys. A: Mater. Sci. Process.*, 2024, **130**, 581.
55. N. A. Pandit and T. Ahmad, Tin oxide-based hybrid nanostructures for efficient gas sensing, *Molecules*, 2022, **27**(20), 7038.
56. S. Sureshkumar, S. Rajakumari and R. Manonmani, Recent advances in the development of carbon/metal oxides nanohybrids for enhanced H_2S detection: a review, *J. Iran. Chem. Soc.*, 2023, **20**(5), 1007–1022.
57. J. Li, H. Zhao, Y. Wang and Y. Zhou, Approaches for selectivity improvement of conductometric gas sensors: an overview, *Sens. Diagn.*, 2024, **3**(3), 336–353.
58. B. Saruhan, R. L. Fomekong and S. Nahirniak, High-sensitivity and --selectivity gas sensors with nanoparticles, nanostructures, and thin films, *Chemosensors*, 2023, **11**(2), 81.
59. S. Yadav and P. Malhotra, Metal–organic frameworks (MOFs) as sensors for environmental monitoring, in *Metal–Organic Frameworks (MOFs) as Catalysts*, Springer Nature Singapore, Singapore, 2022, pp. 393–409.
60. A. Zuliani, N. Khiar and C. Carrillo-Carrión, Recent progress of metal–organic frameworks as sensors in (bio)analytical fields: towards real-world applications, *Anal. Bioanal. Chem.*, 2023, **415**(11), 2005–2023.
61. B. Jia, J. Chen, J. Zhou, Y. Zeng, H.-P. Ho and Y. Shao, Passively and actively enhanced surface plasmon resonance sensing strategies towards single molecular detection, *Nano Res.*, 2022, **15**(9), 8367–8388.
62. V. V. Petrov, V. V. Sysoev, A. P. Starnikova, M. G. Volkova, Z. K. Kalazhokov and V. Y. Storozhenko, *et al.*, Synthesis, characterization and gas sensing study of ZnO-SnO_2 nanocomposite thin films, *Chemosensors*, 2021, **9**(6), 124.
63. M. Kodu, M. Lind, V. Kiisk, I. Renge and R. Jaaniso, Selective detection of toxic gases by arrays of single-layer graphene sensors functionalized with nanolayers of different oxides, in *Eurosensors 2023*, MDPI, Basel Switzerland, 2024.
64. S. R. Benjamin and E. J. Miranda Ribeiro Júnior, Graphene-Based electrochemical sensors for detection of environmental pollutants, *Curr. Opin. Environ. Sci. Health*, 2022, **29**, 100381.
65. S. X. Leong, Y. X. Leong, C. S. L. Koh, E. X. Tan, L. B. T. Nguyen and J. R. T. Chen, *et al.*, Emerging nanosensor platforms and machine learning strategies toward rapid, point-of-need small-molecule metabolite detection and monitoring, *Chem. Sci.*, 2022, **13**(37), 11009–11029.
66. D. Jiang, M. Lian, M. Xu, Q. Sun, B. B. Xu and H. K. Thabet, *et al.*, Advances in triboelectric nanogenerator technology—applications in self-powered sensors, Internet of things, biomedicine, and blue energy, *Adv. Compos. Hybrid Mater.*, 2023, **6**, 57.

67. S. Roy Barman, Y.-J. Lin, K.-M. Lee, A. Pal, N. Tiwari and S. Lee, *et al.*, Triboelectric nanosensor integrated with robotic platform for self-powered detection of chemical analytes, *ACS Nano*, 2023, **17**(3), 2689–2701.
68. T. S. Dhahi, A. K. Yousif Dafhalla, O. E. Tayfour, A. Mubarakali, A. S. Alqahtani and A. E. Tayfour Ahmed, *et al.*, Advances in nano sensors for monitoring and optimal performance enhancement in photovoltaic cells, *iScience*, 2024, **27**(4), 109347.
69. P. Baskaran and M. Rajasekar, Recent trends and future perspectives of thermoelectric materials and their applications, *RSC Adv.*, 2024, **14**(30), 21706–21744.
70. P. Ondrejka and M. Mikolášek, Thin films and coatings for energy storage and conversion: From supercapacitors and batteries to hydrogen generators, *Coatings*, 2023, **13**(4), 742.
71. N. Farooq, Z. Rehman, M. I. Khan, S. Asghar, M. Saleem and R. Irshad, *et al.*, Nanomaterial-based energy conversion and energy storage devices: a comprehensive review, *New J. Chem.*, 2024, **48**(19), 8933–8962.
72. F. Sun, Y. Zheng, E. Liu, T. Chen and Q. Xiao Three-dimensional multi-scale membrane for efficient air filtration performances: Preparation, optimization and characterization. *Res. Square*. 2023.
73. H. S. Hussein, The state of the art of nanomaterials and its applications in energy saving, *Bull. Natl. Res. Cent.*, 2023, **47**, 7.
74. A. M. Oni, A. S. M. Mohsin, M. M. Rahman and M. B. Hossain Bhuian, A comprehensive evaluation of solar cell technologies, associated loss mechanisms, and efficiency enhancement strategies for photovoltaic cells, *Energy Rep.*, 2024, **11**, 3345–3366.
75. M. Noman, Z. Khan and S. T. Jan, A comprehensive review on the advancements and challenges in perovskite solar cell technology, *RSC Adv.*, 2024, **14**(8), 5085–5131.
76. S. Sahai, A. Jangra, L. M. Thomas and V. R. Satsangi, Quantum dots as efficient solar energy absorber: Review on photovoltaics and photoelectrochemical systems, *J. Inst. Eng. (India): Ser. D*, 2024, **105**(1), 553–566.
77. S. M. Kastuar and C. E. Ekuma, Chemically tuned intermediate band states in atomically thin Cu_xGeSe/SnS quantum material for photovoltaic applications, *Sci. Adv.*, 2024, **10**(15), eadl6752.
78. W. Wang, Q. Fu, J. Ge, S. Xu, Q. Liu and J. Zhang, *et al.*, Advancements in thermal insulation through ceramic micro-nanofiber materials, *Molecules*, 2024, **29**(10), 2279.
79. A.-G. Niculescu, D.-I. Tudorache, M. Bocioagă, D. E. Mihaiescu, T. Hadibarata and A. M. Grumezescu, An updated overview of silica aerogel-based nanomaterials, *Nanomaterials*, 2024, **14**(5), 469.
80. Y. Zhang, S. Li, J. Zhang, L.-D. Zhao, Y. Lin and W. Liu, *et al.*, Thermoelectrocatalysis: an emerging strategy for converting waste heat into chemical energy, *Natl. Sci. Rev.*, 2024, **11**(4), nwae036.

81. X. Yu, W. Jin, J. Pang, J. Zuo, X. Kuang and C. Lu, A multifunctional Mg_2Si monolayer with negative Poisson's ratio and ultrahigh thermoelectric performance at room temperature, *J. Mater. Chem. A*, 2024, **12**(3), 1488–1497.
82. Y. Wang, Y. Zhang, X. Xin, J. Yang, M. Wang and R. Wang, *et al.*, In situ photocatalytically enhanced thermogalvanic cells for electricity and hydrogen production, *Science*, 2023, **381**(6655), 291–296.
83. T. Yousefi, F. Veysi, E. Shojaeizadeh and S. Zinadini, An experimental investigation on the effect of Al_2O_3–H_2O nanofluid on the efficiency of flat-plate solar collectors, *Renewable Energy*, 2012, **39**(1), 293–298.
84. H. Das, B. Pathak, S. Khanam, P. K. Kalita and P. Datta, Nanomaterials for next generation energy storage applications, *MRS Commun.*, 2022, **12**(3), 285–294.
85. M. Khan, S. Yan, M. Ali, F. Mahmood, Y. Zheng and G. Li, *et al.*, Innovative solutions for high-performance silicon anodes in lithium-ion batteries: Overcoming challenges and real-world applications, *Nanomicro Lett.*, 2024, **16**, 179.
86. X. Cao, C. Ma, L. Luo, L. Chen, H. Cheng and R. S. Orenstein, *et al.*, Nanofiber materials for lithium-ion batteries, *Adv. Fiber Mater.*, 2023, **5**(4), 1141–1197.
87. Z. Bitew, M. Tesemma, Y. Beyene and M. Amare, Nano-structured silicon and silicon based composites as anode materials for lithium ion batteries: recent progress and perspectives, *Sustainable Energy Fuels*, 2022, **6**(4), 1014–1050.
88. B. B. Gicha, L. T. Tufa, N. Nwaji, X. Hu and J. Lee, Advances in all-solid-state lithium–sulfur batteries for commercialization, *Nanomicro Lett.*, 2024, **16**, 172.
89. L. Du, R. Wu, Z. Wu, H. Huang, Y. Xia and Y. Gan, *et al.*, Research progress of all-solid-state lithium–sulfur batteries with sulfide solid electrolytes: materials, interfaces, challenges, and prospects, *Mater. Chem. Front.*, 2023, 7(22), 5760–5785.
90. M. Tabuyo-Martínez, B. Wicklein and P. Aranda, Progress and innovation of nanostructured sulfur cathodes and metal-free anodes for room-temperature Na–S batteries, *Beilstein J. Nanotechnol.*, 2021, **12**, 995–1020.
91. A. Kausar, I. Ahmad, T. Zhao, O. Aldaghri, K. H. Ibnaouf and M. H. Eisa, Graphene nanocomposites as innovative materials for energy storage and conversion—design and headways, *Int. J. Mol. Sci.*, 2023, **24**(14), 11593.
92. X. Jiao, Y. Liu, T. Li, C. Zhang, X. Xu and O. O. Kapitanova, *et al.*, Crumpled nitrogen-doped graphene-wrapped phosphorus composite as a promising anode for lithium-ion batteries, *ACS Appl. Mater. Interfaces*, 2019, **11**(34), 30858–30864.
93. P. K. Sahoo, N. Kumar, A. Jena, S. Mishra, C.-P. Lee and S.-Y. Lee, *et al.*, Recent progress in graphene and its derived hybrid materials for high-performance supercapacitor electrode applications, *RSC Adv.*, 2024, **14**(2), 1284–1303.

94. Y. Han, H. Ha, C. Choi, H. Yoon, P. Matteini and J. Y. Cheong, *et al.*, Review of flexible supercapacitors using carbon nanotube-based electrodes, *Appl. Sci.*, 2023, **13**(5), 3290.
95. Y. Liang, X. Luo, W. Weng, Z. Hu, Y. Zhang and W. Xu, *et al.*, Activated carbon nanotube fiber fabric as a high-performance flexible electrode for solid-state supercapacitors, *ACS Appl. Mater. Interfaces*, 2021, **13**(24), 28433–28441.
96. J. Jiang, F. Li, L. Ding, C. Zhang, Arramel and X. Li, MXenes/CNTs-based hybrids: Fabrications, mechanisms, and modification strategies for energy and environmental applications, *Nano Res.*, 2024, **17**, 3429–3454.
97. S. Wang, K. Cao, L. Xu, D. Zhao and Y. Tong, Carbon nanotubes/reduced graphene oxide composites as electrode materials for supercapacitors, *Appl. Phys. A: Mater. Sci. Process.*, 2022, **128**, 81.
98. N. Bhadwal, R. Ben Mrad and K. Behdinan, Review of zinc oxide piezoelectric nanogenerators: Piezoelectric properties, composite structures and power output, *Sensors*, 2023, **23**(8), 3859.
99. O. Okhay and A. Tkach, Current achievements in flexible piezoelectric nanogenerators based on barium titanate, *Nanomaterials*, 2023, **13**(6), 988.
100. G. Pranavanathan and S. B. Kandagal, Enhanced energy harvesting application of piezoceramics (PZT) in MEMS devices, in *Mechanisms and Machine Science*, Springer Nature Singapore, Singapore, 2023, pp. 59–69.
101. Y. Mi, Z. Zhao, H. Wu, Y. Lu and N. Wang, Porous polymer materials in triboelectric nanogenerators: A review, *Polymers*, 2023, **15**(22), 4383.
102. K. Lee, S. Mhin, H. Han, O. Kwon, W.-B. Kim and T. Song, *et al.*, A high-performance PDMS-based triboelectric nanogenerator fabricated using surface-modified carbon nanotubes *via* pulsed laser ablation, *J. Mater. Chem. A*, 2022, **10**(3), 1299–1308.
103. P. Hajara, M. R. Shijeesh, K. V. Vijoy, T. P. Rose and K. J. Saji, Harnessing energy through ZnO-based triboelectric nanogenerator: A comparative analysis of polymer materials, with emphasis on PVDF nanofibers, *J. Electron. Mater.*, 2024, **53**(9), 5617–5628.
104. I. Ali, N. Karim and S. Afroj, Textile-based triboelectric nanogenerators integrated with 2D materials, *EcoMat*, 2024, **6**(7), e12471.
105. H. Kim, D. C. Nguyen, T. T. Luu, Z. Ding, Z.-H. Lin and D. Choi, Recent advances in functional fiber-based wearable triboelectric nanogenerators, *Nanomaterials*, 2023, **13**(19), 2718.
106. B. Shan, T. Ai and K. Wang, Triboelectric nanogenerator for ocean energy harvesting: A review of technological advances and future perspectives, *Int. J. Electrochem. Sci.*, 2024, **19**(8), 100694.
107. Y. Tang, H. Fu and B. Xu, Advanced design of triboelectric nanogenerators for future eco-smart cities, *Adv. Compos. Hybrid Mater.*, 2024, **7**, 102.
108. H. Tang, C.-J. Chen, Z. Huang, J. Bright, G. Meng and R.-S. Liu, *et al.*, Plasmonic hot electrons for sensing, photodetection, and solar energy applications: A perspective, *J. Chem. Phys.*, 2020, **152**, 220901.

109. L. Zhang, X. Lu, J. Sun, C. Wang and P. Dong, Insights into the plasmonic "hot spots" and efficient hot electron injection induced by Ag nanoparticles in a covalent organic framework for photocatalytic H_2 evolution, *J. Mater. Chem. A*, 2024, **12**(9), 5392–5405.
110. K. Garg, V. Khullar, S. K. Das and H. Tyagi, Performance evaluation of a brine-recirculation multistage flash desalination system coupled with nanofluid-based direct absorption solar collector, *Renewable Energy*, 2018, **122**, 140–151.
111. M. Khraisheh, S. Elhenawy, F. AlMomani, M. Al-Ghouti, M. K. Hassan and B. H. Hameed, Recent progress on nanomaterial-based membranes for water treatment, *Membranes*, 2021, **11**(12), 995.

Section IV: Challenges, Opportunities and Future Prospects

CHAPTER 12

Challenges in Implementing Nanotechnology Solutions

12.1 Overview of Implementation Challenges

Building on the insights from previous chapters, nanotechnology holds tremendous potential for transforming environmental protection and waste management far beyond the capabilities of traditional technologies. Its ability to manipulate matter at atomic and molecular levels enables innovative approaches to resource recovery, pollutant remediation, and environmental monitoring. Despite these possibilities, scaling nanotechnology from the lab to real-world applications presents significant technical and logistical obstacles.

With global waste production rising and environmental degradation accelerating, the need for innovative solutions is urgent. Nanomaterials, characterized by high surface-area-to-volume ratios, quantum effects, and tunable properties, are particularly well-suited to address these challenges. They offer the ability to enhance pollutant removal efficiency, revolutionize water treatment processes, improve resource recovery, and reduce landfill volumes. Additionally, nanotechnology enables advanced tools for pollution monitoring and ecosystem restoration. However, transitioning these lab breakthroughs to large-scale, practical applications is far from straightforward.

Key challenges include the scalability of production, integration into existing systems, ensuring consistent performance, and addressing safety concerns. For instance, maintaining the unique benefits of nanotechnology while ensuring long-term stability and reliability across various industries, regions, and economies requires innovative strategies. Developed regions may focus on upgrading existing infrastructure to incorporate advanced nanotechnologies, while developing regions might benefit from adopting

Waste Not! How Nanotechnologies Can Increase Efficiencies Throughout Society
By Michael Berger

Published by the Royal Society of Chemistry, www.rsc.org

new nano-enabled solutions directly. Tailoring approaches to different contexts is essential for creating globally relevant applications.

Addressing these challenges requires interdisciplinary collaboration between experts in materials science, chemistry, biology, environmental science, and engineering. While this adds complexity, it also opens up opportunities for innovation. The global nature of environmental issues demands solutions that consider varied economic and industrial landscapes.

This chapter aims to provide a broad understanding of the current state of nanotechnology in waste management, sustainability efforts, and environmental protection, exploring both the hurdles and opportunities involved in realizing its full potential. By examining these issues from a global, interdisciplinary perspective, this overview prepares researchers, policymakers, and industry leaders for the specific challenges in turning nanotechnology's promise into widespread reality.

12.2 Synthesis and Fabrication Challenges

The transition from laboratory-scale synthesis to industrial-scale production of nanomaterials presents a significant hurdle in the implementation of nanotechnology solutions. This section explores the key challenges in scaling up nanomaterial synthesis and fabrication processes.

12.2.1 Scalability of Nanomaterial Synthesis

Scaling up the synthesis of nanomaterials from the laboratory to industrial production presents a complex set of challenges, as illustrated in Figure 12.1. One of the key obstacles lies in **reactor design**. While lab-scale reactors operate under tightly controlled conditions, increasing the reactor size introduces difficulties in maintaining uniform conditions such as temperature, pressure, and concentration gradients. These factors affect heat and mass transfer, making it more difficult to achieve consistent results in larger vessels.[1]

Another major concern is **process control**. Scaling up requires maintaining the same level of precision in controlling reaction conditions, which becomes increasingly challenging as the size of the system grows. This inconsistency can lead to variations in the quality of the nanomaterials produced, which directly impacts their performance and economic viability.[2]

Closely tied to process control is the issue of **yield and efficiency**. While small-scale reactions may yield high-quality products with minimal waste, these efficiencies often decline at larger scales. The loss of yield can result in significant financial drawbacks, undermining the economic feasibility of industrial-scale production.[3]

In many laboratory environments, nanomaterial synthesis is typically conducted in **batch processes**. However, **continuous processing** is often more practical and cost-effective at industrial levels. Adapting batch processes to continuous flow systems presents substantial engineering

Figure 12.1 Key challenges in scaling up nanomaterial synthesis: a visual summary of the major technical and logistical hurdles involved in transitioning from laboratory-scale to industrial-scale production, including reactor design, process control, yield efficiency, and environmental concerns.

challenges, particularly in maintaining consistent quality and purity across large production volumes.

Raw material sourcing also becomes a critical concern when moving to industrial production. The need for high-purity precursor materials on a much larger scale raises both cost and logistical issues, as securing a consistent supply of these materials can be difficult and expensive.[4]

Additionally, **environmental and safety concerns** come into play as production scales up. Handling larger quantities of hazardous materials introduces increased risks of exposure and accidents. Emissions and waste generated during production also grow with scale, necessitating more stringent environmental protocols to ensure safe and sustainable operations.

Achieving **economies of scale** is often the goal of large-scale production, where unit costs can decrease as production volumes rise. However, the initial investments required for larger reactors, specialized equipment, and compliance with industrial regulations can be substantial, potentially running into the tens of millions of dollars. Balancing these costs with potential long-term savings is a critical strategic decision for industries looking to implement nanotechnology solutions.[5]

Finally, implementing **automation and control systems** is crucial for industrial-scale production. Automation allows for precise monitoring and adjustments throughout the process, ensuring consistent product quality and efficiency. However, the complexity and cost of developing and deploying these automated systems represent significant barriers, particularly in terms of ensuring uniformity across large-scale operations. Furthermore, as production scales, ensuring **quality assurance and standardization** across different sites or batches becomes increasingly important to meet the rigorous standards required for industrial applications.

12.2.2 Control of Nanomaterial Properties

Maintaining consistent quality during mass production is crucial for the reliable performance of nanotechnology-based solutions in waste management and environmental protection. This challenge is closely linked to the scalability issues mentioned in Section 12.2.1, as larger-scale synthesis often complicates property control.

As nanomaterial production scales up, maintaining precise control over key properties such as size, shape, composition, and surface characteristics becomes increasingly challenging (see Figure 12.2).[6,7] These properties are critical for ensuring the consistent performance and reliability of nanotechnology applications, particularly in environmental protection and waste management.

One of the primary concerns is **size and shape control**, as the performance of nanomaterials is often highly dependent on these factors. In small-scale synthesis, maintaining uniform size distributions and shapes is manageable. However, as production volumes grow, achieving this consistency becomes difficult. The increased complexity in controlling reaction conditions at larger scales makes it harder to produce nanomaterials with the precise dimensions required for optimal functionality.

Similarly, **ensuring consistent chemical composition and purity** poses a significant challenge during scale-up. In larger reactors, the likelihood of

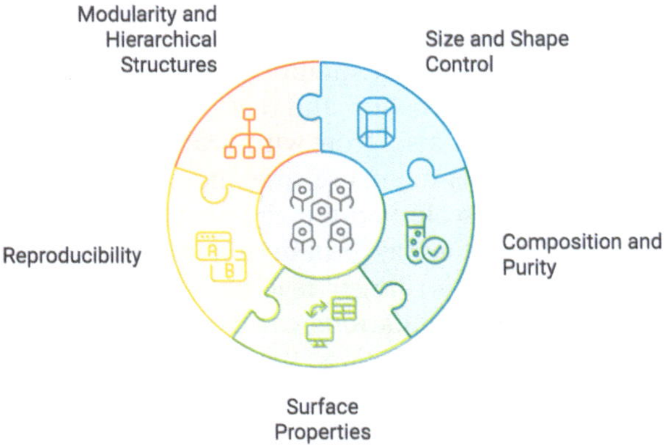

Figure 12.2 Key challenges in controlling nanomaterial properties: a visual representation of the main obstacles in ensuring consistency and quality during large-scale synthesis and fabrication, including size control, composition, surface properties, reproducibility, and modularity.

contamination or unwanted side reactions increases, making it harder to maintain the high levels of purity necessary for many applications. This issue is especially critical for environmental technologies, where impurities could lead to unintended consequences such as secondary pollution.

Another important factor is **surface properties**, as many nanomaterials derive their unique characteristics from their surfaces. Controlling these surface chemistries on an industrial scale is far more complex than in small batches, particularly when it comes to preventing agglomeration or degradation during processing. This is especially important for applications involving pollutant adsorption or catalytic degradation, where the material's surface directly impacts its performance.

In industrial applications, **reproducibility** becomes a crucial requirement. Ensuring that each batch of nanomaterials performs identically is often challenging, as slight variations in synthesis conditions can lead to significant differences in material properties. This issue is particularly relevant for technologies requiring precise and reliable performance, such as environmental sensing and remediation systems.

A promising approach to address these challenges involves the implementation of **modularity and hierarchical structures** in nanomaterial design. Inspired by natural systems, this method allows for the assembly of larger structures from smaller, reliable functional modules. By designing materials with distinct modular units that can be easily assembled and reassembled, manufacturers can achieve greater control over final material properties, simplify troubleshooting during scale-up, and improve reproducibility.[8]

12.2.3 Advanced Fabrication Techniques

Creating functional nanodevices often requires complex fabrication techniques that are difficult to implement at industrial scales. These challenges must be understood to pave the way for broader application of nanotechnology (see Figure 12.3).[9,10]

The fabrication of nanodevices demands a high level of **precision and accuracy**, which becomes increasingly difficult to maintain as production scales up. At the nanoscale, even slight variations in processing conditions can lead to significant differences in the functionality and reliability of the final product. Techniques that offer high precision in laboratory settings may face major hurdles when adapted to industrial production, where maintaining uniform quality and performance across large batches is essential. This challenge is particularly pronounced in fields such as electronics and medical devices, where the intricate structures of nanodevices require exact control over nanoscale dimensions.[11]

One limitation of current nanodevice fabrication lies in the reliance on **planar technologies**, which are highly effective for creating two-dimensional structures, such as those found in modern semiconductor chips.[7] However, the growing complexity of nanodevices necessitates the development of **three-dimensional architectures**, especially in applications where enhanced

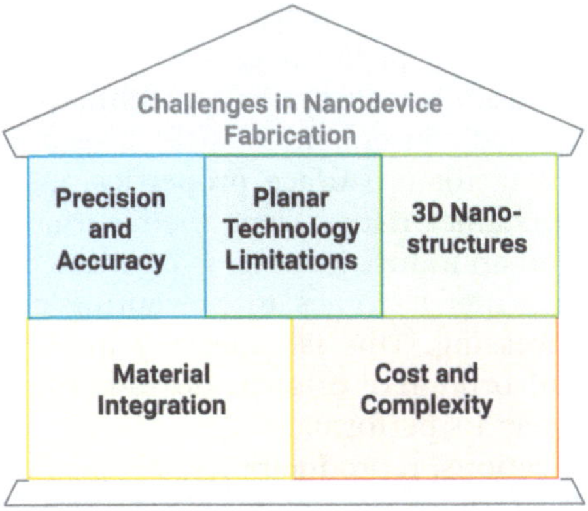

Figure 12.3 Key challenges in nanodevice fabrication: a visual representation of the technical and practical obstacles in producing nanodevices at scale, including precision and accuracy, limitations of planar technologies, integration of materials, and the high cost and complexity of advanced fabrication methods.

connectivity and functionality are required. Shifting from planar to 3D designs offers immense potential for improving the capabilities of nanodevices, but it also presents significant technical challenges.[12] Fabricating intricate 3D nanostructures with the precision required for advanced photonics or biomedical applications, for example, remains an ongoing area of research.

A related challenge is the integration of **multiple materials with different physical or chemical properties**. Many nanodevices combine materials that need to interact seamlessly within a single system, but controlling these interactions—particularly at the interfaces—can be difficult. For example, combining metals, semiconductors, and polymers in a single nanodevice can lead to undesirable reactions or degradation if the interfaces between them are not carefully controlled. This challenge is especially critical in fields like quantum computing and advanced sensing, where the performance of a device hinges on the precise integration of its various material components.[13]

***In situ* characterization** techniques play a crucial role in monitoring and controlling the fabrication process in real-time. As nanodevices are fabricated, observing their behavior at the nanoscale becomes essential for ensuring the quality of the final product. However, current characterization methods are often limited in their ability to provide the level of detail and speed required for real-time monitoring. Developing advanced *in situ* techniques that can capture the dynamic changes during fabrication processes is vital for maintaining precision and ensuring consistency, especially at larger production scales.[14,15]

In addition to these technical barriers, the **cost and complexity** of fabricating nanodevices on a large scale cannot be overlooked. Many advanced fabrication techniques, such as electron-beam lithography or atomic layer deposition, require expensive equipment and highly specialized operators. This not only increases the cost of production but also limits the scalability of these techniques, especially in industries that rely on mass production. The high cost and complexity create barriers to entry for smaller companies or developing economies, making it difficult for them to adopt nanotechnology at scale.[16]

Overcoming these challenges requires ongoing innovation in fabrication methods, as well as a deeper understanding of how materials behave and interact at the nanoscale. While the road to large-scale, cost-effective nanodevice fabrication is complex, advances in 3D nanoprinting, hybrid material integration, and automation offer promising solutions to these barriers.

12.3 Integration and Compatibility Challenges

The successful deployment of nanotechnology in various sectors, including environmental protection, waste management, and industrial manufacturing, hinges not only on overcoming the challenges of synthesis and fabrication but also on the seamless integration of nanomaterials and devices with both existing technologies and market needs. Equally important is ensuring their compatibility with regulatory standards and environmental sustainability, as these factors play an increasing role in nanotechnology's widespread adoption.

Integrating nanotechnology into both microscale and macroscale systems often necessitates significant—and potentially costly—modifications to existing equipment, as well as the development of entirely new characterization and fabrication techniques. Traditional methods like photolithography or reactive-ion etching may require upgrades to handle the precision demanded by nanophotonic and quantum technologies. Advanced techniques such as atomic layer deposition and near-field scanning optical microscopy are becoming critical for ensuring that nanomaterials maintain their integrity during manufacturing processes. These technical advancements must also align with evolving regulatory and environmental standards, which will shape the adoption of nanotechnology.

This is particularly evident in the field of integrated photonics, where traditional fabrication methods for creating photonic integrated circuits on flat substrates are being challenged by the integration of nanophotonic elements.[17] The advancement of quantum technologies in photonics, such as quantum communication and quantum computing, demands precise control of nanoscale photons for the creation and manipulation of quantum states. To achieve this level of precision, significant upgrades to existing equipment are necessary:

- Lithography systems must be enhanced or replaced with more advanced technologies like electron-beam lithography to achieve nanoscale resolution.

- Etching tools need modifications to allow for more precise control at the nanoscale, potentially incorporating atomic layer etching techniques.
- Deposition systems require upgrades to achieve atomic-level precision, such as the implementation of atomic layer deposition systems.
- Characterization equipment needs to be improved to provide nanoscale resolution, possibly incorporating techniques like near-field scanning optical microscopy.

The growing importance of hybrid photonic integration, which allows for the combination of different materials and independent fabrication of devices, further emphasizes the need for significant modifications to existing manufacturing systems. This approach requires equipment capable of handling multiple materials and fabrication processes.

Similarly, in biomedical engineering, microfluidic devices—used for applications such as lab-on-a-chip systems—are fabricated using microscale techniques. Incorporating nanomaterials into these devices to enhance functionality, such as through the addition of nanoparticle-based sensors or drug delivery systems, demands equipment that can handle the integration without disrupting the delicate fluid dynamics within the microchannels.

These examples from photonics and biomedical engineering illustrate a broader trend: as nanotechnology becomes increasingly integrated into various fields, the equipment and processes used in manufacturing and characterization must evolve to meet the demands of working at the nanoscale while maintaining compatibility with existing macro- and microscale systems. This evolution in equipment and processes is crucial for bridging the gap between traditional manufacturing techniques and the precise requirements of nanotechnology integration.

Beyond technical integration challenges, commercializing nanomaterial innovations faces significant economic barriers.[18] The 'Valley of Death' between laboratory success and market viability stems from unclear commercialization pathways, insufficient scale-up funding, and limited industry engagement. Strategic partnerships, intellectual property management, and supportive regulations are crucial for bridging this gap. Joint-development partnerships and licensing agreements can help companies leverage research outputs while sharing development costs.

While these innovations could enable next-generation nanomaterial-enhanced devices across industries, success requires addressing both technical and commercial challenges, including regulatory compliance and environmental concerns. The development of adaptable manufacturing processes, viable business models, and clear regulatory frameworks will be essential for a sustainable nanotechnology industry.

12.4 Performance and Reliability Issues

The successful implementation of nanotechnology in waste management and environmental protection relies not only on overcoming synthesis, fabrication, and integration challenges but also on ensuring consistent, reliable performance over extended periods. This section explores the critical issues of long-term stability, degradation, repeatability, and reliability that must be addressed for nanotechnology solutions to be viable in real-world applications.

12.4.1 Long-term Stability and Degradation

Ensuring the long-term stability of nanomaterials and nanodevices under operational conditions is crucial for their practical implementation in environmental applications. The unique properties that make nanomaterials effective can also make them susceptible to degradation or changes over time, potentially compromising their performance.

Key challenges in maintaining long-term stability include:

12.4.1.1 Chemical Stability

Many nanomaterials are highly reactive due to their large surface area-to-volume ratio. This reactivity can lead to undesired chemical changes when exposed to environmental conditions over extended periods. For instance, perovskite nanocrystals (PNCs) can degrade when exposed to moisture, oxygen, or light, leading to a decrease in their photoluminescence efficiency and overall performance in optoelectronic applications. When PNCs encounter humid conditions, they may undergo a phase transition from a stable cubic structure to a less stable orthorhombic structure, significantly altering their optoelectronic properties.

To address these stability issues, researchers have developed various strategies, including encapsulation methods using organic polymers or inorganic materials.[19] These approaches have shown promise in enhancing the longevity of PNCs, with some encapsulated nanocrystals maintaining a high percentage of their original fluorescence intensity even after extended exposure to challenging environmental conditions.

Such advancements in stability are crucial for the practical application of PNCs in devices like LEDs and solar cells. However, it's important to note that long-term stability and degradation remain significant challenges for nanomaterials in general, including PNCs. Despite ongoing research and improvements, ensuring the sustained performance of nanomaterials over extended periods in real-world applications continues to be an active area of investigation in the field of nanotechnology.

12.4.1.2 Physical Stability

Nanomaterials may undergo physical changes such as agglomeration, sintering, or morphological alterations, which can significantly impact their performance.[20] For example, nanoparticles used in water treatment may aggregate over time, reducing their effective surface area and, consequently, their pollutant removal capacity.[21] Studies on alumina nanoparticles and

CdSe/CdS nanocrystals provide evidence of unintended and undesirable sintering in nanomaterials. Due to their high surface area-to-volume ratio, nanoparticles are prone to sintering over short timescales during processing or use, which can alter their morphology and structure.

For instance, sintering of alumina nanoparticles can lead to substantial changes in their performance characteristics.[22] Similarly, preventing the sintering of CdSe/CdS nanocrystals is crucial to maintaining the thermal stability required for applications in lasers, LEDs, and bio-labeling.[23]

12.4.1.3 Thermal Stability

Temperature fluctuations in operational environments can significantly impact the stability of nanostructures, potentially leading to changes in their physical and chemical properties. This is particularly relevant for applications exposed to harsh conditions or those involving temperature-sensitive processes: In catalysis, nanoparticles like platinum or gold can undergo thermal degradation, reducing their catalytic efficiency.[24] In electronics, silicon nanowires used in transistors might experience morphological changes or phase transformations when exposed to high temperatures, compromising device performance and longevity.[25] In biomedical applications, gold nanorods used for photothermal therapy must maintain their shape and size under varying temperatures to ensure effective treatment.[26]

12.4.1.4 Mechanical Stability

Nanostructured materials or coatings can experience mechanical stress, wear, or deformation over time, which may lead to the degradation of their functional properties. These materials, especially those used in flexible electronics, biomedical implants, or structural applications, are prone to mechanical failures such as cracking, delamination, or abrasion.

For instance, nanocoatings on biomedical devices may face repeated mechanical stresses, resulting in wear and tear that diminishes their protective or functional capabilities. In structural applications, nanoparticles embedded in composites can become dislodged or lose their binding effectiveness due to mechanical impacts or continuous flexing. These mechanical stability issues are critical as they can directly affect the durability and reliability of nanomaterials in practical applications, leading to premature failure or reduced effectiveness.

12.4.1.5 Photostability

Nanomaterials used in applications exposed to light, such as photocatalysts, solar cells, and environmental sensors, face significant challenges in maintaining stability under continuous illumination. Prolonged exposure to light, especially ultraviolet or high-intensity visible light, can trigger photochemical reactions that degrade the material's structure and function.

For example, in solar cells, quantum dots or other nanomaterials may undergo photooxidation or photobleaching, resulting in decreased efficiency and performance over time.[27] In photocatalytic applications, sustained exposure to light can lead to structural breakdowns or the formation of unwanted byproducts, diminishing the catalytic efficiency.[28] The challenge of maintaining photostability is crucial for ensuring that nanomaterials retain their desired properties and performance levels over extended periods in light-exposed environments.

To address these stability issues, researchers are exploring various strategies that target the specific challenges faced by nanomaterials in environmental applications. These approaches aim to enhance the longevity and reliability of nanomaterials while maintaining their unique properties and functionalities. The strategies being developed span a wide range of techniques, from molecular-level modifications to system-level designs, and are often tailored to the specific type of nanomaterial and its intended application. Some of the key approaches include:

12.4.1.6 Surface Modification

Developing coatings or surface treatments to protect nanomaterials from degradation involves creating protective layers or functionalizing the surface of nanomaterials to enhance their resistance to chemical, physical, and environmental stressors. As we have seen above in the example of perovskite nanocrystals, encapsulation methods using organic polymers or inorganic materials have shown promise in improving their stability against moisture, oxygen, and light exposure. These protective coatings can significantly extend the lifespan of nanomaterials while preserving their unique properties and functionalities. However, it is crucial to consider how these modifications might affect other desirable properties of the nanomaterials, such as their optical or electronic characteristics.

A similar approach has been applied to metal–organic frameworks. Researchers have developed superhydrophobic double-layered coatings incorporating MOFs and other materials to protect magnesium alloys from corrosion.[29] In one such design, a layered double hydroxide structure acts as both a protective shield and a means to strengthen the binding between the metal substrate and a top layer of hydrophobic MOF. This type of coating demonstrates how carefully engineered surface treatments can enhance the stability and functionality of nanomaterials like MOFs in challenging environments, while also improving their adhesion to substrates for practical applications in fields such as biomedicine and materials science.

12.4.1.7 Hybrid Systems

Combining nanomaterials with other materials to create more stable composite structures leverages the synergistic effects of different materials to enhance overall stability and performance. For instance, creating hybrid structures by combining different nanomaterials can enhance both photostability and catalytic efficiency of photocatalysts (as is the case with TiO_2/graphene hybrids).[30] These hybrid systems often demonstrate improved resistance to degradation factors such as thermal stress, mechanical wear, and chemical attack, while maintaining or even enhancing the desired functional properties of the nanomaterials. Nevertheless, the complexity of designing and synthesizing these hybrid systems can present challenges, particularly in optimizing the balance between stability and performance.

12.4.2 Repeatability and Reliability

Achieving consistent performance across multiple production batches and over time is essential for the widespread adoption of nanotechnology solutions in environmental applications. Repeatability and reliability issues can arise from various sources, including:

- **Manufacturing variability**: Slight variations in synthesis conditions can lead to significant differences in nanomaterial properties between batches. This is particularly challenging when scaling up production from laboratory to industrial levels.
- **Environmental sensitivity**: The high surface area and reactivity of nanomaterials make them sensitive to environmental conditions during storage, transport, and use. Variations in temperature, humidity, or exposure to contaminants can affect their performance.
- **Measurement and characterization challenges**: Accurately and consistently measuring nanomaterial properties across different batches and over time can be difficult, leading to uncertainties in quality control and performance prediction.
- **Integration issues**: When incorporating nanomaterials into devices or larger systems, ensuring consistent integration and performance across multiple units is crucial.
- **Aging effects**: The performance of nanomaterials and nanodevices may change over time due to various aging mechanisms, even under normal operating conditions.

To address these challenges and enhance the repeatability and reliability of nanotechnology applications, several strategies can be employed:

- **Standardization**: Developing and implementing standardized protocols for synthesis, characterization, and testing of nanomaterials across the industry.
- **Advanced process control**: Implementing sophisticated process control systems in manufacturing to minimize batch-to-batch variations.
- ***In situ* monitoring**: Developing techniques for real-time monitoring of nanomaterial properties during synthesis and application to ensure consistent performance.

- **Predictive modeling**: Utilizing computational models to predict long-term behavior and potential failure modes of nanomaterials in specific applications.
- **Accelerated aging tests**: Designing and conducting accelerated aging tests that can reliably predict long-term performance under various environmental conditions.

Addressing performance and reliability issues is crucial for the successful translation of nanotechnology innovations from the laboratory to real-world environmental applications. As the field advances, ongoing research and development efforts must focus on enhancing the long-term stability, repeatability, and reliability of nanomaterials and nanodevices to fully realize their potential in addressing pressing environmental challenges. These efforts will require interdisciplinary approaches, combining materials science, data analytics, and systems engineering to overcome the complex challenges associated with nanomaterial performance and reliability in environmental applications.

12.5 Measurement and Characterization Challenges

The ability to accurately measure and characterize nanomaterials and nanodevices is crucial for their successful implementation in applications. However, the unique properties of nanoscale materials present significant challenges for traditional characterization techniques. This section explores the limitations of current methods and the emerging approaches for real-time monitoring and quality control in nanomanufacturing.

12.5.1 Limitations of Current Characterization Techniques

Accurately measuring and characterizing nanoscale materials, particularly in dynamic environments, poses several challenges:

Spatial resolution is a critical factor in the characterization of nanomaterials. Many conventional characterization techniques provide limited spatial resolution for analyzing individual nanostructures or nanoscale features. For example, dynamic light scattering (DLS) is sensitive to larger particles and aggregates, which can obscure the contribution from smaller particles in a sample. This limitation necessitates the use of more advanced techniques such as transmission electron microscopy, scanning electron microscopy, or atomic force microscopy for detailed nanoscale analysis. TEM in particular can provide atomic-scale resolution, allowing visualization of crystal structures and individual nanoparticles. Techniques like high-resolution TEM (HRTEM) and scanning transmission electron microscopy (STEM) offer even greater capabilities for studying atomic-level features and structures of nanoparticles. These higher resolution techniques are crucial for characterizing properties like crystal structure, particle size, and morphology at the nanoscale level.[31]

Temporal resolution is another critical factor in nanomaterial characterization, particularly when studying dynamic processes. Capturing rapid dynamic processes at the nanoscale requires extremely high temporal resolution. Many current techniques, such as conventional electron microscopy, often provide excellent spatial resolution but may lack the speed necessary to observe fast chemical reactions or structural changes in real-time. This limitation has driven the development of ultrafast characterization methods, including time-resolved spectroscopy and high-speed imaging techniques.

Recent advancements, such as the use of ultrafast electron diffraction with THz-driven electron bunch compression and time-stamping, have enabled temporal resolutions down to femtoseconds, allowing for the detailed observation of transient phenomena in materials, such as oscillations in THz near fields in gold nanofilms.[32]

Similarly, high-resolution spatiotemporal mapping using ultrafast laser excitation has demonstrated the ability to capture energy transport dynamics in silicon at sub-picosecond temporal resolution, underscoring the importance of these methods in understanding fast processes in nanomaterials.[33]

Sample preparation poses unique challenges in nanomaterial characterization. Many high-resolution characterization techniques, such as TEM, require extensive sample preparation that can potentially alter the properties of nanomaterials. This is particularly problematic for materials intended for environmental applications, where understanding their behavior in natural, unaltered states is crucial. Researchers are continually developing new methods to minimize sample preparation artifacts, such as cryo-electron microscopy techniques that allow for the observation of samples in a near-native state.[34]

***In situ* and *operando* measurements** represent a frontier in the characterization of nanoporous materials. Characterizing these materials under realistic operating conditions (*in situ*) or while they are actively functioning (*operando*) is extremely challenging but crucial for understanding their behavior and optimizing their performance. For example, observing the structural changes of metal–organic frameworks during gas adsorption, or monitoring zeolite catalysts during chemical reactions, requires sophisticated techniques and specialized equipment.

Overcoming these challenges often involves developing custom sample environments, such as high-temperature/high-pressure cells for *in situ* NMR, environmental transmission electron microscopy chambers, or specialized reactors for synchrotron X-ray experiments. These setups aim to maintain realistic conditions while still allowing for high-resolution characterization. The field has seen significant advancements in recent years, with the development of combined techniques like simultaneous *in situ* SAXS/WAXS, time-resolved *in situ* X-ray diffraction, and *operando* spectroscopy methods.

These advanced characterization approaches are essential for establishing structure-property relationships and gaining insights into the dynamic behavior of nanoporous materials during adsorption, catalysis, and other processes. *In situ* and *operando* techniques are at the forefront of efforts to comprehensively understand and optimize the performance of nanoporous materials for various applications.[35]

Multiscale characterization remains a significant challenge in nanotechnology. Nanomaterials often exhibit properties that span multiple length scales, from individual atomic arrangements to macroscale behavior. Developing techniques that can seamlessly bridge these scales remains a significant challenge. This has led to the development of correlative microscopy approaches, where multiple complementary techniques are used to analyze the same sample across different length scales.[36,37] Integrating data from these various techniques and developing a cohesive understanding of material behavior across scales continues to be an active area of research and development.

12.5.2 Real-time Monitoring and Quality Control

Implementing effective real-time monitoring and quality control during nanomanufacturing is essential for ensuring consistent performance and reliability in environmental applications. However, this endeavor presents several unique challenges that researchers and manufacturers must address.

The balance between speed and accuracy is a critical consideration in real-time monitoring. Real-time monitoring requires rapid data acquisition and analysis, but this often comes at the cost of reduced accuracy or resolution. Striking the right balance between speed and precision is crucial for effective quality control. This challenge has spurred the development of advanced sensing technologies and data processing algorithms that aim to maximize both speed and accuracy.

Non-destructive testing presents another significant hurdle in nanomanufacturing quality control. Many traditional quality control methods in manufacturing involve destructive testing of samples. However, for high-value nanomaterials or nanodevices, non-destructive techniques that can be applied to every unit produced are highly desirable but technically challenging to implement. This has led to increased research into non-invasive characterization methods, such as advanced spectroscopic techniques, imaging modalities, and other innovative approaches that can provide detailed information without damaging the sample.

The challenges are particularly pronounced in additive manufacturing of nanocomposites, where conventional testing methods may introduce artifacts or false structural features that can be mistaken for actual defects, compromising the accuracy of defect analysis. As a result, there is a growing emphasis on developing and refining specialized non-destructive testing techniques tailored for nanoscale materials and structures.

Some promising non-destructive methods being explored include X-ray computed tomography, ultrasonic phased array technology, and infrared thermography. These techniques offer the potential for high-resolution, three-dimensional analysis of internal structures and defects without compromising the integrity of the manufactured parts. Implementing these methods effectively at the nanoscale remains a significant technical challenge, requiring ongoing research and development.

The ultimate goal is to develop reliable, efficient, and scalable non-destructive testing methods that can be integrated into nanomanufacturing processes, enabling real-time quality control and ensuring the consistent production of high-quality nanomaterials and nanodevices.[38]

Process variability is a critical concern in nanomanufacturing that complicates real-time monitoring efforts. Nanomanufacturing processes can be highly sensitive to small variations in conditions. Developing monitoring systems that can detect and account for these subtle changes is crucial for maintaining consistent quality. This requires not only highly sensitive detection methods but also sophisticated statistical process control techniques adapted for the nanoscale.

The integration of sensing systems with process control in nanomanufacturing environments presents both technical and logistical challenges. Effective quality control requires not just monitoring but also the ability to make real-time adjustments to the manufacturing process. This necessitates the development of robust feedback systems that can rapidly translate monitoring data into process adjustments, a task that is particularly complex given the sensitivity and precision required in nanomanufacturing.

To address these challenges, several innovative approaches are being developed and refined. In-line spectroscopic techniques, such as Raman spectroscopy or terahertz spectroscopy, are being adapted for in-line monitoring of nanomaterial properties during production. These techniques can provide real-time information on chemical composition and structure without requiring sample preparation, offering a powerful tool for non-destructive, real-time quality control.[39]

Artificial intelligence and machine learning algorithms are being employed to rapidly analyze complex datasets from multiple sensors, enabling real-time detection of process deviations and prediction of product quality. These technologies offer the potential to identify subtle patterns and correlations that might escape human observers, enhancing the precision and reliability of quality control processes.

Machine learning models can be integrated into closed-loop manufacturing systems, allowing for on-demand and reverse process adaptation to improve quality control during production. This capability extends to various manufacturing techniques, including additive manufacturing, where AI-powered systems can perform real-time video monitoring and automated adjustment of processing parameters. By leveraging computer vision and deep learning techniques, these systems can detect defects, anomalies, and quality issues with high accuracy, often surpassing traditional quality control methods.

The integration of AI in nanomanufacturing not only addresses current challenges in product quality variation but also increases confidence in the manufacturing process while reducing unpredictability.[40]

12.6 Safety and Environmental Concerns

As the integration of nanotechnology into waste management and environmental protection advances, it is critical to assess the safety and environmental implications of these innovations. While nanotechnology offers promising solutions for pollution control, resource recovery, and ecosystem restoration, its application raises several concerns related to human health, environmental safety, and regulatory compliance.

12.6.1 Human Health Risks

Nanomaterials exhibit distinct properties which, while beneficial for numerous applications, may also pose significant health risks that are not yet fully understood. Due to their nanoscale dimensions, these materials can easily bypass traditional biological barriers, such as the skin, respiratory tract, and gastrointestinal lining. Once inside the body, their small size enables them to traverse cell membranes, enter the bloodstream, and circulate throughout the body, potentially accumulating in critical organs, such as the lungs, liver, kidneys, and brain.[41]

This ability to penetrate deeply into biological tissues and their persistence in various body compartments can lead to a range of adverse health effects. Inhaled nanoparticles can reach the alveoli in the lungs, where they may cause inflammation, oxidative stress, or even damage to the alveolar cells.[42] The potential for nanomaterials to cross the blood–brain barrier raises concerns about neurotoxicity, as certain types of nanoparticles could interact with neural tissues, potentially leading to neurodegenerative conditions or other neurological impacts.[43]

The high reactivity of nanomaterials may result in the generation of reactive oxygen species, which can lead to oxidative stress, DNA damage, and cellular apoptosis.[44] These processes have been linked to various health issues, including chronic respiratory diseases, cardiovascular problems, and even cancer. Additionally, due to their ability to bypass immune detection and clearance mechanisms, nanoparticles can remain in the body for prolonged periods, leading to chronic exposure and a sustained risk of toxicity.

Given these potential risks, it is crucial to thoroughly evaluate the biocompatibility, bioaccumulation, and long-term health impacts of nanomaterials, especially as their use in industrial, medical, and consumer products continues to grow.

12.6.1.1 Toxicity of Nanoparticles

Nanoparticles exhibit a range of toxicological profiles depending on their chemical composition, size, shape, surface charge, and functionalization (see Figure 12.4).

Different types of nanoparticles demonstrate varying levels of cytotoxicity and can generate ROS that damage cellular components like DNA, proteins,

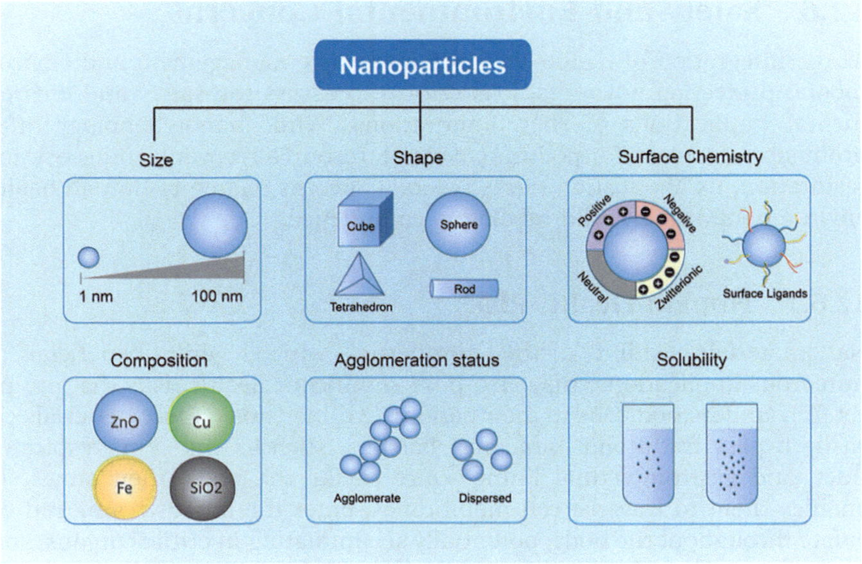

Figure 12.4 Physicochemical properties affecting metal-based nanoparticle toxicity. Different sizes, shapes, chemical compositions, surface chemistry, solubility and agglomeration are closely related to the toxicity induced by metal-based nanoparticles. Reproduced from ref. 45, https://doi.org/10.3389/fbioe.2022.1001572, under the terms of the CC BY 4.0 license, https://creativecommons.org/licenses/by/4.0/.

and lipids. For instance, silver nanoparticles are known for their antimicrobial properties but have also been shown to cause cytotoxic effects in mammalian cells due to their ability to release silver ions, which interact with cellular thiol groups, disrupting cellular respiration and inducing oxidative stress.[45]

Titanium dioxide nanoparticles, commonly used in cosmetics and sunscreens, can also pose significant health risks, particularly when inhaled. Studies have shown that inhaled titanium dioxide nanoparticles can induce pulmonary inflammation and fibrosis, which are linked to oxidative stress and DNA damage.[46] In contrast, carbon nanotubes, due to their needle-like structure and high aspect ratio, have been compared to asbestos fibers in terms of their potential to cause respiratory issues, including inflammation and granuloma formation, when inhaled. Their ability to pierce cell membranes and induce chronic inflammation raises concerns about their carcinogenic potential, especially with long-term exposure.[47]

The conditions under which nanoparticles are synthesized, used, or disposed of can significantly influence their toxicity.[48] For example, nanoparticles that are stable in one environment may become highly reactive and toxic under different environmental or physiological conditions, such as changes in pH or the presence of specific ions. This variability complicates efforts to predict and manage the risks associated with nanomaterials, necessitating comprehensive risk assessments that consider various exposure scenarios.

The interactions of nanoparticles with biological systems are also highly dependent on their surface properties. Surface modifications aimed at improving functionality or biocompatibility can unintentionally increase toxicity. For instance, coating nanoparticles with certain polymers to enhance their stability or targeting capabilities may result in unintended interactions with cell membranes or proteins, potentially leading to adverse health effects.[49] Understanding these nuanced interactions and the mechanisms by which nanoparticles induce toxicity is crucial for developing safer nanomaterials and mitigating potential health risks associated with their use.

12.6.1.2 Occupational Exposure

Occupational exposure to nanomaterials presents a potential risk to workers in industries involved in the production, handling, or application of these materials.[50] The synthesis and manipulation of nanomaterials often involve processes that can release nanoparticles into the air, leading to potential inhalation exposure.

The routes of occupational exposure primarily include inhalation, dermal contact, and, less commonly, ingestion. Inhalation is considered the most significant route of exposure, as nanoparticles can remain suspended in the air and be easily inhaled. Due to their extremely small size, these particles can penetrate deep into the respiratory tract, reaching the alveoli where gas exchange occurs, posing a risk of pulmonary inflammation, oxidative stress, and other respiratory conditions. Dermal exposure, while less understood, can occur when nanoparticles come into direct contact with the skin, potentially leading to localized or systemic effects, particularly if the skin is compromised or damaged.[51]

12.6.1.3 Regulatory Challenges

The rapid advancement of nanotechnology continues to outpace the development of regulatory frameworks designed to ensure safety and environmental protection. One of the significant regulatory challenges is the lack of standardized testing methods for assessing the toxicity and environmental impact of nanomaterials. Traditional toxicological tests, often designed for larger-scale materials, may not adequately capture the unique properties and behaviors of nanoparticles, necessitating the development of new testing protocols tailored specifically to nanoscale materials.[52]

Complicating matters, the environmental fate of nanomaterials is not well understood, which makes regulating their use and disposal more difficult. Nanoparticles can behave differently in various environmental contexts, with their stability, mobility, and persistence being influenced by factors such as temperature, pH, and the presence of other chemicals. This variability poses a challenge for regulators attempting to establish guidelines that account for the wide range of possible scenarios in which nanomaterials may be used or disposed of.

Additionally, there is a significant disparity in how different countries and regions regulate nanotechnology, leading to inconsistencies in safety standards and complicating international trade and collaboration. Some regions have implemented more stringent regulations, while others are still in the process of developing their frameworks. This lack of harmonization poses a challenge for companies operating globally, as they must navigate a complex landscape of varying regulatory requirements.[53,54]

Ongoing regulatory initiatives aim to address these challenges by developing more comprehensive guidelines for nanomaterials, fostering international collaboration to harmonize standards, and investing in research to better understand the health and environmental impacts of nanotechnology.

12.6.2 Environmental Impact

The integration of nanotechnology in various sectors has led to the increased production and use of nanomaterials, raising concerns about their release into the environment. Whether through industrial processes or consumer products, the environmental presence of nanomaterials poses potential risks that are not yet fully understood. The unique properties of nanomaterials significantly influence their behavior, transport, and fate in natural environments. Understanding these dynamics is essential for assessing the ecological impact of nanotechnology and developing strategies to mitigate potential risks.

12.6.2.1 Ecotoxicology

The field of ecotoxicology examines the toxic effects of substances on various ecological entities, and the unique characteristics of nanoparticles present new challenges for researchers. Unlike conventional pollutants, nanoparticles can interact with organisms and ecosystems in complex ways due to their nanoscale dimensions and enhanced reactivity. Studies have shown that certain nanoparticles can adversely affect a wide range of aquatic and terrestrial organisms, including fish, nematodes, and plants.[55]

In aquatic environments, zinc oxide nanoparticles have been found to cause developmental toxicity in zebrafish, affecting processes such as hatching rates and embryonic development. These nanoparticles can also induce changes in gene expression related to inflammation, immune responses, and oxidative stress in fish. For soil organisms like the nematode *Caenorhabditis elegans*, both zinc oxide and titanium dioxide nanoparticles have been shown to alter gene expression patterns associated with stress responses, metabolism, and reproductive functions.

Plants are not immune to the effects of these nanoparticles either. In *Arabidopsis thaliana*, a model plant organism, exposure to zinc oxide and titanium dioxide nanoparticles has been found to trigger changes in gene expression related to stress responses, metabolic processes, and plant development. These molecular-level changes can potentially lead to alterations in plant growth, reproduction, and overall fitness.

The effects of nanoparticles on these diverse organisms highlight the potential for broader ecosystem impacts. Disruptions at various trophic levels, from soil microorganisms to plants and animals, could lead to altered ecosystem functions, including changes in nutrient cycling, primary production, and food web dynamics. As research in this field progresses, it becomes increasingly clear that understanding the ecotoxicological effects of nanoparticles is crucial for assessing their environmental impact and developing appropriate regulatory measures.

In terrestrial environments, nanoparticles can affect plant growth and development. Studies have indicated that nanoparticles like silver and cerium dioxide can interfere with seed germination and root elongation, potentially reducing plant biomass and altering plant community structures.[56] The accumulation of nanoparticles in plants can lead to their entry into the food chain, posing risks to herbivores and, subsequently, to higher trophic levels, including humans. The potential for biomagnification of nanoparticles through the food chain raises additional concerns about long-term ecological and health impacts.[57]

12.6.2.2 Persistence and Bioaccumulation

Nanomaterials' persistence and bioaccumulation potential in the environment are key factors determining their long-term ecological impact. Due to their stability and resistance to degradation, some nanomaterials can persist in soil, water, and sediments for extended periods. For example, carbon-based nanomaterials are known for their durability and resistance to chemical and biological degradation. Once released into the environment, these materials can accumulate in sediments, where they may pose a risk to organisms living on or near the seabed and potentially disrupt communities within the sediment.[58,59]

The small size and high reactivity of nanoparticles also contribute to their bioaccumulation potential in organisms. Once absorbed, nanoparticles can translocate within an organism's tissues, potentially accumulating in specific organs or cellular compartments. Research has shown that certain nanoparticles, such as quantum dots, can accumulate in aquatic organisms, including fish and invertebrates, leading to oxidative stress, inflammation, and other toxic effects.[60,61] The bioaccumulation of nanoparticles in lower levels of the food chain can also lead to biomagnification, where concentrations of nanoparticles increase at each step up, potentially affecting predators and, ultimately, human health.[62]

Understanding the factors influencing the persistence and bioaccumulation of nanomaterials is crucial for assessing their ecological risks. The interactions between nanoparticles and natural organic matter, minerals, and other environmental constituents can alter their stability, mobility, and bioavailability, complicating efforts to predict their behavior and impact in natural settings.

12.6.2.3 Waste Disposal and Recycling

The disposal and recycling of nanomaterial-containing waste present unique challenges that require innovative approaches. Traditional waste treatment methods, such as incineration, landfilling, and conventional water treatment, may not effectively capture or neutralize nanoparticles. Nanoparticles can pass through filtration systems due to their small size, potentially leading to their release into the environment through treated wastewater. Incineration processes may not completely degrade nanoparticles, resulting in the formation of secondary pollutants or the release of nanoparticles into the atmosphere.[63]

To address these challenges, developing waste management strategies specifically tailored for nanomaterials is critical. One approach is to improve the capture and containment of nanoparticles in waste treatment facilities by using advanced filtration technologies, such as ultrafiltration and membrane filtration, which can effectively remove nanoparticles from fluids. In addition, modifying existing incineration processes to achieve higher temperatures and more efficient combustion could help reduce the release of nanoparticles into the atmosphere.

Recycling nanomaterials poses another significant challenge, given their unique properties and potential for contamination. Developing effective recycling methods requires understanding the behavior and interactions of nanoparticles during the recycling process to ensure they do not pose additional risks. Certain nanoparticles could be recovered and reused in industrial processes, reducing the demand for raw materials and minimizing environmental impact. However, these methods need to be carefully designed to prevent the release of nanoparticles during recycling and to ensure that recycled materials do not pose additional risks.[64]

Promoting a circular economy approach, where nanomaterials are reused or recycled within closed-loop systems, can also minimize waste and reduce the environmental footprint of nanotechnology. This approach requires collaboration among industry stakeholders, researchers, and regulators to develop safe and sustainable practices for nanomaterial production, use, and disposal.

12.6.3 Mitigation Strategies

To effectively address the safety and environmental concerns associated with nanotechnology, a multifaceted approach is required. This approach involves the implementation of comprehensive strategies that span across different stages of nanomaterial development and use, from synthesis and application to disposal. By integrating principles of sustainability, safety, and public awareness, these strategies can help mitigate the potential risks posed by nanomaterials and promote the responsible advancement of nanotechnology.[65]

12.6.3.1 Lifecycle Assessment

Lifecycle assessment is a critical tool for evaluating the environmental and health impacts of nanotechnology products throughout their lifecycle (see Figure 12.5). However, current LCA practices for nanomaterials face significant challenges and limitations.[66]

Most assessments are limited to cradle-to-gate analyses, which cover only the extraction of raw materials, manufacturing, and production stages up to the point where the product leaves the factory gate. This approach often neglects the use and end-of-life stages, which may lead to incomplete or misleading conclusions about the overall environmental impact of nanomaterials. In contrast, a cradle-to-grave analysis encompasses the entire lifecycle, including the product's use phase and final disposal or recycling, providing a more comprehensive understanding of its environmental impacts.

Several key challenges persist in conducting LCAs for nanomaterials. For instance, the inadequate definition of functional units often fails to capture the unique properties of nanomaterials. A case in point is the assessment of

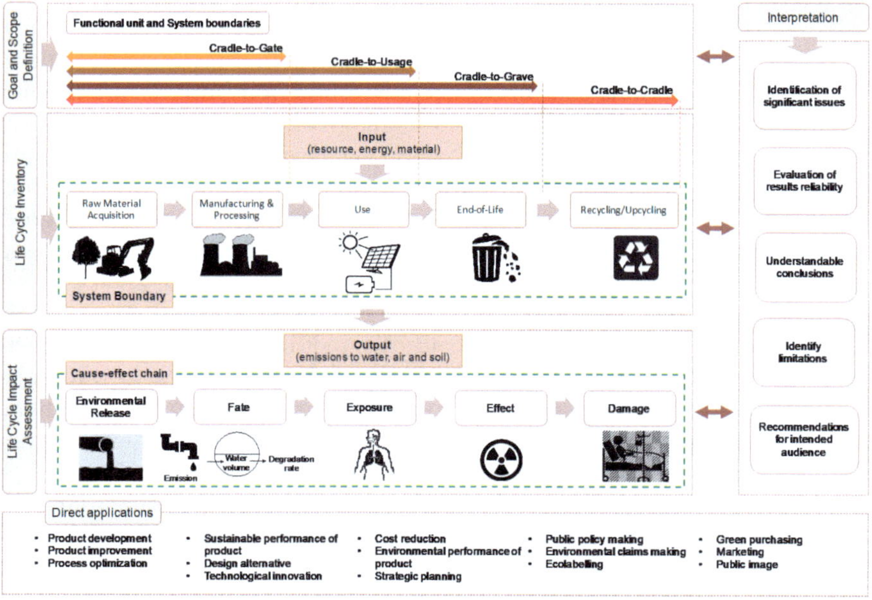

Figure 12.5 A generic life cycle assessment framework. Reproduced from ref. 66, https://doi.org/10.3390/nano11123324, under the terms of the CC BY 4.0 license, https://creativecommons.org/licenses/by/4.0/.

carbon nanotubes, where using weight as the sole functional unit may not accurately represent their enhanced strength and conductivity compared to conventional materials.

Another challenge is the lack of comprehensive and transparent Life Cycle Inventory (LCI) data for nanomaterial production processes. This is exemplified in studies of nano-titanium dioxide, where limited data on the energy and resource inputs for various production methods can lead to inconsistent or unreliable LCA results.

To improve the robustness and reliability of nanomaterial LCAs, future studies need to address these challenges. This involves extending system boundaries from cradle-to-gate to cradle-to-grave analyses, developing more appropriate functional units that reflect the enhanced properties of nanomaterials, improving data collection and sharing for nanomaterial production processes, advancing characterization methods for nanomaterial emissions and their potential impacts, and incorporating uncertainty and sensitivity analyses to address data gaps and variability in nanomaterial properties and behaviors.

By addressing these issues, LCA can provide more accurate and comprehensive assessments of nanomaterial environmental and health impacts. This will enable better-informed decision-making in nanomaterial design, production, use, and regulation, ultimately fostering the development of more sustainable nanotechnology applications.

12.6.3.2 Green Nanotechnology

Green nanotechnology focuses on developing nanomaterials and nanotechnology applications that minimize environmental impact and enhance sustainability. This approach involves using greener synthesis methods, reducing resource consumption, and selecting environmentally benign precursors and solvents. By adopting the principles of green chemistry, which emphasize the reduction or elimination of hazardous substances, green nanotechnology aims to create safer and more sustainable nanomaterial production processes.[67]

Several innovative techniques have emerged under the umbrella of green nanotechnology. For instance, biosynthesis methods using plant extracts, microorganisms, or enzymes have become viable alternatives to traditional chemical synthesis processes. These biological methods can produce nanoparticles under mild conditions without the need for toxic chemicals, significantly reducing the environmental footprint of nanomaterial production.[68,69]

A key aspect of green nanotechnology is the design of nanomaterials with inherent safety in mind, known as the "safe-by-design" approach. This concept involves engineering nanomaterials to minimize potential toxicity and environmental persistence while maintaining their desired functional properties. For example, encapsulation techniques can be employed to reduce the bioavailability or reactivity of active compounds, thereby lowering their potential for bioaccumulation and toxicity.[70]

Green nanotechnology has found applications in various fields, including drug delivery systems, antifouling coatings, corrosion resistance, and environmental remediation. These applications demonstrate the potential of green nanotechnology to address real-world challenges while prioritizing environmental sustainability.[71]

To ensure the responsible development of green nanotechnology, it is crucial to assess the environmental behavior, fate, and effects of new nanomaterials throughout their lifecycle. This comprehensive evaluation helps mitigate potential risks and promotes the safe and responsible use of nanomaterials.

By prioritizing safety and sustainability from the outset, green nanotechnology plays a vital role in addressing societal challenges and supporting sustainable development goals. As the field continues to evolve, it promises to deliver innovative solutions that balance technological advancement with environmental stewardship.

12.6.3.3 Enhanced Regulatory Frameworks

The establishment of robust regulatory frameworks is essential for managing the potential risks associated with nanotechnology and ensuring the safe production, use, and disposal of nanomaterials and other nanostructures. Current regulatory approaches often struggle to keep pace with the rapid development of nanotechnology, resulting in gaps in safety standards and oversight. Enhancing these frameworks requires a coordinated effort among international regulatory bodies, industry stakeholders, and researchers to develop standardized guidelines and best practices for nanomaterial safety.

One of the primary challenges in regulating nanotechnology is the need for standardized definitions and classification systems for nanomaterials. The diverse range of nanomaterials, each with unique properties and behaviors, complicates efforts to establish universal safety standards. Developing standardized testing methods for assessing the toxicity, environmental impact, and fate of nanomaterials is critical to overcoming these challenges. Such methods should account for the unique physicochemical properties of nanoparticles, including their size, shape, surface chemistry, and reactivity.[72]

International cooperation and standardization efforts are also vital for aligning safety guidelines and best practices globally. Initiatives such as the OECD Working Party on Manufactured Nanomaterials aim to harmonize regulatory approaches across countries and promote the safe and sustainable development of nanotechnology. By fostering collaboration and information exchange, these efforts can help ensure that regulatory frameworks are adaptable and responsive to emerging scientific knowledge and technological advancements.[73]

12.6.3.4 Public Awareness and Education

Promoting public awareness and education about nanotechnology's potential risks and benefits is crucial for fostering informed decision-making and encouraging the responsible development and use of nanotechnology applications. Educating stakeholders—industry professionals, policymakers, researchers, and the general public—about the safety, environmental, and ethical considerations associated with nanotechnology can help build trust and facilitate the adoption of best practices for risk management.

Educational initiatives should aim to provide accessible and accurate information about the properties and potential impacts of nanomaterials. This can be achieved through public outreach campaigns, workshops, and educational programs that target specific audiences, such as industry workers, educators, or community members. By raising awareness about the potential risks associated with nanomaterials, these initiatives can empower individuals to make informed choices about the products they use and the practices they adopt in their professional and personal lives.

In addition to raising awareness, fostering a culture of transparency and open communication is essential for addressing public concerns about nanotechnology. Engaging with communities, stakeholders, and civil society organizations through public consultations and participatory decision-making processes can help ensure that diverse perspectives are considered in the development and regulation of nanotechnology. By promoting a collaborative approach to risk management, public awareness, and education efforts can help build a more inclusive and responsible nanotechnology ecosystem.

12.7 Economic and Resource Constraints

Unlocking nanotechnology's full potential for waste management and environmental protection requires more than just technological advancements. Economic and resource constraints present significant challenges that could impede the widespread adoption of these innovative solutions. For nanotechnology to make a meaningful impact in creating a greener, cleaner world, it is essential to address the financial, material, and logistical barriers that currently limit its application.

The integration of nanotechnology into sustainable practices is not just about improving performance or efficiency; it also involves overcoming substantial economic hurdles. The high costs associated with developing and deploying nanomaterials, coupled with scalability issues when moving from laboratory research to industrial-scale production, can deter investment and slow down progress.

In addition to direct costs, the environmental and energy footprint of nanotechnology production must be carefully managed. While the end goal is to enhance sustainability, the processes involved in producing nanomaterials can sometimes be resource-intensive or environmentally taxing, potentially undermining the green credentials of nanotechnology solutions.

To realize the promise of nanotechnology in advancing sustainability goals, it is crucial to develop strategies that address these economic and resource challenges. By innovating production methods, fostering sustainable practices, and promoting supportive policies, stakeholders can work together to overcome these barriers, paving the way for nanotechnology to drive meaningful environmental change.

12.7.1 High Costs and Scalability Issues

Nanotechnology innovations often come with high initial costs due to the need for specialized equipment, materials, and skilled personnel. Developing and implementing these solutions requires significant investment in advanced laboratories, high-precision manufacturing tools, and sophisticated analytical instruments. For instance, the production of nanomaterials often necessitates high-temperature furnaces, cleanroom environments, and precise deposition techniques, which require substantial capital expenditure. Additionally, the development of these technologies involves hiring highly trained scientists, engineers, and technicians, further driving up costs. These upfront investments can be a significant barrier, particularly for smaller companies and organizations in developing regions that may not have the financial resources to support such high-risk, capital-intensive projects.

Scaling up from lab-scale research to industrial-scale production presents additional economic challenges.[18] Many nanomaterial synthesis processes that function well in a controlled laboratory setting do not easily translate to large-scale manufacturing. Maintaining the uniformity and quality of nanomaterials—such as consistent particle size, purity, and defect control—becomes more challenging as production volumes increase. This scaling process often requires redesigning equipment, developing new methods to manage production conditions, and implementing rigorous quality control measures, all of which add to the costs. Transitioning to larger-scale production typically necessitates compliance with stricter regulatory standards related to environmental safety and worker protection, further increasing expenses.

One of the most notable examples of successful commercialization of nanotechnology is in the semiconductor industry, particularly in chip manufacturing.[74] Modern semiconductor fabrication relies heavily on nanoscale processes, with transistors now routinely fabricated at dimensions smaller than 10 nanometers. This industry has managed to overcome many scalability challenges through substantial investments in research and development, advanced lithography techniques, and automated manufacturing processes. The result has been the mass production of increasingly powerful and energy-efficient chips that are integral to nearly all modern electronic devices, from smartphones to supercomputers. This success demonstrates that, despite the high costs and technical hurdles, it is possible to achieve economically viable large-scale production of nanotechnology-based products.

Other nanomaterials have also seen successful commercialization. Titanium dioxide nanoparticles, for instance, are widely used in sunscreens, paints, and coatings, thanks to advancements in scalable production techniques that maintain quality while reducing costs.[75] Silver nanoparticles are used in textiles, healthcare, and consumer products due to efficient manufacturing methods that support large-scale production.

These examples illustrate that, while significant barriers exist, there are viable pathways to commercialization that can lead to economically sustainable production. Strategic investments in infrastructure, collaborative efforts to share risks and resources, and innovative approaches to streamline production processes are essential. By continuing to develop cost-effective, scalable methods for nanomaterial synthesis and ensuring robust supply chains, stakeholders can make nanotechnology solutions more accessible and economically viable, maximizing their potential to drive sustainability and innovation across various sectors.

12.7.2 Resource Availability and Supply Chain Risks

Nanotechnology relies on specific raw materials that are often scarce, costly, or environmentally challenging to produce, posing significant risks to its sustainable deployment. Many applications depend on rare earth elements (REE), a group of 17 metals such as neodymium, cerium, platinum and palladium, which are essential for their unique properties. However, these materials are primarily sourced from a few countries, such as China and Russia, making supply chains vulnerable to geopolitical tensions and trade restrictions. China's control over the REE market means that any export restrictions or geopolitical disputes could lead to supply disruptions and price spikes, as seen in past export bans.[76]

The extraction and refining of these materials are often environmentally damaging and energy-intensive, raising concerns about their sustainability. To mitigate these risks, the sector must explore alternative materials, invest in recycling and recovery initiatives, and develop more sustainable sourcing strategies. By addressing these challenges, nanotechnology can achieve more stable and sustainable growth, ensuring its innovations contribute effectively to environmental protection and sustainability goals.

12.7.3 Energy and Environmental Costs

Producing nanomaterials can be highly energy-intensive, often requiring substantial amounts of electricity, heat, and specialized infrastructure to achieve the precise conditions needed for synthesis. For example, processes such as chemical vapor deposition and high-temperature plasma synthesis, which are commonly used to produce nanomaterials like carbon nanotubes and silicon nanoparticles, consume large amounts of energy. This high energy demand can offset some of the environmental benefits that nanotechnology aims to provide, particularly when the energy is sourced from fossil fuels. The carbon footprint associated with these energy-intensive processes can be substantial, potentially negating the positive impact of nanotechnology applications designed to enhance sustainability, such as improved pollution control or energy storage.[77]

In addition to energy consumption, the production of nanomaterials often involves hazardous chemicals and solvents that can be difficult to manage sustainably. For instance, the synthesis of quantum dots or certain nanocatalysts may require the use of toxic solvents or reagents, such as hydrazine or cadmium compounds, which pose significant health and environmental risks.[78] The disposal of these chemicals is challenging, as they can lead to contamination of soil and water resources if not properly managed. As discussed before, the handling and disposal of nanomaterials themselves can present unique environmental challenges.

Balancing these energy and environmental costs is critical to ensuring that the green potential of nanotechnology is fully realized. This balance involves not only improving the efficiency of nanomaterial production processes but also adopting greener chemistry practices that minimize the use of hazardous substances and reduce waste. Innovations in low-energy synthesis methods, such as microwave-assisted or solvent-free techniques, are emerging as promising alternatives that can reduce both energy consumption and environmental impact. Adopting a lifecycle perspective—considering the energy and environmental impacts of nanomaterials from production through to disposal or recycling—can help identify opportunities for further reducing the overall ecological footprint.[79]

12.8 Interdisciplinary Challenges

This chapter explores two key interdisciplinary challenges in the field: bridging knowledge gaps across disciplines and enhancing communication and collaboration among specialists. These challenges are crucial because they directly impact the effectiveness, safety, and rapid development of nanotechnology applications in real-world settings.

12.8.1 Knowledge Gaps Across Disciplines

One of the fundamental challenges in nanotechnology is the need to bridge knowledge gaps across various scientific and engineering disciplines.[80] Nanotechnology is inherently multidisciplinary, requiring an understanding of physics, chemistry, and materials science.[81] However, to be effectively applied in environmental protection, sustainability and waste management, additional knowledge from biology, toxicology, and environmental science is crucial.

Each discipline brings its own specialized knowledge base, methodologies, and terminologies, making it challenging to integrate these diverse perspectives. For example, materials scientists may develop a nanomaterial with excellent pollutant-adsorbing properties, but they may lack insights into the potential ecological or health impacts of this material once it enters the environment. Conversely, environmental scientists may understand how pollutants behave in natural ecosystems but may not be familiar with the synthetic techniques and mechanisms behind nanomaterial creation and modification.

Bridging these knowledge gaps is essential for several reasons:

- **Integrated solutions**: To develop effective nanotechnology solutions, teams need to integrate expertise from multiple fields to design materials that are not only effective in their intended applications but also safe for humans and the environment. Without this integration, the development of nanotechnology could lead to solutions that are functionally excellent but potentially harmful or unsustainable.
- **Risk assessment and management**: Understanding the full scope of nanomaterial impacts—from their creation to their degradation in natural systems—requires a cross-disciplinary approach. Without it, there is a risk of unintended consequences, such as secondary pollution or unexpected health effects. Effective risk management strategies must consider both the technological capabilities of nanomaterials and their potential long-term impacts on ecosystems and human health.
- **Innovation and optimization**: Cross-disciplinary knowledge can lead to more innovative approaches to problem-solving and optimize existing technologies. For instance, combining insights from materials science with biological perspectives could lead to the development of biocompatible nanomaterials for environmental remediation. This kind of synergy between disciplines can accelerate technological advances and create more holistic solutions to environmental challenges.

12.8.1.1 Educational Aspects: Interdisciplinary Curricula and Training

The challenge of bridging knowledge gaps also extends to education and training. To prepare future generations of scientists and engineers for the complexities of nanotechnology, there is a growing need for interdisciplinary curricula that integrate multiple fields of study. Some universities have already implemented these programs but designing them is challenging due to the depth of knowledge required in each discipline and the limited time available within academic courses. Developing effective interdisciplinary education programs also requires collaboration among faculty from different departments and the creation of joint degree programs or specialized courses in nanotechnology.[82]

12.8.2 Communication and Collaboration

Effective communication and collaboration among specialists from different disciplines are equally crucial but challenging. Nanotechnology projects often require the collaboration of experts from diverse fields who may not share a common language or understanding of each other's methodologies. This can lead to several issues:

- **Terminology barriers**: Each discipline has its own jargon and specialized terms, which can create confusion and miscommunication. For example, terms like "particle size", "reactivity", or "stability" might have different connotations for a materials scientist compared to an environmental toxicologist, leading to misunderstandings about a material's properties or behavior.
- **Methodological differences**: Different disciplines often use varying methodologies to study similar phenomena. For instance, a chemist might focus on the molecular interactions of nanomaterials, while an environmental scientist might be more concerned with their ecological impacts. These differing perspectives can make it difficult to align research goals and experimental designs.
- **Goal alignment**: Interdisciplinary projects often face challenges in aligning goals. While a materials scientist might prioritize enhancing the functionality of a nanomaterial, a toxicologist might focus on minimizing its potential risks. Effective collaboration requires finding a balance between these often-competing objectives to achieve a solution that is both innovative and safe.

Addressing communication and collaboration challenges in nanotechnology is particularly critical due to the unique properties and behaviors of materials at the nanoscale. Unlike other technological fields where materials and methods are often well-defined within specific disciplines, nanotechnology requires integrating expertise from multiple areas—such as materials science, physics, chemistry, biology, environmental science, and toxicology. This integration is essential because the effects of nanomaterials can be highly unpredictable and vary significantly depending on the context in which they are used.

A comprehensive, holistic approach is therefore crucial to account for all potential impacts, from initial development to final deployment. This need for interdisciplinary collaboration is more pronounced in nanotechnology than in other tech areas, where the risks associated with improper deployment—such as unintended ecological damage or health risks from nanoparticle exposure—can be much more severe. Consequently, effective communication and collaboration among disciplines are not just beneficial but necessary to ensure that all aspects of a nanotechnology application are carefully considered, ultimately supporting safe and effective innovation in real-world settings.

12.8.2.1 International Collaboration: Navigating Cultural and Language Differences

Unlike more established technologies, nanotechnology often deals with novel materials and processes for which there are no universally accepted standards or regulations. This lack of standardization can lead to substantial differences in how risk and safety are interpreted and managed across countries. For instance, some countries may have more stringent regulations regarding the environmental release of nanomaterials due to concerns about their long-term ecological impacts, while others may prioritize the economic benefits of rapid nanotechnology development.

These disparities complicate collaborative efforts in developing and implementing nanotechnology solutions, as teams must navigate not only scientific and technical challenges but also a patchwork of regulatory environments that may demand significant adaptations in research design and methodology. Additionally, the high level of uncertainty surrounding the safety and environmental effects of nanomaterials often leads to divergent risk assessments and safety protocols, further complicating international collaborations.

To address these specific challenges, it is crucial to develop clear communication protocols tailored to the complexities of nanotechnology. This may include creating harmonized guidelines for risk assessment, fostering greater regulatory alignment, and utilizing specialized teams that are not only multilingual but also proficient in the regulatory and cultural contexts of the countries involved. Building a culture of mutual respect and understanding is vital, but in the context of nanotechnology, it also requires a shared commitment to adhering to the highest standards of safety and environmental protection, ensuring that collaborative efforts do not compromise on these critical aspects.

12.8.2.2 Emerging Tools and Platforms for Collaboration

Several emerging tools and platforms are designed to facilitate interdisciplinary collaboration in nanotechnology.[83] Databases like the European Union Observatory for Nanomaterials (EUON)[84] provide standardized information on nanomaterial properties, behaviors, and safety aspects, accessible to researchers from various fields. These resources help aggregate and share data on the physical and chemical properties of nanomaterials, as well as their interactions with biological systems and environmental behaviors. This standardized data is crucial for enhancing understanding and promoting consistent methodologies across disciplines.

Additionally, nanotechnology-specific modeling and simulation tools are increasingly used to predict the behaviors and interactions of nanomaterials in different environments. Tools such as NanoHUB,[85] an online platform providing a wide range of simulation tools for nanoscience, allow researchers to model electronic, optical, and mechanical properties of nanomaterials and devices. QuantumATK,[86] a software suite developed by Synopsys, offers atomic-scale modeling of materials, enabling simulations of electronic properties, nano-device characteristics, and molecular dynamics. Similarly, LAMMPS (Large-scale Atomic/Molecular Massively Parallel Simulator)[87] is widely utilized for materials modeling at the atomic scale, particularly for simulating the behavior of nanoparticles, nanotubes, and thin films. These tools help bridge knowledge gaps by allowing researchers to visualize and understand complex nanoscale interactions that would be difficult to study experimentally.

By providing a common framework and language for interpreting these interactions, these platforms help foster more effective communication and collaboration among multidisciplinary teams. In addition, some virtual research environments are tailored specifically to nanotechnology research. These digital platforms provide integrated spaces where researchers can collaborate remotely, share nanomaterial datasets, and access specialized computational tools designed for nanoscience. This targeted approach helps to overcome the communication barriers unique to nanotechnology, enabling researchers to more effectively align their research goals and methodologies.

By leveraging these nanotechnology-specific tools and platforms, the global research community can better address the complex, interdisciplinary nature of nanotechnology challenges, ultimately supporting the development of safe and innovative solutions for environmental protection and waste management.

By addressing these two key interdisciplinary challenges—bridging knowledge gaps and enhancing communication and collaboration—researchers and practitioners can more effectively develop and implement nanotechnology solutions that are innovative, effective, and safe for both humans and the environment. These efforts are critical for advancing nanotechnology's role in creating sustainable solutions for waste management and environmental protection, ultimately contributing to broader societal goals related to sustainability and public health.

12.9 Regulatory and Ethical Challenges

The rapid advancement of nanotechnology offers groundbreaking solutions in many technological areas, presenting unprecedented opportunities to address some of the world's most pressing environmental challenges. However, these advancements also introduce a range of regulatory and ethical issues that are critical to address to ensure safe, responsible, and equitable deployment. This section looks at the regulatory and ethical challenges associated with implementing nanotechnology solutions on a global scale, highlighting why these considerations are vital for the technology's sustainable growth and societal acceptance.

The unique properties of nanomaterials present challenges that existing regulatory frameworks were not designed to handle. For example, traditional safety regulations in the cosmetics industry, like those enforced by the U.S. Food and Drug Administration or the European Medicines Agency, are primarily geared toward assessing the safety of bulk chemical ingredients. However, these regulations fall short when applied to nanoparticles, such as titanium dioxide used in sunscreens. Unlike their bulk counterparts, TiO_2 nanoparticles can penetrate deeper into the skin and generate reactive oxygen species when exposed to UV light, potentially causing oxidative stress and DNA damage—risks that conventional safety assessments do not adequately capture. This gap in regulatory oversight underscores the need for new testing methods and safety standards that specifically address the unique behaviors and risks associated with nanomaterials.

The ethical implications of deploying such advanced technologies, particularly in ways that might exacerbate existing inequalities or cause unintended harm, cannot be overlooked. Addressing these challenges is essential for ensuring that the benefits of nanotechnology are realized in a manner that is safe, fair, and aligned with societal values.

Understanding and navigating these regulatory and ethical landscapes are crucial for several reasons. Firstly, they help mitigate potential risks associated with nanomaterials, protecting both the environment and public health. Secondly, they foster public trust and acceptance by ensuring transparency and ethical considerations in nanotechnology deployment.

Finally, by promoting international cooperation and harmonized standards, these efforts can facilitate more equitable access to the benefits of nanotechnology, particularly for developing regions that might otherwise be left behind. Thus, this section aims to provide a comprehensive overview of the regulatory and ethical challenges facing nanotechnology, underscoring the importance of a proactive and balanced approach to its development and implementation.

12.9.1 Regulatory Challenges

Nanotechnology's unique properties and behaviors at the nanoscale present a range of challenges for existing regulatory frameworks. Traditional regulatory approaches often do not adequately capture the potential risks associated with nanomaterials, leading to gaps in oversight and safety standards.[88,89] Key regulatory challenges include:

- **Lack of standardized definitions and testing protocols**: There is no universally accepted definition of what constitutes a nanomaterial, complicating regulatory efforts.[90,91] Moreover, traditional testing methods for chemical safety may not be appropriate for nanomaterials, which can behave differently due to their size and surface area. This necessitates the development of new testing protocols specifically designed for nanoparticles.
- **Inadequate risk assessment methods**: Current risk assessment frameworks are often ill-equipped to evaluate the unique risks posed by nanomaterials. This includes difficulties in assessing long-term exposure effects, environmental persistence, and bioaccumulation. Improved risk assessment models that account for the distinct characteristics of nanomaterials are urgently needed.

  For instance, silver nanoparticles, widely used in antimicrobial products such as clothing, coatings, and medical devices, pose unique challenges for regulatory bodies. In the United States, the Environmental Protection Agency classifies silver nanoparticles as a pesticide, requiring them to undergo rigorous testing and registration under the Federal Insecticide, Fungicide, and Rodenticide Act (FIFRA).[92] This approach has been criticized for not adequately considering the distinct behaviors and risks of nanoparticles, such as their persistence in the environment and potential to bioaccumulate, which are not typically accounted for in traditional pesticide assessments.

- **Fragmented regulatory landscape**: Different countries and regions have developed their own regulatory standards for nanotechnology, leading to a fragmented global regulatory environment. This inconsistency poses challenges for international trade and cooperation, as companies must navigate a complex array of regulations and compliance requirements.
- **Emerging technologies and unknown risks**: The rapid pace of innovation in nanotechnology often outstrips the ability of regulatory bodies to keep up. Emerging applications, such as nanomaterials used in medicine or food, present unknown risks that current regulations do not adequately address. There is a need for dynamic regulatory approaches that can adapt to new scientific knowledge and technological advancements.

12.9.2 International Cooperation and Ethical Frameworks

The deployment of nanotechnology in environmental and waste management applications raises several ethical concerns that must be carefully considered. Addressing these challenges often requires a mix of international cooperation and context-specific ethical frameworks. While some risks and benefits of nanotechnology, such as the potential for widespread environmental impact or the development of global safety standards, are inherently global in scope, others may be more regional or local, shaped by specific socio-economic, cultural, and environmental conditions. A nuanced approach is needed to ensure that nanotechnology is developed and deployed in a manner that is both globally responsible and sensitive to regional differences, safeguarding human health and the environment while promoting equitable access to its benefits. Key actions include:

- **Harmonization of standards**: Efforts should be made to harmonize regulatory standards and testing protocols across countries and regions to ensure consistent safety and quality standards globally. Organizations such as the OECD and ISO are already working towards developing standardized guidelines for nanomaterials, but more work is needed to ensure global compliance.
- **Ethical guidelines for nanotechnology**: Developing ethical guidelines that govern the use of nanotechnology, particularly in sensitive areas such as environmental protection, is essential. These guidelines should

emphasize the importance of sustainability, equity, and transparency in nanotechnology deployment.[93]

- **Capacity building in developing countries**: Providing technical assistance and capacity-building initiatives for developing countries can help ensure that they are not left behind in the nanotechnology revolution. This includes providing access to knowledge, infrastructure, and financial resources necessary to implement nanotechnology solutions safely and effectively.
- **Promoting public awareness and engagement**: In the context of environmental remediation, sustainability, and resource efficiency, many nanotechnology applications—such as advanced filtration systems, catalytic processes, and high-performance industrial materials—operate behind the scenes within industrial processes or infrastructure. These technologies are not directly visible or apparent to the general public, as they are integrated into systems that manage pollution, enhance material properties, or improve resource use efficiency. In these cases, broad public education may not be essential, as these applications do not involve direct consumer interaction or immediate public visibility.

There is a distinction to be made when considering more consumer-facing nanotechnology products. Applications such as nanomaterials in cosmetics, food packaging, medical devices, and agriculture are directly used or encountered by the public. In these cases, educating the public about potential benefits, risks, and safety considerations is crucial for building trust, ensuring informed consumer choices, and maintaining transparency. Public engagement initiatives in these areas should be an integral part of the regulatory and ethical framework to address any concerns related to health, safety, and ethical use.

By making this distinction, it becomes clear that while public engagement is vital for consumer-facing nanotechnology applications, for environmental and industrial uses, the focus should be on informing and engaging key stakeholders—policymakers, industry leaders, and environmental professionals—to ensure these technologies are safely and effectively integrated into broader environmental management and sustainability efforts.

12.10 Case Studies of Specific Challenges

12.10.1 Scaling Challenges in Graphene Production

Since its isolation in 2004, graphene's remarkable properties have sparked immense interest, yet scaling up production from laboratory to industrial levels remains a significant challenge. This is particularly evident when considering the intricate balance between quality, cost, and environmental impact.[94]

Graphene production can broadly be categorized into top-down and bottom-up methods. In the laboratory, mechanical exfoliation is prized for yielding pristine graphene, but it is impractical for large-scale production. Industrially, top-down methods like physical and chemical exfoliation of graphite dominate due to their scalability. However, these methods often compromise quality. For instance, a survey of 60 commercial producers revealed that most graphene products contain less than 10% single-layer graphene and exhibit carbon purities below 90%, reflecting significant challenges in maintaining quality at scale. Raman spectroscopy data from these producers indicate defect levels, with D/G ratios ranging from 0.08 to 0.65, suggesting that many products are closer to graphite nanoplatelets than true graphene.

On the other hand, bottom-up approaches such as chemical vapor deposition allow for the production of large-area graphene films. CVD can achieve growth rates of up to 25 $cm\ min^{-1}$ on metal substrates, yet scaling this method presents its own set of challenges. Even high-quality CVD graphene typically has a sheet resistance of around 1000 $\Omega\ sq^{-1}$, which is significantly higher than the 60–100 $\Omega\ sq^{-1}$ of indium tin oxide, the material commonly used in applications requiring both conductivity and transparency. Furthermore, the process is resource-intensive, often requiring high-purity gases and extensive post-processing steps, such as the etching of metal substrates, which add to both cost and environmental burden.

Environmental concerns are another critical issue in scaling graphene production. Traditional methods can require up to 1000 liters of solvent per kilogram of graphite processed, raising significant sustainability concerns. Emerging technologies, like flash Joule heating (FJH), offer a promising alternative. FJH can produce graphene at rates of 7.5 $g\ h^{-1}$ while using minimal solvents and only 7.2 kJ of energy per gram of graphene. This method not only reduces the environmental impact but also lowers the overall process mass intensity, making it a more sustainable option for large-scale production.

Despite advances in production techniques, the graphene market faces economic challenges. As of 2022, global production was estimated at 3800 tons per year, yet costs remain prohibitively high, ranging from $50 000 to $200 000 per ton for graphene powders and $45 000 to $100 000 per square meter for graphene films. These prices hinder widespread adoption in many applications, despite the material's potential.

Overcoming these scaling hurdles will be crucial to realizing graphene's full potential in commercial applications. Continued research into methods like FJH, along with improvements in CVD processes and a focus on sustainability, will be key. Furthermore, close collaboration between researchers, manufacturers, and end-users will be essential to tailor production methods to specific applications and drive the graphene industry forward.

12.10.2 Integrating Nanocomposites in Medical Device Manufacturing

The integration of nanocomposites into medical device manufacturing exemplifies the healthcare industry's ongoing struggle with nanoscale precision in production. This technology has rapidly advanced to incorporate various nanoparticles, such as hydroxyapatite, silver, and carbon nanotubes, into conventional materials like polymers and metals, promising enhanced properties such as improved biocompatibility, antimicrobial activity, and mechanical strength.

While offering significant potential for improving medical device performance, the incorporation of nanocomposites presents formidable challenges in maintaining consistent material properties throughout the manufacturing process.[95] As the complexity of nanocomposite formulations increases, so does the difficulty in managing variability across the entire product. This variability manifests most notably in issues like non-uniform dispersion of nanoparticles, which can directly impact the device's reliability and performance.

To illustrate the complexities involved in the integration of nanocomposites into medical device manufacturing, consider the step-by-step process of creating an epoxy matrix composite strengthened with nanoparticles. This manufacturing process begins with the selection and preparation of raw materials, including the base epoxy resin and the chosen nanoparticles. The nanoparticles are then dispersed uniformly within the epoxy matrix using methods such as high-shear mixing, ultrasonication, or ball milling to ensure a homogeneous dispersion. Achieving this uniform dispersion is critical for optimizing the composite's mechanical properties, such as strength and durability. Following the mixing process, the resin is combined with a curing agent, poured into a mold, and subjected to specific temperature and pressure conditions to solidify into a composite material with enhanced performance characteristics. This meticulous process highlights the intricate balance required between precision and scalability in nanocomposite manufacturing, underscoring the significant challenges and innovations needed to achieve consistent material quality at a large scale.

Figure 12.6 visually represents these key steps in the manufacturing process, providing a clear understanding of the complexities involved in producing high-performance nanocomposites for medical applications.

A specific example of this challenge can be seen in the development of nanocomposite-enhanced orthopedic implants. In one case, researchers attempted to improve the osseointegration and antimicrobial properties of titanium dental implants by incorporating both hydroxyapatite and silver nanoparticles into a coating. While initial laboratory results were promising, scaling up the manufacturing process revealed significant inconsistencies in nanoparticle distribution across the implant surface. This led to variations in the implant's performance, with some samples showing excellent osseointegration and antimicrobial activity, while others performed poorly. The challenge of achieving uniform nanoparticle dispersion in the coating required extensive refinement of the manufacturing process, including the development of new deposition techniques and quality control measures.

To combat these precision-related challenges, manufacturers have been forced to implement increasingly complex solutions. These include

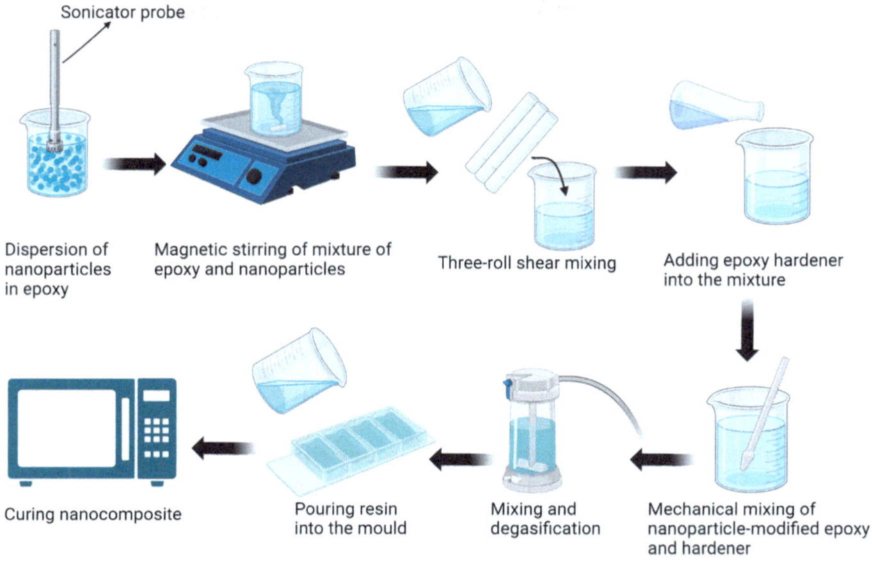

Figure 12.6 Step-by-step manufacturing process for an epoxy matrix composite strengthened with nanoparticles. Reproduced from ref. 95, https://doi.org/10.3390/jcs7050199, under the terms of the CC BY 4.0 license, https://creativecommons.org/licenses/by/4.0/.

advanced processing techniques such as *in situ* polymerization and sonication for improved nanoparticle dispersion, as well as novel coating methods like layer-by-layer deposition. While effective, these approaches add layers of complexity to the production process, inevitably driving up costs.

The push for enhanced functionality and performance has also led to the exploration of "multi-functional nanocomposites," such as those combining drug delivery capabilities with structural support. This shift towards more complex material systems underscores the mounting difficulty and expense involved in maintaining the required level of precision as the technology advances.

Furthermore, as the scale of nanocomposite integration continues to shrink, new challenges emerge. For instance, the need for precise control over nanoparticle release rates in drug-eluting implants becomes critical. Implementing such precise controls consistently across large-scale production represents a significant hurdle for manufacturers.

The case of nanocomposites in medical device manufacturing vividly illustrates the healthcare industry's broader struggle with nanoscale precision manufacturing. The relentless drive for improved device performance, coupled with the need for complex solutions to address variability issues, contributes to escalating production costs and potential inconsistencies in product quality. As the technology continues its advance, these challenges are likely to intensify, highlighting the critical need for breakthrough innovations in manufacturing techniques and materials science within the medical device industry.

This case study underscores the delicate balance manufacturers must strike between leveraging the benefits of nanocomposites and managing the complexities they introduce into the production process. As the field progresses, overcoming these challenges will be crucial for realizing the full potential of nanocomposites in medical devices and improving patient outcomes.

12.10.3 Enhancing Perovskite Stability

Perovskite nanocrystals have become a focal point in optoelectronics due to their remarkable ability to absorb and emit light, making them promising materials for solar cells and LEDs. Despite their potential, these materials face significant challenges related to their stability. Environmental factors such as moisture, oxygen, and light can rapidly degrade PNCs, limiting their practical use. Addressing these stability issues through nanotechnology has brought perovskite-based devices closer to commercial viability.

A major challenge for PNCs, particularly materials like $CsPbI_3$, is their tendency to degrade in humid conditions. Exposure to moisture can cause these nanocrystals to shift from a stable cubic phase to a less stable orthorhombic phase, drastically reducing their efficiency and performance. This transformation significantly impacts the longevity and functionality of perovskite-based devices, making it essential to find ways to stabilize these materials.

To enhance the stability of PNCs, researchers have explored various nanotechnology-driven approaches.[19] One effective method is encapsulation, where the nanocrystals are coated with protective materials. For example, the use of polymers such as polystyrene-*block*-poly(acrylic acid) has been successful in forming protective capsules around the nanocrystals, helping them retain around 60% of their photoluminescence after 180 days in ambient conditions. Another approach involves using silica dioxide as an encapsulating material, which has demonstrated the ability to maintain up to 88% of the fluorescence intensity of PNCs even after 104 days in a high-humidity environment. Although encapsulation techniques can significantly enhance stability, they sometimes introduce challenges, such as reducing light extraction efficiency due to scattering.

In addition to encapsulation, modifying the surface chemistry of PNCs, known as ligand engineering, has proven to be another effective strategy. By designing special organic molecules that attach to the nanocrystal surfaces, researchers can enhance stability by preventing degradation and improving the passivation of surface defects. These ligands help maintain high quantum yields—above 90% in some cases—and ensure the stability of PNCs even after multiple purification processes. For instance, ligands that form strong bonds with the nanocrystal surfaces have been shown to keep the PNCs stable under various environmental conditions, including exposure to light and heat.

Another promising method for improving the stability of PNCs involves doping with metal cations. By introducing small amounts of metal ions, such as zinc or copper, into the PNC structure, researchers have observed substantial improvements in durability. For example, doping $CsPbBr_3$ nanocrystals with zinc ions has been shown to enhance thermal stability, allowing them to maintain 91% of their original luminescence after heating. These doped nanocrystals also exhibit improved resistance to water and oxygen, making them more robust under real-world conditions. However, the effectiveness of doping can vary depending on the type of metal ion used and the specific perovskite material, indicating the need for careful selection and optimization in each case.

Beyond these approaches, optimizing the fabrication process of PNCs has also been a key area of focus. Techniques such as refining the synthesis methods, using advanced post-treatment processes, and developing new materials for encapsulation have shown promise in further enhancing the stability of PNCs. For instance, using a metal–organic framework to encapsulate PNCs has helped create a stable environment that resists photo-induced ion migration and withstands higher laser power irradiation. This method has demonstrated outstanding operational stability in PNC films used for light-emitting devices.

Overall, these nanotechnology strategies—encapsulation, ligand engineering, metal cation doping, and fabrication process optimization—have shown significant promise in overcoming the stability challenges of PNCs. By enhancing resistance to environmental degradation and improving overall durability, these approaches are paving the way for more efficient and longer-lasting perovskite-based solar cells and LEDs. As research continues to evolve, these materials are likely to transition from experimental stages to practical, commercial applications, potentially transforming renewable energy and display technologies.

12.10.4 Size Control of Quantum Dots for Display Performance

In the production of quantum dots for advanced display technologies, precise control over particle size is crucial as it directly impacts optical properties such as emission wavelength, color purity, and efficiency. Quantum dots exhibit size-dependent optical properties, with emission wavelengths that can be tuned across the entire visible spectrum by adjusting particle size. For applications like QD-LED displays, maintaining tight control over these properties is essential to achieve high color purity, a broad color gamut, and optimal efficiency. Narrow emission peaks, with minimal full width at half maximum (FWHM), are critical for delivering vibrant colors and high display performance.

If size control is not adequately maintained during synthesis, quantum dots may exhibit inconsistent emission wavelengths or broader spectral distributions, which can cause color inaccuracies and reduced efficiency in the final display. This variability not only affects color accuracy but also degrades the overall display performance, making it challenging to meet the standards required for high-end applications such as televisions, smartphones, and monitors.

To address these challenges, significant research has focused on developing advanced synthesis methods, such as hot injection and room-temperature synthesis, for quantum dot production.[96] These methods have been tailored to specific materials like cadmium selenide (CdSe), indium phosphide (InP), and more recently, perovskite quantum dots (PQDs). Each of these materials brings its own set of advantages and challenges. For instance, CdSe-based QDs are widely used due to their excellent color performance, but they pose environmental concerns due to cadmium content. InP and perovskites are emerging alternatives, with perovskites offering high photoluminescence quantum yield and tunable emission, though stability issues, particularly thermal and moisture sensitivity, still need to be overcome.

Figure 12.7 visually demonstrates how quantum dots are integrated into micro-LED (μLED) structures, showing a red, green, and blue (RGB) pixel array with nanoring (NR)-μLEDs. The scanning electron microscopy image highlights the close contact between multi-quantum wells and QDs, which is crucial for effective non-radiative resonant energy transfer (NRET) mechanisms. This close interaction facilitates efficient energy transfer, leading to

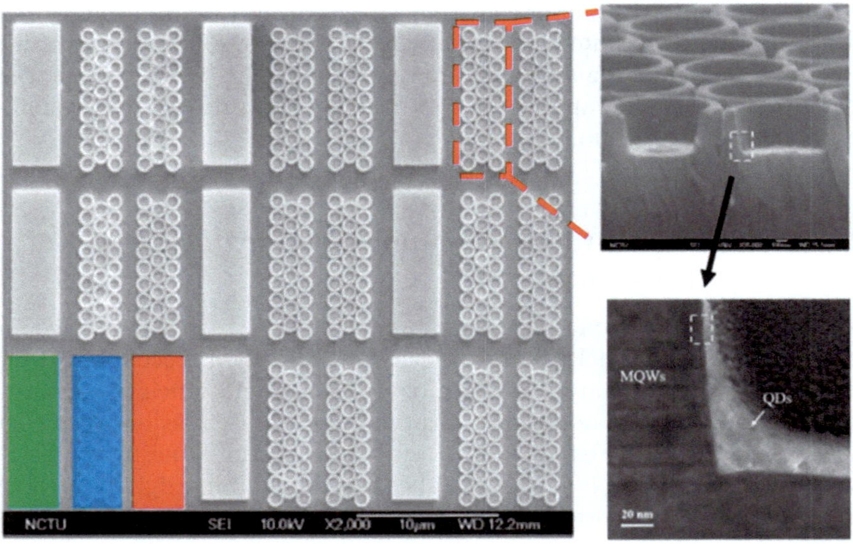

Figure 12.7 Scanning electron microscopy image showing the red, green, and blue (RGB) pixel array and nanoring (NR)-μLED with a 30° tilt. The image highlights the close contact between the multi-quantum wells and quantum dots, which is crucial for efficient energy transfer and color conversion in QD-based displays. Adapted from ref. 96, https://doi.org/10.3390/nano10071327, under the terms of the CC BY 4.0 license, https://creativecommons.org/licenses/by/4.0/.

improved color conversion and high display performance in QD-based technologies.

Process control systems and in-line characterization techniques are being developed to enable real-time feedback during synthesis, allowing for rapid adjustments to maintain the desired quantum dot properties. These systems are essential for ensuring consistent quality in large-scale production, where even minor deviations can lead to significant variations in display performance. However, scaling these solutions to industrial levels presents significant technical and economic challenges, particularly in terms of achieving both high yields and precise control.

The economic implications of stringent control over quantum dot properties are considerable. Achieving high color purity and efficiency often comes at the cost of lower yields and increased production complexity. For example, ensuring that quantum dots meet the narrow emission peaks and color purity required for high-performance displays can result in higher production costs due to the need for more sophisticated synthesis techniques and real-time monitoring systems. Balancing these costs while maintaining commercial viability is a critical challenge for manufacturers.

While the development of quantum dots has enabled significant advances in display technologies, the challenges of controlling quantum dot size and maintaining material stability continue to be focal points of research. The integration of quantum dots into next-generation display technologies will depend not only on overcoming these technical hurdles but also on ensuring that the processes are economically scalable for widespread adoption in consumer electronics.

12.10.5 Challenges in 3D NAND Flash Memory Production

The production of 3D NAND flash memory exemplifies the semiconductor industry's ongoing struggle with nanoscale precision in manufacturing.[97] This technology has rapidly advanced to incorporate up to 176 stacked layers, each requiring exacting control over the deposition and etching of various materials including insulators, charge-trapping layers, and conductive word lines.

This vertical stacking, while enabling significant increases in storage density, presents formidable challenges in maintaining consistent cell characteristics throughout the structure. As the number of layers grows, so does the difficulty in managing variability across the entire pillar. This variability manifests most notably in issues like short-term data retention in charge-trap cells, where threshold voltage distributions can shift and widen shortly after programming, directly impacting the memory's reliability and performance.

To combat these precision-related challenges, manufacturers have been forced to implement increasingly complex solutions. These include advanced programming schemes such as coarse-fine programming and architectural innovations like CMOS under Array. While effective, these approaches add layers of complexity to the production process, inevitably driving up costs.

The push for ever-higher density and performance has also led to the exploration of "disruptive scaling" techniques, such as split cells. This shift away from conventional scaling methods underscores the mounting difficulty and expense involved in maintaining the required level of precision as the technology advances.

Furthermore, as word line pitch continues to shrink, new challenges emerge. For instance, the need for storage layer separation to prevent charge migration between neighboring cells becomes critical. Implementing such precise structural controls consistently across large wafers represents a significant hurdle for manufacturers.

The case of 3D NAND flash memory production vividly illustrates the semiconductor industry's broader struggle with nanoscale precision manufacturing. The relentless drive for higher density and performance, coupled with the need for complex solutions to address variability issues, contributes to escalating production costs and potential inconsistencies in product quality. As the technology continues its advance, these challenges are likely to intensify, highlighting the critical need for breakthrough innovations in manufacturing techniques and materials science within the semiconductor industry.

12.10.6 Bridging Disciplines in Environmental Nanotechnology

A notable example of bridging knowledge gaps in environmental nanotechnology is the collaborative development of nano-enabled water purification systems for heavy metal removal. Interdisciplinary projects involving experts from materials science, surface chemistry, and water technology have been essential for understanding the complex mechanisms of pollutant uptake, fine-tuning the properties of nanoparticles, and adapting these particles for integration into both conventional and novel water treatment schemes.[98]

Materials scientists play a critical role in designing nanoparticles with specific properties suited for targeted pollutant removal. For instance, they developed magnetite (Fe_3O_4) nanoparticles with a size of about 20 nm, optimized to efficiently capture and reduce toxic hexavalent chromium to its less harmful trivalent form. These nanoparticles demonstrated a capacity to remove 1.8 mg of chromium per gram of nanoparticles at neutral pH, a typical condition for drinking water.[99]

Surface chemists further refine these nanoparticles by modifying their surface characteristics to enhance adsorption capacity and selectivity for specific heavy metal ions. This step is crucial for improving the performance of the nanoparticles under varying environmental conditions and for ensuring that they meet safety and regulatory standards.

Environmental engineers contribute by developing practical methods for applying these nanoparticles in real-world water treatment scenarios. They tested the nanoparticles both as dispersed systems and as aggregates in column filters, simulating conditions found in water treatment plants. Notably, the aggregated form in column filters nearly doubled the chromium removal capacity compared to dispersed nanoparticles, showcasing the importance of engineering design in maximizing the efficiency of nanomaterials.

Additionally, environmental considerations were integrated into the research, even though environmental scientists were not part of the core development team. The study evaluated the behavior of these nanoparticles in natural water systems and their potential environmental impact, ensuring a holistic approach to their application.

This collaborative effort not only resulted in a promising solution for chromium removal but also exemplifies how interdisciplinary collaboration can lead to innovative, efficient, and safer methods for addressing drinking water contamination. By bridging the gaps between disciplines, researchers were able to develop a technology that is both scientifically sound and practically viable, underscoring the importance of interdisciplinary approaches in advancing the field of environmental nanotechnology.

References

1. V. Hakke, S. Sonawane, S. Anandan, S. Sonawane and M. Ashokkumar, Process intensification approach using microreactors for synthesizing nanomaterials—A critical review, *Nanomaterials*, 2021, **11**(1), 98.
2. N. Baig, I. Kammakakam and W. Falath, Nanomaterials: a review of synthesis methods, properties, recent progress, and challenges, *Mater. Adv.*, 2021, **2**(6), 1821–1871.
3. C. M. O. González, O. V. Kharissova, L. T. González, M. A. Méndez-Rojas, T. S. Quezada and Y. P. Méndez, Scalable synthesis of nanomaterials, in *Handbook of Nanomaterials and Nanocomposites for Energy and Environmental Applications*, Springer International Publishing, Cham, 2020, pp. 1–24.
4. R. Zhang, C. Tang, W. Ni, J. Yuan, Y. Zhou and X. Liu, Research status and challenges of high-purity quartz processing technology from a mineralogical perspective in China, *Minerals*, 2023, **13**(12), 1505.
5. Nanomaterial scale-up, Cerion Nanomaterials, 2020, [cited 2024 Aug 19]. Available from: https://cerionnano.com/scale-up/
6. V. Harish, M. M. Ansari, D. Tewari, M. Gaur, A. B. Yadav and M.-L. García-Betancourt, *et al.*, Nanoparticle and nanostructure synthesis and controlled growth methods, *Nanomaterials*, 2022, **12**(18), 3226.
7. J. M. Köhler, Challenges for nanotechnology, *Encyclopedia*, 2021, **1**(3), 618–631.
8. J. M. Koehler, What are proteins teaching us on fundamental strategies for molecular nanotechnology?, *Nanotechnol. Rev.*, 2015, **4**(2), 145–160.
9. J. T. Fourkas, J. Gao, Z. Han, H. Liu, B. Marmiroli and M. J. Naughton, *et al.*, Grand challenges in nanofabrication: There remains plenty of room at the bottom, *Front. Nanotechnol.*, 2021, **3**, 700849.

10. U. Tahir, Y. B. Shim, M. A. Kamran, D.-I. Kim and M. Y. Jeong, Nanofabrication techniques: Challenges and future prospects, *J. Nanosci. Nanotechnol.*, 2021, **21**(10), 4981–5013.
11. P. Gargini, F. Balestra and Y. Hayashi, Roadmapping of nanoelectronics for the new electronics industry, *Appl. Sci.*, 2021, **12**(1), 308.
12. K. Muldoon, Y. Song, Z. Ahmad, X. Chen and M.-W. Chang, High precision 3D printing for micro to nano scale biomedical and electronic devices, *Micromachines*, 2022, **13**(4), 642.
13. L. J. Brillson, Defects at nanoscale semiconductor interfaces: Challenges and opportunities, *J. Mater. Res.*, 2024, **39**(2), 177–187.
14. J. Colin, A. Jamnig, C. Furgeaud, A. Michel, N. Pliatsikas and K. Sarakinos, *et al.*, In situ and real-time nanoscale monitoring of ultrathin metal film growth using optical and electrical diagnostic tools, *Nanomaterials*, 2020, **10**(11), 2225.
15. H. Zheng, Imaging, understanding, and control of nanoscale materials transformations, *MRS Bull.*, 2021, **46**(5), 443–450.
16. Navigating the Costly Economics of Chip Making, bcg.com, [cited 2024 Aug 22]. Available from: https://www.bcg.com/publications/2023/navigating-the-semiconductor-manufacturing-costs.
17. F. Lu and Z. Yang, Nanophotonics and integrated photonics, *Appl. Sci.*, 2023, **13**(23), 12605.
18. T. Rambaran and R. Schirhagl, Nanotechnology from lab to industry – a look at current trends, *Nanoscale Adv.*, 2022, **4**(18), 3664–3675.
19. X. Liu and E.-C. Lee, Advancements in perovskite nanocrystal stability enhancement: A comprehensive review, *Nanomaterials*, 2023, **13**(11), 1707.
20. K. Loza, M. Epple and M. Maskos, Stability of nanoparticle dispersions and particle agglomeration, in *NanoScience and Technology*, Springer International Publishing, Cham, 2019, pp. 85–100.
21. X.-H. Zhou, B.-C. Huang, T. Zhou, Y.-C. Liu and H.-C. Shi, Aggregation behavior of engineered nanoparticles and their impact on activated sludge in wastewater treatment, *Chemosphere*, 2015, **119**, 568–576.
22. D. Wang, H. Mei, L. Liu and J. Zhang, The effect of heating rate on sintering mechanism of alumina nanoparticles, *J. Am. Ceram. Soc.*, 2022, **105**(12), 7149–7158.
23. L. Tang, C. Zhang, C. Liao, Y. Liu and Y. Cheng, In situ sintering of CdSe/CdS nanocrystals under electron beam irradiation, *Nanomaterials*, 2023, **13**(24), 3082.
24. A. Aitbekova, C. Zhou, M. L. Stone, J. S. Lezama-Pacheco, A.-C. Yang and A. S. Hoffman, *et al.*, Templated encapsulation of platinum-based catalysts promotes high-temperature stability to 1100 °C, *Nat. Mater.*, 2022, **21**(11), 1290–1297.
25. M. S. R. N. Kiran, T. T. Tran, L. A. Smillie, B. Haberl, D. Subianto and J. S. Williams, *et al.*, Temperature-dependent mechanical deformation of silicon at the nanoscale: Phase transformation *versus* defect propagation, *J. Appl. Phys.*, 2015, **117**, 205901.

26. W. Zhou, Y. Yao, H. Qin, X. Xing, Z. Li and M. Ouyang, *et al.*, Size dependence of gold nanorods for efficient and rapid photothermal therapy, *Int. J. Mol. Sci.*, 2024, **25**(4), 2018.
27. C. Wu, J. Yang, S. Ding, T. Huang, T. Sun and L. Qian, *et al.*, Revealing the origin of the photo-instability to improve the performance of PbS quantum dot solar, *Nano Res.*, 2024, **17**, 10644–10648.
28. S. Zhang, D. Fan, Q. Yan, Y. Lu, D. Wu and B. Fu, *et al.*, Single-molecule fluorescence imaging of photocatalytic nanomaterials, *J. Mater. Chem. A*, 2024, **12**(31), 19627–19662.
29. M. M. Sabzehmeidani and M. Kazemzad, Recent advances in surface-mounted metal–organic framework thin film coatings for biomaterials and medical applications: a review, *Biomater. Res.*, 2023, **27**, 115.
30. A. Nasir, S. Khalid, T. Yasin and A. Mazare, A review on the progress and future of TiO_2/graphene photocatalysts, *Energies*, 2022, **15**(17), 6248.
31. S. Mourdikoudis, R. M. Pallares and N. T. K. Thanh, Characterization techniques for nanoparticles: comparison and complementarity upon studying nanoparticle properties, *Nanoscale*, 2018, **10**(27), 12871–12934.
32. M. A. K. Othman, A. E. Gabriel, E. C. Snively, M. E. Kozina, X. Shen and F. Ji, *et al.*, Improved temporal resolution in ultrafast electron diffraction measurements through THz compression and time-stamping, *Struct. Dyn.*, 2024, **11**, 024311.
33. M. Segovia and X. Xu, Ultrafast, high resolution spatiotemporal mapping of energy transport dynamics for determination of energy transport properties in silicon, *Phys. Rev. B*, 2023, **108**, 125202.
34. Y. Xu and S. Dang, Recent technical advances in sample preparation for single-particle cryo-EM, *Front. Mol. Biosci.*, 2022, **9**, 892459.
35. H. Petersen and C. Weidenthaler, A review of recent developments for the *in situ*/*operando* characterization of nanoporous materials, *Inorg. Chem. Front.*, 2022, **9**(16), 4244–4271.
36. F. Cognigni, M. Rossi, H. Stegmann, G. Sciuto, G. Anastasi, M. Astuto, *et al.* A Multiscale and Multimodal Correlative Microscopy Workflow to Characterize Copper Segregations Identified in Epitaxial Layer of Power MOSFETs, in *International Symposium for Testing and Failure Analysis*, ASM International, 2023.
37. J. Wang, C. Wang, K. Rishi and V. Vogiazi, Direct-on-filter analysis of airborne engineered nanomaterials using correlative microscopy and spectroscopy, *Microsc. Microanal.*, 2024, **30**, ozae044.269.
38. Y. She, J. Tang, C. Wang, Z. Wang, Z. Huang and Y. Yang, Nano-additive manufacturing and non-destructive testing of nanocomposites, *Nanomaterials*, 2023, **13**(20), 2741.
39. IBM.com. TSV Stress Evolution Mapping using in-line Raman Spectroscopy, [cited 2024 Aug 26]. Available from: https://research.ibm.com/publications/tsv-stress-evolution-mapping-using-in-line-raman-spectroscopy.
40. G. Konstantopoulos, E. P. Koumoulos and C. A. Charitidis, Digital innovation enabled nanomaterial manufacturing; Machine learning strategies and green perspectives, *Nanomaterials*, 2022, **12**(15), 2646.

41. M. Awashra and P. Młynarz, The toxicity of nanoparticles and their interaction with cells: an *in vitro* metabolomic perspective, *Nanoscale Adv.*, 2023, **5**(10), 2674–2723.
42. X. Zhou, W. Jin and J. Ma, Lung inflammation perturbation by engineered nanoparticles, *Front. Bioeng. Biotechnol.*, 2023, **11**, 1199230.
43. A. Krsek, A. Jagodic and L. Baticic, Nanomedicine in neuroprotection, neuroregeneration, and blood–brain barrier modulation: A narrative review, *Medicina*, 2024, **60**(9), 1384.
44. T. Wen, J. Liu, W. He and A. Yang, Nanomaterials and reactive oxygen species (ROS), in *Nanotechnology in Regenerative Medicine and Drug Delivery Therapy*, Springer Singapore, Singapore, 2020, pp. 361–387.
45. N. Zhang, G. Xiong and Z. Liu, Toxicity of metal-based nanoparticles: Challenges in the nano era, *Front. Bioeng. Biotechnol.*, 2022, **10**, 1001572.
46. K. Freire, F. Ordóñez Ramos, D. B. Soria, E. Pabón Gelves and A. L. Di Virgilio, Cytotoxicity and DNA damage evaluation of TiO_2 and ZnO nanoparticles. Uptake in lung cells in culture, *Toxicol. Res.*, 2021, **10**(2), 192–202.
47. Z. Nazeri, V. Zarezade, M. Jamalan, M. Cheraghzadeh, S. Azizidoost and A. Kheirollah, Carbon nanotubes induce cytotoxicity and apoptosis through increasing protein levels of Bax and ROS in mouse skin fibroblasts, *Res. Pharm. Sci.*, 2024, **19**(2), 148–156.
48. R. Abbasi, G. Shineh, M. Mobaraki, S. Doughty and L. Tayebi, Structural parameters of nanoparticles affecting their toxicity for biomedical applications: a review, *J. Nanopart. Res.*, 2023, **25**, 43.
49. E. Donadoni, P. Siani, G. Frigerio, C. Milani, Q. Cui and C. Di Valentin, The effect of polymer coating on nanoparticles' interaction with lipid membranes studied by coarse-grained molecular dynamics simulations, *Nanoscale*, 2024, **16**(18), 9108–9122.
50. C. Andraos, M. Gulumian, G. Ichihara, B. Kim and I. J. Yu, Editorial: Occupational exposure to nanomaterials, *Front. Toxicol.*, 2022, **4**, 1014600.
51. S. Omari Shekaftik, N. Mehrparvar, Z. Peivandi and S. J. Shahtaheri, Risk assessment of occupational exposure to engineered and incidental nanomaterials: differences and challenges, *J. Nanopart. Res.*, 2024, **26**, 67.
52. P. O. Isibor, Regulations and policy considerations for nanoparticle safety, in *Environmental Nanotoxicology*, Springer Nature Switzerland, Cham, 2024, pp. 295–316.
53. J. Carter, R. Bjorkland, W. K. Boyes, C. Geraci, V. A. Hackley and J. Howard, *et al.*, U. S. federal perspective on critical research issues in nanoEHS, *Environ. Sci.: Nano*, 2023, **10**(10), 2623–2633.
54. Nanomaterials Regulation – ECHA. Europa.eu, [cited 2024 Aug 27]. Available from: https://euon.echa.europa.eu/regulation.
55. S. Wang, H. Alenius, H. El-Nezami and P. Karisola, A new look at the effects of engineered ZnO and TiO_2 nanoparticles: Evidence from transcriptomics studies, *Nanomaterials*, 2022, **12**(8), 1247.
56. D. K. Verma, S. Patel and K. S. Kushwah, Effects of nanoparticles on seed germination, growth, phytotoxicity and crop improvement, *Agric. Rev.*, 2021, **42**, 1–11.

57. E. A. Kumah, R. D. Fopa, S. Harati, P. Boadu, F. V. Zohoori and T. Pak, Human and environmental impacts of nanoparticles: a scoping review of the current literature, *BMC Public Health*, 2023, **23**, 1059.
58. O. Zaytseva and G. Neumann, Carbon nanomaterials: production, impact on plant development, agricultural and environmental applications, *Chem. Biol. Technol. Agric.*, 2016, **3**, 17.
59. M. Kharlamova and C. Kramberger, Cytotoxicity of carbon nanotubes, graphene, fullerenes, and dots, *Nanomaterials*, 2023, **13**(9), 1458.
60. G. Libralato, E. Galdiero, A. Falanga, R. Carotenuto, E. De Alteriis and M. Guida, Toxicity effects of functionalized quantum dots, gold and polystyrene nanoparticles on target aquatic biological models: A review, *Molecules*, 2017, **22**(9), 1439.
61. F. Wang, W. Guan, L. Xu, Z. Ding, H. Ma and A. Ma, *et al.*, Effects of nanoparticles on algae: Adsorption, distribution, ecotoxicity and fate, *Appl. Sci.*, 2019, **9**(8), 1534.
62. P. O. Isibor, O. S. Adeniyi, O. G. Jerry, A. W. Wurie, A. E. Opeyemi and O. A. Oyewole, Nanoparticles in food chains: Bioaccumulation and trophic transfer, in *Environmental Nanotoxicology*, Springer Nature Switzerland, Cham, 2024, pp. 203–233.
63. Z. Zahra, Z. Habib, S. Hyun and M. Sajid, Nanowaste: Another future waste, its sources, release mechanism, and removal strategies in the environment, *Sustainability*, 2022, **14**(4), 2041.
64. OECD, *Nanomaterials in waste streams: Current knowledge on risks and impacts*, OECD, 2016.
65. US EPA, Advancing Environmental Safety: Celebrating 20 years of Nanotechnology Research at EPA, 2024 [cited 2024 Aug 30]. Available from: https://www.epa.gov/sciencematters/advancing-environmental-safety-celebrating-20-years-nanotechnology-research-epa.
66. N. U. M. Nizam, M. M. Hanafiah and K. S. Woon, A content review of life cycle assessment of nanomaterials: Current practices, challenges, and future prospects, *Nanomaterials*, 2021, **11**(12), 3324.
67. S. S. Salem, A mini review on green nanotechnology and its development in biological effects, *Arch. Microbiol.*, 2023, **205**, 128.
68. N. S. Alsaiari, F. M. Alzahrani, A. Amari, H. Osman, H. N. Harharah and N. Elboughdiri, *et al.*, Plant and microbial approaches as green methods for the synthesis of nanomaterials: Synthesis, applications, and future perspectives, *Molecules*, 2023, **28**(1), 463.
69. A. Q. Malik, T. U. G. Mir, D. Kumar, I. A. Mir, A. Rashid and M. Ayoub, *et al.*, A review on the green synthesis of nanoparticles, their biological applications, and photocatalytic efficiency against environmental toxins, *Environ. Sci. Pollut. Res. Int.*, 2023, **30**(27), 69796–69823.
70. I. Furxhi, A. Costa, S. Vázquez-Campos, C. Fito-López, D. Hristozov and J. A. Tamayo Ramos, *et al.*, Status, implications and challenges of European safe and sustainable by design paradigms applicable to nanomaterials and advanced materials, *RSC Sustainability*, 2023, **1**(2), 234–250.

71. R. Martins and O. B. Kacze–rewska, Green nanotechnology: The latest innovations, knowledge gaps, and future perspectives, *Appl. Sci.*, 2021, **11**(10), 4513.
72. J. Allan, S. Belz, A. Hoeveler, M. Hugas, H. Okuda and A. Patri, *et al.*, Regulatory landscape of nanotechnology and nanoplastics from a global perspective, *Regul. Toxicol. Pharmacol.*, 2021, **122**, 104885.
73. Nanomaterials and advanced materials, OECD, [cited 2024 Aug 30]. Available from: https://www.oecd.org/en/topics/sub-issues/nanomaterials-and-advanced-materials.html.
74. Intel.com, Moore's Law, [cited 2024 Oct 15]. Available from: https://www.intel.com/content/www/us/en/newsroom/resources/moores-law.html.
75. J. Musial, R. Krakowiak, D. T. Mlynarczyk, T. Goslinski and B. J. Stanisz, Titanium dioxide nanoparticles in food and personal care products-what do we know about their safety?, *Nanomaterials*, 2020, **10**(6), 1110.
76. Csis.org, What China's Ban on Rare Earths Processing Technology Exports Means, [cited 2024 Sep 15]. Available from: https://www.csis.org/analysis/what-chinas-ban-rare-earths-processing-technology-exports-means.
77. H. Pedersen, S. T. Barry and J. Sundqvist, Green CVD—Toward a sustainable philosophy for thin film deposition by chemical vapor deposition, *J. Vac. Sci. Technol., A*, 2021, **39**(5), 051001.
78. M. Giroux, Z. Zahra, O. A. Salawu, R. M. Burgess, K. T. Ho and A. S. Adeleye, Assessing the environmental effects related to quantum dot structure, function, synthesis and exposure, *Environ. Sci.: Nano*, 2022, **9**(3), 867–910.
79. S. Alves, M. Gonçalves, H. Monteiro, B. Moura, R. Godina and J. Almeida, Life cycle assessment of nanotechnology: Carbon footprint and energy analysis, in *Environmental Science and Engineering*, Springer Nature Switzerland, Cham, 2023, pp. 233–247.
80. B. Li, S. Chen and V. Larivière, Interdisciplinarity affects the technological impact of scientific research, *Scientometrics*, 2023, **128**(12), 6527–6559.
81. A. L. Porter and J. Youtie, How interdisciplinary is nanotechnology?, *J. Nanopart. Res.*, 2009, **11**(5), 1023–1041.
82. Nanotechnology in curriculum, Nnin.org, [cited 2024 Aug 31]. Available from: https://www.nnin.org/education-training/k-12-teachers/nanotechnology-curriculum.
83. A. Roig, Nanotechnology governance: From risk regulation to informal platforms, *Nanoethics*, 2018, **12**(2), 115–121.
84. National reporting schemes - ECHA [Internet], Europa.eu, [cited 2024 Aug 31]. Available from: https://euon.echa.europa.eu/national-reporting-schemes.
85. Nanohub.org - simulation, education, and community for nanotechnology, Nanohub.org, [cited 2024 Aug 31]. Available from: https://nanohub.org/.
86. Atomistic simulation software, Synopsys.com, [cited 2024 Aug 31]. Available from: https://www.synopsys.com/manufacturing/quantumatk.html.

87. LAMMPS Molecular Dynamics Simulator, Lammps.org, [cited 2024 Aug 31]. Available from: https://www.lammps.org/.
88. M. Ghosh and R. Kumar, Regulatory issues in nanotechnology, in *Livestock Diseases and Management*, Springer Nature Singapore, Singapore, 2024, pp. 765–788.
89. R. Kumari, K. Suman, S. Karmakar, V. Mishra, S. G. Lakra and G. K. Saurav, *et al.*, Regulation and safety measures for nanotechnology-based agri-products, *Front. Genome Ed.*, 2023, **5**, 1200987.
90. K. Rasmussen, J. Riego Sintes and H. Rauscher, How nanoparticles are counted in global regulatory nanomaterial definitions, *Nat. Nanotechnol.*, 2024, **19**(2), 132–138.
91. S. F. Hansen, M. B. Nielsen, O. F. H. Hansen, L. P. W. Clausen, L. M. Skjolding and A. Baun, *et al.*, To be or not to be a nanomaterial, *J. Nanopart. Res.*, 2022, **24**, 235.
92. EPA US, Office of Pesticide Programs, Silver nanoparticles – regulatory actions|chemical search|pesticides|US EPA, [cited 2024 Sep 1]. Available from: https://ordspub.epa.gov/ords/pesticides/f?p=CHEMICALSEARCH:31::::1,3,31,7,12,25:P3_XCHEMICAL_ID:3777.
93. S. Lakshmanan, V. P. Variyam and O. S. N. Ghosh, Making the Research and Innovation Ethical: A Conceptual Framework for Ethics Assessment of Nanotechnology Innovations, Paper presented at: ICN 2024, 2024 May, Kottayam, India. M.G. University [cited 2024 Sep 1]. Available from: https://www.researchgate.net/publication/380457662.
94. K. M. Wyss, D. X. Luong and J. M. Tour, Large-scale syntheses of 2D materials: Flash Joule heating and other methods, *Adv. Mater.*, 2022, **34**(8), 2106970.
95. M. Ramezani and Z. M. Ripin, An overview of enhancing the performance of medical implants with nanocomposites, *J. Compos. Sci.*, 2023, 7(5), 199.
96. Y.-M. Huang, K. J. Singh, A.-C. Liu, C.-C. Lin, Z. Chen and K. Wang, *et al.*, Advances in quantum-dot-based displays, *Nanomaterials*, 2020, **10**(7), 1327.
97. A. Goda, Recent progress on 3D NAND flash technologies, *Electronics*, 2021, **10**(24), 3156.
98. K. Simeonidis, C. Martinez-Boubeta, P. Zamora-Pérez, P. Rivera-Gil, E. Kaprara and E. Kokkinos, *et al.*, Implementing nanoparticles for competitive drinking water purification, *Environ. Chem. Lett.*, 2019, **17**(2), 705–719.
99. K. Simeonidis, E. Kaprara, T. Samaras, M. Angelakeris, N. Pliatsikas and G. Vourlias, *et al.*, Optimizing magnetic nanoparticles for drinking water technology: The case of Cr(VI), *Sci. Total Environ.*, 2015, **535**, 61–68.

CHAPTER 13

Pathways Forward: Advancing Nanotechnology for a Sustainable Future

13.1 Major Nanotechnology Application Areas for Sustainability

Among the many ways nanotechnology contributes to sustainability, three application areas stand out for their potential to drive the most impactful change:

1. Circular economy innovations
2. Energy efficiency and renewable integration
3. Climate change mitigation and adaptation

Current research and implementation indicate these areas have the greatest promise in delivering sustainable solutions through enhanced resource efficiency, energy management, and environmental resilience. These three areas are particularly significant because they are interconnected and mutually reinforcing, addressing both immediate and long-term sustainability challenges. They have demonstrated significant evidence of practical implementation, align with existing policy frameworks and market demands, and offer measurable impacts on sustainability goals. This section examines these high-potential applications, demonstrating how nanotechnology can play a crucial role in advancing global sustainability goals.

Waste Not! How Nanotechnologies Can Increase Efficiencies Throughout Society
By Michael Berger

Published by the Royal Society of Chemistry, www.rsc.org

13.1.1 Circular Economy Innovations

Nanotechnology holds transformative potential in advancing circular economy principles by supporting the development of more circular systems that minimize waste, enhance resource efficiency, and extend the lifecycle of materials. Nanotechnology can support the transition from a linear 'take–make–dispose' model to a more sustainable circular economy by securing resource efficiency and reducing environmental impacts.[1] This transition is further supported by global policy frameworks, such as the European Union's Circular Economy Action Plan, which emphasizes the need for sustainable products and reduced waste generation—objectives that align closely with the capabilities of nanotechnology.[2]

However, realizing this potential requires addressing significant challenges, such as the difficulties in recycling nanomaterials and the need for safe and sustainable design throughout their lifecycle. By overcoming these hurdles, nanotechnology can facilitate the development of innovative solutions that enhance recycling efficiency, enable the reuse and upcycling of materials, and improve the overall sustainability of industrial processes.

Here are several compelling examples, previously discussed in the book, that illustrate the diverse and impactful ways nanotechnology can advance circular economy goals (see Figure 13.1):

13.1.1.1 Enhanced Recycling Processes

Nanotechnology can revolutionize recycling by developing more efficient sorting and separation techniques. For instance, magnetic nanoparticles can be used to separate different types of plastics more effectively, increasing the

Figure 13.1 Nanotechnology advancements for a circular economy.

purity and value of recycled materials. This innovation improves the quality of recycled output and reduces contamination, which often degrades the value of recycled materials (see Section 9.4.1 for detailed discussion).

13.1.1.2 Self-healing Materials for Extended Product Lifespan

Nanotechnology-enabled self-healing materials can autonomously repair small damages, significantly extending the lifespan of products. For example, self-healing concrete infused with nanoparticles can repair small cracks without human intervention, reducing maintenance costs and prolonging infrastructure lifespans. This approach decreases the frequency of replacements and repairs, contributing to more sustainable resource use (see Section 8.4.5).

13.1.1.3 Advanced Filtration and Purification Technologies

Nanomembranes and nanofibers enhance water and air purification systems, allowing for more efficient recovery and reuse of resources that might otherwise be wasted. These technologies filter out contaminants at a molecular level, improving the quality of recycled water and air, and reducing the environmental impact of industrial processes (refer to Sections 11.1 and 11.2).

13.1.1.4 Smart Packaging Solutions to Reduce Waste

Nanotechnology-enhanced smart packaging materials improve food preservation by providing better barrier properties against oxygen, moisture, and microbes, thus reducing spoilage and waste. Moreover, some nanotechnologies enable the creation of biodegradable packaging materials that decompose more efficiently, contributing to waste reduction in landfills (see Section 11.5.1).

13.1.1.5 Energy Efficiency and Renewable Integration

Nanotechnology can enhance the efficiency of energy systems, which is crucial for circular economy development. This includes the improvement of solar cells and batteries through nanomaterials, which increases their efficiency and lifespan. Efficient energy storage solutions are vital for integrating renewable energy sources, such as solar and wind, into the grid (see Section 8.6).

13.1.1.6 Precision Agriculture to Minimize Resource Use

Nanotechnology supports a circular economy by optimizing resource use in agriculture. Nanosensors and smart delivery systems can monitor soil conditions and precisely deliver water, fertilizers, and pesticides, reducing waste and environmental impact. This approach promotes sustainable agricultural practices that maximize crop yields while minimizing resource inputs (see Section 8.5.2).

13.1.1.7 Improved Catalysts for Sustainable Chemical Processes

Nanocatalysts enhance the efficiency of chemical processes by reducing energy consumption and minimizing waste production in industrial applications. These catalysts improve reaction rates and selectivity, leading to less waste generation and more efficient use of raw materials (refer to Sections 8.5.3 and 8.6.5).

13.1.1.8 Material Traceability and Lifecycle Management

Nanotechnology enhances material traceability by using nanoscale markers to track materials throughout their lifecycle. This innovation supports better resource management and enables more effective recycling and reuse strategies, ensuring materials are handled appropriately at the end of their life (see Section 8.5.1).

13.1.1.9 Upcycling Waste to High-value Products

Nanotechnology enables the upcycling of waste materials into new, high-value products rather than simply recycling them into lower-grade materials. For instance, waste plastics can be converted into carbon nanotubes, which have applications across various sectors, such as electronics, construction, and medical devices (see Section 9.5).

13.1.2 Energy Efficiency and Renewable Integration

Nanotechnology holds vast potential to enhance energy efficiency and accelerate the integration of renewable energy sources into our energy infrastructure. While Section 11.3 explored the foundational role of nanotechnology in advancing energy storage, savings, and harvesting, this section focuses on how these applications can further contribute to a more sustainable energy landscape.

This chapter merits a separate discussion because it moves beyond the foundational understanding of nanotechnology's capabilities to explore its future-oriented applications in a comprehensive manner. Unlike Section 11.3, which provided an overview of nanotechnology's basic principles and immediate applications in energy storage and conversion, this chapter delves into emerging innovations and integrative strategies that are shaping the next generation of sustainable energy solutions. It addresses not just the technologies themselves but also the broader impact of these technologies on the energy sector, including their potential to transform the entire energy value chain.

By focusing on advanced and integrated applications, this section emphasizes the forward-looking aspects of nanotechnology. It considers how combining multiple nanotechnological innovations can lead to holistic energy solutions, driving efficiency and sustainability across diverse environments and use cases. This includes hybrid systems that leverage nanostructured materials for enhanced performance in solar cells, batteries, and energy harvesting devices, reflecting a more complex and interconnected approach than what was covered in Section 11.3.

Moreover, this chapter examines the synergistic effects of combining nanotechnology with other cutting-edge fields, such as artificial intelligence and machine learning, to create adaptive and resilient energy systems. This forward-looking perspective is essential for understanding the evolving role of nanotechnology in achieving global sustainability goals and addressing the multifaceted challenges of energy production, storage, and distribution.

By distinguishing between foundational concepts and advanced applications, this chapter allows for a deeper exploration of how nanotechnology is not only a tool for incremental improvements but also a transformative force capable of redefining the future of energy. This separation ensures that readers gain a clear understanding of both the current state and the future potential of nanotechnology, appreciating its role in driving long-term innovation and sustainability in the energy sector.

13.1.2.1 Enhancing Solar Energy Conversion and Harvesting

Nanotechnology plays a critical role in improving solar energy technologies by enhancing the efficiency of energy conversion. Advanced nanostructured materials, such as perovskite crystals, quantum dots, and transition metal dichalcogenides (*e.g.*, MoS_2, WS_2), enable solar cells to capture a broader spectrum of sunlight, significantly improving their efficiency compared to traditional silicon-based cells.[3] For instance, perovskite solar cells incorporating nanoscale crystalline structures can be manufactured at lower temperatures, reducing energy consumption during production while achieving high conversion efficiencies. Quantum dots further extend this efficiency by enabling multiple exciton generation, where a single photon generates multiple electron-hole pairs, thereby surpassing the theoretical efficiency limits of traditional materials.

Metal oxides such as TiO_2, hematite (Fe_2O_3), and tungsten trioxide (WO_3) are increasingly used in photovoltaic devices to enhance their light absorption and charge transfer properties. These materials provide more active sites and improve interactions between active species and electrodes, making them highly suitable for applications in solar energy conversion and storage. Their unique structural and electronic properties allow for the effective conversion of sunlight into electrical energy, broadening the scope of nanotechnology's application in this field.[4]

13.1.2.2 Advancing Energy Storage Solutions

Nanotechnology is also pivotal in advancing energy storage technologies, which are essential for balancing the intermittent nature of renewable energy sources like wind and solar. Nanostructured batteries, such as those using graphene-based composites and silicon-based anodes, offer higher energy density, faster charging capabilities, and improved longevity compared to conventional batteries. For example, the development of nanostructured silicon and phosphorus anodes allows for greater ion transport and capacity, addressing key limitations of traditional lithium-ion batteries. Studies have shown that using nanoscale silicon flakes mixed with conductivity enhancement additives results in high initial coulombic efficiency and better capacity retention over long cycling periods. These advancements are crucial for applications ranging from electric vehicles to grid-scale storage, where high performance and durability are required.[5]

Beyond lithium-ion batteries, the integration of nanowire/graphene hybrids and other metal oxide-based materials, such as cobalt oxides (Co_3O_4) and manganese oxides (MnO_2), enhances the specific capacity and charge-discharge rates, making them attractive for next-generation batteries.[6] These innovations are complemented by the development of supercapacitors using carbon-based nanomaterials like graphene and carbon nanotubes, which provide rapid charge/discharge cycles and long operational lifetimes.[7] Supercapacitors, particularly those based on metal–organic frameworks and covalent organic frameworks, are emerging as high-performance energy storage devices, further expanding the potential of nanotechnology in energy storage applications.[8,9]

13.1.2.3 Improving Thermoelectric and Piezoelectric Energy Harvesting

Nanotechnology also contributes to energy harvesting technologies that capture ambient energy from the environment, such as waste heat, electromagnetic energy, mechanical vibrations, or fluid flows. Nanostructured thermoelectric materials, such as β-$FeSi_2$ and its alloys, are being developed to convert waste heat from industrial processes into electricity. These materials achieve higher thermoelectric performance through nanostructuring, which improves their figure of merit by optimizing both thermal and electrical conductivities. For example, reducing the crystallite size to the nanoscale has been shown to enhance thermoelectric performance at high temperatures.[10]

Piezoelectric and triboelectric nanogenerators are another promising application. These devices use nanoscale materials like zinc oxide nanowires or polymer nanofibers to convert mechanical energy, such as vibrations or human motion, into electrical power. Additionally, the integration of nanomaterials like graphene and carbon nanotubes in rectenna-based devices expands the capability to harvest electromagnetic energy from both the sun and ambient sources, such as wireless communication networks. This technology is particularly suitable for decentralized energy generation in urban environments, where small-scale energy harvesting can power sensors, wearable devices, and other low-energy electronics, contributing to the development of smart cities.[11]

Recent advances also highlight the use of chalcogenides like MoS_2 and WS_2, which are effective in energy harvesting applications due to their ability to undergo multi-electron redox reactions and maintain high catalytic activity. These materials, along with others like vanadium oxides and molybdenum oxides, are being explored for their potential to enhance the efficiency of energy conversion and storage systems.[4]

13.1.2.4 No Single Solution for Energy Challenges

It is crucial to recognize that nanotechnology applications in energy are not a one-size-fits-all solution. The energy sector's complexity—with its varied markets for energy generation, transformation, distribution, storage, and usage—requires a diverse range of nanotechnological innovations. Each nanotechnology application, whether improving solar cell efficiency, enhancing battery storage, or harvesting waste energy, has a unique role to play. The effectiveness of these technologies can vary significantly depending on the specific context and requirements of different energy markets.

For instance, while nanomaterials like quantum dots and plasmonic nanostructures are transforming solar technologies by enabling higher efficiency rates, other materials like nanofluids, MOFs, and nanostructured thermoelectrics target specific aspects of energy conversion and efficiency. Similarly, the deployment of nanostructured materials in batteries addresses the need for higher energy density and faster charging, which is vital for mobile and grid applications, but the same materials might not be as impactful in other energy contexts. The future of energy lies in the continued exploration and combination of these various nanotechnologies to address the diverse challenges in the energy sector.

Additionally, as discussed in Section 11.3, the integration of AI and machine learning with nanosensors could optimize these energy systems in real time, adapting to changing environmental conditions or energy demands. By leveraging the diverse toolkit provided by nanotechnology, it is possible to develop a more resilient and sustainable energy infrastructure that can meet the needs of a rapidly evolving global energy landscape.

13.1.3 Climate Change Mitigation and Adaptation

Nanotechnology offers innovative solutions for both mitigating climate change and adapting to its effects. By harnessing the unique properties of nanomaterials, we can develop technologies that address various climate-related challenges in more efficient and effective ways. Potential applications of nanotechnology span a wide range of sectors, including the development of climate-resilient infrastructure that can withstand extreme weather events, improving water security through advanced desalination and purification techniques, and promoting sustainable agricultural practices with nanosensors and smart delivery systems.

Among these diverse opportunities, three areas stand out for their potential to deliver substantial benefits in the short to mid-term (5–10 years):

1. Carbon capture and sequestration
2. Enhancing solar cell efficiency
3. Improving energy storage for renewable sources

These areas are particularly promising due to their scalability and direct relevance to global efforts to reduce greenhouse gas emissions and promote sustainable energy. By focusing on these strategic advancements, nanotechnology has the potential to drive significant progress toward international climate goals, such as those outlined in the Paris Agreement and the United Nations Sustainable Development Goals.

13.1.3.1 Carbon Capture, Utilization, and Storage

Carbon capture, utilization, and storage (CCUS) are critical technologies for addressing emissions from existing fossil fuel-based infrastructure and industries, which are unlikely to be entirely replaced in the short term. Power supply and carbon-intensive industries—such as cement, steel, limestone, petrochemical and chemical plants, and waste incinerators—account for a large share of CO_2 emissions. CCUS involves two primary pathways: CCS, which captures CO_2 from industrial flue gas for geological storage, and CCU, which converts captured CO_2 into valuable products like chemicals, fuels, construction materials, and biological processes such as algae cultivation.

Several innovative EU projects are advancing CCUS technologies through new materials and techniques. The CARMOF,[12] MOF4AIR,[13] and MEMBER[14] projects are developing more efficient and cost-effective methods through advanced adsorbents like MOFs, modified carbon nanotubes, and innovative membrane-based technologies. MOF4AIR is demonstrating these technologies in power plants and energy-intensive industries, while MEMBER focuses on scaling up manufacturing for pre- and post-combustion capture.

Currently, the most mature CCS technology—post-combustion CO_2 capture using amine scrubbing—faces efficiency and cost-effectiveness challenges. Alternative technologies being explored include oxy-fuel combustion, chilled ammonia technology, adsorptive processes, and calcium looping. However, these conventional methods still face significant challenges in energy requirements, material degradation, and operational costs.

Nanotechnology presents a transformative opportunity to overcome these limitations and enhance CCUS technologies' overall performance. By incorporating advanced materials such as metal–organic frameworks, it is possible to achieve enhanced performance in CO_2 capture, utilization, and storage. MOFs offer exceptional surface areas, tunable pore structures, and versatile chemical functionalities, which enable more selective and efficient CO_2 adsorption and separation. These unique properties can reduce the energy required for CO_2 capture compared to traditional methods like amine scrubbing, potentially lowering operational costs and improving scalability.

Although the specific energy savings are not quantified, the structural and functional advantages of MOFs provide significant potential for more efficient CO_2 management technologies.[15]

Almost all climate plans integrate CCUS as a crucial component for achieving deep emissions reductions, yet significant research and innovation are still needed to make these plans viable. With initiatives like CARMOF, MOF4AIR, and MEMBER, there is a concerted effort to push the boundaries of current technologies by leveraging nanotechnology. These projects are pioneering the use of MOFs and other advanced materials to develop next-generation CCUS solutions that are more efficient, cost-effective, and scalable. By enhancing the performance and scalability of CCUS technologies, these initiatives position nanotechnology as a critical tool in the fight against climate change. These advancements are essential for meeting international climate commitments, including the European Green Deal's goal of achieving climate neutrality by 2050 and a 55% reduction in carbon emissions by 2030.[16]

13.1.3.2 Enhancing Solar Cell Efficiency

Solar energy is one of the most rapidly growing sectors of renewable energy, and its continued expansion is vital for reducing reliance on fossil fuels. In 2023 alone, the global solar capacity surged by 270 gigawatts (GW), representing a 22% growth compared to the previous year. This brought the total installed capacity to over 1 terawatt (TW) and solar power now accounts for 5% of global electricity generation. This rapid growth in solar capacity is expected to accelerate, with projections indicating that it will provide up to 20% of the world's energy by 2050, playing a crucial role in global decarbonization efforts.[17]

Nanotechnology offers significant advancements in this area by enabling more efficient solar cells through materials like quantum dots and perovskites. These materials can enhance light absorption and energy conversion rates, making solar panels more effective and potentially lowering the cost per watt of solar energy. Improvements in durability and performance also extend the lifespan of solar panels, contributing to their economic viability and environmental impact. Given the ongoing investments and research in this area, nanotech-based enhancements in solar cell efficiency are likely to yield substantial benefits in the short to mid-term.

13.1.3.3 Improving Storage for Renewable Energy

Improving energy storage is critical to addressing the variability and intermittency of renewable energy sources such as solar and wind. These energy sources do not produce consistent output, as they depend on environmental conditions, and this makes the development of efficient, scalable, and cost-effective energy storage solutions essential for ensuring a stable energy supply. Advancements in battery technology, especially those driven by nanotechnology, are playing a central role in meeting this challenge.

Nanostructured materials, including graphene-based composites and silicon anodes, have demonstrated promising potential in enhancing the performance of batteries. These materials are contributing to the development of energy storage solutions with higher energy densities, faster charging times, and longer lifespans—three key factors for effective energy storage.[18] Higher energy densities enable batteries to store more energy in a smaller space, making them more practical for applications ranging from electric vehicles to grid-scale storage. Faster charging times reduce the downtime associated with energy transfer, and longer lifespans lower the overall cost of ownership by reducing the need for frequent replacements.

In addition to improving the performance of traditional lithium-ion batteries, researchers are exploring alternative storage technologies, such as solid-state batteries and flow batteries. Solid-state batteries, which replace the liquid electrolyte in traditional batteries with a solid material, promise enhanced safety and energy capacity, while flow batteries offer a flexible solution for large-scale energy storage, as they can be easily scaled to meet the needs of renewable energy systems.

The integration of advanced battery technologies into renewable energy systems also supports the broader transition to a low-carbon economy. Grid-scale energy storage helps balance supply and demand, ensuring that excess energy generated during periods of high production (such as sunny or windy days) can be stored and used during periods of low production. This reduces the need for backup power from fossil fuel-based plants and enhances the reliability of renewable energy sources.

Electric vehicles further benefit from these advancements. With the development of batteries that charge faster and last longer, EVs become more practical for widespread adoption, reducing reliance on internal combustion engines and significantly decreasing greenhouse gas emissions from the transportation sector. As EV charging infrastructure improves in parallel with battery technology, this sector is poised to make a substantial impact on global carbon reduction efforts.

13.1.3.4 Ongoing Research Initiatives

Several international R&D initiatives highlight the critical role of nanotechnology in advancing energy storage, supporting the global transition to renewable energy and reducing environmental impact. Below are some key examples:

13.1.3.4.1 European Union. The European Union is at the forefront of battery innovation, with a focus on sustainability, high energy density, and reducing the environmental impact of energy storage technologies. The NANOBAT Project[19] exemplifies the EU's commitment to advancing lithium-ion batteries by leveraging nanotechnology, such as silicon nanowires and nanoparticles, to enhance anode performance, potentially increasing energy density and improving cycle life. This innovation is part of the EU's broader efforts to develop next-generation battery technologies.

Collaborative projects under the BATT4EU Partnership[20] are central to the EU's research initiatives, aiming to establish a robust European battery value chain by 2030. This partnership focuses on increasing battery performance, improving cycle lifetime, and reducing costs, while also integrating best practices for recycling and reducing the carbon footprint. The European Commission, through programs like Horizon Europe, is dedicating substantial funding to ensure Europe maintains a leadership role in the global transition to clean energy technologies

13.1.3.4.2 United States. In the U.S., battery research is driven by initiatives like the Joint Center for Energy Storage Research (JCESR),[21] led by Argonne National Laboratory, which focuses on beyond-lithium-ion technologies. JCESR explores new chemistries and nanostructured materials to push the boundaries of next-generation energy storage, improving energy density, cycle life, and safety through innovations like solid-ion-conducting materials and advanced electrode designs.

The Battery500 Consortium[22] is another key project, aiming to develop lithium–metal batteries with a target energy density of 500 $W\,h\,kg^{-1}$. This project brings together national laboratories, universities, and industry partners to advance lithium–metal battery technologies using nanostructured electrodes and solid electrolytes, promising significant improvements in performance and energy storage capabilities.

Beyond these projects, the U.S. battery innovation ecosystem is supported by the Federal Consortium for Advanced Batteries (FCAB),[23] a multi-agency collaboration that includes the Department of Energy and the Department of Defense. This consortium ensures a coordinated national strategy to advance battery technologies and secure supply chains. Key research institutions such as the Berkeley Lab Energy Storage Center, Oak Ridge National Lab (ORNL), and universities like UT Austin contribute to breakthroughs across various areas, from solid-state batteries to recycling innovations led by the ReCell Center.

By 2030, the US aims to achieve significant milestones, including a domestic battery supply chain, 90% recycling of key materials, and reducing production cell costs to under \$60 $kW^{-1}\,h^{-1}$. Efforts also focus on eliminating reliance on critical materials like cobalt and nickel, emphasizing sustainability and securing the US's position in the global battery market.

13.1.3.4.3 China. China has made significant strides in battery research, leveraging government initiatives and collaborations across multiple sectors. The country now accounts for a major share of global battery production, focusing on key areas like lithium-ion, solid-state, and alternative battery technologies. Nanotechnology plays a vital role, with research focusing on novel materials such as CO-free cathodes and nano-Si/C anodes. China's strategic goals include breakthroughs in energy density, recycling, and reducing emissions. The government also funds various R&D programs aimed at improving electric vehicle batteries and developing energy storage solutions for industrial and household applications.[24]

13.1.3.4.4 Japan. Japan has a robust battery research ecosystem, focusing on next-generation and solid-state batteries. The government, through organizations like NEDO and METI, has invested heavily in projects aimed at boosting battery production and developing new technologies, such as fluoride and zinc batteries. Japan's strategic goals include increasing domestic production capacity to 150 GW h per year by 2030, advancing recycling technologies, and commercializing solid-state batteries. Research priorities include reducing the need for critical raw materials and decreasing greenhouse gas emissions.[25]

13.1.3.4.5 South Korea. South Korea is a global leader in battery innovation, with a strong focus on next-generation technologies such as solid-state, lithium–sulfur, and lithium–metal batteries. The Korea Institute of Energy Research (KIER) is leading research on nanostructured materials for energy storage. This includes advancements efforts in nanocoating for silicon anode materials and single-walled carbon nanotubes. South Korea aims to commercialize lithium–sulfur and solid-state batteries by 2025 and 2027, respectively. The government, through initiatives like the K-Battery Development Strategy, prioritizes increasing domestic production, recycling capabilities, and global market leadership. The country is also investing in AI and digital twins to optimize battery manufacturing.[26]

13.1.3.4.6 Australia. Australia's battery research is deeply connected to its rich reserves of critical minerals like lithium, cobalt, and nickel, crucial for the global energy transition. Institutions such as the Future Battery Industries Cooperative Research Centre (FBICRC) and StorEnergy, the ARC Training Centre for Future Energy Storage Technologies, are at the forefront of these efforts. The FBICRC focuses on advancing Australia's battery value chain, from raw material extraction to battery recycling, while StorEnergy aims to develop innovative energy storage solutions through cutting-edge research. Nanotechnology plays a key role, with advancements such as super anodes and future electrolyte systems. The Australian government supports these innovations through agencies like ARENA and the CEFC, providing substantial funding for early-stage R&D and large-scale projects. With a focus on expanding domestic manufacturing and recycling capabilities, Australia is positioning itself as a leader in energy storage technologies while targeting net-zero emissions by 2050.[27]

13.2 Policy and Regulation: Setting the Stage for Innovation

Nanotechnology offers immense potential for advancing many aspects sustainability, as illustrated throughout this book, yet its integration into industries relies significantly on the policies and regulations that shape its development. Like any emerging technology, its success depends on navigating frameworks that ensure safety, encourage innovation, and limit environmental impact. This chapter examines the pivotal role of policy and regulation in either accelerating or impeding the adoption of nanotechnologies, emphasizing the importance of flexible regulatory strategies, global cooperation, and incentive mechanisms.

As Figure 13.2 illustrates, in the specific case of nanotechnology-enabled health products, regulatory challenges are particularly significant. This diagram underscores the complexities of establishing clear safety standards and testing methods tailored to these advanced products. These challenges must be addressed to ensure both innovation and the safe application of

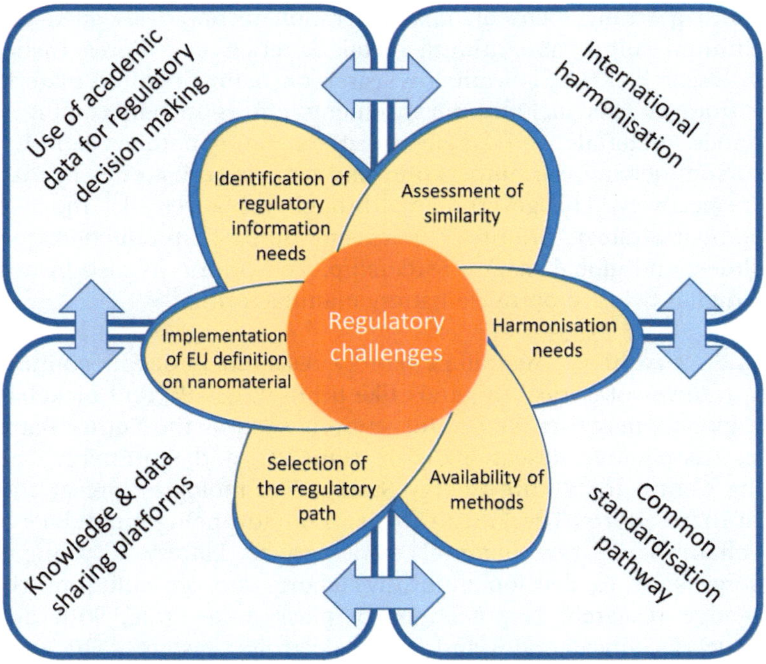

Figure 13.2 Future strategies and perspectives (background squares) in the regulatory science addressing regulatory challenges (flower leaves) for nanotechnology-enabled products. Reproduced from ref. 50, https://doi.org/10.1007/s13346-022-01165-y, under the terms of the CC BY 4.0 license, https://creativecommons.org/licenses/by/4.0/.

nanotechnologies in health-related industries. Similar approaches are necessary in other industries.

Nanotechnology's rapid evolution presents both opportunities and challenges, particularly in terms of regulation. Given the scale at which nanomaterials operate, their behavior can be dramatically different from larger materials, posing unique risks to human health and the environment. At the same time, overly restrictive regulations could stifle innovation and slow the adoption of nanotechnologies that offer transformative potential across industries. Therefore, it is essential to strike a balance—creating regulatory frameworks that protect public health and the environment while fostering innovation.

13.2.1 The Need for Adaptive Regulatory Models

The fast-paced nature of nanotechnology development means that traditional regulatory models, which often involve lengthy approval processes, may not be suited to this field. By the time a regulatory framework is established, new breakthroughs could have rendered the guidelines outdated. An adaptive regulatory framework is therefore critical—one that can evolve with technological advances while providing clear safety guidelines for industries utilizing nanotechnology.

Adaptive regulation should include the following key features:[28]

- **Continuous review**: Regulatory bodies must engage in ongoing research to assess the long-term effects of nanomaterials on health and the environment. This requires collaboration with academic and industrial researchers to ensure that regulations are based on the latest scientific data.
- **Data-driven decision-making**: Regulators should employ real-time data from ongoing research to update and refine guidelines as needed. This could involve creating databases of safety information on nanomaterials, similar to chemical databases used in industries like pharmaceuticals.
- **Tiered risk assessment**: Given the variety of nanomaterials and their uses, a one-size-fits-all regulatory approach is inadequate. Instead, regulators should implement a tiered approach that assesses risks based on the specific characteristics of each nanomaterial, including size, shape, reactivity, and persistence in the environment. Low-risk applications, such as those where the nanomaterial is encapsulated and unlikely to be released, may require less stringent regulations compared to high-risk scenarios like the use of free nanoparticles in medical applications or consumer products.
- **Precautionary principle**: While the promise of nanotechnology is vast, its long-term environmental and health impacts remain uncertain. An adaptive regulatory framework should incorporate the precautionary principle, which advocates for preventative action in the face of potential risks, even if all the scientific data is not yet available. This principle can help mitigate unforeseen consequences, particularly in the early stages of nanotechnology adoption.

13.2.2 Safety Standards for Nanomaterials

Nanomaterials, as highlighted throughout this manuscript, possess distinct properties that set them apart from their bulk counterparts, enhancing their value across various sectors. However, these unique characteristics also pose challenges for safety assessments. As discussed in Sections 12.6.1 and 12.6.2, conventional testing methods for chemicals and materials often fall short in addressing the complex behavior of nanoscale particles.

- **Toxicology of nanomaterials**: The small size and high surface area of nanoparticles can lead to increased reactivity and novel toxicological effects. For instance, nanoparticles can enter the human body through inhalation, ingestion, or skin contact and may pass through biological barriers, such as the blood–brain barrier, more easily than larger particles. Regulatory frameworks must require thorough testing of nanomaterials for toxicological effects, including their potential to cause inflammation, oxidative stress, and DNA damage.
- **Environmental impact**: Nanomaterials can also have unanticipated environmental impacts. When released into air, water, or soil, they may interact with ecosystems in ways that differ from bulk materials. For example, nanoparticles could bioaccumulate in organisms or disrupt microbial communities. Regulations should require environmental impact assessments for nanomaterials, focusing on their persistence in the environment, potential for bioaccumulation, and effects on various ecosystems.
- **International best practices**: Regulatory frameworks should also draw from international best practices. For example, the European Union's REACH (Registration, Evaluation, Authorization, and Restriction of Chemicals) regulation requires companies to provide detailed safety data on nanomaterials. Under REACH, nanomaterials are treated as distinct substances, and manufacturers must provide specific safety information, such as particle size distribution and solubility, to ensure proper risk management.

  Other regions, such as the United States, have begun incorporating nanomaterials into existing regulatory frameworks. The Environmental Protection Agency (EPA) and the Food and Drug Administration (FDA) have both issued guidelines for the safe development and use of nanotechnology. These frameworks can serve as models for countries that have yet to formalize their regulatory approach.

13.2.3 Regulating Innovation Without Stifling It

While safety is paramount, overly burdensome regulations can hinder the development of nanotechnology and reduce its potential to solve pressing global issues, such as climate change and resource scarcity. To encourage innovation while ensuring safety, regulators must design frameworks that are:

- **Proportional to risk**: Regulations should be calibrated to reflect the specific risks associated with a given nanotechnology application. Low-risk uses, such as nanotechnology-enhanced coatings or energy storage, should face less regulatory friction than high-risk applications like nanomaterials in medicine or food products. This proportional approach ensures that innovation is not unnecessarily hindered while maintaining robust safety measures for high-risk uses.
- **Flexible approval processes**: Accelerated approval pathways, similar to those used for pharmaceutical innovations, could be applied to nanotechnology in certain sectors. For example, if a new nanomaterial shows significant promise in energy storage or environmental remediation, regulators could fast-track its approval by requiring interim safety data, allowing it to be deployed more quickly while further research is conducted.
- **Industry involvement**: Policymakers should engage with industry leaders and researchers to create guidelines that are not only enforceable but also practical for businesses to implement. Public-private partnerships can facilitate the development of safe and effective nanotechnology solutions by providing industry players with clarity on regulatory expectations and the resources needed to comply.

The FDA's adaptive approach to regulating nanotechnology in the food and drug industries exemplifies a balanced, innovation-friendly regulatory strategy.[29] The FDA has recognized the unique properties of nanotechnology, such as the ability to manipulate materials at the nanoscale, which introduces novel behaviors that challenge traditional regulatory frameworks. For instance, nanoscale particles can change their properties significantly based on their size, impacting factors like fluorescence, conductivity, and reactivity.

To keep up with the rapid advancements in nanotechnology, particularly in the food and drug sectors, the FDA uses a case-by-case assessment for products involving nanomaterials. This allows the agency to tailor its review process to the specific characteristics of each product, ensuring that regulations evolve alongside technological advancements without stifling innovation. The FDA's guidance documents, such as those addressing the safety of nanomaterials in food and cosmetic products, reflect a proactive stance on regulating emerging technologies.

Moreover, the FDA categorizes nanotechnology products based on their primary mode of action—chemical, mechanical, or biological—and applies the relevant regulatory pathway. For example, nanodrugs like liposomal formulations of cancer treatments and nanoformulated vaccines have already been approved under these flexible regulations.

This adaptive approach aligns with broader global efforts in nanotechnology, where regulatory frameworks are still evolving. The FDA's strategy emphasizes innovation while maintaining safety and efficacy, ensuring that new technologies can reach the market without unnecessary delays.[30]

Similarly, the EPA's approach to regulating nanomaterials has evolved since the launch of the Nanoscale Materials Stewardship Program (NMSP) in 2008. While initial efforts under the NMSP were voluntary, the EPA has shifted towards mandatory reporting and data collection through the Toxic Substances Control Act (TSCA).[31] In 2017, the EPA introduced a final rule requiring manufacturers and processors of nanoscale materials to submit data on chemical identity, production volumes, methods of manufacture, and health and safety information. This rule aims to fill knowledge gaps regarding the environmental and health impacts of nanomaterials, allowing the EPA to determine whether further regulatory action is needed.[32]

Additionally, the EPA has taken more proactive steps in recent years, including developing guidelines for nanomaterials used in pesticides under the Federal Insecticide, Fungicide, and Rodenticide Act (FIFRA), ensuring that nanoscale ingredients in registered pesticides are identified and assessed. These developments reflect a growing focus on balancing innovation with the need for rigorous safety oversight, particularly as nanotechnology becomes more prevalent in commercial products.

For future regulatory directions, the EPA's research initiatives continue to emphasize collaboration with international bodies and domestic stakeholders to improve the understanding and safety of nanomaterials in the environment.

13.2.4 Building Public Trust

A critical element of nanotechnology regulation is public trust. Public perception plays a significant role in the acceptance and widespread adoption of nanotechnology. Transparency in regulatory processes and clear communication about the safety and benefits of nanotechnology are essential for building this trust.

Public trust and consumer awareness are especially crucial in regulating certain nanotechnology applications, particularly in consumer products like sunscreens or food additives, where there may be potential health risks from direct exposure to nanomaterials. However, this manuscript focuses on industrial and circular economy applications, where nanotechnology functions as an integrated, often indiscernible component. In applications like high-performance materials, improved solar cells, or enhanced chemical catalysts, the public does not directly interact with the nanotechnology. Therefore, the regulatory focus here is on industry standards, worker safety, environmental impact, and balancing innovation with safety in these industrial contexts, rather than on consumer-facing issues such as product labeling or public engagement strategies.

13.3 International Collaboration and Interdisciplinary Innovation

The complex nature of nanotechnology and its wide-ranging applications necessitate a collaborative approach that transcends both disciplinary and national boundaries. Nanotechnology, operating at the scale of atoms and molecules, inherently bridges multiple scientific domains, from physics and chemistry to biology and materials science. Its potential to revolutionize fields as diverse as medicine, energy production, environmental remediation, and manufacturing requires expertise that no single discipline or nation can provide in isolation.

The challenges that nanotechnology aims to address—such as climate change, resource scarcity, and global health issues—are inherently global in nature. These problems do not respect national borders, and neither should the solutions. Moreover, the rapid pace of nanotechnology development, coupled with its potential for profound societal impact, demands a coordinated global approach to ensure responsible and equitable progress.

Interdisciplinary collaboration allows for the cross-pollination of ideas and techniques, often leading to breakthrough innovations at the intersections of different fields. For instance, the marriage of nanotechnology with biotechnology has given rise to nanomedicine, offering new possibilities for targeted drug delivery and personalized treatments. Similarly, the integration of nanotechnology with information technology is paving the way for more efficient and powerful computing systems.

International cooperation, on the other hand, enables the pooling of resources, sharing of knowledge, and harmonization of standards and regulations. This global approach not only accelerates scientific progress but also ensures that the benefits of nanotechnology can be realized on a global scale, addressing disparities in technological access and capability between nations.

13.3.1 Bridging Disciplines

Nanotechnology's potential to revolutionize multiple sectors relies heavily on the convergence of various scientific disciplines. Encouraging collaboration between materials science, chemistry, biology, physics, and environmental science can drive innovative solutions that might not be possible within the confines of a single field.

13.3.1.1 Interdisciplinary Synergies

The power of interdisciplinary collaboration in nanotechnology is evident in numerous fields:

13.3.1.1.1 Nanomedicine. Nanomedicine involves the use of nanoscale materials to diagnose, treat, and prevent diseases at the molecular and cellular levels. Unlike traditional medicine, which uses larger, bulk materials, nanomedicine operates at a scale so small that material properties can change significantly. At the nanoscale, materials interact with biological systems in ways that are impossible with larger particles, leading to improved cellular uptake, precise targeting of diseased tissues, enhanced drug delivery, reduced side effects, and new therapeutic possibilities that traditional medicine cannot achieve.

However, nanomedicine's complexity brings challenges that no single discipline can fully address. Interdisciplinary collaboration is crucial to harnessing nanomedicine's full potential and developing safe, effective solutions for real clinical needs. The development of nanoparticles for targeted drug delivery exemplifies this need for interdisciplinary cooperation.[33] Expertise is required from:

- **Materials scientists and chemists** to design and synthesize nanocarriers with specific physical properties, such as inorganic and organic materials, self-assembled polymers, liposomes/lipid vesicles, and drug-polymer conjugates.
- **Chemists** to functionalize nanoparticle surfaces for targeted delivery, controlled release, and optimized drug loading.
- **Biologists** to understand cellular interactions, including adhesion, internalization, and inhibition of intracellular autophagy, as well as potential toxicity.
- **Pharmacologists** to optimize drug loading, release kinetics, and study complex pharmacokinetics, including bioavailability, distribution, and residence time within systemic circulation.
- **Toxicologists** to assess potential toxicity, off-target effects, and altered biological distribution of drugs.
- **Immunologists** to understand interactions with the immune system.
- **Clinical scientists** to ensure relevance to unmet medical needs and guide translation to clinical applications.

No single discipline can fully address the intricate challenges of nanomedicine. Interdisciplinary collaboration is vital for crafting solutions that are not only effective but also safe for clinical application. By drawing on expertise from diverse fields, nanomedicine can be developed to meet pressing clinical needs while overcoming potential hurdles in the translation from research to patient care.

13.3.1.1.2 Nanoelectronics. Nanoelectronics refers to devices where quantum mechanical effects, such as tunneling and quantum confinement, play a crucial role. While modern electronics, like processor chips, may incorporate nanoscale components, they still operate under classical physics principles. In contrast, nanoelectronics operates at a scale where quantum effects become significant, leading to behaviors that don't occur in conventional electronics. This distinction allows nanoelectronics to offer breakthroughs in performance and efficiency that traditional nanoscale electronics cannot achieve.

Because nanoelectronics relies on these unique quantum mechanical properties, its development requires collaboration across multiple scientific fields. No single discipline can fully address the challenges of designing and implementing nanoelectronic devices. Expertise is required from:

- **Materials scientists** to design and fabricate nanoscale electronic components with specific electrical, optical, and magnetic properties.
- **Physicists** to understand and exploit quantum mechanical effects that become prominent at the nanoscale, such as tunneling and quantum confinement.
- **Electrical engineers** to design circuits and systems that integrate nanoscale components and account for their unique behaviors.
- **Chemical engineers** to develop and optimize fabrication processes for nanoscale devices, including etching, deposition, and self-assembly techniques.
- **Computer scientists** to develop new algorithms and architectures that can leverage the capabilities of nanoelectronic devices.
- **Biologists** to explore bio-inspired nanoelectronic systems and potential interfaces between electronic devices and biological systems.
- **Environmental scientists** to assess the potential ecological impacts of nanoelectronic materials and manufacturing processes.

The complexities of nanoelectronics cannot be solved by a single discipline. It is through interdisciplinary efforts that we can pioneer breakthrough technologies capable of addressing the limitations of current electronics. This collaboration paves the way for innovations in computing power, energy efficiency, and device miniaturization, ensuring that nanoelectronics reaches its full potential in real-world applications.

13.3.1.1.3 Energy Applications. Nanotechnology is revolutionizing the energy sector by enabling new materials and technologies that can significantly improve energy generation, storage, and efficiency. At the nanoscale, materials exhibit enhanced electrical, thermal, and chemical properties, making them ideal for energy applications. For example, nanoscale materials can improve the efficiency of solar cells, enhance the storage capacity of batteries, and reduce energy loss in power transmission. These advances have the potential to address current energy challenges by making renewable energy more efficient, energy storage more reliable, and power grids more resilient.

However, the complexity of these energy applications requires an interdisciplinary approach to realize their full potential. The development of nanomaterials and devices for energy generation and storage requires expertise from:

- **Materials scientists** to design and fabricate nanomaterials with specific properties, such as high surface area, conductivity, and chemical stability, for use in solar cells, batteries, fuel cells, and supercapacitors.
- **Chemists** to develop new materials and chemical processes that enhance energy conversion efficiency, such as catalysts for fuel cells and nanostructured materials for solar energy harvesting.
- **Physicists** to study the fundamental interactions of light, heat, and electricity at the nanoscale and exploit quantum mechanical effects to improve energy efficiency.
- **Electrical engineers** to design energy storage systems and power grids that can integrate nano-enhanced materials and devices for improved performance and stability.
- **Environmental scientists** to evaluate the sustainability and environmental impact of nanomaterials used in energy technologies, from production to disposal.
- **Mechanical engineers** to optimize the thermal management of energy devices, ensuring that heat is effectively dissipated in systems like batteries and solar cells.

The development of advanced energy technologies through nanotechnology requires a collective effort across multiple fields. Only by working together can experts create solutions that not only enhance performance but also minimize environmental impact. Interdisciplinary cooperation is essential for meeting the global demand for clean, sustainable energy and for creating long-term, impactful solutions to today's energy challenges.

13.3.1.2 Emerging Trends

As nanotechnology advances, new trends are emerging that further emphasize the importance of interdisciplinary collaboration. These trends not only shape the future of nanotechnology but also require new and evolving partnerships across multiple disciplines:

- **Convergence of nanotechnology with AI and big data:** Nanotechnology is increasingly intersecting with artificial intelligence and big data analytics. AI techniques such as machine learning, deep learning, and neural networks are being employed to optimize the discovery of new nanomaterials, predict their properties, and enhance their applications in fields like medicine and electronics. This synergistic relationship explores how these AI methods improve the efficiency, precision, and scalability of nanotechnology applications. As a result, collaboration between nanoscientists and data scientists is becoming essential to harness AI for materials discovery and optimization. Researchers in computer science, materials science, and chemistry are working together to develop data-driven approaches that accelerate innovation.[34]

- **Nano-bio interface**: The integration of nanomaterials with biological systems, known as the nano-bio interface, is becoming a focal point for future applications in healthcare and biotechnology.[35] This area includes the development of nanoscale devices for drug delivery, biosensors, and tissue engineering. As these technologies advance, even closer collaboration between physical scientists, life scientists, and clinicians is essential. Expertise in materials science, biology, and medicine will be crucial to ensuring the safety, effectiveness, and ethical use of nano-bio technologies.
- **Sustainable/green nanotechnology**: The environmental implications of nanotechnology are gaining increasing attention, with a growing emphasis on developing sustainable and eco-friendly nanomaterials. Green nanotechnology focuses on minimizing the environmental impact of nanomaterials throughout their life cycle, from production to disposal.[36] This trend requires collaboration between ecologists, toxicologists, environmental engineers, and nanomaterial developers to create safer, more sustainable technologies. These efforts ensure that nanotechnology can contribute to solving environmental challenges, rather than creating new ones.

13.3.2 Institutional Approaches

Universities and research institutions around the world play a critical role in fostering the interdisciplinary collaboration necessary for advancements in nanotechnology. Many have established dedicated nanotechnology centers that bring together researchers from diverse fields, enabling breakthroughs in areas such as healthcare, electronics, and energy. Below is a partial list of prominent institutions that are leading the way in interdisciplinary nanotechnology research:

- **Center for Nanoscale Systems (CNS) at Harvard University**:[37] As a shared user facility, the CNS at Harvard facilitates collaboration across multiple departments, including physics, chemistry, biology, and engineering, enabling breakthroughs in areas such as quantum computing, biomimetic materials, and nanomedicine. Researchers from various fields work together to develop nanostructures and nanoscale devices that are used in cutting-edge applications. This center provides state-of-the-art equipment and resources to support interdisciplinary projects.
- **International Iberian Nanotechnology Laboratory (INL) in Portugal**:[38] The INL is a unique international research organization that brings together researchers from various countries and disciplines to address global challenges in health, environment, and renewable energy using nanotechnology. It fosters collaboration between physicists, chemists, engineers, biologists, and environmental scientists, enabling the development of innovative solutions in areas like cancer therapy, air quality monitoring, and energy-efficient materials.

- **Swiss Nanoscience Institute (SNI) at the University of Basel:**[39] SNI promotes interdisciplinary collaboration through its innovative PhD program, where each doctoral student is supervised by two faculty members from different disciplines. This approach ensures that students gain expertise in both the scientific and practical applications of nanotechnology. The institute focuses on areas like nanomedicine, quantum devices, and energy solutions, making it a hub for cross-disciplinary research.
- **Kavli Nanoscience Institute (KNI) at Caltech:**[40] The Kavli Nanoscience Institute at the California Institute of Technology is dedicated to advancing nanoscale research by integrating physics, chemistry, biology, and engineering. KNI researchers work on a broad range of interdisciplinary projects, including nanophotonics, quantum information, and nanofabrication. The center also emphasizes the importance of collaboration between academia and industry to accelerate the commercialization of nanotechnology.
- **Nanotechnology Characterization Lab (NCL) at the National Cancer Institute, USA:**[41] The NCL is a critical resource for nanomedicine, working at the interface of nanotechnology and cancer research. It helps researchers from diverse fields—such as medicine, pharmacology, and materials science—evaluate the safety and efficacy of nanomedicines. The NCL's goal is to bridge the gap between laboratory discoveries and clinical applications, providing the infrastructure necessary for collaboration between basic scientists, clinicians, and regulatory agencies. According to the site, The NCL has tested more than 500 unique nanomaterials and worked with more than 125 investigators worldwide.
- **London Centre for Nanotechnology (LCN), UK:**[42] A partnership between University College London, King's College London and Imperial College London, the LCN is a multidisciplinary research center that focuses on the intersections of nanotechnology with life sciences, quantum technologies, and information processing. It brings together physicists, chemists, engineers, and biologists to work on nanomaterials and devices that address challenges in healthcare, energy, and security.
- **National Center for Nanoscience and Technology (NCNST), China:**[43] Located in Beijing, NCNST is one of the leading centers in China for nanotechnology research and interdisciplinary collaboration. Researchers from physics, chemistry, materials science, biology, and engineering work together to explore nanotechnology applications in healthcare, energy, and environmental protection. The center also collaborates with industry to ensure that new technologies are scalable and market ready.
- **Centre for Nano Science and Engineering (CeNSE), Indian Institute of Science:**[44] The Centre for Nanoscience and Engineering at IISc Bangalore is one of Asia's leading institutions for interdisciplinary research in nanotechnology. CeNSE fosters collaboration between fields such as materials science, electrical engineering, and biology to develop

technologies for energy harvesting, medical devices, and environmental sustainability. The center provides world-class infrastructure for nanoscale research and frequently collaborates with industry and global research organizations.
- **National Institute for Materials Science (NIMS), Japan:**[45] NIMS, located in Tsukuba, is one of Japan's leading research institutions focused on materials science and nanotechnology. Researchers from chemistry, physics, biology, and engineering collaborate on projects involving nanomaterials for energy, electronics, and healthcare applications. NIMS plays a significant role in Japan's nanotechnology innovation, and its interdisciplinary teams are instrumental in the development of new materials and devices for industrial and medical uses.

13.3.3 Cross-sector Knowledge Transfer

Nanotechnology innovations often transcend their original sectors, offering immense potential when applied to different industries. This cross-pollination of ideas and technologies can lead to breakthroughs that might not be immediately apparent within a single field. Facilitating knowledge transfer across sectors is therefore critical to unlocking the full potential of nanotechnology.

One of the key advantages of nanotechnology is its ability to address problems across a broad range of applications. By repurposing existing technologies, industries can accelerate development, reduce costs, and foster innovation. Here are a few examples of how cross-sector knowledge transfer in nanotechnology can drive advancement:

- **Medical diagnostics to environmental monitoring**: Nanosensors, originally developed with a strong focus on medical diagnostics to detect subtle biological changes, have a wide range of applications beyond healthcare. These sensors can be used not only to identify specific cells and deliver targeted treatments by measuring various physical characteristics, such as concentration, movement, and temperature, but also for monitoring environmental pollutants and pathogens. Their sensitivity enables real-time tracking of pollutants in air or water, as well as monitoring processes in manufacturing and transportation systems, making them a valuable tool for addressing challenges across multiple sectors.
- **Nanocoatings across industries**: Nanocoatings developed to enhance fuel efficiency in the automotive sector by reducing friction and wear are now being adapted in construction to improve insulation, provide UV and corrosion resistance, and reduce maintenance. Beyond these areas, nanocoatings are also widely applied in textiles for water and stain resistance, pharmaceuticals for controlled drug delivery, paints for self-cleaning and anti-corrosion properties, electronics for moisture protection, food packaging to extend shelf life, and the energy sector to

improve efficiency in solar panels and wind turbines. This cross-industry transfer showcases the versatility of nanocoating technologies, promoting resource efficiency across multiple sectors.

- **Lightweight composites for diverse applications**: Nanomaterials originally designed for aerospace applications, such as lightweight, high-strength composites, are now being utilized in industries as diverse as sports equipment manufacturing and medical prosthetics. These materials reduce weight without sacrificing durability, making them ideal for applications that require both strength and lightness. The repurposing of aerospace technology into everyday items demonstrates how advances in one field can ripple into others, leading to widespread benefits.

13.3.3.1 Case Study: Graphene Flagship

The Graphene Flagship, launched in 2013 with a €1 billion budget, set out to make Europe a global leader in graphene research and commercialization.[46] The initiative embodied the core principle that nanotechnology's progress requires international and interdisciplinary collaboration. With graphene's potential in sectors like energy, healthcare, and electronics, the Flagship highlights how pooling expertise across disciplines and countries accelerates innovation and addresses global challenges. However, the initiative also provides valuable lessons on the difficulties of commercialization and the need for better integration between research and market demands.[47]

The Graphene Flagship brought together over 170 partners from 21 countries, demonstrating the necessity of collaboration in addressing the complex technical challenges associated with nanotechnology. The project facilitated interdisciplinary synergies by uniting scientists, engineers, biologists, and industry leaders to collectively solve problems that no single discipline could tackle alone.

This broad collaboration enabled breakthrough innovations in diverse sectors, highlighting how combining different areas of expertise is critical to realizing nanotechnology's full potential. For example, chemists working on material properties joined forces with engineers focusing on production scalability, while biologists and medical researchers collaborated on healthcare applications. The success of the Flagship underscores that only through interdisciplinary teamwork can complex challenges be overcome.

- **Biomedical innovations**: One of the most significant successes of the Flagship was its contribution to nanomedicine, particularly through graphene-based brain implants and biosensors. These advances were only possible through close collaboration between materials scientists, biologists, and clinicians, all working to harness graphene's properties for precise diagnostics and targeted treatments. By integrating expertise across multiple fields, the project made strides in addressing major healthcare challenges, such as developing more effective treatments for neurological disorders.

- **Scalable manufacturing**: A major focus of the Flagship was on scaling up graphene production to meet industrial needs. This required collaboration between physicists, chemists, and engineers to solve the technical challenges of manufacturing high-quality graphene at a larger scale. The creation of the 2D Experimental Pilot Line, which sought to mass-produce graphene-based devices, exemplifies the importance of interdisciplinary cooperation. By combining knowledge of graphene's fundamental properties with practical manufacturing techniques, the Flagship advanced the potential for industrial-scale production—an essential step toward broader commercialization.
- **Energy solutions with global impact**: Graphene's potential in energy storage was another key area of the Flagship's focus. Collaborative efforts between materials scientists, chemists, and engineers led to innovations in supercapacitors and batteries that have the potential to address global energy challenges. These advancements could lead to more efficient renewable energy systems and longer-lasting energy storage solutions, demonstrating how interdisciplinary collaboration can generate technologies with global significance. For example, by improving the efficiency and lifespan of energy storage devices, graphene-based technologies could contribute to reducing carbon emissions and enhancing energy sustainability worldwide.

13.3.3.2 Challenges in Commercialization

While the Graphene Flagship made significant progress, it also faced substantial barriers to widespread market adoption. One of the main challenges was the lack of a "killer application" for graphene, which made it difficult to compete with established technologies in industries like electronics. Despite its remarkable properties, graphene struggled to find its niche in consumer markets, particularly due to high production costs and complex fabrication processes.

This difficulty in translating research breakthroughs into commercially viable products highlights the gap between academic research and industrial needs. The Flagship's experience illustrates the need for deeper collaboration between academia and industry from the outset. By involving industry partners more directly in the early stages of research, future projects could ensure that the technologies developed are better aligned with market demands and have a clearer path to commercialization.

13.3.3.3 Lessons for Future Nanotechnology Initiatives

The Graphene Flagship offers important lessons for future large-scale projects in nanotechnology. First, the initiative's success in research and innovation underscores the importance of international and interdisciplinary collaboration. By bringing together experts from various fields, the Flagship was able to address complex challenges that would have been insurmountable within isolated disciplines. This model of collaboration should be a blueprint for future projects, especially as nanotechnology continues to expand into new sectors.

However, the Flagship also highlights the need for better integration between research and commercialization. The challenges in scaling up graphene production and finding a market-ready application suggest that future projects should foster closer ties with industry early on. More targeted efforts to bridge the gap between research labs and the marketplace—such as involving industry partners in designing research goals—could help ensure that nanotechnology innovations are both technically feasible and commercially viable.

Additionally, future initiatives must prioritize addressing global challenges, such as climate change and healthcare disparities, through interdisciplinary research. The Flagship's advancements in energy storage and biomedical technologies show how nanotechnology can contribute to global sustainability efforts. By focusing on how their innovations can address pressing global issues, future projects can not only drive technological progress but also ensure that their work has a broader societal impact.

13.3.3.4 Outcome

The Graphene Flagship demonstrated that international and interdisciplinary collaboration is essential for advancing nanotechnology. The project's breakthroughs in areas like healthcare and energy storage highlight how pooling expertise across fields can lead to significant innovations. However, the Flagship also reveals the importance of better integrating research with market demands, as well as aligning technological advancements with global challenges. Future nanotechnology initiatives should learn from the Graphene Flagship's successes and challenges, using collaboration to drive innovation while ensuring that their developments are both commercially viable and socially impactful.

13.4 A Vision for the Future—"NanoLEAP"

As we approach the conclusion of this exploration into nanotechnology's transformative potential, a simple summary wouldn't do justice to this fast-moving field, where advancements are constantly reshaping what is possible. So, instead of merely recapping the findings, this final section offers a bold vision for an initiative to mobilize all that nanotechnology has to offer to solve critical sustainability challenges.

This final section sketches out a high-level concept for a hypothetical yet feasible initiative—let's call it the "**Nano**technology for **L**everaging **E**merging **A**dvances and **P**artnerships—NanoLEAP". While the details would need to be refined, this preliminary blueprint, modeled after successful large-scale efforts such as the Graphene Flagship, illustrates how targeted funding, stakeholder collaboration, and adaptive governance could accelerate progress in key areas.

Having demonstrated nanotechnology's potential throughout this book, this scenario showcases how a coordinated, strategic "Moonshot" effort, combined with ambitious yet achievable goals, could significantly impact global sustainability. It exemplifies how innovative research can be transformed into actionable, real-world solutions that drive meaningful change.

This approach reinforces the underlying message of this book: that the choices we make today—in research, investment, and collaboration—will shape the future of our planet and our ability to confront the urgent sustainability challenges we face.

13.4.1 Mission

The NanoLEAP Fund is designed to accelerate the development and large-scale deployment of nanotechnology solutions that directly address the urgent global challenges of climate change mitigation, resource efficiency, and environmental protection. By focusing on the areas where nanotechnology holds the greatest potential for transformative, game-changing impacts, the fund aims to bridge the gap between research innovation and real-world implementation, ensuring that cutting-edge technologies move swiftly from the lab to industrial applications.

13.4.1.1 Objectives of the Fund

The fund's objectives focus on harnessing nanotechnology to address three critical environmental challenges (see Figure 13.3):

Environmental protection leverages nanotechnology's precision at the atomic and molecular levels to develop highly effective environmental solutions. The fund will focus on advancing water purification systems, air

Figure 13.3 Nanotechnology's role in environmental objectives.

quality monitoring, and pollution control technologies that can drastically reduce contaminants in the environment. Emphasis will be placed on scalable solutions that can be rapidly integrated into industrial processes and infrastructure, thereby protecting ecosystems and human health.

Climate change mitigation represents a key investment priority. The fund will support projects that leverage nanomaterials and nanoscale applications to develop breakthrough solutions for reducing greenhouse gas emissions, improving energy efficiency, and enhancing carbon capture technologies. These innovations aim to mitigate the effects of climate change and create adaptive technologies capable of withstanding environmental stresses.

Resource efficiency addresses the challenges of accelerating resource scarcity through revolutionary approaches to resource management. The fund will support nano-enabled innovations in recycling, energy-efficient manufacturing, and smart materials designed to reduce waste. By advancing circular economy models, the fund seeks to promote the development of products with longer lifecycles and materials that can be reused or upcycled. This approach aims to reduce dependency on raw resources and minimize environmental footprints.

13.4.1.2 The Vision

The fund seeks to foster a future in which nanotechnology serves as a driving force behind Europe's leadership in global sustainability. By investing in key areas where nanotechnology can outpace conventional approaches, the fund aims to set the foundation for a low-carbon, resource-efficient, and resilient economy. This vision will be achieved through coordinated efforts that harness Europe's existing strengths in research, innovation, and industrial capacity.

13.4.1.3 Key Principles of the Fund

- **Scalability and integration**: The fund will prioritize solutions that have the potential for wide-scale industrial integration and can be rapidly deployed. It will emphasize innovations that not only push the boundaries of research but also demonstrate a clear path to commercialization and large-scale adoption.
- **Interdisciplinary collaboration**: Given the complexity of sustainability challenges, the fund will encourage interdisciplinary collaboration between scientists, engineers, economists, and policymakers. It will foster a holistic approach to solution development, ensuring that technical breakthroughs align with economic and social objectives.
- **Adaptation and flexibility**: The fund recognizes the fast-paced nature of both technological advancements and global environmental challenges. As such, it will adopt an adaptive management framework, allowing for regular reassessment of priorities and reallocating resources as needed to address emerging challenges and capitalize on new opportunities.

- **Global leadership and collaboration**: While initially focused on Europe, the fund will actively encourage international collaboration, partnering with global innovators to share knowledge and solutions. This international dimension will help scale promising technologies worldwide, ensuring that Europe plays a leading role in driving global sustainability efforts.

13.4.2 Set-up and Identification of Key Impact Areas

The establishment of NanoLEAP will follow a structured process that ensures the initiative is both grounded in expert knowledge and responsive to evolving technological advancements and societal needs. Similar to the Graphene Flagship, NanoLEAP will begin with the formation of a core consortium tasked with setting the strategic direction, identifying priority areas, and managing the initial phase of the initiative. The consortium will be responsible for aligning NanoLEAP's objectives with broader sustainability and policy goals and ensuring that the initiative remains adaptable and future oriented.

13.4.2.1 Formation of the Core Consortium

The creation of NanoLEAP will begin with a call for proposals by the European Commission (or equivalent governing body). Research institutions, industry stakeholders, and academic leaders with expertise in nanotechnology, environmental science, and sustainability will be invited to form partnerships and submit proposals outlining their vision for NanoLEAP's role in addressing sustainability challenges.

Once submitted, these proposals will be evaluated by an independent panel of experts to ensure that the selected consortium demonstrates a comprehensive understanding of nanotechnology's transformative potential and a clear pathway for moving from research to commercialization.

The core consortium will consist of:

- **Research Institutions**: Universities and research organizations with leading nanotechnology and sustainability programs.
- **Industry Stakeholders**: Key industrial partners from sectors such as energy, materials, and manufacturing who can facilitate the rapid deployment and commercialization of nanotechnology solutions.
- **Policymakers and NGOs**: Representatives from governmental bodies and environmental organizations to ensure alignment with sustainability and regulatory frameworks.

This multidisciplinary consortium will guide NanoLEAP's strategy, ensuring that the initiative targets high-impact areas and contributes to both the European and global sustainability agenda.

13.4.2.2 Expert Advisory Panels

NanoLEAP will establish rotating expert advisory panels to provide ongoing guidance in identifying the most promising opportunities for nanotechnology to address critical sustainability challenges. These panels will consist of specialists from fields such as:

- Materials science and nanotechnology
- Environmental engineering and climate policy
- Economics and industrial integration

The advisory panels will be responsible for regularly reviewing emerging research, technological developments, and market needs. They will provide data-driven recommendations on which sectors and projects should receive funding, ensuring that NanoLEAP's investments are directed toward the most impactful and scalable technologies.

13.4.2.3 Stakeholder Engagement

NanoLEAP's stakeholder engagement strategy forms the foundation of its approach to developing nanotechnology solutions that address both scientific innovation and real-world challenges. This strategy is built on four key pillars, each representing a crucial stakeholder group (see Figure 13.4).

Research organizations form the first pillar, bringing cutting-edge scientific discoveries to the table. Their involvement ensures that NanoLEAP's

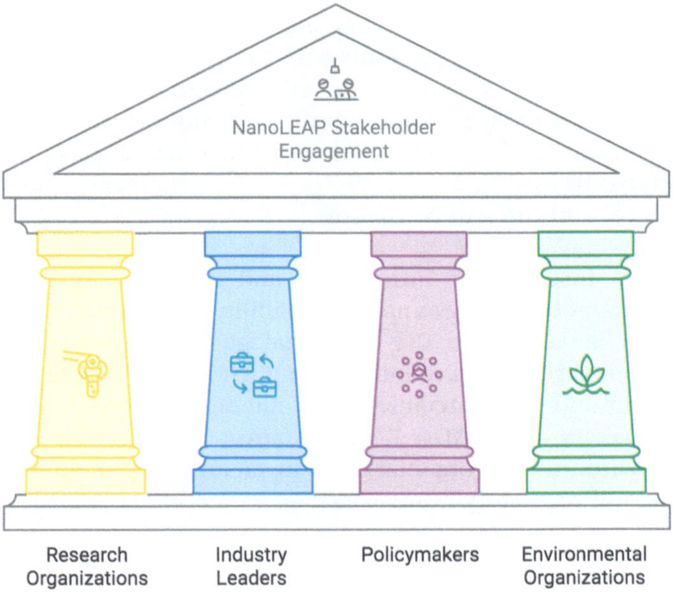

Figure 13.4 NanoLEAP Stakeholders.

projects are grounded in the latest advancements and helps identify emerging nanotechnologies with high-impact potential and scalability.

Industry leaders constitute the second pillar, providing invaluable insights into market demands and practical considerations. Their input is crucial for ensuring that nanotechnology applications are designed for easy scalability and seamless integration into industrial processes. These partners also offer feedback on the commercial viability of potential solutions, bridging the gap between laboratory innovations and real-world applications.

Policymakers represent the third pillar, playing a vital role in aligning NanoLEAP's initiatives with broader sustainability goals. Their involvement ensures that projects are developed in harmony with national, regional, and global policies, such as the European Green Deal and the Paris Agreement. By providing insights into regulatory frameworks, policymakers help shape the development and deployment of nanotechnology solutions in a way that meets both scientific and governance standards.

Environmental organizations complete the structure as the fourth pillar, bringing a crucial perspective on ecological and societal impacts. Their participation ensures that NanoLEAP's projects contribute meaningfully to environmental protection, resource efficiency, and sustainability goals. These organizations offer a broader view of public and ecological benefits, helping to assess and validate the positive impact of funded projects.

Through regular consultations and collaborative efforts with these diverse stakeholder groups, NanoLEAP aims to remain responsive to both industrial and societal needs. This comprehensive engagement strategy ensures that the initiative's nanotechnology solutions are not only scientifically innovative but also practically applicable and aligned with broader environmental and societal objectives.

13.4.2.4 Data-driven Assessments

To identify the areas where nanotechnology can have the greatest impact, NanoLEAP will employ data-driven assessments. These will include:

- **Lifecycle analysis**: Evaluating the long-term environmental and economic benefits of nanotechnology applications.
- **Cost-benefit assessments**: Identifying projects that offer the greatest return on investment, both financially and environmentally.
- **Big data analytics**: Utilizing machine learning and other advanced techniques to assess the scalability and market readiness of emerging technologies.

These assessments will allow NanoLEAP to make informed decisions about which projects to fund, ensuring that its investments lead to measurable, scalable outcomes.

13.4.2.5 *Alignment with Global and Regional Sustainability Goals*

A core aspect of NanoLEAP's mission is to contribute directly to the achievement of global sustainability goals, such as those outlined in the Paris Agreement and the European Green Deal. The consortium and advisory panels will work closely with policymakers to ensure that all target areas and funded projects are aligned with climate change mitigation, resource efficiency, and environmental protection goals.

This alignment will position NanoLEAP as a key driver of Europe's leadership in sustainability, ensuring that its projects have global relevance and impact.

13.4.2.6 *Technology Readiness and Scalability*

A key criterion for selecting NanoLEAP's target areas will be the technology readiness level of nanotechnology solutions. NanoLEAP will prioritize projects that are close to commercialization but will also invest in emerging technologies with long-term potential. Each project will be evaluated based on its:

- **Scalability**: Can the technology be integrated into existing industrial processes?
- **Market readiness**: Is there a clear path to commercialization?
- **Societal benefits**: Does the technology contribute to global sustainability goals?

By focusing on projects with both short-term and long-term impact, NanoLEAP will ensure that it supports a balanced portfolio of research, development, and commercialization.

13.4.2.7 *Regular Review and Adaptation*

Recognizing the rapid pace of technological advancement and the evolving nature of global sustainability challenges, NanoLEAP will implement a regular review process. The Steering Committee and Expert Advisory Panels will conduct annual reviews of NanoLEAP's target areas and funded projects. This process will include:

- **Reallocation of funding**: Ensuring that resources are directed toward the most promising and impactful projects.
- **Assessment of emerging opportunities**: Identifying new sectors and technologies that may have emerged since the initial target areas were identified.
- **Adapting to global challenges**: Pivoting to address new environmental or economic challenges as they arise.

This adaptive management framework will ensure that NanoLEAP remains flexible and responsive, positioning it to capitalize on new opportunities while staying focused on its mission of driving transformative sustainability outcomes through nanotechnology.

13.4.3 Likely Candidates for Key Impact Areas

At the time of writing this manuscript, the following areas appear to be the most likely candidates where nanotechnology could have the greatest transformative impact. However, this list should be considered an educated guess, as it reflects current research, technological advancements, and societal needs. These examples illustrate where nanotechnology could potentially provide high-impact solutions to sustainability challenges.

13.4.3.1 Carbon Capture and Utilization

Nanotechnology can significantly enhance carbon capture technologies by developing advanced nanomaterials, such as metal–organic frameworks and carbon nanotubes, to capture CO_2 more efficiently from industrial emissions and the atmosphere. Given that carbon emissions are a primary driver of climate change, improving the economic and technical feasibility of CO_2 capture and reuse is crucial. In the broader sustainability picture, CCU offers the potential to reduce atmospheric carbon levels while transforming captured CO_2 into valuable products like fuels and chemicals, contributing both to climate mitigation and the circular economy.

13.4.3.2 Energy Efficiency

Energy consumption remains one of the largest sources of greenhouse gas emissions globally. Nanotechnology can offer breakthrough solutions by enhancing materials used in buildings, vehicles, and industrial processes. For example, nanotech-enhanced materials can improve insulation, reduce friction, and enable the creation of lighter and more energy-efficient products. In the overall sustainability context, energy efficiency is often considered the "low-hanging fruit" because it can deliver immediate reductions in energy use and emissions without requiring major behavioral or infrastructure changes, making it a critical target area.

13.4.3.3 Renewable Energy

The transition to renewable energy is a cornerstone of global efforts to reduce dependency on fossil fuels and mitigate climate change. Nanotechnology can improve the performance and cost-efficiency of renewable energy systems like solar cells, wind turbines, and energy storage technologies. For example, using nanomaterials in solar cells can significantly increase their efficiency while lowering production costs, making solar power more accessible and competitive with fossil fuels. In the big picture, scaling up renewable energy is essential for achieving a low-carbon future, and nanotechnology could play a pivotal role in making this transition more rapid and cost-effective.

13.4.3.4 Water Purification

Access to clean water is an increasing global challenge, exacerbated by climate change, population growth, and pollution. Nanotechnology offers powerful solutions for water purification and desalination through nanomembranes and nanofilters, which can remove contaminants with higher efficiency and lower energy requirements than conventional methods. In terms of sustainability, ensuring a reliable supply of clean water is fundamental to both human health and ecosystem stability, especially in regions suffering from water scarcity. By improving the energy efficiency and scalability of water treatment, nanotechnology can make a substantial contribution to global water security.

13.4.3.5 Sustainable Agriculture

Global food production faces significant sustainability challenges due to resource depletion, climate change, and the growing demand for agricultural products. Nanotechnology can enhance precision agriculture by developing smart materials that deliver water and nutrients more efficiently, reducing waste while increasing yields. Nano-sensors can monitor soil and crop health in real time, allowing farmers to use resources more efficiently. In the larger sustainability context, improving the efficiency of agriculture is essential for feeding a growing population while reducing the environmental impacts of farming, such as water use and soil degradation.

13.4.3.6 Circular Economy Initiatives

The shift toward a circular economy—where products are designed to last longer, be reused, and eventually recycled—is key to reducing waste and conserving natural resources. Nanotechnology can facilitate this transition by enabling more efficient recycling processes, creating self-healing materials, and producing high-value products from waste materials. For example, nano-enabled recycling can improve the sorting and recovery of valuable materials from electronic waste. In the broader picture, circular economy initiatives help to close the resource loop, minimizing the environmental footprint of industrial processes and promoting sustainable resource management.

While these candidates represent a well-informed estimation of the most promising areas for nanotechnology's application in sustainability, they are by no means exhaustive or static. As new research and technologies emerge, the focus areas for NanoLEAP will need to adapt. Nevertheless, each of these fields plays a critical role in the broader sustainability agenda, offering the potential for significant environmental, societal, and economic impacts.

13.4.4 Structure and Implementation

The success of NanoLEAP will depend on a well-defined structure and a strategic approach to implementation. This section outlines the key components that will guide the allocation of resources, the fostering of interdisciplinary collaboration, the involvement of industry partners, and the establishment of measurable goals to ensure that the initiative delivers impactful results within its 10-year timeframe.

13.4.4.1 Funding Allocation

NanoLEAP will allocate €1 billion over 10 years, distributed through competitive grants to university labs, research institutions, and startups across the EU. Funding will prioritize emerging technologies, underfunded areas, and promising innovations with clear scalability potential. A portion of the budget may be reserved for early-stage seed funding to support high-risk, high-reward projects. Additionally, a mid-term review process (at year 5) will reassess funding distribution to ensure that successful projects receive continued support and new priorities are addressed.

13.4.4.2 Interdisciplinary Approach

Every funded project must involve collaboration across at least three disciplines (*e.g.*, materials science, environmental engineering, economics) to ensure a holistic approach to sustainability challenges. This fosters innovation by encouraging researchers to work beyond traditional boundaries, combining technical expertise with economic, policy, and social perspectives. Annual conferences, workshops, and knowledge-sharing platforms will further encourage collaboration between interdisciplinary teams.

13.4.4.3 Industry Partnerships

Projects must include at least one industry partner to ensure the rapid commercialization and real-world application of nanotechnology solutions. Industry partners will co-invest in promising technologies, ensuring that innovations are aligned with market demands and have clear pathways to commercialization. Early access to research outputs will provide industry partners with a competitive edge in adopting sustainable technologies, promoting faster transitions from lab to market.

Inspired by Denmark's Kalundborg Symbiosis model,[48] NanoLEAP will go beyond traditional partnerships to foster industrial symbiosis, where companies collaborate by sharing resources, by-products, and nanotechnology innovations. In this model, businesses will cooperate to minimize waste and create closed-loop systems, where the waste or output of one process becomes the input for another. This symbiotic approach will encourage companies to adopt nanotechnology-enabled processes that improve resource efficiency and promote sustainability, while reducing costs and accelerating the commercialization of circular economy solutions.

By adopting the Kalundborg model, NanoLEAP will enable industries to work together on nanotech innovations such as CO_2 capture, waste recycling, and energy-efficient materials, ensuring that solutions are scalable and have broad industrial applications across sectors.

13.4.4.4 Adaptive Management

An annual review process will reassess priorities, funding allocations, and goals based on progress, technological advancements, and global needs. This will be conducted by an expert panel of scientists, policymakers, and industry leaders to ensure that NanoLEAP remains agile and responsive to external changes. The fund will pivot towards emerging technologies or global challenges as necessary, while phasing out underperforming projects. Input from research institutions and industry stakeholders will further inform adaptive decisions.

NanoLEAP will employ AI-powered analytics and real-time data to monitor project performance and environmental impact. By incorporating advanced data-driven tools inspired by Denmark's eco-innovation framework, the initiative will maintain flexibility, ensuring that resources are dynamically allocated to the most impactful projects and that progress is continuously tracked.

13.4.4.5 Ambitious but Realistic Goals

NanoLEAP will define a set of clear, measurable objectives across various areas of focus, including CO_2 reduction, energy efficiency improvements, and the percentage of circular economy processes and applications developed using nanotechnology. These objectives will be broken down into:

- **Short-term goals (1–3 years)**: The initial phase will focus on identifying low-hanging fruit—existing, well-advanced research projects with clear commercialization potential. NanoLEAP will prioritize fast-tracking these technologies, facilitating their transition from research to small-scale pilot projects and early market deployment. The goal is to achieve quick wins by identifying technologies with near-term impact potential and pushing them toward commercialization.
- **Medium-term goals (3–7 years)**: Expanded deployment and industrial integration of nanotechnology solutions, focusing on scaling successful pilot projects into wider use. By this phase, the goal will be to see broad adoption of nanotechnology applications across industries, contributing to meaningful CO_2 reduction, improved resource efficiency, and circular economy innovations.

- **Long-term goals (7–10 years):** By the final phase, NanoLEAP will aim to achieve full commercialization of successful technologies within target sectors. The focus will be on ensuring widespread adoption of nanotechnology solutions at scale, delivering measurable improvements in areas such as carbon capture, energy efficiency, and waste reduction. The ultimate goal will be to see these technologies contributing significantly to Europe's sustainability agenda and global climate targets.

The goals will remain adaptive, allowing NanoLEAP to reassess and realign based on progress, emerging technologies, and shifting global priorities, ensuring that the initiative remains relevant and impactful throughout its 10-year span.

13.4.4.6 Performance Metrics

To ensure NanoLEAP delivers measurable impact, clear and quantifiable performance targets will be set. These targets will focus on both environmental and economic performance, enabling NanoLEAP to contribute to sustainable development in line with global climate goals. While specific targets will evolve, potential examples could include:

- **Carbon capture**: A potential target could be a 50% improvement in CO_2 capture efficiency by 2030, significantly reducing industrial emissions.
- **Energy efficiency**: Materials that reduce energy consumption in buildings by 30% by 2035 could provide significant cost savings and environmental benefits.
- **Renewable energy**: **Increasing solar cell efficiency by 25%** through nanotechnology innovations by 2032 could help renewable energy compete with fossil fuels.

In addition to environmental performance, NanoLEAP should adopt economic metrics based on frameworks like the OECD's Green Growth Indicators.[49] These metrics will help track how nanotechnology contributes to green growth by quantifying the economic impact of innovation:

- **Job creation**: Assessing the number of jobs generated in emerging nanotechnology industries.
- **Economic returns**: Measuring financial benefits from nanotechnology solutions, such as energy cost savings or the value generated through circular economy processes.
- **Market growth**: Tracking the expansion of new markets and industries benefiting from nanotechnology commercialization.

By incorporating both environmental and economic metrics, NanoLEAP will not only measure its contributions to sustainability but also ensure that it drives economic prosperity in tandem with global climate goals.

These performance metrics will be tracked through regular reporting from funded projects, and the outcomes will be published publicly to ensure transparency and accountability. Projects that meet or exceed these metrics may receive additional funding for further development and scaling.

13.4.4.7 Stakeholder Involvement

Regular consultations with environmental NGOs, industry associations, policymakers, and research organizations will ensure that NanoLEAP's priorities are aligned with broader sustainability goals and regulatory frameworks. This multi-stakeholder engagement process will help prevent potential bottlenecks, such as regulatory delays or market resistance, and will create a collaborative ecosystem to scale successful innovations.

13.4.4.8 Public Engagement

A robust public communication strategy will be implemented to educate and engage the public on nanotechnology's potential to address climate change and sustainability challenges. This will include public demonstrations, interactive workshops, and online platforms for knowledge sharing. Citizen science initiatives will encourage public participation in tracking the impact of nanotechnology-based solutions, while partnerships with the media will highlight success stories and maintain transparency.

NanoLEAP will extend its public engagement strategy by introducing citizen science initiatives, encouraging communities to actively participate in nanotechnology trials and monitoring. This approach will democratize innovation, allowing citizens to contribute to sustainability efforts by testing nano-enabled technologies such as water purification and air quality monitoring systems.

13.4.4.9 International Collaboration and Standardization

Although EU-focused, NanoLEAP will seek international collaboration to share breakthroughs and technologies with global partners, especially in regions where sustainability challenges are more severe. Partnerships with global standardization bodies will help harmonize nanotechnology standards, ensuring that solutions are scalable across different markets. Collaborative projects with non-EU countries will support the global climate agenda, particularly in areas like energy storage, water management, and circular economy innovations.

13.4.5 Realizing Nanotechnology's Potential for Sustainability

This book has demonstrated how nanotechnologies can play a pivotal role in addressing some of the world's most pressing environmental challenges.

However, the true success of NanoLEAP hinges on the collective efforts of policymakers, researchers, industry leaders, and citizens. The strategic framework set by NanoLEAP offers a blueprint for action, showing how coordinated investment, research, and innovation can accelerate progress toward a sustainable, resilient future.

As we conclude this exploration, it is clear that the choices we make today, especially through initiatives like NanoLEAP, will significantly shape our environmental legacy. Now is the time for all stakeholders to embrace this vision and work together to unlock nanotechnology's full potential in building a sustainable future for generations to come.

References

1. S. F. Hansen, R. Arvidsson, M. B. Nielsen, O. F. H. Hansen, L. P. W. Clausen and A. Baun, *et al.*, Nanotechnology meets circular economy, *Nat. Nanotechnol.*, 2022, **17**(7), 682–685.
2. EUR-Lex – 52015DC0614 – ES – EUR-Lex, Closing the Loop: An EU Action Plan for the Circular Economy, European Commission, 2015, [cited 2024 Sep 3]. Available from: https://eur-lex.europa.eu/legal-content/EN/TXT/?uri=CELEX:52015DC0614.
3. D. Dragoman, Nanomaterials for energy harvesting, *Nanomaterials*, 2023, **13**(7), 1154.
4. N. Farooq, Z. Rehman, M. I. Khan, S. Asghar, M. Saleem and R. Irshad, *et al.*, Nanomaterial-based energy conversion and energy storage devices: a comprehensive review, *New J. Chem.*, 2024, **48**(19), 8933–8962.
5. Y. Tzeng, C.-Y. Jhan, Y.-C. Wu, G.-Y. Chen, K.-M. Chiu and S. Y.-E. Guu, High-ICE and high-capacity retention silicon-based anode for lithium-ion battery, *Nanomaterials*, 2022, **12**(9), 1387.
6. M. Imran, A. M. Afzal, M. W. Iqbal, H. H. Hegazy, M. Z. Iqbal and S. Mumtaz, *et al.*, Manganese (Sulfide/Oxide) based electrode materials advancement in supercapattery devices, *Mater. Sci. Semicond. Process.*, 2023, **158**, 107366.
7. X. Chen, R. Paul and L. Dai, Carbon-based supercapacitors for efficient energy storage, *Natl. Sci. Rev.*, 2017, **4**(3), 453–489.
8. S. Xu, J. Wu, X. Wang and Q. Zhang, Recent advances in the utilization of covalent organic frameworks (COFs) as electrode materials for supercapacitors, *Chem. Sci.*, 2023, **14**(47), 13601–13628.
9. B. Yang, B. Li and Z. Xiang, Advanced MOF-based electrode materials for supercapacitors and electrocatalytic oxygen reduction, *Nano Res.*, 2023, **16**(1), 1338–1361.
10. L. Abbassi, D. Mesguich, D. Berthebaud, S. Le Tonquesse, B. Srinivasan and T. Mori, *et al.*, Effect of nanostructuring on the thermoelectric properties of β-$FeSi_2$, *Nanomaterials*, 2021, **11**(11), 2852.
11. M. Dragoman, M. Aldrigo, A. Dinescu, D. Vasilache, S. Iordanescu and D. Dragoman, Nanomaterials and devices for harvesting ambient electromagnetic waves, *Nanomaterials*, 2023, **13**(3), 595.

12. CARMOF, [cited 2024 Sep 16]. Available from: https://carmof.eu/.
13. MOF4AIR, [cited 2024 Sep 16]. Available from: https://www.mof4air.eu/.
14. MEMBER, [cited 2024 Sep 16]. Available from: https://member-co2.com/.
15. J. Gandara-Loe, L. Pastor-Perez, L. F. Bobadilla, J. A. Odriozola and T. R. Reina, Understanding the opportunities of metal–organic frameworks (MOFs) for CO_2 capture and gas-phase CO_2 conversion processes: a comprehensive overview, *React. Chem. Eng.*, 2021, **6**(5), 787–814.
16. European climate law, Climate Action, [cited 2024 Sep 5]. Available from: https://climate.ec.europa.eu/eu-action/european-climate-law_en.
17. World Energy Outlook 2023, IEA, [cited 2024 Sep 6]. Available from: https://www.iea.org/reports/world-energy-outlook-2023.
18. A. Machín and F. Márquez, The next frontier in energy storage: A game-changing guide to advances in solid-state battery cathodes, *Batteries*, 2023, **10**(1), 13.
19. Optimised battery production through nanotechnology, Nanobat.eu, [cited 2024 Sep 9].
20. BATT4EU partnership – BATT4EU [Internet]. Bepassociation.eu, [cited 2024 Sep 9].
21. Joint center for energy storage research, [cited 2024 Sep 9]. Available from: https://www.jcesr.org/.
22. The innovation center for Battery500 consortium. PNNL, [cited 2024 Sep 9]. Available from: https://www.pnnl.gov/innovation-center-battery500-consortium.
23. Federal Consortium for Advanced Batteries (FCAB). Energy.gov, [cited 2024 Sep 9]. Available from: https://www.energy.gov/mesc/federal-consortium-advanced-batteries-fcab.
24. Battery Innovation System of China, Batterieseurope.eu, [cited 2024 Sep 9]. Available from: https://batterieseurope.eu/wp-content/uploads/2024/06/3-Battery-Innovation-China_250424.pdf.
25. Battery Innovation System of Japan, Batterieseurope.eu, [cited 2024 Sep 9]. Available from: https://batterieseurope.eu/wp-content/uploads/2024/06/2-Battery-Innovation-JAPAN_210524.pdf.
26. Battery Innovation System of South Korea, Batterieseurope.eu, [cited 2024 Sep 9]. Available from: https://batterieseurope.eu/wp-content/uploads/2024/06/4-Battery-Innovation-South-Korea_250424.pdf.
27. Battery Innovation System of Australia, Batterieseurope.eu, [cited 2024 Sep 9]. Available from: https://batterieseurope.eu/wp-content/uploads/2024/06/Battery-Innovation-Australia_310520245.pdf.
28. Wiener J. and Bennear L., Adaptive Regulation: A Framework for Policy Learning over Time, 2019, [cited 2024 Sep 9]. Available from Harvard University: https://www.hks.harvard.edu.
29. FDA.gov, FDA's Approach to Regulation of Nanotechnology Products, [cited 2024 Sep 10]. Available from: https://www.fda.gov/science-research/nanotechnology-programs-fda/fdas-approach-regulation-nanotechnology-products.

30. J. Paradise, Regulating nanomedicine at the food and drug administration, *AMA J. Ethics*, 2019, **21**(4), E347–E355.
31. US EPA O, Control of nanoscale materials under the toxic substances control act, 2015, [cited 2024 Sep 10]. Available from: https://www.epa.gov/reviewing-new-chemicals-under-toxic-substances-control-act-tsca/control-nanoscale-materials-under.
32. US EPA O, Advancing Environmental Safety: Celebrating 20 years of Nanotechnology Research at EPA, 2024, [cited 2024 Aug 30]; Available from: https://www.epa.gov/sciencematters/advancing-environmental-safety-celebrating-20-years-nanotechnology-research-epa.
33. A. Owen, S. Rannard, R. Bawa and S.-S. Feng, Interdisciplinary nanomedicine publications through interdisciplinary peer-review, *J. Interdiscip. Nanomed.*, 2016, **1**(1), 4–8.
34. G. G. Naik and V. A. Jagtap, Two heads are better than one: Unravelling the potential Impact of Artificial Intelligence in nanotechnology, *Nano TransMed*, 2024, **3**, 100041.
35. Y. Wang, X. Zhang and H. Yue, Two-dimensional nanomaterials induced nano-bio interfacial effects and biomedical applications in cancer treatment, *J. Nanobiotechnol.*, 2024, **22**, 67.
36. L. S. Alqarni, M. D. Alghamdi, A. A. Alshahrani and A. M. Nassar, Green nanotechnology: Recent research on bioresource-based nanoparticle synthesis and applications, *J. Chem.*, 2022, **2022**, 1–31.
37. Harvard Center for Nanoscale Systems, [cited 2024 Sep 12]. Available from: https://cns1.rc.fas.harvard.edu/.
38. International Iberian Nanotechnology Laboratory, [cited 2024 Sep 12]. Available from: https://www.inl.int/.
39. Swiss Nanoscience Institute, [cited 2024 Sep 12]. Available from: https://nanoscience.unibas.ch/en/.
40. The Kavli Nanoscience Institute at Caltech, [cited 2024 Sep 12]. Available from: https://kni.caltech.edu/.
41. Nanotechnology Characterization Laboratory, [cited 2024 Sep 12]. Available from: https://www.cancer.gov/nano/research/ncl.
42. London Centre for Nanotechnology [cited 2024 Sep 12]. Available from: https://www.london-nano.com/.
43. National Center for Nanoscience and Technology, China [cited 2024 Sep 12]. Available from: http://english.nanoctr.cas.cn/.
44. Centre for Nano Science and Engineering, [cited 2024 Sep 12]. Available from: http://www.cense.iisc.ac.in/.
45. National Institute for Materials Science, [cited 2024 Sep 12]. Available from: https://www.nims.go.jp/eng/.
46. Graphene Flagship, [cited 2024 Sep 13]. Available from: https://graphene-flagship.eu/.
47. A mixed legacy, *Nat. Mater.*, 2023, **22**(7), 793.
48. Kalundborg Symbiosis, [cited 2024 Sep 15]. Available from: https://www.symbiosis.dk/en/.

49. Green growth indicators (Edition 2023), [cited 2024 Sep 15]. Available from: https://www.oecd-ilibrary.org/environment/data/oecd-environment-statistics/green-growth-indicators-edition-2023_8fbb3633-en.
50. B. Halamoda-Kenzaoui, R. Geertsma, J. Pouw, A. Prina-Mello, M. Carrer and M. Roesslein, *et al.*, Future perspectives for advancing regulatory science of nanotechnology-enabled health products, *Drug Delivery Transl. Res.*, 2022, **12**(9), 2145–2156.

Subject Index